Student Solutions Manual to accompany

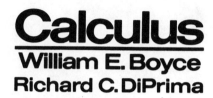

Calculus
William E. Boyce
Richard C. DiPrima

Prepared by

Charles W. Haines & Thomas C. Upson
Rochester Institute of Technology

John Wiley & Sons
New York   Chichester   Brisbane   Toronto   Singapore

ISBN 0 471 85450-6

Printed in the United States of America

10 9 8 7 6 5 4 3

PREFACE

This supplement has been prepared for use in conjunction with CALCULUS, by William E. Boyce and Richard C. DiPrima. The supplement contains the outline of the solution to all odd problems in the text. We have chosen not to include all steps to the solutions since that would have resulted in a cumbersome document. Also, we feel that one learns mathematics by doing mathematics. You will notice that there is more detail in the first chapters than in the latter. For problems that are similar, details are usually more complete for the first or second of that type so you may find it helpful to look at the answers to these for a better understanding.

Students should be aware that following these solutions is very different from designing and constructing one's own solution. Using this supplemental resource appropriately for learning calculus is outlined as follows:

1. Make an honest attempt to solve the problem without using the guide.

2. If needed, glance at the beginning of the solution in the guide and then try again to generate the complete solution. Continue using the guide for hints when you reach an impasse.

3. Compare your final solution with the one provided to see whether yours is more or less efficient than the guide, since there is frequently more than one correct way to solve a problem.

4. Ask yourself why that particular problem was assigned.

We wish to express our appreciation to Mrs. Susan A. Hickey, Mrs. Susan E. Menrisky, and Ms. Beth E. Strothmann for their excellent typing of a very difficult manuscript and to Mr. Jeffery W. Benck for his proofreading assistance.

Charles W. Haines
Thomas C. Upson
Rochester Institute of Technology
Rochester, New York
May, 1988

CONTENTS

CHAPTER 1

Section 1.1, Page 6

1. $A \cup B = (-1,4)$ since all points in this interval lie in either A or B or both. $A \cap B = [0,2)$ since all points in this interval lie in both A and B. Note that the point 0 lies in both A and B and thus $A \cap B$ includes 0 and the square bracket is used for the left end point.

3. Note that all real points, except zero, are contained in either A or B but not both. Thus $A \cup B = R^1 - \{0\}$, while $A \cap B = \Phi$ (the empty set).

5a. Since 2 is an element of A we have $2 \in A$.

 b. Since 2 is an element and not a set, the statement $2 \subset B$ is false.

 c. Since the set A is not an element of B, $A \in B$ is false.

 d. Since the elements of A are also in B we know A is a subset of B and thus $A \subset B$.

7. Since the points 2 and 3 are the only points in A that are not in B we have $A - B = \{2,3\}$.

9. Since the interval $[1/4, 3/4)$ and the point 1 are in A but not B, we have $A - B = [1/4, 3/4) \cup \{1\}$.

11. Since S has only a finite number of points, it is bounded above and bounded below.

13. Since S contains all real numbers less than 1, the set S will be bounded above but not bounded below.

15. Since $1/n$ is always less than 1 for $n = 2, 3, \ldots$, we see that the elements of S are always less than 1 and hence S is bounded above. Since $1/n$ decreases as n increases and for $n = 2$, $1 - 1/2 = 1/2$, we see that S is bounded below.

17. We need to find s so that $x^2 + x - s > 0$ for all real values of x. This means the equation $x^2 + x - s = 0$ has no real solutions. Using the quadratic formula, the roots of this last equation are $x = (-1 \pm \sqrt{1+4s})/2$. Thus, if $1 + 4s < 0$ there will be no roots and $S = \{s \mid s < -1/4\}$.

19.   Using Eq.(8) we have $|2x+3|$ = $2x+3$ = 1 for $2x+3 \geq 0$. Thus
       x = -1 is one solution. If $2x+3 < 0$, then $|2x-3|$ =
       $-(2x+3)$ = 1 and thus x = -2.

21.   Since $|1-x|$ is always greater than or equal to zero we must
       have $x \geq 1$ for the equation to be defined. In this case
       $|1-x|$ = x-1 for all $x \geq 1$.

23.   Using Eq.(13) we have $|(3-x)/(3+x)|$ = $|3-x|/|3+x|$, $x \neq -3$.
       Thus the given equation becomes $|3-x|$ = $4|3+x|$, $x \neq -3$ and
       we may follow the example of the text. If both 3-x and 3+x
       are non-negative, $x \in [-3,3]$, then we have 3-x = 4(3+x) or
       x = -9/5. If either 3-x or 3+x are negative,
       $x \in (-\infty,-3) \cup (3,\infty)$, then 3-x = -4(3+x) or x = -5.

25.   Using Eq.(8), if $x \geq 0$ then $|x|$ = x so that $(1/2)(|x|+x)$ =
       $(1/2)(x+x)$ = x. If x < 0, then $|x|$ = -x so that
       $(1/2)(|x|+x)$ = $(1/2)(-x+x)$ = 0.

27.   To prove the given statement we must consider the following
       four cases:
       1. $x \geq 0$, $y \geq 0$. In this case $xy \geq 0$ and $|xy|$ = xy, $|x|$ = x
       and $|y|$ = y by Eq.(8). Thus $|xy|$ = xy = $|x||y|$.
       2. x < 0, $y \geq 0$. If y = 0, then both sides of the equation
       are zero and thus the equality holds. Otherwise xy < 0 and
       hence $|xy|$ = -xy, $|x|$ = -x and $|y|$ = y. Thus
       $|xy|$ = -xy = $|x||y|$.
       3. $x \geq 0$, y < 0. If x = 0 then the equation holds as in
       case (2). Otherwise $|xy|$ = -xy, $|x|$ = x and $|y|$ = -y. Thus
       $|xy|$ = -xy = x(-y) = $|x||y|$.
       4. x < 0, y < 0. In this case xy > 0 and $|xy|$ = xy,
       $|x|$ = -x and $|y|$ = -y. Thus $|xy|$ = xy = (-x)(-y) = $|x||y|$.

29.   S may be written as S = $\{0,1/3,1/2,3/5,...\}$. Thus the
       greatest lower bound is 0 and the least upper bound is 1,
       since (n-1)/(n+1) is always less than 1.

31.   Since y = $(x^2 + 2x + 1) - 1$ = $(x+1)^2 - 1$ we see that the
       greatest lower bound is -1, which occurs for x = -1. The
       least upper bound is 3, for the x range given.

Section 1.2, Page 13

1.   Following Ex.(6) we see that x is in the set S if the
        distance from -3 to x is less than 3.   Thus x is in the
        interval (-6,0).   An alternate approach is to use the
        extension of Theorem 1.2.1, in which case $|x+3| < 3$ if and
        only if $-3 < x + 3 < 3$.   Adding -3 to each part yields
        $-6 < x < 0$, which is the interval (-6,0).   As stated
        following Ex.(1) we should also prove the converse.   That
        is, if x is in the interval (-6,0), then it is in S.   This
        can be done by reversing the last several steps since they
        involve addition only.   As in the text, we will not
        ordinarily include this portion of the problem.

3.   Using Theorem 1.2.1 and its extension we see that t is in S
        if and only if $-1 < t-1 < 0$ or $0 < t-1 < 1$.   Adding 1 to all
        terms yields $0 < t < 1$ or $1 < t < 2$.   Thus $S = (0,1) \cup (1,2)$.

5.   x is in S if and only if $-\varepsilon < x-a < \varepsilon$.   Adding a to all terms
        yields $a-\varepsilon < x < a+\varepsilon$ and hence S is the interval $(a-\varepsilon, a+\varepsilon)$.

7.   To isolate x, first add 7 to both sides of the inequality to
        obtain $3x \geq 9-x$.   Next add x to both sides to find $4x \geq 9$
        and finally divide by 4 to yield $x \geq 9/4$ or $[9/4, \infty)$.

9.   Using the second approach shown in Ex.(2) we first add 1 to
        each term of the inequality to obtain $4 < 4x \leq 7$.   Dividing
        all terms by 4 then yields $1 < x \leq 7/4$ or $(1, 7/4]$.

11.   The quotient is positive if and only if $x-2 > 0$ and $x-3 > 0$
        or $x-2 < 0$ and $x-3 < 0$.   The first case is satisfied by
        $x > 3$ since then $x-3 > 0$ and $x-2 > 1 > 0$.   The second case
        is satisfied if $x < 2$ since then $x-2 < 0$ and $x-3 < -1 < 0$.
        Thus $S = (-\infty, 2) \cup (3, \infty)$.   A line drawing similar to Fig.1.2.1
        may be helpful in this case.

13.   By adding -3 and then $2/x$ to both sides we obtain $4 > 2/x$.
        Multiplying by 1/2 then yields $2 > 1/x$.   If $x > 0$, then we
        may multiply by x to find $2x > 1$ or $x > 1/2$.   If $x < 0$ then
        $1/x < 0$ and hence $1/x < 2$ is satisfied for all $x < 0$.   Thus
        $S = (-\infty, 0) \cup (1/2, \infty)$.

15.   Following Ex.(5) we first consider 2 - 1/x < 1. Adding 1/x
        and then -1 to both sides yields 1 < 1/x. If x > 0, then
        multiply both sides by x to obtain x < 1. If x < 0 we have
        a contradiction since 1/x < 0 which can't be greater than 1.
        Thus 0 < x < 1 are the only values to satisfy the right
        inequality. The left inequality, -1 < 2 - 1/x, yields
        1/x < 3 using similar steps to above. Again, if x > 0 then
        1 < 3x, or x > 1/3. If x < 0 then 1/x < 0 for all x and
        hence 1/x < 3. Finally, the only values of x which satisfy
        both sides of the given inequality is the intersection of
        the intervals $(-\infty,0)$, $(0,1)$ and $(1/3,\infty)$ and thus S =(1/3,1).

17.   Since the numerator can be written as $(x-1)^2$, which is
        always greater than or equal to zero, we may conclude that
        the quotient is negative only when $x^2 - 2x - 3 =$
        $(x+1)(x-3) < 0$. This inequality holds if x+1 > 0 and
        x-3 < 0 or if x+1 < 0 and x-3 > 0. There are no values of x
        satisfying the latter conditions. The former conditions are
        both satisfied for -1 < x < 3. Finally, the given quotient
        is zero only when the numerator is zero, which occurs at
        x = 1 and is already in the interval (-1,3).

19.   Using the extension of Theorem 1.2.1 we have -1 < 4-1/x < 1.
        The right inequality yields 3 < 1/x by adding 1/x and -1 to
        both sides. If x > 0 then multiplication by x gives 3x < 1
        or x < 1/3 and thus 0 < x < 1/3. Since 3 is a positive
        number, x < 0 (and therfore 1/x < 0) is a contradiction.
        Likewise the left inequality yields 1/x < 5. If x is
        positive we have 1 < 5x or x > 1/5. For x negative, 1/x < 0
        and hence 1/x < 5 for all x < 0. Thus the original
        inequalities hold only for the intersection of the three
        intervals (0,1/3), $(1/5,\infty)$ and $(-\infty,0)$ which is the interval
        (1/5,1/3).

21a.   For x ≥ 0 and y ≥ 0 we have x+y ≥ 0. Thus |x+y| = x+y,
        |x| = x, |y| = y and hence |x+y| = |x| + |y|.
  b.   For x < 0 and y < 0 we have x+y < 0. Thus |x+y| = -(x+y) =
        -x-y, |x| = -x, |y| = -y and hence |x+y| = |x| + |y|.
  c.   For x > 0 and y < 0 we have |x| = x and |y| = -y. In this
        case, however, we must consider x+y > 0, x+y < 0 and x+y = 0
        separately. For the latter case |x+y| = 0 which is clearly
        less than |x|+|y|, which must be a positive number.
        For x+y > 0 we have |x+y| = x+y < x+|y| since y < 0. Since
        |x| = x, this last inequality yields |x+y| < |x|+|y|. For
        x+y < 0 we have |x+y| = -(x+y) = -x-y = -|x|+|y| < |x|+|y|.

21d. For $x < 0$ and $y > 0$ we have $|x| = -x$ and $|y| = y$. As in
      part (c) we have three cases:
      $x+y > 0$ and hence $|x+y| = x+y < |x|+y = |x|+|y|$;
      $x+y = 0$ and hence $|x+y| = 0 < |x|+|y|$;
      $x+y < 0$ and hence $|x+y| = -(x+y) = -x-y = |x|-y < |x|+|y|$.

23.   Theorem 1.2.2 says that if x and w are real numbers then
      $|x+w| \leq |x|+|w|$. If we let $w = -y$, then we have

      $|x-y| \leq |x|+|-y|$. But $|-y| = |y|$ and thus $|x-y| \leq |x|+|y|$.

      From Prob.21 we see that equality holds when $x \geq 0$ and

      $-y \geq 0$ or when $x \leq 0$ and $-y \leq 0$ and thus equality holds for

      $x \geq 0$ and $y \leq 0$ or $x \leq 0$ and $y \geq 0$.

25.   Using the hint we have $|x| = |(x-y)+y| \leq |x-y|+|y|$ by

      Thm. 1.2.2. Thus $|x|-|y| \leq |x-y|$. From Problem 21,

      equality holds when $x-y \geq 0$ and $y \geq 0$ or when $x-y \leq 0$ and

      $y \leq 0$ and thus for $x \geq y \geq 0$ or $x \leq y \leq 0$ equality holds.

27.   If $b > a$ then $b-a > 0$. Multiplying both sides by $b+a$, which
      is positive, then yields $(b-a)(b+a) > 0$. Since $(b-a)(b+a) = b^2-a^2$ we have $b^2-a^2 > 0$ if $b > a$ and $b+a > 0$.

29.   The desired inequality may be established by expanding the
      left side and then adding and subtracting appropriate terms
      to obtain the right side as shown:

      $$(a_1b_1 + a_2b_2)^2 = a_1^2b_1^2 + 2a_1a_2b_1b_2 + a_2^2b_2^2 =$$

      $$a_1^2b_1^2 + a_1^2b_2^2 + a_2^2b_1^2 + a_2^2b_2^2 - (a_1^2b_2^2 - 2a_1a_2b_1b_2 + a_2^2b_1^2) =$$

      $$(a_1^2 + a_2^2)(b_1^2 + b_2^2) - (a_1b_2 - a_2b_1)^2 \leq (a_1^2 + a_2^2)(b_1^2 + b_2^2).$$
      Equality occurs when $a_1b_2 - a_2b_1 = 0$ or $a_1b_2 = a_2b_1$.

31. $S = \{x| (x-3)(x+2)<0 \} = \{x| -2<x<3 \}$, so the l.u.b. is 3
      and the g.l.b. is -2.

33. $S = \{x| (x-3)(x+2)>0 \} = \{x| x<-2$ or $x>3 \}$, so no g.l.b. or
      l.u.b. exist.

Section 1.3, Page 24

1.   Using Eq.(2), the distance between the points is given by
      $d = \sqrt{[2-(-1)]^2 + [-3-4]^2} = \sqrt{9+49} = \sqrt{58}$.

3.   If $t > 0$ then the points are (t,t) and $(t,-t)$ and Eq.(2)
      gives $d = \sqrt{(t-t)^2 + (t+t)^2} = 2t$.   If $t < 0$ then the points are
      $(t,-t)$ and $(-t,-t)$ and we have $d = \sqrt{(t+t)^2 + (-t+t)^2} = 2|t|$.   If
      $t = 0$, the points are identical and hence $d = 0$.   Thus
      $d = 2|t|$ for all t.

5.   Eq.(2) yields $d(P_1,P_2) = \sqrt{(x_2-x_1)^2 + (y_2-y_1)^2}$,

      $d(P_1,P_3) = \sqrt{\left(\dfrac{x_1+x_2}{2}-x_1\right)^2 + \left(\dfrac{y_1+y_2}{2}-y_1\right)^2} = (1/2)d(P_1,P_2)$, and

      $d(P_2,P_3) = \sqrt{\left(\dfrac{x_1+x_2}{2}-x_2\right)^2 + \left(\dfrac{y_1+y_2}{2}-y_2\right)^2} = (1/2)d(P_1,P_2)$.

      Since P_3 is equidistance from P_1 and P_2 and since that
      distance is one-half of the distance from P_1 to P_2 we may
      conclude that P_3 is the midpoint of the line segment joining
      the points P_1 and P_2.

7.   If P_1 is $(-3,-2)$, P_2 is $(-6,9)$ and P_3 is $(3,2)$ then Eq.(2)
      yields $d(P_1,P_2) = \sqrt{130}$, $d(P_2,P_3) = \sqrt{130}$ and $d(P_1,P_3) = \sqrt{52}$ so
      that the given triangle is isosceles.

9.   If P_1 is $(-2,1)$, P_2 is $(2,1)$ and P_3 is $(0,1+2\sqrt{3})$ then Eq.(2)
      yields $d(P_1,P_2) = 4$, $d(P_2,P_3) = 4$ and $d(P_1,P_3) = 4$ so that
      the given triangle is equilateral.

11.   We use Eq.(7) with the slope $m = 1$ and the point
      $(x_1,y_1) = (0,0)$ to obtain $y-0 = 1(x-0)$ or $y = x$.

13.   Using Eq.(4) we have $m = \dfrac{3-(-1)}{2-(-1)} = 4/3$ and thus Eq.(7) yields

      $y-(-1) = (4/3)[x-(-1)]$ which reduces to $y = (4/3)x + 1/3$.

15.   If we let x = 0 then 2y = -7 or
       y = -7/2, which is the y axis
       intercept. Letting y = 0 we find
       0 = 4x-7 or x = 7/4, which is the
       x axis intercept. The graph is
       then the straight line through
       these two points.

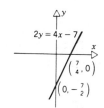

17.   Letting x = 0 we find that
       y = 2 is the y intercept and
       letting y = 0 we find that
       x = 2 is the x intercept.
       Drawing a line through these
       two points yields the graph shown.

19a.   From Fig.1.3.11 we see that the angle between the line with
       slope m_2 and the x axis is α +90°. Therefore,

$$m_2 = \tan(\alpha+90°) = \frac{\sin(\alpha+90°)}{\cos(\alpha+90°)} = \frac{\sin\alpha\cos 90° + \cos\alpha\sin 90°}{\cos\alpha\cos 90° - \sin\alpha\sin 90°}$$

$$= -\frac{\cos\alpha}{\sin\alpha} = -\frac{1}{\tan\alpha} = -\frac{1}{m_1} \ .$$

b.   Assume $A_1 \neq 0$, $B_1 \neq 0$, $A_2 \neq 0$, and $B_2 \neq 0$, then
      y = (A_1/B_1)x - C_1/B_1 and y = (A_2/B_2)x - C_2/B_2 and thus
      m_1 = A_1/B_1 and m_2 = A_2/B_2. If the lines are perpendicular
      part(a) tells us that m_1 = $-1/m_2$, or
      A_1/B_1 = $-1/(A_2/B_2)$ = $-B_2/A_2$ and hence $A_1A_2 + B_1B_2 = 0$.
      Conversely, if $A_1A_2+B_1B_2 = 0$ then A_1/B_1 = $-B_2/A_2$ = $-1/(A_2/B_2)$
      so that m_1 = $-1/m_2$ and the lines are perpendicular.

      If one of the coefficients is zero, A_1 for instance, then B_1
      cannot be zero (since there would be no line) and thus
      y = $-C_1/B_1$ which has a zero slope. If the second line is
      perpendicular to the first line then it must have a vertical
      slope. Writing the second line as x = $- (B_2/A_2)$y $-C_2/A_2$ we
      then conclude that B_2/A_2 = 0 or B_2 = 0, in which case
      $A_1A_2 + B_1B_2 = 0$ since both A_1 and B_2 are zero. Conversely,
      if $A_1A_2 + B_1B_2 = 0$ we have that $B_1B_2 = 0$, since we have
      assumed A_1 = 0. However $B_1 \neq 0$, since then there would be
      no line, and hence we conclude that B_2 = 0 making the
      second line x = $-C_2/A_2$, which is vertical and hence
      perpendicular to y = $-C_1/B_1$.

21.   Solving for y we obtain $y = -(1/2)x - 5/2$ so that $m_1 = -1/2$.
From Prob.18b we have $m_2 = m_1$ for parallel lines and thus
$m_2 = -1/2$. Using Eq.(7) then yields $y-2 = -(1/2)[x-(-2)]$ or
$x+2y = 2$.

23.   The slope of the given line is $m_1 = -1/2$ and thus the slope
of the desired line is $m_2 = -1/m_1 = 2$ from Prob.19a. Thus
Eq.(7) yields $y-2 = 2(x+2)$ or $2x-y = -6$.

25.   We must write the equation in the form of Eq.(16) by
completing the squares in x and y. Thus $x^2+y^2-6x-4y = -8$
becomes $x^2-6x + (-6/2)^2 + y^2-4y + (-4/2)^2 = -8+9+4$, or
$(x-3)^2 + (y-2)^2 = 5$. Thus the center is at $(3,2)$ and the
radius is $\sqrt{5}$.

27.   Multiplying the equation by -1 yields $x^2+y^2+4x-4y = 0$.
Completing the squares then gives $x^2+4x+4+y^2-4y+4 = 4+4$ or
$(x+2)^2 + (y-2)^2 = 8$. Thus the center is at $(-2,2)$ and the
radius is $\sqrt{8} = 2\sqrt{2}$.

29.   Setting $y = 2x$ in the second equation yields $x^2+(2x)^2 = 9$ or
$5x^2 = 9$. One root of this equation is $x = 3/\sqrt{5}$ which means
$y = 6/\sqrt{5}$ from $y = 2x$. Similarly the second root is $x = -3/\sqrt{5}$
which gives $y = -6/\sqrt{5}$. Thus the graphs intersect at
$(3/\sqrt{5}, 6/\sqrt{5})$ and at $(-3/\sqrt{5}, -6/\sqrt{5})$.

31.   Solving the first equation for x gives $x = y^2-2$, which when
substituted in the second equation yields $2y-(y^2-2) = 3$.
Simplifying this last equation we get $y^2-2y+1 = 0$ or
$(y-1)^2 = 0$. The only solution of this is $y = 1$.Substituting
$y = 1$ into either of the given equations yields $x = -1$ and
thus the graphs intersect at $(-1,1)$ only. If one were to
actually draw the graphs, the line $2y-x = 3$ is tangent to
the parabola $x+2 = y^2$ at $(-1,1)$.

33.   First draw the line x+3y = 8,
        which has the intercepts
        (8,0) and (0,8/3). Now if we
        start from a point on the line
        and increase either x or y then
        x+3y > 8 and the relation is
        satisfied. If we decrease either
        x or y then x+3y < 8 and the relation is not satisfied.
        Thus the shaded region shown, including the line itself, is
        the graph of the given relation.

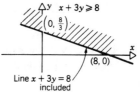

35.   Use the definition of |x| and consider the four cases: (1)
        y-x > 0 and y+x > 0.   Then |y-x| = y-x, |y+x| = y+x and thus
        y-x = y+x, or x = 0.   (2) y-x > 0 and y+x < 0.   Then
        |y-x| = y-x, |y+x| = -(y+x) and y-x = -(y+x), or y = 0.
        (3) y-x < 0 and y+x > 0. Then
        -(y-x) = y+x, or y = 0.
        (4) y-x < 0 and y+x < 0. Then
        -(y-x) = -(y+x) or x = 0.
        Thus the graph consists
        of the coordinate axis only.

37.   The given relation says that
        the distance from (x,y) to (-3,5)
        must be greater than 4 [using
        Eq.(2)]. Thus the graph consists
        of all points outside the circle
        of radius 4 with center at (-3,5).

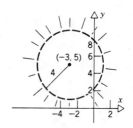

39.   If x ≥ 0 and y ≥ 0 then the
        equation reduces to y-y =
         x+x or x = 0. Since y ≥ 0
        this is the positive y axis
        only. If x ≥ 0 and y < 0 the
        equation reduces to y-(-y) =
        x+x or y = x. However, since
        x ≥ 0 and y < 0 there are no points on this line. If
        x < 0 and y ≥ 0 the equation becomes y-y = x+(-x) or 0 = 0
        That is, all points in the second quadrant, including the x
        axis, satisfy the equation. Finally, if x < 0 and y < 0 the
        equation becomes y-(-y) = x+(-x), or y = 0. This is
        excluded here since y < 0, but note that y = 0 for x < 0 is
        included above.

41.   If (x,y) is any point less than two units from (-1,-3), then
        Eq. (2) gives us $\sqrt{[x-(-1)]^2+[y-(-3)]^2}$ < 2. Since both sides are
        positive, we may square both sides to obtain
        $(x+1)^2+(y+3)^2$ < 4.

43.   Let (x,y) be any point which is equidistant from the two
        points. Then Eq. (2) gives us $\sqrt{(x+1)^2+(y-3)^2} = \sqrt{(x-4)^2+(y-2)^2}$.
        Expanding the terms under the square roots and then squaring
        both sides yields $(x^2+2x+1)+(y^2-6y+9) = (x^2-8x+16)+(y^2+4y+4)$
        which reduces to 5x-y = 5.

45.   A graphical appproach is helpful
        in solving the given inequality.
        To plot y = |4-x| we have y = 4-x
        for 4-x > 0, or x < 4, and
        y = x-4 for 4-x < 0, or x > 4.
        Likewise, for y = |2x|, y = 2x for
        x > 0 and y = -2x for x < 0. These
        four line segments are shown on
        the graph, where it is now seen
        that |2x| < |4-x| for -4 < x < 4/3.

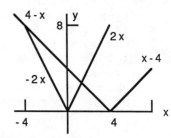

        The end points are the points of intersection of the
        indicated lines.

Section 1.4, Page 31

1a.   Since f(x) = 2-3x, then f(3) = 2-3(3) = 2-9 = -7.

 b.   f(-3) = 2-3(-3) = 2+9 = 11.

 c.   -f(3) = -[2-3(3)] = -(2-9) = 7.

 d.   -f(-3) = -[2-3(-3)] = -(2+9) = -11.

3.   Since the function is a quotient, it will be defined for all
        real x except the values for which the denominator is zero.
        Thus the domain is all real x except x = 3 and
        X = $(-\infty,3)\cup(3,\infty)$. Since the numerator is never zero and
        the denominator can take on all real values, the range will
        be all reals except 0 and hence f(X) = $(-\infty,0)\cup(0,\infty)$.

5.   As in Prob.3, the denominator is zero for $x = \pm 3$ and thus
      $X = \{x \mid x \neq \pm 3\}$. To determine the range we consider the
      denominator, x^2-9, which is always greater than or equal to
      -9 for all real x. For $x^2-9 > 0$, the reciprocal, $1/(x^2-9)$,
      can take any positive value for $|x| > 3$. For $x^2 \cdot 9 < 0$, the
      reciprocal takes on all negative values less than or equal
      to $-1/9$ for $|x| < 3$. The value $-1/9$ occurs at $x = 0$. Thus
      $f(X) = (-\infty, -1/9] \cup (0, \infty)$.

7.   In order that the function be defined we must have $9-x^2 \geq 0$,
      or $x^2 \leq 9$. Thus $-3 \leq x \leq 3$ and hence $X = [-3,3]$. Since
      $f(0) = 3$, and $f(\pm 3) = 0$ and since x^2 is always subtracted
      from 9 we have $0 \leq f(x) \leq 3$ and hence $f(X) = [0,3]$.

9a.   If $f(x) = \sqrt{1-x^2}$, $-1 \leq x \leq 1$, then $f(s-1) = \sqrt{1-(s-1)^2} =$
      $\sqrt{2s-s^2}$ for $-1 \leq s-1 \leq 1$, or $0 \leq s \leq 2$.

 b. $f(-t) = \sqrt{1-(-t)^2} = \sqrt{1-t^2}$ for $-1 \leq -t \leq 1$, or $-1 \leq t \leq 1$.

 c. $f(1/u) = \sqrt{1-(1/u)^2} = \sqrt{1-1/u^2}$ for $-1 \leq 1/u < 0$ and
      $0 < 1/u \leq 1$, or $-u \geq 1$ and $u \geq 1$ which can be written as
      $|u| \geq 1$.

 d. $f(w) = \sqrt{1-w^2}$ so that $f[f(w)] = \sqrt{1-\left(\sqrt{1-w^2}\right)^2} = \sqrt{1-1+w^2} =$
      $\sqrt{w^2} = |w|$. To find the domain, we must have $f(w)$ defined,
      i.e., $-1 \leq w \leq 1$, and $f(f(w))$ defined, ie., $0 \leq \sqrt{1-w^2} \leq 1$.

      Both of these are satisfied for $-1 \leq w \leq 1$.

11.   If $f(x) = 2x-3$, then $f(x+h) = 2(x+h)-3 = 2x+2h-3$ and
      $f(x+h) - f(x) = 2h$. Thus $[f(x+h) - f(x)]/h = 2$.

13.   Again, if $f(x) = 2x^2+3x-4$ then
      $f(x+h) = 2(x+h)^2 + 3(x+h) - 4 = 2x^2+3x-4 + 4xh+3h+2h^2$ and
      $[f(x+h) - f(x)]/h = [(4x+3)h+2h^2]/h = 4x+3+2h$.

12

Section 1.4

15.   In this case f(x+h) = 1/x+h so that

$$[f(x+h)-f(x)]/h = \frac{\frac{1}{x+h} - \frac{1}{x}}{h} = \frac{x-(x+h)}{hx(x+h)} = \frac{-1}{x(x+h)}.$$

17. $\phi(-3) = |-3+2|+|-3-2| = |-1|+|-5| = 6.$ In a similar fashion
       $\phi(-1) = |1|+|-3| = 4,$ $\phi(1) = |3|+|-1| = 4$ and
       $\phi(3) = |5|+|1| = 6.$ If $x \geq 2$ then x+2 $\geq$ 0 and x-2 $\geq$ 0 and
       $\phi(x) = (x+2)+(x-2) = 2x.$ If $-2 \leq x < 2$ then x+2 $\geq$ 0

and x-2 < 0 so that
$\phi(x) = (x+2)-(x-2) = 4.$
Finally,if x < -2 then
x+2 < 0 and x-2 < 0 so that
$\phi(x) = -(x+2) -(x-2) = -2x.$
Thus the graph is as shown.

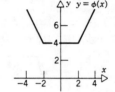

19.   The relation of $y^2 = x$ does not define y as a function of x
       since for each positive x value there are two possible
       values for y.

21.   The relation y = |2x| does define y as a function of x since
       for each x value there is precisely one y value.

23.   As in Prob.19, for each x value in the interval [-1,1] there
       are two possible values for y. Thus the relation does not
       define y as a function of x.

25.   The simplest function to look for would be a polynomial.
       Since the function must pass through the three given points
       it must contain three unknown constants and therefore it
       must be at least quadratic. Thus we assume $f(x) = ax^2+bx+c$
       which means f(-2) = 4a-2b+c = 0, f(3) = 9a+3b+c = 0 and
       f(0) = c = -36. Solving the three equations yields a = 6,
       b = -6 and c = -36 and hence $f(x) = 6x^2-6x-36 = 6(x-3)(x+2).$
       Higher order polynomials with three unknown constants can
       also be used. For instance we can assume $f(x) = x^3+ax^2+bx+c$
       and follow the same procedure as above to find a = 5,
       b = -12 and c = -36. Thus $f(x) = x^3+5x^2-12x-36$ is also a
       possible answer.

Section 1.5, Page 40

1.   If $3+2x \geq 0$ then $f(x) = 3+2x$.
        If $3+2x < 0$ then
        $f(x) = -(3+2x)$. Thus we graph
        $f(x) = 3+2x$ for $x \geq -3/2$ and
        $f(x) = -3-2x$ for $x < -3/2$.
        From the graph we see that the
        range is all reals greater than
        or equal to zero, $[0,\infty)$.

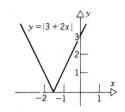

3.   If $x \geq 1$, $f(x) = (x+2)-(x-1) = 3$,
        if $-2 \leq x < 1$ then
        $f(x) = (x+2)+(x-1) = 2x+1$,
        if $x < -2$ then
        $f(x) = -(x+2)+(x-1) = -3$.
        Graphing each of these segments
        yields the graph shown, from
        which it can be seen that the
        range is $[-3,3]$.

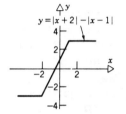

5.   Since $g(x)$ is a multiple of the
        function $f(x)$ of Eq.(15), its
        graph can be obtained from Fig.1.5.11
        by multiplying each value of $f(x)$
        by 2, obtaining the graph shown.
        The range is then seen to be all
        positive and negative even
        integers and zero.

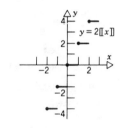

7.   Again, the graph for $g(x)$ is
        obtained from Fig.1.5.11 by
        multiplying each value of $f(x)$ by
        -1 to obtain the graph shown. The
        range is then all positive
        and negative integers and zero.

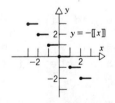

9.   As in Ex.3, $f(x)$ is a parabola.
        If we set $f(x)$ equal to zero we
        find $x = \pm 1$, which are the x
        intercepts. Setting $x = 0$ we find
        that $f(0) = -1$, which is the y
        intercept and thus the graph is as
        shown. From the graph we see that the range is $[-1,\infty)$.

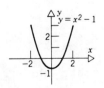

11.   Since $f(x) = 2x^2+x-6 = (2x-3)(x+2)$
        we see that the x intercepts are
        3/2 and -2. Also, since $f(0) = -6$,
        the y intercept is -6. To find
        the vertex in this case we must
        complete the square as follows:
        $f(x) = 2(x^2+x/2+1/16) - 6 - 1/8$
            $= 2(x+1/4)^2 - 49/8$. Thus the
        vertex is at (-1/4,-49/8) as shown
        on the graph. From the graph the range is $[-49/8,\infty)$.

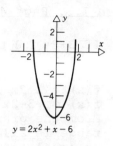

13.   Setting $g(x) = 0$ we find the x
        intercepts are -2,0,1 and letting
        $x = 0$ we find the y intercept is 0.
        Since $g(x) = x(x-1)(x+2) = x^3+x^2-2x$,
        we see that $g(x)$ behaves like x^3
        when $|x|$ is large and hence the
        graph is as shown. The range is all reals, $(-\infty,\infty)$.

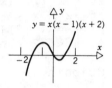

15.   If we let $y = \sqrt{9-x^2}$ and square
        both sides we obtain $x^2+y^2 = 9$
        which is a circle with center at
        (0,0) and radius 3. Since $\phi(x)$
        must be positive we conclude that
        the graph of ϕ is the upper half circle as shown. The range
        is then [0,3].

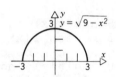

17.   Since $\psi(0) = 0$ the y intercept is

        0 and since $\dfrac{x}{1-x^2} = 0$ has the

        solution $x = 0$, the x intercept is
        also the origin. For $0 < x < 1$,
        $\psi(x) > 0$ and for $-1 < x < 0$,
        $\psi(x) < 0$. For $x > 1$, $\psi(x) < 0$ and
        for large x, $\psi(x)$ behaves like $-1/x$
        since the 1 in the denominator is negligible. Likewise, for
        $x < -1$, $\psi(x) > 0$ and for large values of $-x$, $\psi(x)$ behaves
        like $1/x$. Plotting some individual points then yields the
        graph shown, which indicates that the range is $(-\infty,\infty)$.

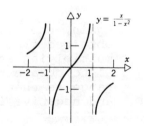

19.   Setting $\psi(x) = 0$ we find $x = 1$
and since $\psi(0) = 1$ we conclude
that the x and y intercepts are
each 1. For large positive values
of x, $\psi(x)$ will be slightly larger
than -1, while for large negative
values of x, $\psi(x)$ will be slightly
less than -1. For x slightly
greater than -1, $\psi(x)$ will have large positive values while
for x slightly less than -1, $\psi(x)$ will have large negative
values. Plotting some individual points will then assist in
obtaining the graph shown. Since $\psi(x)$ never takes on the
value -1, we see that the range is $(-\infty,-1)\cup(-1,\infty)$.

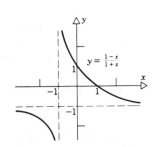

21.   Since $f(x) = \begin{cases} 0 & x < 0 \\ \sqrt{x} & x \geq 0 \end{cases}$ we see that $f(x+2) = \begin{cases} 0 & x+2 < 0 \\ \sqrt{x+2} & x+2 \geq 0 \end{cases}$

or $f(x+2) = \begin{cases} 0 & x < -2 \\ \sqrt{x+2} & x \geq -2 \end{cases}$.

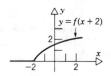

Also $f(x-2) = \begin{cases} 0 & x-2 < 0 \\ \sqrt{x-2} & x-2 \geq 0 \end{cases}$

or $f(x-2) = \begin{cases} 0 & x < 2 \\ \sqrt{x-2} & x \geq 2 \end{cases}$.

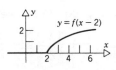

23.   From the definition of H in Prob.22 we see that the product
H(x)H(2t-x) will be non-zero as long as x and 2t-x are
greater than or equal to zero. Thus we must have $x \geq 0$ and
$2t-x \geq 0$ or $0 \leq x \leq 2t$, if we assume $t > 0$. Thus the
intervals are [0,1/2], [0,1], [0,2] and [0,4] respectively
for $t = 1/4$, 1/2, 1, and 2. Since t is multiplied by 2 the x
interval is expanding twice as fast as t is increasing.

25.   If $f(x) = -3x$, then $f(-x) = -3(-x) = 3x = -(-3x) = -f(x)$.
Thus $f(x)$ is an odd function.

27.   If $f(x) = 2x^3+3x^2-7x+4$, then $f(-x) = -2x^3+3x^2+7x+4$, which is
       neither $f(x)$ nor $-f(x)$ and thus $f(x)$ is neither even nor
       odd. In this case f involves both even and odd powers of x.

29.   In this case $f(-x) = \phi(|-x|) = \phi(|x|) = f(x)$ and thus $f(x)$
       is an even function.

31.   Let $h(x) = f(x) + g(x)$. If f and g are even functions then
       $h(-x) = f(-x) + g(-x) = f(x) + g(x) = h(x)$ and thus h is
       also even. If f and g are odd functions then
       $h(-x) = f(-x) + g(-x) = -f(x) - g(x) = -h(x)$ and thus h is
       also odd. If f is even and g is odd then
       $h(-x) = f(-x) + g(-x) = f(x) - g(x)$, which is neither $h(x)$
       nor $-h(x)$ unless one of the functions f or g is everywhere
       zero.

33.   Assume g is an even function and h is an odd function. The
       following steps show how to find g and h so that
       $f(x) = g(x) + h(x)$. If this latter equation is true, then
       $f(-x) = g(-x) + h(-x) = g(x) - h(x)$. Thus
       $f(x) + f(-x) = 2g(x)$, or $g(x) = [f(x) + f(-x)]/2$. Likewise
       $f(x) - f(-x) = 2h(x)$, or $h(x) = [f(x) - f(-x)]/2$. Note that
       $g(-x) = g(x)$ and $h(-x) = -h(x)$ and that $g(x) + h(x) = f(x)$.
       Thus we have found an even function, g, and an odd function,
       h, which add together to give f.

Section 1.6, Page 48

1.   Using Eq.(19) we have

$$\sin\left(\frac{\pi}{2} + \frac{\pi}{4}\right) = \sin\frac{\pi}{2}\cos\frac{\pi}{4} + \cos\frac{\pi}{2}\sin\frac{\pi}{4} = \cos\frac{\pi}{4} = \frac{\sqrt{2}}{2}\ .$$

3.   Using Eq.(20) we have $\sin\left(\frac{\pi}{3} - \frac{5\pi}{4}\right) = \sin\frac{\pi}{3}\cos\frac{5\pi}{4} - \cos\frac{\pi}{3}\sin\frac{5\pi}{4}$.

       In order to use Table 1.1 we must write $5\pi/4 = \pi + \pi/4$ and

       use Eqs.(17) and (19) to obtain $\sin\left(\frac{\pi}{3} - \frac{5\pi}{4}\right) =$

$$\left(\frac{\sqrt{3}}{2}\right)\left[\cos\pi\cos\frac{\pi}{4} - \sin\pi\sin\frac{\pi}{4}\right] - \left(\frac{1}{2}\right)\left[\sin\pi\cos\frac{\pi}{4} + \cos\pi\sin\frac{\pi}{4}\right] =$$

$$-\left(\frac{\sqrt{3}}{2}\right)\left(\frac{\sqrt{2}}{2}\right) + \left(\frac{1}{2}\right)\left(\frac{\sqrt{2}}{2}\right) = \left(\frac{-\sqrt{6}+\sqrt{2}}{4}\right).$$

5. Let $\phi = \pi/4$ and use Eq.(25) to obtain $\sin\dfrac{\pi}{8} = \sin\left(\dfrac{\pi/4}{2}\right) =$

$+\sqrt{\dfrac{1-\cos\pi/4}{2}} = \sqrt{\dfrac{1-\sqrt{2}/2}{2}} = \dfrac{\sqrt{2-\sqrt{2}}}{2}$. We use the + sign in

Eq.(25) since $\pi/8$ lies in the first quadrant.

7. Using Eq.(26) with $\phi = \pi/6$ we obtain

$$\cos\frac{\pi}{12} = \cos\frac{(\pi/6)}{2} = \sqrt{\frac{1+\cos(\pi/6)}{2}} = \sqrt{\frac{1+\sqrt{3}/2}{2}} = \frac{\sqrt{2+\sqrt{3}}}{2} .$$

The + sign is used since $\pi/12$ lies in the first quadrant.

9. The graph of f(x) will be
similar to Fig.1.6.6 . Since
the coefficient of sine is 1,
the amplitude of f(x) is 1.
The maximum values of f(x),1,
occur at the points where
$2x = \pi/2$ and $5\pi/2$, which are

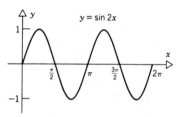

$x = \pi/4$ and $5\pi/4$. Likewise the minimum values, -1, occur at
$2x = 3\pi/2$ and $7\pi/2$, or $x = 3\pi/4$ and $7\pi/4$. Finally f(x) = 0
when $2x = 0$, π, 2π, 3π, and 4π, or $x = 0$, $\pi/2$, π, $3\pi/2$ and
2π. Other points may be plotted in a similar fashion, using
values of $\sin x$ from Table 1.1.

11. This graph will be related to
that shown in Fig.1.6.7. The
y intercept, in this case, is
$f(0) = -2\cos(-\pi/4) = -\sqrt{2}$
and the amplitude is 2. In this
case the peak values, ±2, occur
at $x-\pi/4 = 0$ and π, or $x = \pi/4$ and

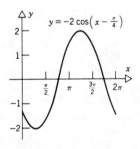

$5\pi/4$ and the x intercepts occur at
$x-\pi/4 = \pi/2$ and $3\pi/2$, or $x = 3\pi/4$ and $7\pi/4$.Finally, use of
Table 1.1 yields the graph shown.

13.   Since this is a sine function
        we find the peak values by
        setting $3x = \pi/2,\ 3\pi/2,\ 5\pi/2,$
        $7\pi/2,\ 9\pi/2$ and $11\pi/2$, which
        yield $x = \pi/6,\ \pi/2,\ 5\pi/6,\ 7\pi/6,$
        $3\pi/2$ and $11\pi/6$. Since the
        amplitude is two, $f(x)$ will be 2
        at $x = \pi/6,\ 5\pi/6$ and $3\pi/2$ and will

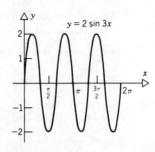

be -2 at the remaining three. The x intercepts occur at $3x$
$= n\pi$, or $x = n\pi/3$, for $n = 0,1,2\ldots6$. Other points may be
plotted by the use of Table 1.1.

15.   Using Eqs.(6b) and (9) we have have
        $\sec(-\theta) = 1/\cos(-\theta) = 1/\cos\theta = \sec\theta$ and thus $\sec\theta$ is an even
        function. Using Eqs.(6c) and (10) we have
        $\csc(-\theta) = 1/\sin(-\theta) = -1/\sin\theta = -\csc\theta$ and thus $\csc\theta$ is an
        odd function. Finally, Eqs.(6a) and (11) yield
        $\cot(-\theta) = 1/\tan(-\theta) = -1/\tan\theta = -\operatorname{ctn}\theta$ and thus $\cot(\theta)$ is an
        odd function.

17.   Using Eq.(5c) we have $\tan\theta = y/x$
        for $P(\theta)$ and $\tan(\theta+\pi) = -y/-x$
        for $P(\theta+\pi)$. Thus $\tan\theta = \tan(\theta+\pi)$.
        Likewise Eq.(6a) gives us
        $\cot\theta = x/y$ and $\cot(\theta+\pi) = -x/-y$
        and thus $\cot\theta = \cot(\theta+\pi)$.

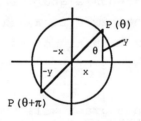

19.   We look for values of P for which $\sin[(x+P)/2] = \sin(x/2)$.
        Using Eq.(19) we have
        $\sin[(x+P)/2] = \sin(x/2)\cos(P/2) + \sin(P/2)\cos(x/2)$. The
        right side will reduce to $\sin(x/2)$ when $\cos(P/2) = 1$ and
        $\sin(P/2) = 0$, which are both satisfied when P/2 is a
        multiple of 2π. Thus the fundamental period is $P = 4\pi$.

21.   Again we look for values of P for which
        $2\cos[(x+P)/3 + 4] = 2\cos(x/3 + 4)$. Using Eq.(17) we have
        $2\cos[(x+P)/3 + 4] = 2\cos[(x/3 + 4)+P/3]$
                $= 2\cos(x/3 + 4)\cos(P/3) - 2\sin(x/3 + 4)\sin(P/3)$. The
        right side will reduce to $2\cos(x/3 + 4)$ provided $\cos(P/3) =$
        1 and $\sin(P/3) = 0$, which are both satisfied when $P/3 =$
        $n(2\pi)$, $n = 1,2\ldots$. Thus the fundamental period is $P = 6\pi$.

23.   Multiplying both sides of Eq.(27) by two yields
       $2\cos\theta\cos\phi = \cos(\theta+\phi) + \cos(\theta-\phi)$. If $\theta = (x+y)/2$ and
       $\phi = (x-y)/2$ then $\theta+\phi = x$ and $\theta-\phi = y$. Substituting these
       four relations into the first equation then yields
       $2\cos[(x+y)/2]\cos[(x-y)/2] = \cos x + \cos y$.

25.   Multiplying both sides of Eq.(29) by two yields
       $2\sin\theta\cos\phi = \sin(\theta+\phi) + \sin(\theta-\phi)$. Let θ and ϕ be the same as
       in Prob.23 and thus $2\sin[(x+y)/2]\cos[(x-y)/2] = \sin x + \sin y$.

27.   We have $R\sin(\theta+\delta) = R\sin\theta\cos\delta + R\cos\theta\sin\delta$
                           $= (R\cos\delta)\sin\theta + (R\sin\delta)\cos\theta$. Equating this
       to $a\sin\theta + b\cos\theta$ we obtain $a = R\cos\delta$ and $b = R\sin\delta$, so that
       $a^2 + b^2 = R^2(\cos^2\delta + \sin^2\delta) = R^2$ or $R = (a^2 + b^2)^{1/2}$,
       $\cos\delta = a/R$, and $\sin\delta = b/R$.

29.   Comparing $\sin\theta + \sqrt{3}\cos\theta$ to
       $a\sin\theta + b\cos\theta$ we see that $a = 1$
       and $b = \sqrt{3}$. Thus, from the
       paragraph just ahead of Prob.27,

$y = 2\sin(\theta + \frac{\pi}{3})$

       we have $R = \sqrt{1^2 + (\sqrt{3})^2} = 2$ and
       $\cos\delta = 1/2$, $\sin\delta = \sqrt{3}/2$. Since both cosine and sine are
       positive, δ is a first quadrant angle such that $\tan\delta = \sqrt{3}$.
       Thus $\delta = \pi/3$ and $\sin\theta + \sqrt{3}\cos\theta = 2\sin(\theta+\pi/3)$. As discussed
       in the text, the graph will then be a sine curve with
       amplitude 2 and shifted to the left $\pi/3$ units. Substituting
       these values into Fig.1.6.14 yields the graph above.

31.   As in Prob.27 we have $a=3$ and
       $b=-2$ and thus $R = \sqrt{(3)^2 + (-2)^2} = \sqrt{13}$.

$y = \sqrt{13}\sin(\theta + \delta)$

       Hence $\cos\delta = 3/\sqrt{13}$ and
       $\sin\delta = -2/\sqrt{13}$ and therefore δ
       must be a fourth quadrant angle
       such that $\tan\delta = -2/3$, or $\delta = 2\pi - .588 \cong 5.6952$ radians.
       Substituting these values for R and δ into the graph of
       Fig.1.6.14 yields the graph shown.

33.   The perpendicular line of Fig.1.6.13 divides side a into two
       segments. Using trigonometry, the left segment has length
       $b\cos\theta$ and thus the right segment has length $a-b\cos\theta$. The
       Pythagorean Theorem applied to the right triangle then

       yields $c^2 = h^2+(a-b\cos\theta)^2 = h^2+a^2-2ab\cos\theta+b^2\cos^2\theta$. Likewise

       for the left triangle we have $b^2 = h^2+(b\cos\theta)^2$, or

       $h^2 = b^2-b^2\cos^2\theta$. Substituting this into the first equation

       yields $c^2 = b^2 + a^2 - 2ab\cos\theta$, which is the law of cosines.

35.   If we replace ϕ of Prob.32 by $\phi+\pi/2$ we then have
       $\cos[\theta-(\phi+\pi/2)] = \cos\theta\cos(\phi+\pi/2) + \sin\theta\sin(\phi+\pi/2)$

                        $= -\cos\theta\sin\phi + \sin\theta\cos\phi$, using Eqs.(14) and

       (15).   However, $\cos[\theta-(\phi+\pi/2)] = \cos[(\theta-\phi)-\pi/2] =$

       $\cos(\theta-\phi)\cos\pi/2 + \sin(\theta-\phi)\sin\pi/2 = \sin(\theta-\phi)$, where we have
       again utilized Prob.32.   Substituting this last result for
       the left side of the first equation gives the desired
       result.

37.   The given identity is derived by the following steps, which
       are justified by the given equations.

$$\frac{\cos(x+h)-\cos x}{h} = \frac{(\cos x\,\cos h-\sin x\,\sin h)-\cos x}{h};\quad \text{Eq. (17)}.$$

$$= \frac{\cos x(\cos h-1)-\sin x\,\sin h}{h}$$

$$= \frac{\cos x\left(-2\sin^2\frac{h}{2}\right)-\sin x\left(2\sin\frac{h}{2}\cos\frac{h}{2}\right)}{h};\quad \text{Eqs. (23), (22)}.$$

$$= -\frac{\sin(h/2)}{h/2}\left[\sin x\,\cos(h/2)+\cos x\,\sin(h/2)\right]$$

$$= -\frac{\sin(h/2)}{h/2}\,\sin(x+h/2)\quad \text{Eq. (19)}.$$

Chapter 1 Review, Page 50

1.   The graph is sketched using
      the intercepts (4,0) and
      (0,-6).

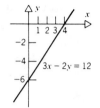

3.   Completing the square gives
      $2(x^2-2x+1) + 2(y^2+4y+4) = 9$ or
      $(x-1)^2 + (y+2)^2 = 9/2$, which
      is a circle with center at
      $(1,-2)$ and radius $3/\sqrt{2}$.

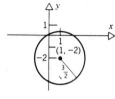

5. (x,y) in the first quadrant
      gives $y = x$. (x,y) in the
      second quadrant gives the
      negative x-axis. (x,y) in the
      third quadrant gives all points.
      (x,y) in the fourth quadrant
      gives the negative y-axis.

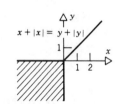

7.   For $0 \leq x \leq 2\pi/3$, $y = 1+2\cos x$.
      For $2\pi/3 < x < 4\pi/3$,
      $y = -(1+2\cos x)$. For $4\pi/3 \leq x \leq 2\pi$,
      $y = 1+2\cos x$. The function is
      periodic with period 2π.

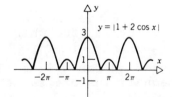

9. $X = (-\infty,-1)\cup(-1,\infty)$. With
      $f(x) = 1+2/(x+1)$, $y = 1$ is a
      horizontal asymptote.
      $x = -1$ a vertical asymptote
      and $f(X) = (-\infty,1)\cup(1,\infty)$.

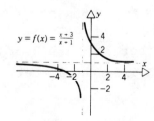

11. The graph is bounded by |x|, that
     is −x ≤ xsin2x ≤ x, with x-axis
     intercepts the same as those of
     sin2x. The graph also has y-axis
     symmetry. X = (−∞,∞) and
     f(X) = (−∞,∞).

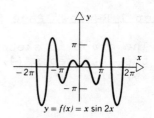

13. For x < 0, f(x) = −1, for
     0 ≤ x ≤1, f(x) = 2x−1 and for
     1 < x, f(x) = 1,
     X = (−∞,∞), f(X) = [−1,1].

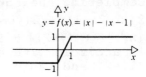

15. f(x) = −(x+2)(x−1) is a parabola
     with x-intercepts at −2 and 1
     and vertex at (−1/2, 9/4).
     X = (−∞,∞), f(X) = (−∞,9/4).

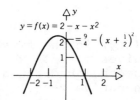

17. $x^2 \geq 16$ is equivalent to x ≤ −4 and x ≥ 4. Thus
     S = (−∞,−4]∪[4,∞).

19. For x > 0 the inequality becomes −7x < 6+x^2 < 5x. The left
     equality is true when (x+1)(x+6) > 0, which is satisfied for
     x in (−∞,−6)∪(−1,∞). Similarly, the right inequality is
     satisfied for x in (2,3). Thus for x in (2,3) both
     inequalities are satisfied. Now, for x < 0, the inequality
     becomes 5x < 6+x^2 < −7x. The left inequality is satisfied for
     all x < 0, while the right inequality is satisfied for x in
     (−6,−1). Thus S = (−6,−1)∪(3,2).

21. Either 3x + 1 ≥ 2 or 3x + 1 ≤ −2. The former gives
     x ≥ 1/3, the latter, x ≤ −1. Thus S = (−∞,−1]∪[1/3,∞).

23. By periodicity, we need examine only $0 \leq x \leq \pi$. In that
    interval, $\cos 2x \geq 1/2$ for $0 \leq x \leq \pi/6$ and $5\pi/6 \leq x \leq \pi$. Thus
    $S = [0,\pi/6] \cup [5\pi/6,\pi]$, as well as sets of the form
    $[k\pi - \pi/6, k\pi + \pi/6]$, $k = 0, \pm 1, \pm 2, \ldots$. Hence
    $S = \bigcup\limits_{k=-\infty}^{\infty} [k\pi - \pi/6, \ k\pi + \pi/6]$.

25. The slope of the given line is $m = -1/2$, as is the slope of
    the required line, whose equation is $y-5 = (-1/2)(x-2)$ or
    $x+2y = 12$.

27. The radius of the circle is $[(2+3)^2 + (3-1)^2]^{1/2} = 29^{1/2}$ so the
    equation is $(x+3)^2 + (y-1)^2 = 29$.

29. Let $A = (-1,3)$, $B = (4,2)$ and P be a point satisfying the
    conditions $d(P,A) = 2d(P,B)$ or
    $[(x+1)^2 + (y-3)^2]^{1/2} = 2[(x-4)^2 + (y-2)^2]^{1/2}$. Squaring and
    simplifying gives $3x^2 + 3y^2 - 34x - 10y + 74 = 0$.

31. $X = (-\infty,\infty)$ and
    $f(X) = \bigcup\limits_{k=-\infty}^{\infty} [2k,2k+1]$.

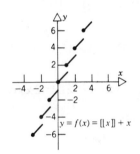

33. $f(-x) = -x[-x-4][-x+4]/4$
    $\qquad = -x(x+4)(x-4)/4,$
    $\qquad = -f(x)$ so f is odd.
    $X = (-\infty,\infty)$ and $f(X) = (-\infty,\infty)$.

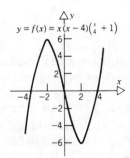

35.   X = [1,∞) and
      f(X) = {y|y = 1/k, k=1,2,3,...}.

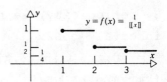

37.   Write f(x) = (4/x) - x. Then
      X = (-∞,0)∪(0,∞) and
      f(X) = (-∞,∞).

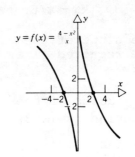

39.   f(x) = |cos2(x-π/4)| is a
      translation by π/4 of |cos2x|,
      and so has period π.

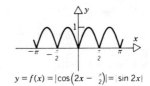

41.   Write f(x) = 2sin2x/cos2x.

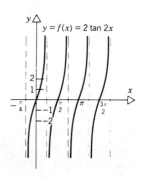

43.   From tanx = sinx/cosx, tan²x = sin²x/(1-sin²x). Solving for
      sin²x gives sin²x = (tan²x)/(1+tan²x).

45.   With a,b > 0 and from b > a, we find b-a > 0 and b(b-a) > 0
      so b² > ab. Similarly a(b-a) > 0 so ab > a². From the second
      property of inequalities in Section 1.2, we have from a < b,
      (1/a)a < (1/a)b or 1 < b/a and (1/b)(1) < (1/b)(b/a) or
      1/b < 1/a. These results are not always true if a < 0. For
      instance if a < 0 and b > 0, then b > a and 1/b > 1/a also.
      The first result is clearly not true if a < 0.

47. Now, $(\sqrt{a} + \sqrt{b})^2 = a + 2\sqrt{ab} + b > a+b$ as $\sqrt{ab} > 0$. But $a+b$ may be seen as $(\sqrt{a+b})^2$. With this in mind, then from Prob.46, $\sqrt{a} + \sqrt{b} > \sqrt{a+b}$.

CHAPTER 2

Section 2.1, Page 60

1. From Eq. (7), with $f(t) = 5t^2-3$ and $t_0 = 1$ we have

$$v_{av}(1,t) = \frac{f(t)-f(1)}{t-1} = \frac{(5t^2-3)-2}{t-1} = \frac{5(t^2-1)}{t-1} = 5(t+1), t \neq 1.$$ From Eq. (8)

$$\lim_{t \to t_0} v_{av}(t_0,t) = \lim_{t \to 1} v_{av}(1,t) = \lim_{t \to 1} 5(t+1).$$ For t close to 1, $5(t+1)$ is close to 10 so that $v(1) = 10$.

3. Again, from Eq. (7), $v_{av}(2,t) = \frac{f(t)-f(2)}{t-2} = \frac{(2t^3-t^2)-12}{t-2} = \frac{2t^3-t^2-12}{t-2}$

so $v_{av}(2,t) = \frac{(2t^2+3t+6)(t-2)}{t-2} = 2t^2+3t+6; \ t \neq 2.$ Finally,

$v(2) = \lim_{t \to 2} v_{av}(2,t) = \lim_{t \to 2} (2t^2+3t+6).$ Thus, since $2t^2+3t+6$ is close to 20 when t is close to 2, $v(2) = 20$.

5. By Eq. (11), the slope of the tangent line to the graph of

$y = f(x)$ at $(1,4)$ is $m = \lim_{x \to 1} \frac{(3x^2+1)-4}{x-1} = \lim_{x \to 1} \frac{3(x^2-1)}{x-1} = \lim_{x \to 1} 3(x+1) = 6.$

Using the point slope form, $y-y_0 = m(x-x_0)$, the equation of

the tangent line is $y-4 = 6(x-1)$ or $y = 6x-2$.

7. Since $f(x) = 2x^2-3x+4$, $x_0 = 2$, we have

$m = \lim_{x \to 2} \frac{(2x^2-3x+4)-6}{x-2} = \lim_{x \to 2} \frac{(2x+1)(x-2)}{x-2} = \lim_{x \to 2} (2x+1) = 5.$ The tangent

line then is given by $y-6 = 5(x-2)$ or $y = 5x-4$.

9. $m = \lim_{x \to 2} \frac{5/(x+3)-1}{x-2} = \lim_{x \to 2} \frac{2-x}{(x-2)(x+3)} = \lim_{x \to 2} \frac{-1}{x+3} = -1/5.$

The tangent line is then $y-1 = (-1/5)(x-2)$ or $x+5y = 7$.

11a. First determine the average velocity over the time interval

from t_1 to t_2: $v_{av}(t_1, t_2) = \dfrac{16t_2^2 - 16t_1^2}{t_2 - t_1} = 16(t_1 + t_2)$. Now the

midpoint is $t_0 = (t_1+t_2)/2$ and thus:

$$v(t_0) = \lim_{t \to t_0} \frac{16t^2 - 16t_0^2}{t - t_0} = \lim_{t \to t_0} 16(t+t_0) = 32t_0 = 16(t_1+t_2) = v_{av}(t_1, t_2).$$

b. $v(t_0) = \displaystyle\lim_{t \to t_0} \frac{t^3 - t_0^3}{t - t_0} = \lim_{t \to t_0} (t^2 + t_0 t + t_0^2) = 3t_0^2.$

$$v_{av}(t_0 - \tau, t_0 + \tau) = \frac{(t_0 + \tau)^3 - (t_0 - \tau)^3}{(t_0 + \tau) - (t_0 - \tau)} = \frac{6\tau t_0^2 + 2\tau^3}{2\tau} = 3t_0^2 + \tau^2.$$

13a. For $s(t) = -\dfrac{1}{2} gt^2 + v_0 t$ we have

$$v(t_0) = \lim_{t \to t_0} \frac{\left(-\frac{1}{2}gt^2 + v_0 t\right) - \left(-\frac{1}{2}gt_0^2 + v_0 t_0\right)}{t - t_0} =$$

$$\lim_{t \to t_0} \frac{-\left(\frac{1}{2}gt^2 - \frac{1}{2}gt_0^2\right) + (v_0 t - v_0 t_0)}{t - t_0} = \lim_{t \to t_0} \frac{-\frac{1}{2}g(t - t_0)(t + t_0) + v_0(t - t_0)}{t - t_0} =$$

$$\lim_{t \to t_0} \left[-\frac{1}{2} g(t + t_0) + v_0\right] = -gt_0 + v_0, \text{ or } v(t) = -gt + v_0.$$

b. Intuitively, when the object is at its greatest height its
instantaneous velocity is zero so $-gt + v_0 = 0$ or $t = v_0/g$.

15. $D(x) = \dfrac{f(x) - f(2)}{x - 2} = \dfrac{\sqrt{x^2 + 5} - 3}{x - 2}$. Thus $D(2.1) = .676$, $D(2.01) = .668$,

$D(1.9) = .657$, $D(1.99) = .666$ and hence we may estimate
$\displaystyle\lim_{x \to 2} D(x) \cong .667 \cong 2/3.$

17. $D(x) = \dfrac{(x^2 - 5)^{3/2} - 8}{x - 3}$. Thus $D(2.9) = 17.03$, $D(2.99) = 17.90$,

$D(3.1) = 18.98$, $D(3.01) = 18.09$ so $\displaystyle\lim_{x \to 3} D(x) \cong 18.00.$

19. $D(x) = \dfrac{\tan x + 1}{x - 3\pi/4}$.   Thus $D(2.3) = 2.12$, $D(2.35) = 2.01$,

$D(2.36) = 1.99$, $D(2.4) = 1.92$ and $\lim\limits_{x \to 3\pi/4} D(x) = 2.00$.

21. $D(x) = \dfrac{\sin 2x - \sqrt{3}/2}{x - \pi/3}$.   Thus $D(1.1) = -1.09$, $D(1.05) = -1.005$,

$D(1.04) = -.987$, $D(1.00) = -.917$ and $\lim\limits_{x \to \pi/3} D(x) \cong -1.000$,

where we have used the middle two values weighted towards -
1.005 since $\pi/3$ is closer to 1.05 than to 1.04.

Section 2.2, Page 72

1. Note that $f(-2)$ is undefined, so that the limit cannot be
calculated as $f(-2)$. In fact, at $x = -2$ $f(x)$ has the form
$0/0$. When this happens an algebraic or other technique might
be useful. In this case the numerator can be factored so we

have $f(x) = \dfrac{(x^2-4)}{(x+2)} = \dfrac{(x+2)(x-2)}{x+2} = x-2 \ (x \ne -2)$.   From this it is

apparent that $f(x)$ is close to -4 when x is close to -2.

3. f can be simplified for $x \ne 2$ as $f(x) = \dfrac{x^2-x-2}{x-2} = \dfrac{(x-2)(x+1)}{x-2} =$

$x+1; \ (x \ne 2)$.   Therefore when x is close to 2, $f(x)$ is close
to 3.

5. $f(x) = \dfrac{x-1}{x^2-4x+3} = \dfrac{x-1}{(x-1)(x-3)} = \dfrac{1}{x-3} ; (x \ne 1,3)$.   When x is close to 1,
$f(x)$ is close to -1/2. However, when x is close to 3, $f(x)$
becomes unbounded, and thus has no limit.

7. Simplify $f(x) = \dfrac{x^2+x-6}{x^2-x-2} = \dfrac{(x+3)(x-2)}{(x+1)(x-2)} = \dfrac{x+3}{x+1} ; \ (x \ne -1,2)$.   When x is
close to 2, $f(x)$ is close to 5/3 but when x is close to -1,
$(x+3)/(x+1)$ becomes unbounded. Thus f has no limit at -1.

9. Using the hint $f(x) = \dfrac{\sqrt{x}-1}{x-1} = \dfrac{\sqrt{x}-1}{x-1} \cdot \dfrac{\sqrt{x}+1}{\sqrt{x}+1} = \dfrac{x-1}{(x-1)(\sqrt{x}+1)}$ so
$f(x) = \dfrac{1}{\sqrt{x}+1} (x \ne 1)$. Now, for $x > 0$ and close to 1, $f(x)$ is
close to 1/2. Alternatively, observing that $x-1 = (\sqrt{x}-1)(\sqrt{x}+1)$
leads to the same simplified form for $f(x)$.

11.   In this problem f(x) is defined for all x and thus for x
      close to 0, f(x) is close to 1.

13.   If 0<x<1/2, then 0<2x<1 so -3 < 2x-3 < -4 and f(x) = -3.
      If 1/2<x<1, then 1<2x<2 so -2 < 2x-3 < -1 and f(x) = -2.
      Therefore, there is no single L such that when x is close to
      1/2, f(x) is close to L, showing f has no limit at x = 1/2.
      The situation is different for x close to 1/4. In this case,
      f(x) is close to (actually equals) -3, showing f has limit -3
      for x = 1/4.

15.   For f(x) = $\sqrt{|x|}$, the limit as
      x→0 can be determined
      geometrically. It is seen
      that for x close to 0, f(x)
      is close to 0.

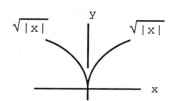

17.   Note that f(x) = f(-x) and that f(.5) = .4897, f(.1) = .4996,
      and f(.05) = .4999. Thus we estimate the limit as 1/2.

19.   We have f(.9) = 2.270, f(.8) = 2.030, f(.75) = 1.932,
      f(.65) = 1.771 and thus we estimate the limit as 2.00 since
      π/4 is closer to .8 than to .75.

21.   Note that f(-x) = f(x) and that f(.5) = .1646, f(.1) = .1666,
      and f(.01) = .1667 and thus we estimate the limit as 1/6.

23.   The slope of the tangent line at x = 0 is given by
      $\lim_{x\to 0} \dfrac{f(x)-f(0)}{x-0}$ = $\lim_{x\to 0} |x|/x$. Now for x < 0, |x|/x = -x/x = -1 and
      for x > 0, |x|/x = x/x = 1 and thus there is no single value
      L such that when x is close to 0, |x|/x is close to L. The
      conclusion is that f(x) = |x| does not possess a tangent line
      at x = 0.

25.   Suppose $\lim_{x\to c} f(x) = L$. From the definition, given ε > 0 there is
      a δ > 0 such that |f(x) - L| < ε when 0 < |x-c| < δ.
      Replacing x with (c+h) yields |f(c+h) - L| < ε when
      0 < |h| < δ (x-c = c+h-c = h). But this last statement
      means $\lim_{h\to 0} f(c+h) = L$, giving: $\lim_{x\to c} f(x) = \lim_{h\to 0} f(c+h)$.

27a. We have $|f(x)-L| = |(2x+1)-3| = |2x-2| = |2(x-1)|$. So if
      $|f(x)-L| < 1/10$ then $|2(x-1)| < 1/10$ so $|x-1| < 1/20$. Thus
      choose $0 < \delta_1 \leq 1/20$.

  b. To guarantee $|f(x)-L| = |2(x-1)| < 1/100$ choose $0 < \delta_2 \leq 1/200$
      so $0 < |x-1| < 1/200$.

  c. For $|f(x)-L| = 2|x-1| < \varepsilon$, choose $0 < \delta(\varepsilon) \leq \varepsilon/2$. Then when
      $0 < |x-1| < \delta(\varepsilon)$ we have $|f(x)-L| < \varepsilon$.

29.   We are to show $\lim_{x \to 2}(3x-8) = -2$ or given $\varepsilon > 0$, we must find
      $\delta > 0$ such that $|(3x-8)-(-2)| < \varepsilon$ whenever $0 < |x-2| < \delta$. But
      $|(3x-8)-(-2)| = |3x-6| = 3|x-2| < \varepsilon$ whenever
      $|x-2| < \varepsilon/3$. So by choosing $0 < \delta(\varepsilon) \leq \varepsilon/3$ we are guaranteed
      $|(3x-8)-(-2)| < \varepsilon$ when $|x-2| < \delta(\varepsilon)$. Thus $\delta_1 \leq 1/30$ and $\delta_2 \leq 1/300$.

31.   Since $|f(x)-(ax_0+b)| = |(ax+b)-(ax_0+b)| = |a||x-x_0|$ we have
      that $|f(x)-(ax_0+b)| < \varepsilon$ whenever $|a||x-x_0| < \varepsilon$ or
      $|x-x_0| < \varepsilon/|a|$. Thus $\delta(\varepsilon) \leq \varepsilon/|a|$, $\delta_1 \leq 0.1/|a|$ and $\delta_2 \leq 0.01/|a|$.

33.   If we restrict $\delta < 1$ then $|x-2| < \delta < 1$ means that x must satisfy
      $1 < x < 3$. Now $|f(x) - 1/2| = |1/x - 1/2| = |(2-x)/2x| =$
      $|1/2||1/x||x-2| \leq |1/2||x-2|$ since $|1/x| < 1$. Thus $\delta(\varepsilon) \leq 2\varepsilon$,
      $\delta_1 \leq 1/5$ and $\delta_2 \leq 1/50$.

Section 2.3, Page 82

1. $\lim_{x \to 3}(2x^2 + x - 5) = \lim_{x \to 3} 2x^2 + \lim_{x \to 3} x + \lim_{x \to 3}(-5)$ Thm 2.3.1a

          $= 2\lim_{x \to 3} x^2 + \lim_{x \to 3} x - \lim_{x \to 3} 5$ Thm 2.3.1b

          $= 2 \lim_{x \to 3} x \lim_{x \to 3} x + \lim_{x \to 3} x - \lim_{x \to 3} 5$ Thm 2.3.1c

          $= 2(3)(3) + 3 - 5 = 16$ Eq.(10).

3.   By Eq.(8), $\lim_{x \to 1}(x-2)^5 = \lim_{x \to 1}(x-2) \lim_{x \to 1}(x-2) \ldots \lim_{x \to 1}(x-2)$ (5 factors)

          $= \left[\lim_{x \to 1} x - \lim_{x \to 1} 2 \right] \ldots \left[\lim_{x \to 1} x - \lim_{x \to 1} 2 \right]$ (by Eq.5)

          $= (1-2) \ldots (1-2) = (-1)^5 = -1$ (by Eq.10).

5. $\lim\limits_{x\to 2}\dfrac{x^2-3x-4}{x+2}=\dfrac{\lim\limits_{x\to 2}(x^2-3x-4)}{\lim\limits_{x\to 2}(x+2)}=\dfrac{2^2-3\cdot 2-4}{2+2}=-\dfrac{3}{2}$, by Thm.2.3.1d and Eq.(11).

7. $\lim\limits_{x\to 2}\left(x^2+\dfrac{1}{x}\right)=\lim\limits_{x\to 2}x^2+\lim\limits_{x\to 2}\dfrac{1}{x^2}$,   by Thm 2.3.1b.

$=\lim\limits_{x\to 2}x^2+\dfrac{\lim\limits_{x\to 2}(1)}{\lim\limits_{x\to 2}x^2}=4+1/4=17/4$,   by Thm 2.3.1d, Eq.(10).

9. $\lim\limits_{x\to 2}\sqrt{1+2x^2}=\sqrt{\lim\limits_{x\to 2}(1+2x^2)}=\sqrt{1+8}=3$,   by Thm 2.3.2, Eq.(11).

11. $\lim\limits_{x\to 2}\left(\sqrt{x}+\dfrac{1}{\sqrt{x}}\right)=\lim\limits_{x\to 2}\sqrt{x}+\lim\limits_{x\to 2}\dfrac{1}{\sqrt{x}}$ Thm 2.3.1b

$=\lim\limits_{x\to 2}\sqrt{x}+\dfrac{\lim\limits_{x\to 2}1}{\lim\limits_{x\to 2}\sqrt{x}}$ Thm 2.3.1d

$=\sqrt{\lim\limits_{x\to 2}x}+\dfrac{1}{\sqrt{\lim\limits_{x\to 2}x}}=\sqrt{2}+1/\sqrt{2}=3\sqrt{2}/2$, by Eq.(10) and

Thm 2.3.2.

13. $\lim\limits_{x\to 1}\left(\dfrac{2x^3-4x^2+3}{6x^2+4x-1}\right)^{1/2}=\left[\lim\limits_{x\to 1}\dfrac{2x^3-4x^2+3}{6x^2+4x-1}\right]^{1/2}$ Thm 2.3.2

$=\left[\dfrac{\lim\limits_{x\to 1}(2x^3-4x^2+3)}{\lim\limits_{x\to 1}(6x^2+4x-1)}\right]^{1/2}=\left(\dfrac{1}{9}\right)^{1/2}=\dfrac{1}{3}$,   by Thm 2.3.1d and Eq.(11).

15. $\lim\limits_{x\to c}\sin x=\lim\limits_{h\to 0}\sin(c+h)=\lim\limits_{h\to 0}(\sin c\cos h+\cos c\sin h)=\sin c$

by Eqs.(18).

17. $\lim\limits_{x\to \pi/4}(1-\sin^3 x)=1-\lim\limits_{x\to \pi/4}\sin^3 x=1-(1/\sqrt{2})^3=(4-\sqrt{2})/4$, by Eq.(24).

19. $\lim\limits_{x\to \pi/2}(x\sin x-\cos x)=\lim\limits_{x\to \pi/2}x\sin x-\lim\limits_{x\to \pi/2}\cos x$

$=\lim\limits_{x\to \pi/2}x\lim\limits_{x\to \pi/2}\sin x-\lim\limits_{x\to \pi/2}\cos x=(\pi/2)\sin(\pi/2)-\cos(\pi/2)=\pi/2$.

21. First recall that $\sin 2x = 2\sin x \cos x$ so that

$$\lim_{x\to 0}\frac{\sin 2x}{x} = \lim_{x\to 0}\frac{2\sin x \cos x}{x} = 2\lim_{x\to 0}\frac{\sin x}{x}\lim_{x\to 0}\cos x \; = \; (2)(1)(1) \; = \; 2.$$

23. We must first rewrite $\dfrac{\sin x^2}{x}$ as $\dfrac{\sin x^2}{x^2}\cdot x$ so that

$$\lim_{x\to 0}\frac{\sin x^2}{x} = \lim_{x\to 0}\frac{x\sin x^2}{x^2} = \lim_{x\to 0}x\cdot\lim_{x\to 0}\frac{\sin x^2}{x^2}. \text{ Now let } z = x^2 \text{ so}$$

$$\lim_{x\to 0}\frac{\sin x^2}{x^2} = \lim_{z\to 0}\frac{\sin z}{z} = 1. \text{ Thus } \lim_{x\to 0}\frac{\sin x^2}{x} = 0\cdot 1 = 0.$$

25. $\lim\limits_{x\to 0}\dfrac{\tan x}{x} = \lim\limits_{x\to 0}\dfrac{\sin x}{x\cos x} = \lim\limits_{x\to 0}\dfrac{\sin x}{x}\lim\limits_{x\to 0}\dfrac{1}{\cos x} = 1\cdot 1 = 1.$

27. Noting that $|\sin 1/x| \le 1$ for all $x \ne 0$ we can write
$0 \le |x\sin 1/x| = |x|\,|\sin 1/x| \le |x|$ for all $x \ne 0$. Now,
by the Sandwich Principle with $g(x) = 0$, $h(x) = |x|$,

$f(x) = |x\sin 1/x|$ we have $\lim\limits_{x\to 0}|x\sin 1/x| = 0$ since $\lim\limits_{x\to 0}|x| = 0$.

Finally, by Theorem 2.3.3b $\lim\limits_{x\to 0}x\sin 1/x = 0$.

29. One pssibility is given in the text. Another is to choose

$f(x) = \dfrac{5}{(x-2)^2}$. Then for x satisfying $0 < |x-2| < 2$, we

have $f(x) \ge 1 = g(x)$. However, $\lim\limits_{x\to 2}f(x)$ does not exist.

31a. If $f(x) = g(x) = \operatorname{sgn} x$, then $f(x)g(x) = \operatorname{sgn}^2 x = 1$ for
$x \ne 0$ and thus $\lim\limits_{x\to 0}f(x)g(x) = 1$ while neither f nor g has a limit
at $x = 0$.

  b. If $f(x) = x$ and $g(x) = 1/x$ then $\lim\limits_{x\to 0}f(x)g(x) = 1$ and $\lim\limits_{x\to 0}f(x) = 0$,
while g does not have a limit at $x = 0$. Another example is
given in the text.

33a. $|f(x)+g(x) - (A+B)| = |[f(x)-A] + [f(x)-B]| \le |f(x)-A| + |g(x)-B|$.

  b. If $0 < |x-c| < \delta$, with δ the minimum of δ_1 and δ_2, then from
part (a), $|f(x)+g(x) - (A+B)| \le |f(x)-A| + |g(x)-B| < \varepsilon' + \varepsilon' = \varepsilon.$

35.   Since $g(x) \le f(x) \le h(x)$ then $g(x)-L \le f(x)-L \le h(x)-L$. Thus,
       for $0 < |x-c| < \delta$, $-\varepsilon < g(x)-L \le f(x)-L \le h(x)-L < \varepsilon$, where
       δ is the minimum of δ_1 and δ_2. Therefore $|f(x) - L| < \varepsilon$.

Section 2.4, Page 91

1.   Since $f(x)$ is defined only for $4-x^2 \ge 0$ or $-2 \le x \le 2$ we have

      $\lim\limits_{x \to 2-} \sqrt{4-x^2} = 0$, $\lim\limits_{x \to -2+} \sqrt{4-x^2} = 0$ and by Thm.2.4.1

      $\lim\limits_{x \to 1-} \sqrt{4-x^2} = \lim\limits_{x \to 1+} \sqrt{4-x^2} = \lim\limits_{x \to 1} \sqrt{4-x^2} = \sqrt{3}$. All the other required

      limits are meaningless.

3. $\lim\limits_{x \to 2+} (4-x^2)^{1/3} = \left[\lim\limits_{x \to 2+} (4-x^2)\right]^{1/3} = 0^{1/3} = 0$

      $\lim\limits_{x \to 2-} (4-x^2)^{1/3} = \left[\lim\limits_{x \to 2-} (4-x^2)\right]^{1/3} = 0^{1/3} = 0$ and so by Thm 2.4.1

      $\lim\limits_{x \to 2} (4-x^2)^{1/3} = 0$.

5. $\lim\limits_{x \to -1-} \sqrt{|x^2-1|} = \sqrt{\lim\limits_{x \to -1-} |x^2-1|} = \sqrt{0} = 0$. Similarly $\lim\limits_{x \to 1+} \sqrt{|x^2-1|} = 0$

      and thus $\lim\limits_{x \to 1} \sqrt{|x^2-1|} = 0$.

7. $\lim\limits_{x \to 0-} f(x) = \lim\limits_{x \to 0-} (1+x^2) = 1$ since $f(x) = 1+x^2$ for $x < 0$

      $\lim\limits_{x \to 0+} f(x) = \lim\limits_{x \to 0+} (1-x^2) = 1$ since $f(x) = 1-x^2$ for $x > 0$. Thus, by

      Thm.2.4.1 $\lim\limits_{x \to 0} f(x) = 1$.

9. $\lim\limits_{x \to 0+} x^2 = 0$, but $\lim\limits_{x \to 0-} f(x) = \lim\limits_{x \to 0-} (1/x^2) = \infty$, since $1/x^2$ is unbounded on

      any interval $a < x < 0$. Thus $\lim\limits_{x \to 0} f(x)$ does not exist.

11. $\lim\limits_{x \to 0-} f(x) = \lim\limits_{x \to 0-} \sin x = 0$ from the properties of $\sin x$. Also

      $\lim\limits_{x \to 0+} f(x) = \lim\limits_{x \to 0+} \sin 3x = 0$ and thus $\lim\limits_{x \to 0} f(x) = 1$.

13. $\lim\limits_{x \to \infty} \dfrac{1-x}{1+x} = \lim\limits_{x \to \infty} \dfrac{\frac{1}{x}-1}{\frac{1}{x}+1} = \dfrac{\lim\limits_{x \to \infty}\left(\frac{1}{x}-1\right)}{\lim\limits_{x \to \infty}\left(\frac{1}{x}+1\right)} = \dfrac{\lim\limits_{x \to \infty}\frac{1}{x}-\lim\limits_{x \to \infty} 1}{\lim\limits_{x \to \infty}\frac{1}{x} + \lim\limits_{x \to \infty} 1} = \dfrac{-1}{1} = -1$.

15. $\displaystyle \lim_{x\to\infty} \frac{3x^2+4x-5}{2x^3-6x^2+x+1} = \lim_{x\to\infty} \frac{3+\dfrac{4}{x}-\dfrac{5}{x^2}}{2x-6+\dfrac{1}{x}+\dfrac{1}{x^2}}$. $\dfrac{4}{x}, \dfrac{5}{x^2}, \dfrac{1}{x}, \dfrac{1}{x^2}$ all go to 0 as

x goes to ∞.   Therefore, $\displaystyle \lim_{x\to\infty}\frac{3x^2+4x-5}{2x^3-6x^2+x+1} = \lim_{x\to\infty}\frac{3}{2x-6} = 0$.

17. $\displaystyle \lim_{x\to\infty}\frac{\sqrt{1+x^2}}{x} = \lim_{x\to\infty}\frac{x\sqrt{1/x^2+1}}{x} = \lim_{x\to\infty}\sqrt{1/x^2+1} = \sqrt{1} = 1$.

19. $\displaystyle \lim_{x\to-\infty}\left(\frac{2x^2-x+1}{x^2+4}\right)^{1/2} = \lim_{x\to-\infty}\frac{-x}{-x}\left(\frac{2-\dfrac{4}{x}+\dfrac{1}{x^2}}{1+\dfrac{4}{x^2}}\right)^{1/2} = \sqrt{2}$.

21. Since $-1/x \le (\sin x)/x \le 1/x$ and $\displaystyle\lim_{x\to\infty}(-1/x) = \lim_{x\to\infty}(1/x) = 0$ we have

by the Sandwich Principle, Thm.2.3.4 for $c = \infty$, that

$\displaystyle\lim_{x\to\infty}\frac{\sin x}{x} = 0$.

23. $\displaystyle \lim_{x\to-\infty}\frac{x^3+2x-4}{x^2+1} = \lim_{x\to-\infty}\frac{x\left(1+\dfrac{2}{x^2}-\dfrac{4}{x^3}\right)}{\left(1+\dfrac{1}{x^2}\right)} = \lim_{x\to-\infty}x = -\infty$.

25. $\displaystyle \lim_{x\to 0+}x\sqrt{1+\frac{1}{x}} = \lim_{x\to 0+}\sqrt{x^2\left(1+\frac{1}{x}\right)} = \lim_{x\to 0+}\sqrt{x^2+x} = 0$.

27. $\displaystyle \lim_{x\to 1-}\frac{(1-x^2)}{\sqrt{1-x^4}} = \lim_{x\to 1-}\frac{(1-x^2)}{\sqrt{(1-x^2)(1+x^2)}} = \lim_{x\to 1-}\sqrt{\frac{(1-x^2)}{1+x^2}} = 0$.

29. Using trigonometric identities $\displaystyle \lim_{x\to 0}\frac{\csc x - \cot x}{x} = \lim_{x\to 0}\frac{1-\cos x}{x\sin x}$

$\displaystyle = \lim_{x\to 0}\frac{(1-\cos x)}{x\sin x}\frac{(1+\cos x)}{(1+\cos x)} = \lim_{x\to 0}\frac{1-\cos^2 x}{x\sin x(1+\cos x)}$

$\displaystyle = \lim_{x\to 0}\frac{\sin^2 x}{x\sin x(1+\cos x)} = \lim_{x\to 0}\left(\frac{\sin x}{x}\cdot\frac{1}{1+\cos x}\right) = (1)(1/2) = 1/2$.

31.   The example shown satisfies
       conditions (i) through (v).

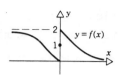

33.   To show $\lim\limits_{x \to 0+} \sqrt{x} = 0$ we must find $\delta(\epsilon)$ so that if $0 < x < \delta$
       then $0 < \sqrt{x} < \epsilon$. Note, $\sqrt{x} < \epsilon$ whenever $(\sqrt{x})^2 < \epsilon^2$, or
       $x < \epsilon^2$. Thus, choosing $\delta \leq \epsilon^2$ guarantees that if $0 < x < \delta$
       (or $0 < x < \epsilon^2$), then $0 < \sqrt{x} < \epsilon$.

35.   If $|x| < 1/\sqrt{M}$, then $x^2 < 1/M$ so that $1/x^2 > M$. This means a
       good choice for δ would be $\delta(M) = 1/\sqrt{M}$. Thus $\lim\limits_{x \to 0} \dfrac{1}{x^2} = \infty$ since
       if $0 < x < 1/\sqrt{M}$, then $1/x^2 > M$.

37a.   With δ the minimum of δ_1 and δ_2, then $|f(x)-L| < \epsilon$ when
       $c-\delta < x < c$ or $c < x < c+\delta$. That is, $|f(x)-L| < \epsilon$ when
       $-\delta < x-c < 0$ or $0 < x-c < \delta$. But the last inequalities are
       equivalent to $0 < |x-c| < \delta$.

  b.   The inequality $0 < |x-c| < \delta$ is equivalent to the two
       inequalities $c-\delta < x < c$ and $c < x < c+\delta$ [see part (a)]. Thus
       $|f(x)-L| < \epsilon$ when $c-\delta < x < c$,   so $\lim\limits_{x \to c-} f(x) = L$, and
       $|f(x)-L| < \epsilon$ when $c < x < c+\delta$,   so $\lim\limits_{x \to c+} f(x) = L$.

Section 2.5, Page 98

1. $f(x) = x/(x^2-1) = x/(x-1)(x+1)$ has an infinite discontinuity
       at $x = 1$ and $x = -1$ since at these values the denominator is
       zero.

3. $f(x) = x^2/(x^2-3x+2) = x^2/(x-1)(x-2)$ has an infinite
       discontinuity at $x = 1$ and $x = 2$.

5.   f has a jump discontinuity at $x = 0$ with a jump of 2.

7.   f has jump discontinuities (of 1) at $x = 0, \pm 3, \pm 6, \ldots, \pm 3n, \ldots$.

9.   f has jump discontinuities at each x for which $1-x^2 = 0$.
      That is, when $x = \pm 1$.

11.   f has removable discontinuities at each x for which
      $1+\sin x = 0$. That is, for $x = -\pi/2, 3\pi/2, 7\pi/2, \ldots 3\pi/2 \pm 2n\pi$.

13.   f has an infinite discontinuity at $x = 0$ since $\lim_{x \to 0-} f(x)$ is
      unbounded.

15. $f(x) = \dfrac{x}{|x|} = \begin{cases} -1 & x < 0 \\ 1 & x > 0 \end{cases}$ has a jump discontinuity of 2 at

      $x = 0$. It is not possible to define $f(0)$ to remove the
      discontinuity since the limit does not exist at $x = 0$.

17. $f(x) = x+1$ for $x \neq 0$, so $\lim_{x \to 0} f(x) = 1$. Thus by defining
      $f(0) = 1$, the discontinuity would be removed.

19. $f(x) = x\left(1 + \dfrac{1}{\sqrt{|x|}}\right) = x + \dfrac{x}{\sqrt{|x|}}$. $f(x) = \begin{cases} x + \sqrt{x} & x > 0 \\ x - \sqrt{|x|} & x < 0 \end{cases}$ so

      $\lim_{x \to 0} f(x) = 0$. Defining $f(0) = 0$ removes the discontinuity.

21.   f will be continuous at $x = 1$, if $\lim_{x \to 1-} f(x) = \lim_{x \to 1+} f(x)$. Since

      $\lim_{x \to 1-} f(x) = \lim_{x \to 1-}(ax+b) = a+b$ and $\lim_{x \to 1+} f(x) = \lim_{x \to 1+}(cx^2 + dx + e) = c + d + e$;
      the function is continuous at $x = 1$ if $a+b = c+d+e$.

23.   If $\varepsilon \neq 0$, $f(\varepsilon) = 2$, while if $\varepsilon = 0$, $f(\varepsilon) = 1$. Thus, f is
      discontinuous at $\varepsilon = 0$.

25.   Let $f(x) = g(x) = \begin{cases} -1 & x \geq 0 \\ 1 & x < 0 \end{cases}$. Both f and g are discontinuous at

      $x = 0$, but $f(x)g(x) = 1$ for all x is continuous at $x = 0$.

27.   The statement of the problem assumes that f is defined on
      (a,b). Further $0 \leq |f(x)-f(t)| < |x-t|$ for each pair of
      points x and t in the interval. Finally, since $\lim_{x \to t} |x-t| = 0$,

we have that $\lim\limits_{x\to t} |f(x)-f(t)| = 0$ by Theorem 2.3.4 (Sandwich

Principle). Thus $\lim\limits_{x\to t} f(x) = f(t)$, and hence for each t in the
interval (a,b), f is continuous, or f is continuous on (a,b).

Section 2.6, Page 106

To facilitate your answering problems 1-10 draw a graph of f(x)
in each case.

1.   The given interval is not closed and f(x) is not continuous
      on the closed bounded interval $[0,\pi/2]$ since $f\to\infty$ as $x\to\pi/2$.
      f(x) is not bounded and has no maximum or minimum.

3.   f is defined on the open interval (-1,1) and is not
      continuous at x = 0. f is bounded since $|f(x)| \le 1$ for all
      $x\in (-1,1)$. f has a minimum of 0 but no maximum on the interval
      and the intermediate value property fails on some
      subintervals.

5.   f is not continuous at x = 2 on the closed interval, but f is
      bounded, has a maximum, but no minimum and the intermediate
      value property fails on some subintervals.

7.   The domain of f is not bounded. f is not continuous at x=3,
      is not bounded, has no maximum, and the intermediate value
      property fails on some subintervals.

9.   The domain of f is not closed, but f is bounded above by 1/2
      and below by 0. f is not continuous at $\pi/2$, has no minimum,
      and the intermediate value property fails on some
      subintervals.

11.   We have f(-3)=-11,f(-2)=4,f(-1)=-3,f(3)=-11,f(4)=52 and thus
      from Bolzano's Theorem, the roots lie on the intervals
      (-3,-2), (-2,-1), and (3,4) since f(-3)f(-2) = (-11)(4) < 0;
      f(-2)f(-1) = (4)(-3) < 0; and f(3)f(4) = (-11)(52) < 0.

13.   We have f(-1)=-14.7,f(0)=45,f(1)=0.34,f(2)≅-100,f(6)≅-103,
      and f(7)≅236 so f(x) has roots on each of the intervals
      (-1,0), (1,2) and (6,7).

15. Let $f(x) = a_{2k+1}x^{2k+1} + a_{2k}x^{2k} + \cdots + a_1 x + a_0$. Further, we may

assume $a_{2k+1} > 0$. Since $\lim_{x \to \infty} f(x) = +\infty$, there is some x^* for

which $f(x) > 0$ for all $x > x^*$. Let x_1 be chosen so that

$x_1 > x^*$. Then $f(x_1) > 0$. Similarly $\lim_{x \to -\infty} f(x) = -\infty$ so that for some

x^{**}, $f(x) < 0$ for all $x < x^{**}$. Choose any $x_2 < x^{**}$ and thus

$f(x_2) < 0$. Finally, $f(x)$ has a root between x1 and x2 since

$f(x1) f(x2) < 0$. A slight modifcation of the proof yields the

same result if $a_{2k+1} < 0$.

17. To find the desired interval we start with $x = 1$ and $x = 2$ to
     to find $f(1) = 0.6829$ and $f(2) = -0.1814$. Thus there is a
     root on the interval $(1,2)$. We now begin the bisection:

Int. size	x	$f(x) = 2\sin x - x$
	1	0.6829
1	2	-0.1814
1/2	1.5	0.4950
$1/2^2$	1.75	0.2180
$1/2^3$	1.875	0.0332
$1/2^4$	1.9375	-0.0705
$1/2^5$	1.90625	-0.0177
$1/2^6$	1.890625	0.0079
$1/2^7$	1.8984375	-0.0048

Thus $f(x) = 0$ for some x in the interval $(1.890625, 1,8984375)$.

19. There is a root between $x = 10$ and $x = 11$ since $f(10) = 0.0377$
     and $f(11) = -0.134$. Now, as in Prob. 17, we have
     $f(10.500) = -0.050$, $f(10.250) = -0.006$, $f(10.125) = 0.016$,
     $f(10.1875) = 0.005$, $f(10.21875) = -0.001$, $f(10.203125) = 0.002$,
     $f(10.2109375) = 0.0004$, and thus the root must lie in
     $(10.2109375, 10.21875)$.

21. The area of any rectangle inscribed in a circle of radius r
     is bounded below by 0 and above by πr^2, the area of the
     circle. The area of the rectangle
     is $A = xy$ where $y = \sqrt{4r^2 - x^2}$.

     Thus $A(x) = x\sqrt{4r^2 - x^2}$, which is
     continuous in $0 \le x \le 2r$ and
     hence by Theorem 2.6.2(b) there
     is a $c \in [0, 2r]$ such that $A(c)$ is a maximum.

23. In many cases like this, where the object is to show two
     quantities equal, it is useful to consider their difference.
     In this case, consider the function $h(x) = f(x) - g(x)$. Then
     from the given conditions $h(0) = 2$ and $h(1) = -4$. Since $h(x)$
     is continuous on $[0,1]$, then by Theorem 2.6.2(c), h takes on
     all values between 2 and -4 at least once on $[0,1]$. In
     particular, for some $c \in [0,1]$, $h(c) = 0 = f(c) - g(c)$. That is
     $f(c) = g(c)$.

25. Suppose the rim of the canyon is H feet above the floor. Let
     $h_1(t)$ be the hiker's height above the canyon floor at time t
     ($t \in [8am, 4pm]$) on the first day and $h_2(t)$ his height on the
     second day. (Both are continuous functions of t.) Finally,
     as in Prob.23 above, let $h(t) = h_1(t) - h_2(t)$. Then
     $h(8am) = H - 0 = H$ and $h(4pm) = 0 - H = -H$. Thus by Theorem
     2.6.2(c), there is some $t^* \in [8am, 4pm]$ for which $h(t^*) = 0$.
     That is $h_1(t^*) - h_2(t^*) = 0$ or $h_1(t^*) = h_2(t^*)$. So at the same time,
     t^*, the hiker will be at the same point on the trail.

27a.   b.

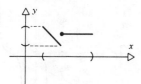   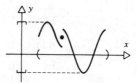

29.

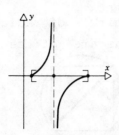

31. If $|x-c| < \delta$ then, for $\varepsilon = f(c)/2$, $|f(x)-f(c)| < \varepsilon$ so
     $-\varepsilon < f(x)-f(c) < \varepsilon$. Thus, from the left inequality,
     $f(c) - \varepsilon < f(x)$ or $f(c) - [f(c)/2] = f(c)/2 < f(x)$, and hence
     $f(x) > 0$. If $f(c) < 0$, then choose $\varepsilon = -f(c)/2$ so
     $f(x) - f(c) < \varepsilon$ and $f(x) < f(c) + \varepsilon = f(c)/2 < 0$.

Chapter 2 Review, Page 107

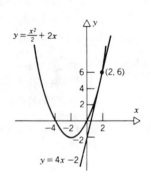

1. As in Ex.2 of Section 2.1, we find the
    slope of the secant line joining the
    point $(2,6)$ to any point (x,y) to be
    $m_S = [f(x)-f(2)]/(x-2) =$

    $(x^2+4x-12)/2(x-2) = (x+6)/2$. Thus the
    slope of the tangent is $m_T = \lim\limits_{x\to 2} m_S = 4$

    and hence the equation of the tangent
    line is $(y-6) = 4(x-2)$ or $y = 4x-2$.

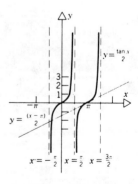

3. We have $m_S = (1/2)(\tan x - \tan\pi)/(x-\pi) =$
    $(1/2)(1/\cos x)[\sin x/(x-\pi)] =$
    $-(1/2)(1/\cos x)[\sin(x-\pi)/(x-\pi)]$. Thus
    $m_T = -(1/2)\lim\limits_{x\to\pi}(1/\cos x)\lim\limits_{x\to\pi}\sin(x-\pi)/(x-\pi)$
    $= (1/2)\lim\limits_{t\to 0}\sin t/t = 1/2$ from Section 2.3,
    and hence the equation of the tangent
    line is $y = (1/2)(x-\pi)$.

5. $f(2) = 5$ so $[f(x)-f(2)]/(x-2) = (x^2-4)/(x-2) = x+2$. Thus
    $\lim\limits_{x\to 2}[f(x)-f(2)]/(x-2) = 4$.

7. $f(1) = -1/2$ so $[f(x)-f(1)]/(x-1) = 3/2(x+1)$ and
      $\lim_{x \to 1} 3/2(x+1) = 3/4$.

9. $\lim_{x \to 0} (x^2+2)/(x-1) = (2)/(-1) = -2$.

11. $\lim_{x \to 3} (x+3)(x-3)(x-2)/(x^2-9) = \lim_{x \to 3} (x-2) = 1$.

13.   Since $1/[\![x]\!] = -1$ for $-1 \le x < 0$ and $1/[\![x]\!]$ does not exist
      for $0 \le x < 1$, the required limit does not exist.

15. $\lim_{x \to -1} f(x) = \lim_{x \to -1} (x^2) = 1$.

17. $\lim_{x \to 1} \dfrac{x^{3/2} - 3x + 3x^{1/2} - 1}{(\sqrt{x} - 1)^3(x+2)} = \lim_{x \to 1} \dfrac{(\sqrt{x} - 1)^3}{(\sqrt{x} - 1)^3 (x + 2)} = 1/3$.

19. $\lim_{x \to 4} \dfrac{x^2 - 16}{x - \sqrt{x} - 2} = \lim_{x \to 4} \dfrac{(\sqrt{x} - 2)(\sqrt{x} + 2)(x+4)}{(\sqrt{x} - 2)(\sqrt{x} + 1)} = 32/3$.

21. $\lim_{x \to 0} \dfrac{\sin^2 2x}{2x^2} = 2 \lim_{x \to 0} \left(\dfrac{\sin 2x}{2x} \right)^2 = 2$.

23. $\lim_{x \to 0} \dfrac{x}{\sin 3x} = \dfrac{1}{3} \lim_{x \to 0} \dfrac{3x}{\sin 3x} = \dfrac{1}{3}$.

25.   Noting that $\sin 3x = -\sin 3(x-\pi)$ we have
      $\lim_{x \to \pi} \dfrac{\sin 3x}{3(x - \pi)} = \lim_{x \to \pi} \dfrac{-\sin 3(x-\pi)}{3(x - \pi)} = -\lim_{w \to 0} \dfrac{\sin w}{w} = -1$.

27.   The limit does not exist since the numerator has limit -2
      and the denominator has limit 0.

29. $\lim_{x \to \infty} \dfrac{x^3 + 2x + 1}{3x^3 - 4} = \lim_{x \to \infty} \dfrac{1 + 2/x^2 + 1/x^3}{3 - 4/x^3} = 1/3$.

31. $\lim_{x \to 0} \dfrac{x}{x^2 - 4} = 0$.

33. $f(x)$ is discontinuous at $x = 1$, but by defining $f(1) = -\pi$,
      the discontinuity is removed. (See Prob.26).

35.   Since sin(1/x) is undefined at x = 0, f is discontinuous
           there. However, since $\lim_{x\to 0+}$ xsin(1/x) = $\lim_{v\to\infty}$(sinv)/v = 0 and

           $\lim_{x\to 0-}$ xsin(1/x) = $\lim_{y\to -\infty}$ (siny)/y = 0, the discontinuity may be

           removed by defining f(0) = 0.

37a.   Yes, for any function that has a jump discontinuity such
           as f(x) = [|x|] for x in [-1/2,1/2]. However, if $\lim_{x\to c}$f(x) = ∞,
           for example f(x) = 1/x² for x in [-1,1], the boundedness
           property would not hold.

  b.   Yes, for example the function $f(x) = \begin{cases} x & a \le x \le c \\ (x-c)^{-1} & c < x \le b. \end{cases}$

39.   There is a root between x = 3 and x = 4 since f(3) = -7,
           f(4) = 216 so f(3)f(4) < 0. Similarly the root lies
           between 3.2 and 3.3 as f(3.2) = -2.432 and f(3) = .137.
           Continuing, we find the root lies between 3.294 and 3.295.
           Thus, to two decimal places, the root is x = 3.29.

41a.   If f(x) = 1/x then $\lim_{x\to 0}$ x(1/x) = $\lim_{x\to 0}$ x/x = 1.

  b.   Let g(x) = 2/x so $\lim_{x\to 0}$ 2x/x = 2.

  c.   If h(x) = 1/x² then $\lim_{x\to 0}$ x/x² does not exist.

43.   One such function is
           f(x) = (x+1)³(x-2)/(x²-2x).

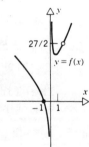

45.   One such function is
           f(x) = (x²-25)/(x-2)², x ≠ 2.

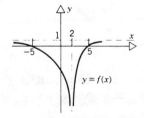

47.   One such function is

$$f(x) = \begin{cases} -1 - x, & x < 0 \\ x^2/(x-2)^2, & x \geq 0 \end{cases}.$$

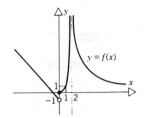

49.   One such function is
         $f(x) = 2(x^2-1)/(x^2-x)$.

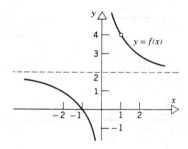

CHAPTER 3

Section 3.1, Page 120

Problems 1 through 12 use Eq.(3).

1. $f'(x) = \lim\limits_{h\to 0}\dfrac{(x+h)^2-x^2}{h} = \lim\limits_{h\to 0}\dfrac{x^2+2xh+h^2-x^2}{h} = \lim\limits_{h\to 0}2x+h = 2x.$

3. $g'(x) = \lim\limits_{h\to 0}\dfrac{\left(\dfrac{1}{x+h}\right)^2-\dfrac{1}{x^2}}{h} = \lim\limits_{h\to 0}\dfrac{x^2-(x+h)^2}{h(x+h)^2x^2} = \lim\limits_{h\to 0}\dfrac{-(2x+h)}{(x+h)^2x^2} = \dfrac{-2}{x^3}.$

5. $g'(s) = \lim\limits_{h\to 0}\dfrac{(s+h)^4+(s+h)^2-s^4-s^2}{h}$

    $= \lim\limits_{h\to 0}\dfrac{(s^4+4s^3h+6s^2h^2+4sh^3+h^4)+(s^2+2sh+h^2)-s^4-s^2}{h}$

    $= \lim\limits_{h\to 0}4s^3+2s+h(6s^2+4sh+h^2+h) = 4s^3+2s.$

7. $f'(x) = \lim\limits_{h\to 0}\dfrac{x+h-\dfrac{1}{x+h}-\left(x-\dfrac{1}{x}\right)}{h} = \lim\limits_{h\to 0}\dfrac{h-\left[\dfrac{x-(x+h)}{x(x+h)}\right]}{h}$

    $= \lim\limits_{h\to 0}\left[1+\dfrac{1}{x(x+h)}\right] = 1+\dfrac{1}{x^2}.$

9. $h'(u) = \lim\limits_{h\to 0}\dfrac{\dfrac{1}{1+\sqrt{u+h}}-\dfrac{1}{1+\sqrt{u}}}{h} = \lim\limits_{h\to 0}\dfrac{1+\sqrt{u}-(1+\sqrt{u+h})}{h(1+\sqrt{u})(1+\sqrt{u+h})}$

    $= \lim\limits_{h\to 0}\dfrac{\sqrt{u}-\sqrt{u+h}}{h(1+\sqrt{u})(1+\sqrt{u+h})}\cdot\dfrac{\sqrt{u}+\sqrt{u+h}}{\sqrt{u}+\sqrt{u+h}}$

    $= \lim\limits_{h\to 0}\dfrac{u-(u+h)}{h(1+\sqrt{u})(1+\sqrt{u+h})(\sqrt{u}+\sqrt{u+h})} = \dfrac{-1}{2\sqrt{u}(1+\sqrt{u})^2}.$

11. $f'(x) = \lim\limits_{h\to 0}\dfrac{\sqrt{x+h}-\dfrac{1}{(x+h)^2}-\left(\sqrt{x}-\dfrac{1}{x^2}\right)}{h} = \lim\limits_{h\to 0}\left[\dfrac{\sqrt{x+h}-\sqrt{x}}{h}-\dfrac{x^2-(x+h)^2}{hx^2(x+h)^2}\right]$

    $\lim\limits_{h\to 0}\left[\dfrac{h}{h(\sqrt{x}+\sqrt{x+h})}-\dfrac{2x+h}{x^2(x+h)^2}\right] = \dfrac{1}{2\sqrt{x}}+\dfrac{2}{x^3}.$

13. If $f(x) = x^6$ then $f'(x) = 6x^{6-1} = 6x^5$, which is defined for all x.

15. $f'(x) = (3/7)x^{3/7-1} = (3/7)x^{-4/7}$, which is defined for all $x \neq 0$.

17. $f'(x) = -4x^{-4-1} = -4x^{-5}$, which is defined for all $x \neq 0$.

19. $f'(x) = (3/5)x^{3/5-1} = (3/5)x^{-2/5}$, $x \neq 0$.

21. Note that $f(x) = x^2\sqrt{x} = x^{5/2}$ and thus $f'(x) = (5/2)x^{3/2}$, $x > 0$.

23. Since $y' = 3x^2$ we see that the slope at $x = 2$ is 12 and thus the tangent line is given by $y-8 = 12(x-2)$, or $12x-y = 16$.

25. Since $y' = -3x^{-4}$ we see that the slope at $x = -2$ is $-3/16$ and thus the tangent line is given by $y+1/8 = (-3/16)(x+2)$, or $3x+16y = -8$.

27. For $y = x^2$ we have $y' = 2x$ and thus the tangent line at $(1,1)$ is $y-1 = 2(x-1)$, or $2x-y = 1$. For $y = 1/x$ we have $y' = -1/x^2$ and thus the tangent line at $(1,1)$ is $y-1 = -(x-1)$, or $x+y = 2$. Since the respective slopes (2 and -1) are not negative reciprocals, the lines are not perpendicular.

29. We have $y' = -1/x^2$ and thus the tangent line at $(1/2,2)$ is given by $y-2 = (-4)(x-1/2)$, or $4x+y = 4$. Setting $y = 0$ we find $(1,0)$ is the x intercept and setting $x = 0$ we find $(0,4)$ is the y intercept.

31. Setting $x = -a$ in Eq.(3) we have

$$f'(-a) = \lim_{h\to 0}\frac{f(-a+h)-f(-a)}{h} = \lim_{h\to 0}\frac{f(a-h)-f(a)}{h}, \text{ since } f(x) \text{ is an}$$

even function. Setting $w = -h$, and hence $w\to 0$ as $h\to 0$, we

have $f'(-a) = \lim_{w\to 0}\dfrac{f(a+w)-f(a)}{-w} = -\lim_{w\to 0}\dfrac{f(a+w)-f(a)}{w} = -f'(a)$. Thus

$f'(-1) = -2$ and $f'(-a) = -b$.

33. The slope of $x-6y+9 = 0$ is 1/6 for all x and the slope of $y = \sqrt{x}$ is $1/2\sqrt{x}$ for each x. Thus the slopes are equal only for $x = 9$ (since $1/2\sqrt{9} = 1/6$). Since $(9,3)$ lies on both the line and the curve $y = \sqrt{x}$ we conclude that the straight line is tangent to the curve at $(9,3)$.

35.   Since y' = 2x we see that the tangent line to y = x² has
       slope 2a at the point (a,a²). For distinct values of a,
       then, the slope is unique and the tangent lines cannot be
       parallel.

37a.   For x < 0 we have $f'(x) = \lim\limits_{h\to 0} \dfrac{x+h-x}{h} = 1$ and for x > 0 we have

       $f'(x) = \lim\limits_{h\to 0} \dfrac{(x+h)^2-x^2}{h} = 2x.$

  b.   For x = 0 we have $f'(0) = \lim\limits_{h\to 0} \dfrac{f(h)-f(0)}{h} = \begin{cases} \lim\limits_{h\to 0} \dfrac{h-0}{h} = 1 & h < 0 \\[2ex] \lim\limits_{h\to 0} \dfrac{h^2-0}{h} = 0 & h > 0 \end{cases}$

       and thus f'(0) does not exist.

  c.

                    $y = f(x)$ $y = f'(x)$

39a.   For x < 0 we have $f'(x) = \lim\limits_{h\to 0} \dfrac{-(x+h)^2+x^2}{h} = -2x$ and for x > 0

       we have $f'(x) = \lim\limits_{h\to 0} \dfrac{(x+h)^2-x^2}{h} = 2x.$

  b.   For x = 0 we have $f'(0) = \lim\limits_{h\to 0} \dfrac{f(h)-f(0)}{h} = \left.\begin{cases} \lim\limits_{h\to 0} \dfrac{-h^2+0}{h} & h < 0 \\[2ex] \lim\limits_{h\to 0} \dfrac{h^2-0}{h} & h > 0 \end{cases}\right\} = 0.$

  c.

                    $y = f(x)$ $y = f'(x)$

Section 3.2, Page 128

1.   Using Eqs.(1) and (2) we have
      $f'(x) = (3x^2-4x)' + (1)' = 3(x^2)' - 4(x)' + 0 = 6x-4.$

3.   Using Eqs.(1) and (2) we have
      $f'(x) = (4x^5+7x^3)' - (6x)' = 4(x^5)' + 7(x^3)' - 6 = 20x^4+21x^2-6.$

5.   Writing $f(x) = (2x^2+1)(2x^2+1)$ and using Eq.(3) we have
      $f'(x) = (4x)(2x^2+1)+(2x^2+1)(4x) = 16x^3+8x.$

7.   Writing $f(x) = (x+2)[(x-1)(x+3)]$ and using Eq.(3) twice we
      have $f'(x) = (1)[(x-1)(x+3)] + (x+2)[(1)(x+3)+(x-1)(1)]$
      $= 3x^2+8x+1.$

9. $f'(x) = (2x)(x^2+4) + (x^2+1)(2x) = 4x^3+10x$ by Eq.(3).

11. $f(x) = x-6 + 8/x - 5/x^2$ and thus differentiating each term
      yields $f'(x) = 1-8/x^2+10/x^3.$

13.   Using Eq.(4) we have $f'(x) = \dfrac{(x^2-5x+6)(4x^3-6x)-(x^4-3x^2+2)(2x-5)}{(x^2-5x+6)^2}$

      $= \dfrac{2x^5-15x^4+24x^3+15x^2-40x+10}{(x^2-5x+6)^2}, \ x \neq 2,3.$

15.   Differentiating each term yields
      $f'(x) = 6x+(4/3)x^{1/3} - 1/x^2+6/x^4, \ x \neq 0.$

17.   Eq.(4) yields $f' = \dfrac{(x+1)(3/2)x^{1/2}-x^{3/2}(1)}{(x+1)^2} = x^{1/2}(x+3)/2(x+1)^2, x>0.$

19.   Differentiating each term using the appropriate equation
      yields $f'(x) = $
      $3+(1/3)x^{-2/3}(x^{2/3}-1)+(x^{1/3}+1)(2/3)x^{-1/3}+[(1+x^2)7-7x(2x)]/(1+x^2)^2$
      $= 4 + (2/3)x^{-1/3} - (1/3)x^{-2/3} + 7(1-x^2)/(1+x^2)^2, \ x \neq 0.$

21a. Differentiating each term we find $y' = 2x-6.$

  b. At $x = 1$ $y' = -4$ and thus the desired tangent line is
      $y+3 = -4(x-1)$, or $4x+y = 1.$

21c.   The tangent line will have zero
       slope when y' = 0, that is when
       2x-6 = 0, or x = 3, y = -7.
       Since y = x²-6x+9+2-9 = (x-3)²-7,
       we see that (3,-7) is the vertex
       of the parabola and hence is the
       lowest point on the graph. The
       tangent line at this point is y = -7.

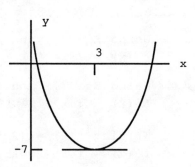

23a. $y' = \dfrac{(x^2+1)3-(3x)(2x)}{(x^2+1)^2} = \dfrac{3(1-x^2)}{(x^2+1)^2}$.

   b.   The tangent line will be horizontal when y' = 0, which

        occurs when 1-x² = 0 or x = ± 1.   For x = 1, y = 3/2 and for
        x = -1, y = -3/2 and thus (1,3/2) and (-1,-3/2) are the
        desired points.

   c.   The graph of f(x) is
        shown and thus we see
        that (1,3/2) is the
        greatest value and
        (-1,-3/2) is the least
        value of f(x).

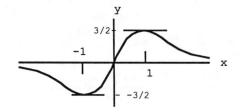

25a.   For y = x²-3x-1 we have y' = 2x-3 = 5 at x = 4 and thus
       y-3 = 5(x-4) or 5x-y = 17 is its tangent line at (4,3).   For
       y = √x + 1 we have y' = (1/2)x⁻¹/² = 1/4 at x = 4 and thus
       y-3 = (1/4)(x-4) or x-4y = -8 is its tangent line at (4,3).

   b.   From the graph we see that
        tanφ = tan(A-B) =
        $\dfrac{\tan A - \tan B}{1+\tan A \tan B} = \dfrac{5-(1/4)}{1+5/4} = \dfrac{19}{9}$

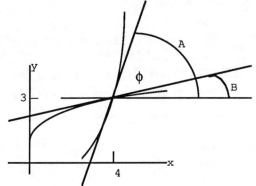

27. Since y' = 1-4x we see that the slope of the tangent line at
     any point (a,5+a-2a^2) is 1-4a. Thus the tangent line is
     y-(5+a-2a^2) = (1-4a)(x-a). This line must pass through
     (-1,10) and thus 10-(5+a-2a^2) = (1-4a)(-1-a), which reduces
     to 2(a^2+2a-3) = 0, or a = -3,1. Thus (-3,-16) and (1,4) are
     the desired points.

29. The slope of the given curve is 3x^2+4x-6 at any point (x,y)
     and the slope of the line 2x+y = 3 is -2 at all points.
     Thus we need 3x^2+4x-6 = -2 or 3x^2+4x-4 = 0, which has roots
     x = -2, 2/3. Hence (-2,16) and (2/3,32/27) are the desired
     points.

31. Since y' = 2x we find that (y-a^2) = 2a(x-a) is the tangent
     line to y = x^2 at the point (a,a^2). The tangent line must
     pass through (4,7) and thus (7-a^2) = 2a(4-a) which yields
     a^2-8a+7 = 0 or a = 1,7. Hence y-1 = 2(x-1), or 2x-y = 1, is
     tangent to y = x^2 at (1,1) and y-49 = 14(x-7), or 14x-y = 49
     is tangent to y = x^2 at (7,49).

33a. From Section 2.1 we have v = s' = 3t^2-9t+15 and thus
     3t^2-9t+15 = 9, or 3(t^2-3t+2) = 0, which yields t = 1,2.
  b. Since v = 3(t^2-3t+9/4+5-9/4) = 3[(t-3/2)^2+11/4] we see that
     the smallest value of v occurs when t = 3/2.

35. Since $f' = \begin{cases} 2x & x < 2 \\ a & x > 2 \end{cases}$ we must have $\lim\limits_{x\to2-} f'(x) = \lim\limits_{x\to2+} f'(x)$,
     or 4 = a, in order for f(x) to be differentiable at x = 2.
     In addition, by Thm.3.2.2, f(x) must be continuous at x = 2
     and thus $\lim\limits_{x\to2-}$ f(x)(=4) must equal $\lim\limits_{x\to2+}$ f(x)(=2a+b) and hence 4 =
     2a+b, which yields b = -4.

37. Since A = πr^2 we have A' = 2πr.

39. Let x denote the side of the cube. Then V = x^3 and S = 6x^2,
     thus V = (S/6)$^{3/2}$ = $\dfrac{S^{3/2}}{6^{3/2}}$. Hence V' = $\dfrac{(3/2)S^{1/2}}{6\cdot6^{1/2}}$ = $\sqrt{S}/4\sqrt{6}$.

41. Applying Eq.(3) twice we have
     f'(x) = u'(x)v(x)w(x) + u(x)[v(x)w(x)]'
           = u'(x)v(x)w(x)+u(x)[v'(x)w(x)+v(x)w'(x)]
           = u'(x)v(x)w(x)+u(x)v'(x)w(x)+u(x)v(x)w'(x).

43.   Let $u_1(x) = u_2(x) = \ldots = u_n(x) = u(x)$, then Prob.42 yields

$$\frac{f'(x)}{f(x)} = \frac{u'(x)}{u(x)} + \frac{u'(x)}{u(x)} + \ldots + \frac{u'(x)}{u(x)} \quad \text{where } f(x) = [u(x)]^n \text{ and there}$$

are n terms in the sum on the right.   Thus
$$f'(x) = n[u(x)]^n \frac{u'(x)}{u(x)} = n[u(x)]^{n-1}u'(x).$$

Section 3.3, Page 137

1. $f'(x) = (2x)(x^2-2x+4)+(x^2+1)(2x-2)$
$$= 2x^3-4x^2+8x+2x^3-2x^2+2x-2 = 4x^3-6x^2+10x+2$$
and thus $f''(x) = 12x^2-12x+10$.

3. $Df = (1/3)x^{-2/3}+2x^{-3}, x \neq 0$, so $D^2f = -(2/9)x^{-5/3}-6x^{-4}; x \neq 0$.

5.   Writing $y = 1/x + x^{-5/3}$ we have $y' = -x^{-2}-(5/3)x^{-8/3}$, $x \neq 0$;
$y'' = 2x^{-3}+(40/9)x^{-11/3}$, $x \neq 0$; and thus
$y''' = -6x^{-4}-(440/27)x^{-14/3}$, $x \neq 0$.

7. $y' = 4x^3-6x$ and thus $y'' = 12x^2-6$.

9. $y' = 2x+(3/2)x^{-1/2}+x^{-2}-2x^{-3/2}$, $x > 0$;
$y'' = 2-(3/4)x^{-3/2}-2x^{-3}+3x^{-5/2}$, $x > 0$ and thus
$y''' = (9/8)x^{-5/2}+6x^{-4}-(15/2)x^{-7/2}$, $x > 0$.

11.   We have $D^{-1}(x-4) = D^{-1}(x) - D^{-1}(4)$ by Eq.(18),
$D^{-1}(x) = x^2/2+c_1$ by Eq.(17), and $D^{-1}(4) = 4D^{-1}(1) = 4x+c_2$ by
Eqs.(19) and (17).   Thus $D^{-1}(x-4)=x^2/2-4x+c$, where $c= c_1+c_2$.

13. $D^{-1}(x^2+x^{-2}) = D^{-1}(x^2)+D^{-1}(x^{-2}) = x^3/3 - x^{-1} + c$.

15. $D^{-1}(3x^{2/3}-4x^{1/3}) = 3D^{-1}(x^{2/3})-4D^{-1}(x^{1/3}) =$
$3[x^{5/3}/(5/3)]-4[x^{4/3}/(4/3)]+c = (9/5)x^{5/3}-3x^{4/3}+c$.

17. $D^{-1}(3/\sqrt{x}) = 3D^{-1}(x^{-1/2}) = 3x^{1/2}/(1/2)+c = 6\sqrt{x}+c$.

19.   Since $D\left(\dfrac{A}{2+x}\right) = \dfrac{-A}{(2+x)^2}$ we choose A = -1 and thus

$$D^{-1}\left(\frac{1}{(2+x)^2}\right) = -1/(2+x)+c.$$

21.   For the given f'(x) we must have f(x) = x^2-3x+c, which
        yields f(0) = c. Thus c = 4 and hence f(x) = x^2-3x+4.

23.   For the given f'(x) we must have f(x) = x^4-x^2+c, which
        yields f(2) = 16-4+c. Thus c = 4-(12) = -8 and hence
        f(x) = x^4-x^2-8.

25.   Since f(x) = $x^{3/2}$/(3/2) + c we have f(1) = 2/3 + c. Thus
        c = 1-2/3 = 1/3 and hence f(x) = (2/3)$x^{3/2}$ + 1/3.

27.   Assume P(x) = ax^2+bx+c and therefore P'(x) = 2ax+b and
        P"(x) = 2a. Thus P(2) = 4a+2b+c = 1, P'(0) = b = -3, and
        P"(4) = 2a = 6. Solving, we find a = 3, b = -3 and c = -5
        and hence P(x) = $3x^2$-3x-5.

29.   Let s be measured positively downward with the origin at the
        edge of the cliff. Then, as in Ex.5, we have Newton's Law
        F = ma, where the only force is due to gravity and F = mg
        since the force now acts in the positive direction. Thus
        a = v' = g so that v = gt+c_1, where c_1 = 0 since there is no
        initial velocity [v(0) = 0]. Continuing, we then have
        s' = v so that s = gt^2/2 + c_2, where c_2 = 0 since s(0) = 0
        is the top of the cliff. To find the height of the cliff we
        now set t = 3 to find s(3) = $32(3)^2$/2 = 144 ft.

31a.   Let s be measured positively upward, with the origin at the
        top of the building. This is the same setup as Ex.6 and
        thus from Eq.(25) we have v = -9.8t+20 (since v_0 = 20 for this
        problem). Thus the maximum height occurs for v = 0, or
        t = 20/9.8 = 2.04 sec. s(t) is then given by Eq.(26), which
        yields s(2.04) = $-4.9(2.04)^2$+20(2.04) = 20.4 meters above
        the building, or 50.4 meters above the ground.

   b.   When the ball hits the ground s(t) = -30 and thus we must
        solve -30 = $-4.9t^2$+20t, which has a positive root at
        t = 5.25 sec., using the quadratic formula.

33.   (f'g-fg')'= (f'g)'-(fg')'= (f"g+f'g')-(f'g'+fg")= f"g-fg".

35.   There are two steps in a mathematical induction proof. The
        first is to show that the desired result holds for at least
        one value of n, which was done in Prob.34 for n= 2 and n= 3.
        The second step is to assume the desired result holds for
        n = N and then to show this implies the result holds for
        n = N+1. Therefore we assume Eq.(i) holds for n = N:

$$(fg)^{(N)} = f^{(N)}g + Nf^{(N-1)}g' + \ldots + \binom{N}{k-1} f^{(N-[k-1])}g^{(k-1)} +$$

$\binom{N}{k} f^{(N-k)}g^{(k)} + \ldots + fg^{(N)}$. Now we differentiate both sides to

obtain $(fg)^{(N+1)} = f^{(N+1)}g + f^{(N)}g' + Nf^{(N)}g' + Nf^{(N-1)}g'' + \ldots +$

$\binom{N}{k-1} f^{(N-(k-1)+1)}g^{(k-1)} + \binom{N}{k-1} f^{(N-(k-1))}g^{(k)} + \binom{N}{k} f^{(N-k+1)}g^{(k)} +$

$\binom{N}{k} f^{(N-k)}g^{(k+1)} + \ldots + f'g^{(N)} + fg^{(N+1)}$. Combining similar terms we

then obtain $(fg)^{(N+1)} = f^{(N+1)}g + (N+1)f^{(N)}g' + \ldots +$

$\left[\binom{N}{k-1} + \binom{N}{k}\right] f^{(N-k+1)}g^{(k)} + \ldots + fg^{(N+1)}$. Using Eq.(ii) we have

$\binom{N}{k-1} + \binom{N}{k} = \binom{N+1}{k}$ and thus $(fg)^{(N+1)} = f^{(N+1)}g +$

$(N+1)f^{(N)}g' + \ldots + \binom{N+1}{k} f^{(n+1-k)}g^{(k)} + \ldots + fg^{(N+1)}$, which is

        Eq.(i) for n = N+1 and thus Eq.(i) holds for all n by
        induction.

Section 3.4, Page 146

1.   Since y' = -2x, Eq.(1) yields the tangent line
        y = 3+(-2)(x-1) = 5-2x.

3. $y' = \dfrac{(2x^2+3x+4)(6)-(6x)(4x+3)}{(2x^2+3x+4)^2}$ which has the value y' = 4/3 at
        x = -1. Thus y = (-2)+(4/3)(x+1), or 4x-3y = 2, is the
        desired tangent line.

5.   y' = 2x-3 and thus $y = y_0 + (2x_0 - 3)(x - x_0)$ is the tangent line
        at (x_0, y_0).

7.   f(2) = 5 and f'(2) = 4 and thus T(x;2) = 5+4(x-2) = 4x-3.
        In this case we have f(x) = T(x;2) and hence the left side
        of Eq.(3) is zero and therefore r(x;2) = 0.

9. $f(0) = -4$ and $f'(0) = 2$ and thus $T(x;0) = -4+2(x-0) = 2x-4$.
    Eq.(3) then yields $r(x;0)(x-0) = (x^2/3+2x-4)-(2x-4) = x^2/3$
    and thus $r(x;0) = x/3$.

11. $f(2) = 2/3$, $f'(x) = \dfrac{(x+1)-x}{(x+1)^2} = 1/(x+1)^2 = 1/9$ at $x = 2$.   Thus
    $T(x;2) = 2/3 + (1/9)(x-2) = (1/9)x + 4/9$ and hence
    $r(x;2)(x-2) = \dfrac{x}{x+1} - \dfrac{(x+4)}{9} = \dfrac{-(x-2)^2}{9(x+1)}$. Thus
    $r(x;2) = -(x-2)/9(x+1)$, $x \neq -1$.

13. Following Ex.3 we choose $f(x) = x^{1/3}$ and $x_0 = 27$. Then we have
    $f'(27) = (1/3)(27)^{-2/3} = 1/27$ and $T(x,27) = 3+(1/27)(x-27)$.
    Now we let $x = 27.16$ to obtain $T(27.16,27) = 3.00593$. Now
    $(27.16)^{1/3} = 3.00591$, so that the linear approximation is
    correct to four decimal places, since $.00002 < 5/10^5$.

15. Let $f(x) = x^{1/4}, g(x) = x^{-1/2}$ and $x_0 = 16$. Since these are each
    evaluated at different points we must find a linear
    approximation to each one and then add them. For $f(x)$ we
    have $T_1(x,16) = 2+(1/32)(x-16) = 2-(.18/32) = 1.99438$ when
    $x = 15.82$. Likewise for $g(x)$ we have
    $T_2(x,16) = 1/4-(1/128)(x-16) = 1/4-(.12/128) = .24906$ when
    $x = 16.12$. Adding these two we obtain 2.24344 as the
    desired linear approximation, which is correct to four
    decimal places since the exact value is 2.24342.

17. We know that $V(r) = (4/3)\pi r^3$ so that
    $T(r,1) = V(1)+V'(1)(r-1) = 4\pi/3+4\pi(r-1)$. Thus
    $T(.96,1) = 4\pi/3-.16\pi = 3.69$ ft^3. Since $V(.96) = 3.71$ ft^3,
    T is correct to one decimal place.

19. $T(x;1) = f(1)+f'(1)(x-1) = 8+28(x-1)$ and thus
    $T(1.01,1) = 8.280$, which is correct to two decimal places
    since $f(1.01) = 8.282$.

21. Since $dy/dx = y'$ and $d^2y/dx^2 = y''$ we have $dy/dx = 4x^3+4x$ and
    $d^2y/dx^2 = 12x^2+4$.

23. $\dfrac{dy}{dx} = \dfrac{(x^2+1)(2x)-(x^2-1)(2x)}{(x^2+1)^2} = \dfrac{4x}{(x^2+1)^2}$ and
    $\dfrac{d^2y}{dx^2} = \dfrac{(x^2+1)^2(4)-(4x)[2x(x^2+1)+(x^2+1)2x]}{(x^2+1)^4} = \dfrac{4-12x^2}{(x^2+1)^3}$.

25. $dy/dx = 6x - (4/3)x^{1/3} - 1/(x+1)^2$; $x \neq -1$.

$$d^2y/dx^2 = 6 - (4/9)x^{-2/3} - \left[\frac{(x+1)^2(0) - 1(2x+2)}{(x+1)^4}\right]$$

$$= 6 - (4/9)x^{-2/3} + 2/(x+1)^3; x \neq -1. \text{Note that}$$

$[(x+1)^2]' = [x^2+2x+1]' = 2x+2$ in the middle equation.

27. $\dfrac{dy}{dx} = \dfrac{(x+2)(1)-(x+1)(1)}{(x+2)^2} = \dfrac{1}{(x+2)^2}$, $x \neq -2$.

$\dfrac{d^2y}{dx^2} = \dfrac{(x+2)^2(0)-1(2x+4)}{(x+2)^4} = \dfrac{-2}{(x+2)^3}$, $x \neq -2$.

29a. Since $\lim\limits_{x \to x_0} f(x) = \lim\limits_{x \to x_0} A(x_0) + \lim\limits_{x \to x_0} B(x_0)(x-x_0) + \lim\limits_{x \to x_0} r(x; x_0)(x-x_0) =$

$A(x_0) + B(x_0) \cdot 0 + 0 \cdot 0 = A(x_0)$ and since $f(x)$ is continuous means $\lim\limits_{x \to x_0} f(x) = f(x_0)$, we conclude that $f(x_0) = A$.

  b. From Eq.(i) we have $f(x) - A = B(x_0)(x-x_0) + r(x; x_0)(x-x_0)$.
Substituting $A = f(x_0)$ and dividing both sides by $(x-x_0)$ we
obtain Eq.(iii).

  c. From Eq.(iii) we have $\lim\limits_{x \to x_0} \dfrac{f(x)-f(x_0)}{x-x_0} = B$ using Eq.(ii). If we

let $x = x_0 + h$, then $x \to x_0$ implies $h \to 0$ and thus

$\lim\limits_{x \to x_0} \dfrac{f(x)-f(x_0)}{x-x_0} = \lim\limits_{h \to 0} \dfrac{f(x_0+h)-f(x_0)}{h} = f'(x_0)$. Thus $f'(x_0) = B$.

31a. For $V(x) = x^3$ we have

$V(x_0+\Delta x) - V(x_0) = (x_0+\Delta x)^3 - x_0^3 = 3x_0^2 \Delta x + 3x_0 \Delta x^2 + \Delta x^3$. Setting
$x_0 = 3$ ft. and $\Delta x = 0.02$ ft. we find $V(3.02) - V(3) = .5436$ ft^3.

  b. The relative error is $.5436$ ft^3/27 ft$^3 \cong .0201$.

  c. By Eq.23 $dV = 3x^2 dx = 3(3)^2(.02) = .54$ ft^3. Thus
$dV/V = .54/27 = .02$

Section 3.5, Page 155

1. $f \circ g = 3(2x+7) - 2 = 6x + 19$, with domain $-\infty < x < \infty$.

    $g \circ f = 2(3x-2) + 7 = 6x + 3$, with domain $-\infty < x < \infty$.

3. $f \circ g = 1 - (\sqrt{x+1})^2 = -x$, with domain $-1 \leq x < \infty$ since that is

the domain of $g(x)$. $g \circ f = \sqrt{(1-x^2)+1} = \sqrt{2-x^2}$, with domain
$-\sqrt{2} \leq x \leq \sqrt{2}$, which is the set of values for which $f(x) = 1-x^2$ lies in the domain of $g(x) = \sqrt{x+1}$.

5. $f \circ g = \sqrt{(x^2-3x+2)}$, which has domain all x for which
$x^2-3x+2 = (x-2)(x-1) \geq 0$, which yields $x \geq 2$ or $x \leq 1$.

$g \circ f = (\sqrt{x})^2 - 3\sqrt{x} + 2 = x - 3\sqrt{x} + 2, \ x \geq 0$.

7. $f \circ g = \sqrt{-(x^2-3x+6)}$, which does not exist since
$-(x^2-3x+6) < 0$ for all x. $g \circ f = -(x - 3\sqrt{x} + 6)$ for $x \geq 0$.

9. $f \circ g = \cos 2\left(\sqrt{x^2-4}\right)$ for $|x| \geq 2$. $g \circ f = \sqrt{\cos^2 2x - 4}$, however,
does not exist since $\cos^2 2x - 4 < 0$ for all x.

11.   We have $dy/dx = (dy/dz)(dz/dx) = (2z+2)(2x) = 4(z+1)x$, by
Eq.(14). To verify this answer we see that
$y = (x^2-7)^2 + 2(x^2-7) - 5 = x^4 - 12x^2 + 30$ and thus
$y' = 4x^3 - 24x = 4x(x^2-6) = 4x(z+1)$.

13.   As in Prob.11 we have $\dfrac{dy}{dx} = \dfrac{(z^2+1)(2z)-(z^2-1)(2z)}{(z^2+1)^2}(3) = \dfrac{12z}{(z^2+1)^2}$.

To verify this, find y' from $y = \dfrac{(3x-2)^2-1}{(3x-2)^2+1} = \dfrac{9x^2-12x+3}{9x^2-12x+5}$.

15. $\dfrac{dy}{dx} = (2z-3)\dfrac{(x+1)(1)-x(1)}{(x+1)^2} = (2z-3)\left(\dfrac{1}{x+1}\right)^2$, $x \neq -1$. To verify this,

find y' from $y = \dfrac{x^2}{(x+1)^2} - \dfrac{3x}{x+1} + 4 = \dfrac{2x^2+5x+4}{(x+1)^2}$.

17. $dy/dx = (3u^2)[(1/2)(x+1)^{-1/2}] = (3/2)(x+1)(x+1)^{-1/2} = (3/2)(x+1)^{1/2}, \ x > -1$.

19. $dy/dx = [(7/3)z^{4/3}](2x) = (14/3)x(x^2+5)^{4/3}$, all x.

21. $dy/dx = (-2z)[(1/2)(x+4)^{-1/2}] = -1, \ x > -4$, since that is the
domain of $z = \sqrt{x+4}$.

23.   From Eq.(13) we have $D_x(f \circ g) = (D_z f)(D_x g) = (2z+2)(2x)$.   Now
      $x = 2 \Rightarrow z = g = -3$ and thus $D_x(f \circ g) = (-6+2)(4) = -16$.

25.   Again $D_x(f \circ g) = (D_z f)(D_x g)$.   To find D_z let $w = z^2-16$, then
      $df/dz = (df/dw)(dw/dz) = (1/2)w^{-1/2}(2z)$.   Thus
      $D_x(f \circ g) = (1/2)(z^2-16)^{-1/2}(2z)(2x+2)$.   If $x = 1$, then
      $z = g = 5$ and thus $D_x(f \circ g) = (5/3)(4) = 20/3$.

27.   Let $w = 2x+1$.   Then $\dfrac{d}{dx}[w(x)]^3 = 3w^2 \cdot \dfrac{dw}{dx} = 3(2x-1)^2(2) =$
      $6(2x-1)^2$.

29.   Following the pattern of Prob.27 we have
      $$\frac{d}{dx}\left(\frac{x+1}{2}\right)^{100} = 100\left(\frac{x+1}{2}\right)^{99}\left(\frac{1}{2}\right) = 50\left(\frac{x+1}{2}\right)^{99}.$$

31. $\dfrac{d}{dx}[(x-1)^4\sqrt{2x-3}] = 4(x-1)^3(1)(2x-3)^{1/2} + (x-1)^4(1/2)(2x-3)^{-1/2}(2)$

      $$= (x-1)^3\left[4(2x-3)^{1/2} + \frac{x-1}{(2x-3)^{1/2}}\right] = \frac{(x-1)^3(9x-13)}{\sqrt{2x-3}}.$$

33. $\dfrac{d}{dx}\left(\dfrac{2x-1}{x}\right)^{3/2} = \dfrac{3}{2}\left(\dfrac{2x-1}{x}\right)^{1/2}\left[\dfrac{x(2)-(2x-1)(1)}{x^2}\right] = \dfrac{3}{2}x^{-5/2}(2x-1)^{1/2}.$

35. $\dfrac{dy}{dx} = (2z+2)(2x)$; $\dfrac{d^2y}{dx^2} = \left[\dfrac{d}{dx}(2z+2)\right](2x) + (2z+2)\dfrac{d}{dx}(2x)$

      $= \left[\dfrac{d}{dz}(2z+2)\dfrac{dz}{dx}\right](2x) + (2z+2)(2) = [(2)(2x)](2x)+4(z+1)$

      $= 8x^2+4(z+1).$

37.   Let $z = x^3+1$, then $dy/dx = (1/2)z^{-1/2}(3x^2) = (3/2)x^2(x^3+1)^{-1/2}$.
      $d^2y/dx^2 = 3x(x^3+1)^{-1/2} + (3/2)x^2(-1/2)(x^3+1)^{-3/2}(3x^2)$
      $= 3x/(x^3+1)^{1/2} - (9/4)x^4/(x^3+1)^{3/2} = (3/4)x(x^3+4)(x^3+1)^{-3/2}$.

39. $\dfrac{dy}{dx} = \dfrac{(x^2+1)(2x) - (x^2-1)(2x)}{(x^2+1)^2} = \dfrac{4x}{(x^2+1)^2}$;

      $\dfrac{d^2y}{dx^2} = \dfrac{(x^2+1)^2(4) - (4x)(2)(x^2+1)(2x)}{(x^2+1)^4} = \dfrac{4(1-3x^2)}{(x^2+1)^3}.$

41. Let $z = -x$, then $\dfrac{d}{dx} f(z) = \dfrac{df}{dz}\dfrac{dz}{dx} = f'(z)(-1) = -f'(-x)$.

43. Let $z = ax$, then $\dfrac{d}{dx} f(z) = \dfrac{df}{dz}\dfrac{dz}{dx} = f'(z)(a) = af'(ax)$.

45. From Prob.43 we have $\dfrac{d^2}{dx^2} f(ax) = a\dfrac{d}{dx} f'(ax) = a[af''(ax)] = a^2 f''(ax)$.
     Now use induction (see Sec.6.1) and assume
     $\dfrac{d^n}{dx^n} f(ax) = a^n f^{(n)}(ax)$. Taking the derivative of both sides then
     yields $\dfrac{d^{n+1}}{dx^{n+1}} f(ax) = a^n \dfrac{d}{dx} f^{(n)}(ax) = a^n [af^{(n+1)}(ax)] = a^{n+1} f^{(n+1)}(ax)$.

47. $d/dx\ [u'(x)]^2 = 2[u'(x)]\ d/dx\ u'(x) = 2u'(x)u''(x)$.

49. $\dfrac{d}{dx}\{[u'(x)]^2 u''(x)\} = [2u'(x)u''(x)]u''(x) + [u'(x)]^2 u'''(x)$
     $= 2u'(x)[u''(x)]^2 + [u'(x)]^2 u'''(x)$.

51. In this case choose $u(x) = x-2$, then $u'(x) = 1$ and thus
     Eq.(16) yields $D^{-1}[(x-2)^4] = (x-2)^5/5 + c$.

53. Choose $u(x) = 2x-1$, then $u'(x) = 2$ and $D^{-1}[2(2x-1)^5] =$
     $D^{-1}[(2x-1)^5(2)] = (2x-1)^6/6 + c$.

55. $D^{-1}[(2x+1)^{4/3}] = D^{-1}[(1/2)(2x+1)^{4/3}(2)] = 3(2x+1)^{7/3}/14 + c$.

57. Let $u(x) = x^2+4$, then $u'(x) = 2x$ and $D^{-1}[3x(x^2+4)^{1/2}] =$
     $D^{-1}[(3/2)(x^2+4)^{1/2}(2x) = (3/2)(x^2+4)^{3/2}/(3/2) + c =$
     $(x^2+4)^{3/2} + c$.

59. Let $w = h(x)$ and $z = g(w)$, then $\dfrac{dF}{dx} = \dfrac{df}{dz}\dfrac{dz}{dx} = \dfrac{df}{dz}\left[\dfrac{dz}{dw}\dfrac{dw}{dx}\right] =$
     $f'\{g[h(x)]\}\ g'[h(x)]\ h'(x)$.

61. $\dfrac{dy}{dx} = \dfrac{g(x)f'(x) - f(x)g'(x)}{[g(x)]^2} = \dfrac{gf'-fg'}{g^2}$;
     $\dfrac{d^2y}{dx^2} = \dfrac{g^2[g'f' + gf'' - (f'g' + fg'')] - [gf' - fg']2gg'}{g^4}$
     $= [g(gf''-fg'') - 2g'(gf'-fg')]/g^3$.

63.   Since $V(t) = (4/3)\pi r^3(t)$ we have $dV/dt = 4\pi r^2(t)\, dr/dt$.
       However $dr/dt = -2$ and thus $dV/dt = -8\pi r^2$.

65.   In this case $V(x) = x^3$ and thus $dV/dt = 3x^2\, dx/dt$. We are
       given that $dx/dt = 2$ in./sec. (rate of change of the edge)
       and thus when $x = 6$ we have $dV/dt = 3(6)^2(2) = 216$ in.^{3}/sec.

67.   From Eq.(i) we have

$$m_0 \frac{d}{dt}\left(\frac{v}{(1-v^2/c^2)^{1/2}}\right) = \frac{(1-v^2/c^2)^{1/2}v' - v(1/2)(1-v^2/c^2)^{-1/2}(-2v/c^2)v'}{(1/m_0)\,(1-v^2/c^2)}$$

$$= m_0\,\frac{(1-v^2/c^2)\,v' + v^2v'/c^2}{(1-v^2/c^2)^{3/2}} = \frac{m_0 a}{(1-v^2/c^2)^{3/2}} \quad \text{where } v' = a.$$

Section 3.6, Page 165

Problems 1 through 18 make use of Eq.(17).

1. $(\cos 4x)' = (-\sin 4x)(4) = -4\sin 4x$.

3. $(\sin 4x - 3\cos 2x)' = \cos 4x(4) - 3(-\sin 2x)(2) = 4\cos 4x + 6\sin 2x$.

5. $(x^2\sin 2x)' = (2x)(\sin 2x) + x^2[(\cos 2x)(2)] = 2x\sin 2x + 2x^2\cos 2x$.

7. $(\sin^2 2x)' = 2(\sin 2x)(\sin 2x)' = 2\sin 2x(\cos 2x)(2) = 4\sin 2x\cos 2x$.

9. $[x(\sec x - \tan x)]' = (1)(\sec x - \tan x) + x(\sec x\tan x - \sec^2 x)$
                    $= (1 - x\sec x)(\sec x - \tan x)$.

11. $\dfrac{d}{dx}\left(\dfrac{1+\sin x}{x-\cos x}\right) = \dfrac{(x-\cos x)(\cos x) - (1+\sin x)(1+\sin x)}{(x-\cos x)^2} =$

$$= \frac{x\cos x - 2\sin x - 2}{(x-\cos x)^2}.$$

13. $$\frac{d}{dx}\left(\frac{1-\cos x}{1+\cos x}\right)^{1/2} = \frac{1}{2}\left(\frac{1-\cos x}{1+\cos x}\right)^{-1/2}\frac{d}{dx}\left(\frac{1-\cos x}{1+\cos x}\right)$$

$$= \frac{1}{2}\left(\frac{1-\cos x}{1+\cos x}\right)^{-1/2}\frac{(1+\cos x)(\sin x)-(1-\cos x)(-\sin x)}{(1+\cos x)^2}$$

$$= \left(\frac{1+\cos x}{1-\cos x}\right)^{1/2}\frac{\sin x}{(1+\cos x)^2} = \frac{\sin x}{(1+\cos x)^{3/2}(1-\cos x)^{1/2}} = \frac{\text{sgn}(\sin x)}{1+\cos x}.$$

15. $[\sin(\cos x)]' = \cos(\cos x)(\cos x)' = -\sin x \cos(\cos x)$.

17. $[\tan(\sec^2 x)]' = \sec^2(\sec^2 x)(\sec^2 x)' = \sec^2(\sec^2 x)[2\sec x(\sec x)']$
       $= \sec^2(\sec^2 x)(2\sec x)(\sec x \tan x) = 2\tan x \sec^2 x \sec^2(\sec^2 x)$.

19.   Let $u = 2x$, then Eq.(28) yields
       $D^{-1}[\sin 2x] = D^{-1}[(1/2)\sin 2x(2)] = (-1/2)\cos 2x + c$.

21.   Let $u = 3x$ then Eq.(27) yields
       $D^{-1}[2\cos 3x] = D^{-1}[(2/3)\cos 3x(3)] = (2/3)\sin 3x + c$.

23. $D^{-1}[\sin(x+\pi/2)] = -\cos(x+\pi/2) + c$.

25.   Let $u = x/2$, then $D^{-1}[\sec(x/2)\tan(x/2)] =$
       $D^{-1}[2\sec(x/2)\tan(x/2)(1/2)] = 2\sec(x/2) + c$.

27.   Since $\sin 2x = 2\sin x\cos x$ we have $D^{-1}[\sin 2x\cos 2x] =$
       $D^{-1}[(1/2)\sin 4x] = D^{-1}[(1/8)\sin 4x(4)] = -(1/8)\cos 4x + c$.   We
       could also let $u = \sin 2x$, then $u' = 2\cos 2x$ and
       $D^{-1}[\sin 2x\cos 2x] = D^{-1}[(1/2)uu'] = (1/2)u^2/2 + c_1 =$
       $(1/4)\sin^2 2x + c_1 = (1/8)(1-\cos 4x) + c_1 = -(1/8)\cos 4x + c_2$.

29. $D^{-1}[\cos^2 x] = D^{-1}[1/2+(1/2)\cos 2x] = D^{-1}[1/2+(1/4)\cos 2x(2)] =$
       $x/2+(1/4)\sin 2x + c$.

31.   Let $u = x^2$ then $u' = 2x$ and
       $D^{-1}[x\sin x^2] = D^{-1}[(1/2)\sin x^2(2x)] = (-1/2)\cos x^2 + c$.

33.   Let $u = (x-\pi)^2$ then $u' = 2(x-\pi)$ and $D^{-1}[(x-\pi)\cos(x-\pi)^2] =$
       $D^{-1}[(1/2)\cos(x-\pi)^2 \, 2(x-\pi)] = (1/2)\sin(x-\pi)^2 + c$.

35.   From Eq.(24) we have $D^8 \sin x = D^4(D^4 \sin x) = D^4 \sin x = \sin x$.
      Thus $D^9 \sin x = D(D^8 \sin x) = D \sin x = \cos x$.

37.   Using Eqs.(17),(24) and (25) we have $D^3 \sin(x/2) =$
      $-(1/2)^3 \cos(x/2)$ and $D^4 \cos(x/2) = (1/2)^4 \cos(x/2)$. Thus
      $D^{11} \sin(x/2) = D^4\{D^4[D^3 \sin(x/2)]\} = D^4\{D^4[-(1/2)^3 \cos(x/2)]\} =$
      $D^4\{-(1/2)^7 \cos(x/2)\} = -(1/2)^{11} \cos(x/2)$.

39. $d(\tan ax)/dx = \sec^2 ax(a) = a\sec^2 ax$;
      $d^2(\tan ax)/dx^2 = (2a \sec ax)[\sec ax \tan ax(a)] = 2a^2 \sec^2 ax \tan ax$.

41. $D(\sec x) = \sec x \tan x$ and thus $D^2(\sec x) = D(\sec x \tan x) =$
      $(\sec x \tan x)\tan x + \sec x(\sec^2 x) = \sec x(\tan^2 x + \sec^2 x)$.

43.   By Eq.(17e) we have $D[\cot 2x] = -\csc^2 2x(2) = -2\csc^2 2x$. Thus
      $D^2[\cot 2x] = -4\csc 2x D[\csc 2x] = -4\csc 2x(-2\csc 2x \cot 2x)$, using
      Eq.(17f). Thus $D^2[\cot 2x] = 8\csc^2 2x \cot 2x$.

45. $y' = -4\sin 2x(2) = -8\sin 2x$. Thus $y' = -8\sqrt{3}/2 = -4\sqrt{3}$ at
      $x = \pi/6$ and hence $y-5 = -4\sqrt{3}(x-\pi/6)$.

47.   We have
      $v = ds/dt = 2\cos 3t(3) - 3(-\sin 2t)(2) = 6\cos 3t + 6\sin 2t$. Also
      $a = dv/dt = 6(-\sin 3t)(3) + 6(\cos 2t)(2) = 12\cos 2t - 18\sin 3t$.

49. $f(x) = D^{-1}(\sin 2x - 3\cos x) = D^{-1}(\sin 2x) - 3D^{-1}(\cos x) =$
      $D^{-1}[(1/2)\sin 2x(2)] - 3D^{-1}[\cos x] = -(1/2)\cos 2x - 3\sin x + c$.
      Thus $f(\pi/4) = 0 - 3/\sqrt{2} + c$ which must have the value -1 yielding
      $c = 3/\sqrt{2} - 1$ and hence $f(x) = -(1/2)\cos 2x - 3\sin x + (3/\sqrt{2} - 1)$.

51.   By Theorem 3.2.2 $f(x)$ must be continuous at $x = \pi/4$ and thus
      $a\pi/4 + b = \cos \pi/4 = 1/\sqrt{2}$. In addition, we have
      $$f'(x) = \begin{cases} a & x < \pi/4 \\ -\sin x & x > \pi/4 \end{cases}$$ which means $a = -\sin(\pi/4) = -1/\sqrt{2}$ in
      order for $f'(x)$ to exist at $x = \pi/4$. Substituting this
      value for a in the earlier equation yields
      $b = 1/\sqrt{2} - (-1/\sqrt{2})\pi/4 = (1+\pi/4)/\sqrt{2}$ and thus $f(x)$ is
      differentiable at $x = \pi/4$ with $f'(\pi/4) = -1/\sqrt{2}$.

53. $D(\cos x) = \lim\limits_{h\to 0}\dfrac{\cos(x+h)-\cos x}{h} = \lim\limits_{h\to 0}\dfrac{\cos x\cos h-\sin x\sin h-\cos x}{h}$

$= \lim\limits_{h\to 0}\cos x\,\dfrac{(\cos h-1)}{h} - \lim\limits_{h\to 0}\sin x\,\dfrac{\sin h}{h} = -\sin x$ using Eqs.(4),(5).

55.   Since 1.6 is close to $\pi/2 \cong 1.571$ we have
$\sin x = \sin(\pi/2) + \cos(\pi/2)(x-\pi/2) = 1.000$. The exact value
is .9996 and thus we have three decimal place accuracy.

57.   Since 1.0 is close to $\pi/3 \cong 1.047$ we have
$\tan x = \tan(\pi/3) + \sec^2(\pi/3)(x-\pi/3) = 1.7321+4(1-1.0472) = 1.5433$.
The exact value is 1.5574 and thus the approximation has one
decimal place accuracy since the difference is $.0141 < 5/10^2$.

59.   From Fig.3.6.1, the bisector of the angle θ intersects PR at
right angles, so the straight line distance PR is given by
$2r\sin(\theta/2) = 4\sin(\theta/2)$. Thus

$d(PR)/dt = 4\cos(\theta/2)(1/2)d\theta/dt = 2\cos(\theta/2)d\theta/dt$. To

determine $d\theta/dt$ we recall that the arc length PR is given by

$s = r\theta$ and thus $v = ds/dt = rd\theta/dt$, where v is given as

3mi/hr and thus $d\theta/dt = 3/2$ radians/hr. Using this in the

above equation with $\theta = \pi/3$ yields
$d(PR)/dt = 2\cos(\pi/6)(3/2) = 3\sqrt{3}/2$ mi/hr.

61.   From the definition of the derivative we have

$f'(0) = \lim\limits_{h\to 0}\dfrac{f(0+h)-f(0)}{h} = \lim\limits_{h\to 0}\dfrac{h\sin(1/h)-0}{h} = \lim\limits_{h\to 0}\sin(1/h)$,

which does not exist by Prob.24 Section 2.2.

Section 3.7, Page 171

1a.   Since $(-1)^2-(-1)(-1)+(-1)^2 = 1-1+1 = 1$ we see that $(-1,-1)$
satisfies the given equation.

b.   Differentiating both sides of the equation we obtain
$2x-[(1)y+xdy/dx] + 2ydy/dx = 0$ or $(2x-y)+(2y-x)dy/dx = 0$ so
that $dy/dx = (y-2x)/(2y-x)$.

c.   Setting $x = -1$ and $y = -1$ in the above we obtain
$dy/dx = [-1-2(-1)]/[2(-1)-(-1)] = 1/-1 = -1$ at $(-1,-1)$.

3a.   Since $(1)^2+(1)(2)+(2)^2 = 7$ and $12-(1)^2-(2)^2 = 7$ the equation is satisifed at $(1,2)$.

b.   Differentiating both sides of the equation yields $2x+(y+xdy/dx)+2ydy/dx = 0-2x-2ydy/dx$. Solving for dy/dx we obtain $dy/dx = -(4x+y)/(x+4y)$.

c. $dy/dx = -(4+2)/(1+8) = -2/3$ at $(1,2)$.

5a.   Since $(1)^3+(1)-(1) = 1$, the equation is satisfied at $(1,1)$.

b.   Differentiating both sides we get $3y^2dy/dx + dy/dx-1 = 0$, or $dy/dx = 1/(1+3y^2)$.

c. $dy/dx = 1/(1+3) = 1/4$ at $(1,1)$.

7a.   Since $[(3/4)^2+(\sqrt{3}/4)^2]^{3/2} = 3\sqrt{3}/8$ and $2(3/4)(\sqrt{3}/4)=3\sqrt{3}/8$, the equation is satisfied at the given point.

b.   Differentiating both sides we obtain $(3/2)(x^2+y^2)^{1/2}(2x+2ydy/dx) = 2y+2xdy/dx$, which can be solved to yield $dy/dx = -[2y-3x(x^2+y^2)^{1/2}]/[2x-3y(x^2+y^2)^{1/2}]$.

c. $dy/dx = -[2\sqrt{3}/4 - 3(3/4)(3/4)^{1/2}/[2(3/4)-3(\sqrt{3}/4)(3/4)^{1/2}] = 5\sqrt{3}/3$.

9a. $[\sin(\pi/4)]^2 + [\cos(\pi/4)]^2 = 1/2 + 1/2 = 1$. Actually, by Eq.(12) of Sect.1.6 the equation holds for all points (x,y).

b. $2\sin x\cos x+2\cos y(-\sin y)dy/dx = 0$, or $dy/dx =\sin x\cos x/\sin y\cos y$.

c. $dy/dx = (1/\sqrt{2})(1/\sqrt{2})/(1/\sqrt{2})(1/\sqrt{2}) = 1$ at $(\pi/4, \pi/4)$.

11.   From Prob.1 we have $dy/dx = (y-2x)/(2y-x)$ and thus
$$\frac{d^2y}{dx^2} = \frac{(2y-x)(dy/dx -2) - (y-2x)(2dy/dx -1)}{2(y-x)^2}.$$
Substituting $x = -1$, $y = -1$, $dy/dx = -1$ (from Prob.1) we
find $\dfrac{d^2y}{dx^2} = \dfrac{(-2+1)(-1-2)-(-1+2)(-2-1)}{(-2+1)^2} = 6$ at $(-1,-1)$.

13.   From Prob.5 we have $dy/dx = 1/(3y^2+1)$ and thus
$$\frac{d^2y}{dx^2} = \frac{(3y^2+1)(0) - (1)(6y\,dy/dx)}{(3y^2+1)^2} = \frac{-6(1)(1/4)}{[3(1)^2+1]^2} = \frac{-3}{32}$$ at $(1,1)$. Note
that, from Prob.5, $dy/dx = 1/4$ at $(1,1)$.

15. Note that (2,1) does satisfy the equation and hence we may
     differentiate both sides to obtain
     4x-5y-5xdy/dx + 6ydy/dx = 0 which yields
     dy/dx = (5y-4x)/(6y-5x) = 3/4 at (2,1). Thus the tangent
     line is (y-1) = (3/4)(x-2) or 3x-4y = 2.

17. Note that $(\pi/4,\pi/4)$ does satisfy the equation and hence we
     have siny + xcosy(dy/dx) + (dy/dx)cosx - ysinx = 0 which
     yields dy/dx = (ysinx - siny)/(xcosy + cosx) = $(\pi-4)/(\pi+4)$
     at $(\pi/4,\pi/4)$. Thus the tangent line is
     $y-\pi/4 = [(\pi-4)/(\pi+4)](x-\pi/4)$ or $(4-\pi)x + (4+\pi)y = 2\pi$.

19. Since (1,-1) satisfies the equation we may differentiate to
     obtain 6x-4y-4xdy/dx + 4ydy/dx = 0. Solving for dy/dx we
     find dy/dx = (4y-6x)/(4y-4x) = 5/4, which is the slope of
     the tangent line at (1,-1). Thus the desired line is given
     by y-(-1) = (-4/5)(x-1), or 4x+5y = -1.

21. The given point satisfies the given equation and hence
     differentiating yields
     $2\pi cosxsiny + 2\pi sinx(cosy)dy/dx = \sqrt{3}(1+dy/dx)$, or
     dy/dx = $(\sqrt{3}-2\pi cosxsiny)/(2\pi sinxcosy - \sqrt{3})$. Thus the slope
     of the tangent line at $(\pi/6,\pi/3)$ is given by
     dy/dx = $(2\sqrt{3}-3\pi)/(\pi-2\sqrt{3})$ and hence the desired line is

     $(y-\pi/3) = [(\pi-2\sqrt{3})/(3\pi-2\sqrt{3})](x-\pi/6)$, or

     $(\pi-2\sqrt{3})(x-\pi/6)-(3\pi-2\sqrt{3})(y-\pi/3) = 0$.

23. Assume the given equation defines a function $A(\omega)$ and then

     differentiate both sides with respect to ω to obtain

     $2\omega = (1/4)AdA/d\omega + (1/A^2)dA/d\omega$ or $dA/d\omega = 8\omega A^2/(A^3+4)$.

25. If x = f(y), then 1 = (df/dy)(dy/dx) = f'(y)dy/dx and thus
     dy/dx = 1/(f'(y). Also y = g(x) so dy/dx = g'(x) and hence
     g'(x) = 1/f'(y).

27. We have f(y) = $\sqrt{y+2}$ so that f'(y) = $(1/2)(y+2)^{-1/2} = 1/2\sqrt{y+2}$.
     From Prob.25 we then obtain dy/dx = 1/f'(y) = $2\sqrt{y+2}$.
     However, x = $\sqrt{y+2}$ and thus dy/dx = 2x.

29. We have f(y) = tan y so that f'(y) = $sec^2 y$ and thus
     dy/dx = $1/sec^2 y$ = $cos^2 y$. Since x = tany we have
     cosy = $1/\sqrt{1+x^2}$ from the definition of the trigonometric
     functions and hence dy/dx = $(1/\sqrt{1+x^2})2 = 1/(1+x^2)$.

Chapter 3 Review, Page 172

1. $f'(x) = \lim\limits_{\Delta x \to 0} \dfrac{(x + \Delta x + \sqrt{x + \Delta x}) - (x + \sqrt{x})}{\Delta x} = \lim\limits_{\Delta x \to 0} \left[1 + \dfrac{\sqrt{x + \Delta x} - \sqrt{x}}{\Delta x} \right]$

$= \lim\limits_{\Delta x \to 0} \left[1 + \dfrac{\Delta x}{\Delta x \, (\sqrt{x + \Delta x} + \sqrt{x})} \right] = 1 + 1/2\sqrt{x}.$

3. $h'(x) = \lim\limits_{\Delta x \to 0} \dfrac{1/(x + \Delta x + 1) - 1/(x + 1)}{\Delta x}$

$= \lim\limits_{\Delta x \to 0} \dfrac{-1}{(x + \Delta x + 1)(x + 1)} = -\dfrac{1}{(x + 1)^2}$

5. Combining results of Probs. 1 and 2, we have

$f'(t) = \lim\limits_{\Delta t \to 0} \dfrac{\sqrt{t + \Delta t} - \sqrt{t}}{\Delta t} + \lim\limits_{\Delta t \to 0} \dfrac{1/(t + \Delta t) - 1/t}{\Delta t} = \dfrac{1}{2\sqrt{t}} - \dfrac{1}{t^2}$.

7a. $f'(x) = 6x - (5/2)x^{3/2}$.
 b. $f'(4) = 4$ and $f(4) = 16$, so $y - 16 = 4(x-4)$, or $y = 4x$.

9a. $f'(x) = 3x^2 - 2x + 1$.
 b. $f'(4) = 41$ and $f(4) = 51$, so $y - 51 = 41(x-4)$, or
$y = 41x - 113$.

11a. $f'(x) = (1-x^2)/(1+x^2)^2$.
  b. $f'(4) = -15/289$ and $f(4) = 4/17$, so
$y - 4/17 = -(15/289)(x-4)$, or $15x + 289y - 128 = 0$.

13a. $f'(x) = x^{1/2}(3-x^2)/2(x^2+1)^2$.
  b. $f'(4) = -13/289$ and $f(4) = 8/17$. Thus
$y - 8/17 = -(13/289)(x-4)$ or $y = -(13/289)x + 188/289$.

15. $f'(x) = 4x(x^2+1)$, which is 0 only when $x = 0$.

17. $f'(x) = (7x^2 - 32x + 28)/(x^2-4)^2 = 0$ for $x = (16 \pm 2\sqrt{15})/7$.

19a. $F(x) = x^4/4 + 3x + c$. With $F(1) = -2$, $c = -21/4$.
  b. $f'(x) = 3x^2$, so $f'(4) = 48 = m$.

21a. $F(x) = (3/5)x^{5/3} + 1/x + c$. With $F(1) = -2$, $c = -18/5$.
  b. $f'(x) = (2/3)x^{-1/3} + 2/x^3$, so $f'(4) = (2/3)4^{-1/3} + 1/32 = m$.

23a. $f(x) = ax^3+bx^2+cx+d$, $d = f(0) = 2$, $c = f'(0) = -4$,
         $f''(x) = 6ax+2b$ and $f'''(x) = 6a$. With $f'''(1) = 12$, $a = 2$.
         Thus $f''(1) = 12+2b = 2$ so $b = -5$ and $f(x) = 2x^3-5x^2-4x+2$.

  b.   We have $\lim\limits_{x \to -2-} g(x) = f(-2) = -26 = 4-2p+q$ and
         $\lim\limits_{x \to -2-} g'(x) = f'(-2) = 40 = -4+p$ or $p = 44$ and $q = 58$, so
         $g(x) = x^2+44x+58$.

25. $f'(x) = \dfrac{2(2x)(x^2-1)(x^{1/2}+3) - (1/2)x^{-1/2}(x^2-1)^2}{[x^{1/2}+3]^2}$

         $= (x^2-1)(7x^2 + 24x\sqrt{x} + 1) / [2\sqrt{x}(\sqrt{x}+3)^2]$.

27. $\dfrac{dg}{dx} = \dfrac{dg}{dy}\dfrac{dy}{dx} = [-6y\sin(3y^2 - 1)][1/2x^{1/2}]$

         $= 3(x^{-1/2} - 1)\sin[3(x^{1/2} - 1)^2 - 1]$.

29. $f'(x) = -(3/2)(x^{1/2})\sin(x^{3/2} + 1)$ so
         $f''(x) = -(3/4)(x^{-1/2})\sin(x^{3/2} + 1) - (9/4)x\cos(x^{3/2} + 1)$.

31. $\dfrac{dh}{dt} = \dfrac{dh}{dx}\dfrac{dx}{dt} = (1+1/x^2)(2t) = 2t(t^4+2t^2+2)/(t^2+1)^2$.

33. $f'(x) = [-2ax(x^2-1)\sin(ax^2+b) - 2x\cos(ax^2+b)]/(x^2-1)^2$.

35. $\dfrac{dh}{ds} = \dfrac{dh}{dy}\dfrac{dy}{ds} = -3y^2\sin(y^3-1)[2\sin s \cos s]$
         $= 6\sin s \cos s[4+\sin^2 s]^2\sin[(4+\sin^2 s)^3 - 1]$.

37. $Dg(x) = 2(x-\pi)\cos(3x^{3/2})\sec[(x-\pi)^2]\tan[(x-\pi)^2]$
                  $- (9/2)\sqrt{x}\sin(3x^{3/2})\sec[(x-\pi)^2]$.

In Prob.39-45 let F be an antiderivative of the given function f.

39.   With $f(x) = 2\sin 3x$, $F(x) = (-2/3)\cos 3x + c$.

41. $f(x) = \sin 2x[2\cos 2x]$ so $F(x) = (1/2)\sin^2 2x + c$.

43. $F(x) = -\cos(x^3+1) + c$.

45. $f(x) = (\sin x)^{-2}[-\cos x]$ so $F(x) = [\sin x]^{-1} + c$.

47a.   Since $f(x_0) = x_0^3 + x_0^{1/2}$ and $f'(x_0) = 3x_0^2 + (1/2)x_0^{-1/2}$,

$T(x;x_0) = x_0^3 + x_0^{1/2} + [3x_0^2 + (1/2)x_0^{-1/2}](x-x_0)$ and

$$r(x;x_0) = \frac{(x^3 + x^{1/2}) - (x_0^3 + x_0^{1/2})}{x - x_0} - [3x_0^2 + (1/2)x_0^{-1/2}].$$

b.   With $x_0 = 1$ and $x = 15/16$,
$T(15/16;1) = 2 + (7/2)(-1/16) = 57/32$.

c.   Error $= r(15/16;1)(-1/16) = 0.01097$, by Eq.3, Section 3.4.
Relative error $= |[f(15/16)-T(15/16;1)]/f(1)| = 0.5485\%$.

49a. $f(x_0) = x_0^{1/2} + x_0^{-1/2}$, $f'(x_0) = (1/2)(x_0^{-1/2} - x_0^{-3/2})$ so

$T(x;x_0) = (x_0^{1/2} + x_0^{-1/2}) + (1/2)(x_0^{-1/2} - x_0^{-3/2})(x-x_0)$ and

$r(x;x_0) = [(x^{1/2}+x^{-1/2}) - (x_0^{1/2} + x_0^{-1/2})]/(x-x_0) - (x_0^{-1/2} - x_0^{-3/2})/2$.

b.   With $x_0 = 9$,
$T(80/9;9) = (10/3)+(1/2)[(1/3)-(1/27)]/(-1/9) = 806/243$.

c.   Error $= f(80/9) - T(80/9;9) = -3.83 \times 10^{-5}$.
Relative error $= |(\text{error})/f(9)| = 1.15 \times 10^{-3}\%$.

51a. $(1)^2 + (1)(2) = 3$, $(3)^2 + (3)(-\sqrt{2})^2 \neq 3$, and

$4^2 + (4)(\sqrt{13}/2)^2 \neq 3$, so the latter two points do not satisfy the equation.

b.   Differentiating implicitly, $2x + 2xyy' + y^2 = 0$, and again
$2 + 2yy' + 2x(y')^2 + 2xyy'' + 2yy' = 0$. Thus
$y' = -(2x+y^2)/2xy$ and $y'' = -[1+2yy'+x(y')^2]/xy$ so at
$(1,\sqrt{2})$, $y' = -\sqrt{2}$, and $y'' = 1/\sqrt{2}$.

53a. $(1)^2(-1)+(1)(-1)^2 \neq 2$, $(1)^2(1)+(1)(1)^2 = 2$, and

$(-1)^2(-1)+(-1)(-1)^2 \neq 2$, so only $(1,1)$ satisfies the
equation.

b.   Differentiating implicitly, $(2xy+y^2)+(x^2+2xy)y' = 0$ and
$2y+4xy'+4yy'+2x(y')^2 + (x^2+2xy)y'' = 0$. These give, for
$(1,1)$, $y' = -1$ and $y'' = 4/3$.

CHAPTER 4

Section 4.1, Page 184

1.   f(x) is continuous on
      [-1,2] and differentiable on
      (-1,2) so there is a c in
      (-1,2) such that f(2)-f(-1) =
      f'(c)(2+1). Since f'(x) = 2x we
      obtain c = 1/2, f(c) = 1/4.

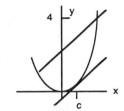

3.   f(x) = |x²-1| is continuous on
      [-2,2] but not differentiable at
      x = -1 or x = 1. Thus the Mean
      Value Theorem does not apply.
      However, Eq.(5) is satisfied for
      c = 0 since f'(0) = 0 and
      f(2) - f(-2) = 0.

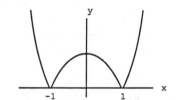

5.   f(x) is not continuous at x = 0,
      which is in the interval [-1,1],
      so that the Mean Value Theorem
      does not apply. There is no value
      of c for which Eq.(5) is
      satisfied.

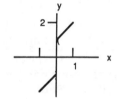

7a.   Following the technique of Ex.1 suppose $f(x) = x^3-3x+b$ is
      zero at x_1 and x_2 so $0 = f(x_1) = f(x_2)$. Further, suppose
      $-1 < x_1 < x_2 < 1$. By Rolle's Theorem there is a point c in
      the interval (x_1,x_2) where f'(c) = 0. However, f'(x) =
      $3x^2-3 = 3(x^2-1) = 0$ only for x = 1 and x = -1. Thus there
      cannot be more than one root of $f(x) = x^3-3x+b$ in the
      interval [-1,1].

  b. From the sketch, which is f(x) for
      b = 0, the curve will continue to
      have a root in [-1,1] for any
      vertical translation between -2
      units and +2 units. That is,
      whenever $-2 \le b \le 2$, $f(x) = x^3-3x + b$
      will be 0 for some x in [-1,1].

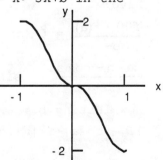

9.   Following the method of Ex.1, suppose there are two values
      of x, say $x_1 < x_2$ for which $f(x_1) = f(x_2) = 0$. Then by
      Rolle's Theorem there is a c in (x_1,x_2) for which $f'(c) = 0$.
      But $f'(x) = 3x^2 + a$ which is never less than a for all x.
      Since a > 0 we have $f'(x) > 0$ for all x. Thus there is no c
      in (x_1,x_2) for which $f'(c) = 0$, so the assumption that $f(x_1)$
      $= f(x_2) = 0$ for $x_1 \neq x_2$ is impossible. We conclude that
      there is at most one root of $f(x) = x^2 + ax + b$ with a > 0.

11a. Suppose there were two x in [a,b] for which $f(x_1)=f(x_2) = 0$.
      Without loss of generality we may take $a \leq x_1 < x_2 \leq b$. By
      Rolle's Theorem there is a c in (x_1,x_2) for which $f'(c) = 0$.
      This contradicts the assumption that $f'(x) > 0$ for all x in
      (a,b). Thus f can have no more than one root in (a,b).

  b. Suppose there are $x_1 < x_2 < x_3$, all in (a,b) such that
      $f(x_1) = f(x_2) = f(x_3) = 0$. Then by Rolle's Theorem there are
      c_1 and c_2 such that $x_1 < c_1 < x_2$ and $x_2 < c_2 < x_3$ and
      $f'(c_1) = 0$ and $f'(c_2) = 0$. Since c_1 is in (x_1,x_2) then c_1 is
      in (a,b) and since c_2 is in (x_2,x_3) then c_2 is in (a,b) also.
      Now, applying Rolle's Theorem to $f'(x)$ on $[c_1,c_2]$, there is
      a c_3 in (c_1,c_2) [and hence in (a,b)] for which $f''(c_3) = 0$,
      which contradicts the hypothesis that $f''(x) > 0$ for all x in
      (a,b). Thus f can have no more than two zeroes in (a,b).

  c. A generalization of parts (a),(b) is: Let $f,f',f'',..f^{(n-1)}$ be
      continuous on [a,b] and $f^{(n)}(x) > 0$ for each x in (a,b).
      Then $f(x)$ has no more than n roots in (a,b).

13.   In a manner similar to Ex. 2, let $f(x) = \cos x$. On any
      interval [a,b], $f(x)$ is continuous on [a,b] and
      differentiable on (a,b). Applying the Mean Value Theorem we
      have:

      $$\frac{f(b) - f(a)}{b-a} = f'(c) \text{ for some c in (a,b). Thus}$$

      $$\frac{\cos b - \cos a}{b-a} = -\sin c \text{ or } |\cos b - \cos a| = |\sin c|\ |b-a|.$$

      Now letting b = 0, and since $|\sin c| \leq 1$ for all c, we have
      $|1 - \cos a| \leq |a|$ or $|1 - \cos x| \leq |x|$.

15.   For any x with $0 \leq x < \pi/2$, f(x) = tanx is continuous on
        [0,x] and differentiable on (0,x) with (tanx)' = sec^{2}x.
        Hence, by the Mean Value Theorem, $\frac{\tan x}{x}$ = sec^{2}c for some c
        in (0,x). Thus |tanx| = |x|sec^{2}c $\geq$ |x| since sec^{2}c $\geq$ 1. A
        similar argument shows |tanx| $\geq$ |x| for $-\pi/2 < x < 0$.
        Therefore |tanx| $\geq$ |x| for |x| < $\pi/2$.

17.   Since f is differentiable in $(0,\infty)$, then f is continuous in
        $(0,\infty)$. Thus the Mean Value Theorem applies on any interval
        [a,b] with 0 < a < b. In particular $\frac{f(x) - f(1)}{x-1}$ = f'(c) for
        some c between x and 1, provided x > 0.
 a.   Setting x = 3 we have [f(3) - f(1)]/(3-1) = 1/c for some c
        in (1,3) so f(3) = 2/c; but for 1 < c < 3, 1/3 < 1/c < 1 so
        2/3 < 2/c < 2 and 2/3 < f(3) < 2.
 b.   In like manner, there is a c in (1,a) for which
        f(a) - f(1) = (1/c)(a-1). Since 1 < c < a, then
        1/a < 1/c < 1 so (a-1)/a < (a-1)/c = f(a) < (a-1).
 c.   Similarly, there is a c satisfying a < c < 1 such that
        f(a) = (a-1)/c. Now since a < c < 1; 1 < (1/c) < (1/a) and
        (a-1)/a < (a-1)/c = f(a) < (a-1) because (a-1) < 0.

19a.   Since f' exists on any interval (a,b) with -1 < a < b < 1, f
        is continuous on [a,b]. In particular, f is continuous on
        [0,x] for 0 < x < 1 and differentiable on (0,x). By the
        Mean Value Theorem [f(x) - f(0)]/(x-0) = f'(c) for some c
        in (0,x). Thus, since f(0) = 0, f(x) = $x/\sqrt{1-c^2}$. Finally,
        since $\sqrt{1-c^2}$ < 1, $1/\sqrt{1-c^2}$ > 1 and thus f(x) = $x/\sqrt{1-c^2}$ > x.
 b.   For the case x < 0, f(x) = $x/\sqrt{1-c^2}$ < x, as the inequality
        $1/\sqrt{1-c^2}$ > 1 is multiplied by a negative number.

21.   Under the stated conditions, the function
        F(x) = [f(b) - f(a)]g(x) - [g(b) - g(a)]f(x) satisfies the
        conditions of Theorem 4.1.3. Thus for some c in (a,b),
        F'(c) = [F(b) - F(a)]/(b - a). By direct calculation
        F(b)=F(a) so F'(c) = [f(b)-f(a)]g'(c)-[g(b)-g(a)]f'(c) = 0.
        If g' is never zero in (a,b) then g(a) $\neq$ g(b) and we have
        $\frac{f(b) - f(a)}{g(b) - g(a)} = \frac{f'(c)}{g'(c)}$.

Section 4.2, Page 196

1a.   f'(x)= 2(x+2) so x = -2 is the only critcal point.
 b.   Since f'(x) < 0 for x < -2, f is decreasing on $(-\infty,-2]$.
        Since f'(x) > 0 for x > -2, f is increasing on $[-2,\infty)$.
 c.   Since f"(x)=2 > 0 for all x, f is concave upward on $(-\infty,\infty)$.
 d.   f"(-2)> 0, so by Theorem 4.2.4, f has a local minimum there.

3a.   The only critical points are for
        f'(x) = $3x^2-6x-9$ = 3(x+1)(x-3)= 0; that is x = -1, x = 3.
 b.   f'(x) > 0 for x < -1 and x > 3 so f increases for x $\leq$ -1 and
        x $\geq$ 3. f'(x) < 0 for -1 < x < 3 so f is decreasing for
        $-1 \leq x \leq 3$.
 c.   Since f"(x)=6(x-1) > 0 for x > 1, f is concave up for x $\geq$ 1.
        Since f"(x) < 0 for x < 1, f is concave down for x $\leq$ 1.
 d.   f"(-1) = -12 < 0 so x = -1 is a local maximum point and
        f"(3) = 12 > 0, so x = 3 is a local minimum point.

5a.   f'(x) = $-1/(x+1)^2$ if x $\neq$ -1. Thus the only critical point is
        x = -1, where f' fails to exist.
 b.   f'(x) < 0 for $(-\infty,-1)$ and $(-1,\infty)$ so f is decreasing for
        $(-\infty,-1)$ and $(-1,\infty)$.
 c.   f"(x) = $2/(x+1)^3$ if x $\neq$ -1 so f"(x) < 0 for x < -1 and f is
        concave down on $(-\infty,-1)$. Likewise f"(x) > 0 for x > -1 so f
        is concave up on $(-1,\infty)$.
 d.   Since for a < -1 < b, f(a) < f(-1) < f(b), and x = -1 is the
        only critical point, f has no local maximum points or local
        minimum points.

7a.   f'(x) = $(-2/3)x^{-1/3}$ for x $\neq$ 0 so the only critical point is
        at x = 0, since f'(0) does not exist.

 b.   f is increasing on $(-\infty,0)$ since f'(x) > 0 there. f is
        decreasing on $(0,\infty)$ since f'(x) < 0 there.
 c.   f"(x) = $(2/9)x^{-4/3}$ > 0 for all x $\neq$ 0 so f is concave up on
        $(-\infty,0]$ and $[0,\infty)$.
 d.   Since f(x) < f(0) for all x $\neq$ 0, we have that x = 0 is a
        local maximum point.

9a.   The critical points are the endpoints x = $-\pi$, π and when
        f'(x) = 1 + 2sinx = 0 or sinx = -1/2 so x = $-5\pi/6,-\pi/6$.
 b.   f is increasing for $[-\pi,-5\pi/6]$ and $[-\pi/6,\pi]$ since f'(x) > 0,
        there and f is decreasing for $[-5\pi/6,-\pi/6]$ since f'(x) < 0
        there.

9c.   f is concave upward for x in $[-\pi/2, \pi/2]$ since
       $f''(x) = 2\cos x > 0$ there. f is concave downward for x in
       $[-\pi, -\pi/2]$ and $[\pi/2, \pi]$ as $f''(x) < 0$ there.

d. $f''(x) > 0$ for $x = -\pi/6$ so f has a local minimum there and
       $f''(x) < 0$ for $x = -5\pi/6$ so f has a local maximum there. From
       part (c) we conclude from the concavity of the function that
       $x = -\pi$ is a local minimum point and that $x = \pi$ is a local
       maximum point.

11.   To analyze $f(x) = |x^3 - 12x|$ on $-4 \leq x \leq 4$ it is convenient to
       express $f(x)$ without the absolute values. We begin by
       observing: $x^3 - 12x = (x + 2\sqrt{3})x(x - 2\sqrt{3})$ which changes sign
       at $x = -2\sqrt{3}$, $x = 0$, and $x = 2\sqrt{3}$. Thus

$$f(x) = \begin{cases} x^3 - 12x & \text{for } -2\sqrt{3} < x < 0 \text{ and } 2\sqrt{3} < x \leq 4 \\ 0 & \text{for } x = -2\sqrt{3},\ x = 0 \text{ and } x = 2\sqrt{3} \\ 12x - x^3 & \text{for } -4 \leq x < -2\sqrt{3} \text{ and } 0 < x < 2\sqrt{3} \end{cases}$$

so

$$f'(x) = \begin{cases} 3x^2 - 12 = 3(x+2)(x-2), & \text{for } -2\sqrt{3} < x < 0,\ 2\sqrt{3} < x < 4 \\ \text{does not exist} & \text{for } x = \pm 4, \pm 2\sqrt{3},\ 0 \\ 12 - 3x^2 = 3(2+x)(2-x), & -4 < x < -2\sqrt{3},\ 0 < x < 2\sqrt{3} \end{cases}$$

$$f''(x) = \begin{cases} 6x & \text{for } -2\sqrt{3} < x < 0,\ 2\sqrt{3} < x < 4 \\ \text{does not exist} & \text{for } x = \pm 4,\ \pm 2\sqrt{3},\ 0 \\ -6x & \text{for } -4 < x < -2\sqrt{3},\ 0 < x < 2\sqrt{3} \end{cases}$$

a.   Thus the critical points are (i) the end points $x = \pm 4$; (ii)
       points for which $f'(x)$ does not exist, $x = \pm 2\sqrt{3}$, 0; (iii)
       points for which $f'(x) = 0$: $x = \pm 2$.

b.   Let $f_1(x) = f(x)$ on $-2\sqrt{3} < x < 0$, $2\sqrt{3} < x < 4$ and
       $f_2(x) = f(x)$ on $-4 < x < -2\sqrt{3}$, $0 < x < 2\sqrt{3}$. Then
       $f_1'(x) > 0$ for $-2\sqrt{3} < x < -2$ and $2\sqrt{3} < x < 4$ and $f_2' > 0$ for $0 < x < 2$.

       Thus f is increasing for $-2\sqrt{3} \leq x \leq -2$, $0 \leq x \leq 2$ and $2\sqrt{3} \leq x \leq 4$.

       Also $f_1' < 0$ for $-2 < x < 0$ and $f_2' < 0$ for $-4 < x < -2\sqrt{3}$ and $2 < x < 2\sqrt{3}$.
       Thus f is decreasing for $-4 \leq x \leq -2\sqrt{3}$, $-2 \leq x \leq 0$ and $2 \leq x \leq 2\sqrt{3}$.

11c. $f_1''(x) > 0$ for $2\sqrt{3} < x < 4$; $f_2''(x) > 0$ for $-4 < x < -2\sqrt{3}$.

Thus f is concave upward for $-4 \le x \le -2\sqrt{3}$ and $2\sqrt{3} \le x \le 4$.
Similarly $f_1''(x) < 0$ for $-2\sqrt{3} < x < 0$, $f_2''(x) < 0$ for $0 < x < 2\sqrt{3}$ so f is concave downward for $-2\sqrt{3} \le x \le 0$ and $0 \le x \le 2\sqrt{3}$.

  d.   Using the results of parts (b) and (c) we have that f has a local maximum of 16 at the points $x = \pm2, \pm4$ and a local minimum of 0 at the points $x = 0, \pm 2\sqrt{3}$.

13a. $f'(x) = -x/\sqrt{4-x^2}$ so $x = 0$ is a critcal point as are the end points $x = \pm2$.
  b. $f'(x) > 0$ for $-2 < x < 0$, so f increases for $-2 \le x \le 0$. $f'(x) < 0$ for $0 < x < 2$, so f decreases for $0 \le x \le 2$.
  c. $f''(x) = -4(4-x^2)^{-3/2} < 0$ for $-2 < x < 2$, so f is concave downward there.
  d. $x = \pm 2$ are local minima from the concavity of f and $f''(0) < 0$, so f has a local maximum at $x = 0$.

15a.   From Prob.1 the only global minimum is at $x = -2$ and the global maximum is at the end point $x = -4$.
  b.   On $(-\infty,\infty)$ the only global minimum is $x = -2$, and there is no global maximum.

17a.   From Prob.3, the critical points for f on $(-\infty,\infty)$ are $x = -1$, $x = 3$. For the domain $[-1,5]$, the only additional critical point is $x = 5$. Since $f(-1) = 9$, $f(3) = -23$, $f(5) = 9$. We have $x = -1,5$ are the global maximum points and $x = 3$ is the global minimum point.

  b.   The only critical points of f on $(-\infty,\infty)$ are $x = -1$, and $x = 3$, which are only local maximum and minimum respectively since f is unbounded

19a.   From Prob.5, the only critical points are $x = -1$ and the endpoint $x = 1$. Since $f(-1) = 0$ and $f(x) > 0$ otherwise, $x = -1$ is the global minimum point and f has no maxima on $[-1,1]$.
  b.   On the interval $[0,2]$ the only critical points are $x = 0,2$. As f decreases on $[0,2]$, $x = 0$ is the global maximum point and $x = 2$ is the global minimum point.

21a.   From Prob.7, the only critical point is $x = 0$. As $f'(x) > 0$ for $x < 0$ and $f'(x) < 0$ for $x > 0$, $x = 0$ is a global maximum point.
  b.   On $[-1,1]$, the global maximum point is $x = 0$ and the global minimum points are $x = \pm1$ since $f(-1) = f(1) = 0$.

23a.   From Prob.9 f has critical points $x = -\pi, -5\pi/6, -\pi/6$, and π.
       $f(-\pi) = -\pi + 2$, $f(-5\pi/6) = -5\pi/6 + \sqrt{3}$, $f(-\pi/6) = -\pi/6 - \sqrt{3}$,
       $f(\pi) = \pi + 2$. Thus $x = -\pi/6$ is the global minimum point and
       $x = \pi$ is the global maximum point.

  b.   On $[-\pi/2, \pi/2]$, f has critical points $x = -\pi/2, -\pi/6$, and $\pi/2$.
       $f(-\pi/2) = -\pi/2$, $f(-\pi/6) = -\pi/6 - \sqrt{3}$, and $f(\pi/2) = \pi/2$. Thus
       $x = -\pi/6$ is the global minimum point and $x = \pi/2$ is the
       global maximum point.

25a.   From Prob.11 the global maximum points on $[-4,4]$ are
       $x = \pm4, \pm2$, at each of which $f(x) = 16$. The global minimum
       points on $[-4,4]$ are $x = \pm2\sqrt{3}, 0$ at each of which $f(x) = 0$.

  b.   The only change from part (a) is that on $(-3,3)$ $x = \pm4, \pm2\sqrt{3}$
       are no longer in the domain, hence they are not considered.
       Thus the global maximum points are $x = \pm2$ and the global
       minimum point is $x = 0$.

27a.   From Prob.13 the global minimum points are $x = \pm2$ and the
       global maximum point is $x = 0$.

  b.   On the interval $(0,1)$, f has no critical points and hence no
       global maximum or minimum.

29.   31.

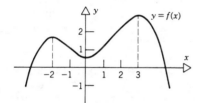   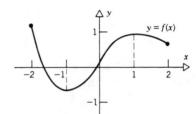

33.

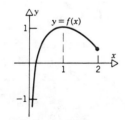

35.   To determine a, b, c so that the graph of $y = ax^2 + bx + c$
       passes through $(0,1)$ and $(3,0)$ and possesses a maximum at
       $x = 1$, we have the three equations $f(0) = 1 = c$, $f(3) = 0 =$
       $9a + 3b + c$ and $f'(1) = 0 = 2a + b$. The third equation
       arises from the fact that since f is continuous and

       differentiable on $(-\infty, \infty)$, then the only maxima and minima
       can occur when $f'(x) = 0$. The three equations reduce to
       $c = 1$, $9a+3b = -1$, $2a+b = 0$, which yield $a = -1/3$ and $b = 2/3$.

37.   Assuming c in (a,b) for which f'(c) < 0, then there is some
       h > 0 for which [f(c+h) − f(c)]/h < 0 or f(c+h) − f(c) < 0.
       Thus c + h > c and f(c+h) < f(c) so f is decreasing at c,
       which contradicts the hypothesis.

Section 4.3, Page 206

1.   Denote the product by P and the sum by s. Then P = xy and
       s = x+y, with $0 \leq x < \infty$, $0 \leq y < \infty$. The other restriction
       is that P = xy = k, a constant. Before beginning, note that
       for large x and small y we can have xy = k, but that
       s = x+y would be large. Similarly, large y and small x
       satisfying xy = k yields a large sum. Thus we expect some
       balance between x and y. The problem is formulated as:
       minimize s = x+y subject to xy = k, $0 \leq x$, $0 \leq y$. Solving
       xy = k for y gives y = k/x so s = x+(k/x) and s' = 1−(k/x²).
       Thus the only critical points are the endpoint x = 0, and x
       such that s'(x) = 0, or (x²−k)/x² = 0 or $x = \sqrt{k}$. Since
       s"(x) = 2k/x³ > 0 the critical point $x = \sqrt{k}$ is a local
       minimum. Further, since s'(x) < 0 for $x < \sqrt{k}$ and s'(x) > 0
       for $x > \sqrt{k}$, we have $x = \sqrt{k}$ as a global minimum. From
       y = k/x we have $y = \sqrt{k}$, also. Thus for fixed product, xy,
       the sum x+y, will be minimized when $x = y = \sqrt{k}$, which agrees
       with our intitial observation.

3.   This problem is a geometrical application of Ex.1, as we
       seek a rectangle of dimensions x by y for which the
       perimeter P = 2x + 2y is a constant and the area A = xy, is
       a maximum. Now let the fixed perimeter be 2s so 2x+2y = 2s,
       or x + y = s. Thus y = s − x and A = x(s−x). We
       are asked to maximize A = x(s−x) for $0 \leq x \leq s$. The critical
       points are the endpoints x = 0 and x = s and all x for which
       A'(x) = 0. Since A(x) = sx − x², A'(x) = s−2x = 0 for
       x = s/2. Further A"(x) = −2 < 0, so x = s/2 is a local
       maximum point. Finally, since A'(x) = s −2x > 0 for x < s/2
       and A'(x) = s − 2x < 0 for x > s/2, x = s/2 is the global
       maximum point. From P = 2x + 2y = 2s, y = s/2 when x = s/2,
       so the area A = s²/4 is a maximum when the sides are equal
       and thus the rectangle must be a square.

5.   Let the circle have diameter d
       and the rectangle have sides of
       length x and y. Then the
       rectangle has area A = xy, with

       $0 \leq x \leq d$ and $0 \leq y \leq d$. The
       endpoints x = 0,d and y = 0,d   give area

zero, which clearly is not the largest possible. To apply our techniques for determining a maximum, we must have A as a function of one variable. Referring to the sketch we see $d^2 = x^2 + y^2$ or $y = \sqrt{d^2-x^2}$. Thus $A = xy = x\sqrt{d^2-x^2}$,

so $A'(x) = \sqrt{d^2-x^2} - x^2/\sqrt{d^2-x^2}$ or $A'(x) = (d^2-2x^2)/\sqrt{d^2-x^2} = 0$ for $x = d\sqrt{2}/2$. Since $A''(x) = -x(3d^2-2x^2)/(d^2-x^2)^{3/2} < 0$ for all $0 < x < d$, then $x = d\sqrt{2}/2$ is the global maximum

point. From $y = \sqrt{d^2-x^2}$, we see that $y = d\sqrt{2}/2$ also, so that the largest rectangle is indeed a square.

7.   The total surface area is made up of the lateral area plus the top and bottom. The lateral area uses a complete rectangle of area $2\pi rh$ and the top and bottom circles need to be cut from squares of side $2r$ and area $4r^2$. Thus the total material needed is $A = 2\pi rh + 8r^2$. The volume is fixed at V so that $V = \pi r^2 h$ or $h = V/\pi r^2$. Substituting, we have $A = (2V/r) + 8r^2$, so A is given as a function of the single variable r. The only critical points are the values of r for which $A'(r) = 0$. Since $A'(r) = 2(8r^3-V)/r^2$, $r_c = V^{1/3}/2$ gives $A'(r_c) = 0$. Also $A''(r) = (22r^3+4V)/r^3 > 0$, so that $r_c = V^{1/3}/2$ is a global minimum. Since $h = V/\pi r^2$,

$h = 4V^{1/3}/\pi$, and $d = 2r_c = V^{1/3}$ shows $h/d = 4/\pi \cong 1.27$.

9a.   The surface area in this case is the lateral area plus the top and bottom: $S = 4xh + 2x^2$. The fixed volume is $V = hx^2$, so $h = V/x^2$ and $S = 4V/x + 2x^2$ for $x > 0$. The only critical point is the value of x for which $S'(x) = 0$. Since $S'(x) = 4(x^3-V)/x^2$, $S'(x) = 0$ for $x = V^{1/3}$. Also, $S'(x) < 0$ for $x < V^{1/3}$ and $S'(x) > 0$ for all $x > V^{1/3}$, giving $x = V^{1/3}$ as a global minimum. From $h = V/x^2$, $h = V^{1/3}$ when $x = V^{1/3}$. This is interpreted as follows. A rectangular box of fixed volume having square, closed top and bottom, has least surface area when it is a cube.

b.   With open top, the volume remains $V = hx^2$, but the surface area is now $S = 4xh + x^2 = (4v/x) + x^2$. Again the only critical point is the x for which $S'(x) = 0$. Now $S'(x) = 2(x^3-2V)/x^2 = 0$ when $x = (2V)^{1/3}$. Again if $x < (2V)^{1/3}$, then $S'(x) < 0$ and if $x > (2V)^{1/3}$ $S'(x) > 0$ so $x = (2V)^{1/3}$ is a global minimum point. Again, since $h = V/x^2$, $h = (2V)^{1/3}/2 = x/2$.

11.   From Fig.4.3.8 the length of the pipe is given by
$l^2 = (b+y)^2 + (a+x)^2$, where x is the horizontal distance to
the right from a and y is the vertical distance down from b.
From similar triangles $b/x = y/a$, so $y = ab/x$ and
$l^2 = (b+ab/x)^2 + (a+x)^2$, which

simplifies to $l = (a/x+1)\sqrt{b^2+x^2}$. This gives

$l'(x) = (-a/x^2)\sqrt{b^2+x^2} + (a/x+1)x/\sqrt{b^2+x^2}$

       $= (x^3-ab^2)/x^2\sqrt{b^2+x^2}$. The only critical point is the x
for which $l'(x) = 0$ so $x^3-ab^2 = 0$, or $x = (ab^2)^{1/3}$. Now
$l'(x) < 0$ when $x < (ab^2)^{1/3}$ and $l'(x) > 0$ when $x > (ab^2)^{1/3}$,
so $x = (ab^2)^{1/3}$ is a global maximum point. Since $y = ab/x$,
$y = (a^2b)^{1/3}$ and $l^2 = [b+(a^2b)^{1/3}]^2 + [a+(ab^2)^{1/3}]^2$ or
$l^2 = b^2 + 3a^{2/3}b^{4/3} + 3a^{4/3}b^{2/3} + a^2$ so $l^2 = (b^{2/3} + a^{2/3})^3$.
Therefore the longest pipe which can pass horizontally
around the corner is $l = (b^{2/3}+a^{2/3})^{3/2}$ feet.

13   Referring to Fig.4.3.10, the base of the box will be a
square with sides of length a-2x and the height of the box
will be x. Thus the volume of the box will be $V = x(a-2x)^2$,
with $0 < x < a/2$. Being a polynomial, V is a continuous
differentiable function throughout its domain so the only
critical points are those x for which $V'(x) = 0$. But
$V'(x) = (a-2x)(a-6x)$ and thus $x = a/2, a/6$, the first of
which gives 0 volume, and is not even in the domain.
$V''(x) = 8(3x-a)$ which is negative for $x = a/6$ and hence
$x=a/6$ is a global maximum and $V(a/6)=(a/6)[a-a/3]^2 = 2a^3/27$.

15.   Recall that the area of an equilateral triangle with side of
length s is $A = s^2\sqrt{3}/4$. Now suppose the wire is cut into
pieces of length x and 1-x. The first is bent to form an
equilateral triangle with side of length $s = x/3$. The
second is bent to form a square with side of length
$w = (1-x)/4$. The total area is the sum of the two so
$A(x) = x^2\sqrt{3}/36 + (1-x)^2/16$ with $0 \le x \le 1$. Since $A(x)$ is a
polynomial, it is continuous and differentiable for all x so
the only critical points are for $A'(x) = 0$, or $x = 0,1$. Now
$A'(x) = \sqrt{3}x/18 - (1-x)/8$ or $A'(x) = [(4\sqrt{3}+9)x - 91]/72 = 0$
for $x = 91/(4\sqrt{3}+9) = x^*$. For $x < x^*$, $A'(x) < 0$ and for
$x > x^*$, $A'(x) > 0$, so x^* is a local minimum. Further,
$A''(x) = (4\sqrt{3}+9)/72 > 0$ for all x, so x^* is a global
minimum. Also $A(0) = 1^2/16$ and $A(1) = 1^2\sqrt{3}/36$ or

       $A(0) \cong (.063)1^2$, $A(1) \cong (.048)1^2$. Thus the total area will

Section 4.3 — page 77

be a minimum if the perimeter of the triangle is x* and the total area will be a maximum when the entire length is used for the square.

17. If we let x be the position of ship A and y the position of ship B at time t, then $x = 30-15t$, $y = 20-15t$ and the distance between them is $d = \sqrt{x^2+y^2} = \sqrt{(30-15t)^2 + (20-15t)^2}$ or $d = \sqrt{450t^2 - 1500t + 1300}$ for $t \geq 0$. This function is continuous and differentiable on $t > 0$, so the only critical points are those t for which $d'(t) = 0$. Now

$d'(t) = (900t-1500)/2\sqrt{450t^2 - 1500t + 1300}$ which is zero for $t = 5/3$ hours or 1 hour and 40 minutes. Also $d'(t) < 0$ for $t < 5/3$ and $d'(t) > 0$ for $t > 5/3$, showing $t = 5/3$ is a global minimum. Thus $d(5/3) = \sqrt{x^2 + y^2} = \sqrt{5^2 + (-5)^2} = 5\sqrt{2}$.

Hence at 1:40 am the ships are $5\sqrt{2}$ miles apart.

19. If the third camper is located at $P(x,y)$ on the circle then the sum of the distances from P to A and B is

$d = \sqrt{(x-1)^2 + y^2} + \sqrt{x^2 + (y-a)^2}$. Since (x,y) is on the circle, $x^2+y^2 = 1$, so d can be expressed as $d = \sqrt{2}[\sqrt{1-x} + \sqrt{1-y}]$ with $-1 \leq x \leq 1$ and $-1 \leq y \leq 1$. It will be simpler to differentiate implicitly rather than substituting $y = \sqrt{1-x^2}$. Since d is continuous and differentiable for $-1 < x < 1$ the only critical points are the endpoints, $x = \pm 1$, and those for which $d'(x) = 0$. Now $d'(x) = \sqrt{2}[(-1/2\sqrt{1-x}) - (y'/2\sqrt{1-y})] = \sqrt{2}[-\sqrt{1-y} - y'\sqrt{1-x}]/2\sqrt{1-x}\sqrt{1-y}$. So $d'(x) = 0$ when $\sqrt{1-y} + y'\sqrt{1-x} = 0$ or $\sqrt{1-y} = -y'\sqrt{1-x}$ or $1-y = (y')^2(1-x)$. Now we use the relationship $x^2+y^2 = 1$ to find $2x + 2xyy' = 0$ or $y' = (-x/y)$ so $(y')^2 = x^2/y^2$. Thus $1-y = (x^2/y^2)(1-x)$ and $y^2 - y^3 = x^2 - x^3$ or $(y^2 - x^2) - (y^3 - x^3) = 0 = (y-x)(y+x) - (y-x)(y^2+xy+x^2)$.

Therefore $(y-x)(y+x-y^2-xy-x^2) = 0$. Since $x^2+y^2 = 1$, $(y-x)(y-xy+x-1) = 0$ or $(y-x)(y-1)(x-1) = 0$. Thus the critical points are $x = 1$, $x = y$, or $y = 1$. The points $x = 1$ and $y = 1$ correspond to camping at site A or B, which minimize the total distance. The condition $x = y$ has two solutions $x = y = 1/\sqrt{2}$ or $x = y = -1/\sqrt{2}$. Clearly, the latter gives a greater total distance. Thus the third camper should camp at $(-1/\sqrt{2}, -1/\sqrt{2})$.

21a.   We have $d = \sqrt{(x-a)^2 + (y-b)^2} = \sqrt{(x-a)^2 + (f(x)-b)^2}$.
       Since f is differentiable for $-\infty < x < \infty$, d(x) is also.
       Thus if there is a minimum for (x_0, y_0), x_0 must satisfy
       $d'(x_0) = 0$. But $d'(x) = \{(x-a) + f'(x)[f(x)-b]\}/\sqrt{(x-a)^2 + [f(x)-b]^2}$
       from which it follows that if (x_0, y_0) is the point nearest
       (a,b), then $d'(x_0) = (x_0-a) + f'(x_0)(f(x_0)-y) = (x_0-a) + f'(x_0)(y_0-b) = 0$.
   b. The equation of the normal at point
       (x_0, y_0) is: $y-y_0 = -(x-x_0)/f'(x_0)$, or $(x-x_0) + f'(x_0)(y-y_0) = 0$;   from
       part (a) this is satisfied when x = a and y = b, since
       $(a-x_0) + f'(x_0)(b-y_0) = 0$. Thus (a,b) is on the normal to the
       curve at (x_0, y_0). In this sense the minimal distance is the
       "perpendicular" distance.

23a.   The relationship between profits, sales, and costs is
       $P(x) = S(x) - C(x)$, where each expression represents the
       annual figure based on production and sales of x houses,
       measured in kilo-dollars. Thus $S(x) = 80x$ and
       $C(x) = 25+65x+x^2/3$ so $P(x) = 80x-(25+65x+x^2/3) = -x^2/3+15x-25$.
       Since P(x) is a polynomial it is continuous and
       differentiable for x > 0 so the only critical points are
       those x for which $P'(x) = 0$. $P'(x) = -2x/3 + 15 = 0$ for
       x = 22.5. $P''(x) = -2/3$ for all x and thus the production of
       22 or 23 houses annually will generate the greatest profit.
   b. In this case the 25 is replaced by α in C(x) giving

       $P(x) = -x^2/3 + 15x - \alpha$. Again $P'(x) = -2x/3 + 15 = 0$ for

       x = 22.5, which is independent of α. To guarantee

       $P(x) \geq 125$, we have $P(22) = -(22)^2/3 + 15(22) - \alpha \geq 125$ or

       $168.67 - \alpha \geq 125$ so that $\alpha \leq 43.67$ kilodollars.
   c. In this case the income from sales is $S(x) = p(x)x$, where
       x = 4(100-p) so $p(x) = 100 - x/4$ and the profit is
       $P(x) = x(100-x/4) - (25+65x+x^2/3)$. Thus
       $P'(x) = 100 -x/2 - 65 - 2x/3 = 35 - 7x/6 = 0$ for x = 30,
       which is the optimum production of houses per year.

25.   If the average per engineer is w hours per day, then if x
       engineers are employed, the total number of hours is T = wx.
       From the given expression $T = 8[x - x^3/(x^2+130)]$ which is
       continuous for $x \geq 1$ and differentiable for x > 1. The
       critical points are x = 1 and those x for which $T'(x) = 0$.

But $T'(x) = 8[1-(x^4+390x^2)/(x^2+130)^2] = 0$ for
$x^4+260x^2+(130)^2-x^4-390x^2 = -130x^2+(130)^2 = 0$, which gives
$x^2 = 130$. If $x^2 < 130$, then $T'(x) > 0$ and if $x^2 > 130$,
$T'(x) < 0$, so $T(x)$ has a maximum value for $x^2 = 130$, that is
$x = \sqrt{130} \cong 11$. (We can't hire 11.4 engineers).

Section 4.4, Page 212

1.   The volume of a sphere of radius r is given by the function
     $V(r) = 4\pi r^3/3$. To find how fast the radius is increasing is
     to ask for the value of dr/dt. Now $dV/dt = (dV/dr)(dr/dt)$
     and $dV/dr = 4\pi r^2$, so $dV/dt = (4\pi r^2)(dr/dt)$. We are given
     that $dV/dt = 2$ ft^3/min and thus
     $dr/dt = 2/4\pi r^2 = 2/4\pi(3)^2 = 1/18\pi$ ft/min when r = 3.

3.   The area of an equilateral triangle with side of length s,
     is given by $A(s) = \sqrt{3}\,s^2/4$. This yields
     $dA/dt = (dA/ds)(ds/dt) = (\sqrt{3}\,s/2)ds/dt$. When s = 7 and
     $ds/dt = 2$, $dA/dt = 7\sqrt{3}$ in^2/min.

5.   Since we are asked to find dy/dt we simply take the
     derivative of both sides of the equation with respect to t
     to get $(2x)dx/dt + (8y)dy/dt = 0$. With x = -2, y = 1, and
     $dx/dt = 3$ we have $(-4)(3) + 8dy/dt = 0$ or $dy/dt = 3/2$.

7.   From $pV^{7/5} = c$ we are asked to find dp/dt when V = 32 in^3,
     p = 20 lb/in^2 and $dV/dt = -0.1$ in^3/sec. Now
     $d(pV^{7/5})/dt = dc/dt = 0$ so that
     $V^{7/5}dp/dt + (7/5)pV^{2/5}(dV/dt) = 0$. Substituting the given
     values and solving yields $dp/dt = 7/80$ lb/in^{2}sec.

9.   Let x be the length and w the width then the perimeter
     $P = 2(x+w) = 6w$ and the area $A = xw = 2w^2$. Thus
     $dA/dt = 4w(dw/dt)$ and $dP/dt = 6(dw/dt)$ so
     $dw/dt = (dP/dt)/6 = 1/2$. Now $A = 24 \Rightarrow w = 2\sqrt{3}$. Thus
     $dA/dt = 4(2\sqrt{3})(1/2) = 4\sqrt{3}$ in^2/min.

11.   If the man is x feet from the
       street light and his shadow is
       s feet long, then from similar
       triangles we have (x+s)/20 = s/6
       so s = 3x/7 and
       ds/dt = (3/7)dx/dt = 12/7 ft/sec,
       independent of where the man is.

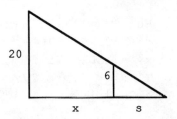

13.   From the equation for the volume of a sphere, $V = (4/3)\pi r^3$,

       we have $dV/dt = 4\pi r^2 (dr/dt)$. Since $dr/dt = 1/(1+r^2)$, then

       $dV/dt = 4\pi r^2/(1+r^2)$. When r=3, $dV/dt = 36\pi/10 = 18\pi/5$ cm³/min.

15.   For a square inscribed in a circle
       of radius r, the sides have length
       $r\sqrt{2}$, since the triangle on the
       diagonal of the square is a
       45,45,90 right triangle. Thus
       the area of the square is
       $A = (r\sqrt{2})(r\sqrt{2}) = 2r^2$ and so
       $dA/dt = 4r(dr/dt)$. When r=10 and dr/dt=3, dA/dt=120 cm/min.

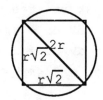

17.   The volume of a cone is

       $V = (1/3)\pi r^2 h$ and from similar

       triangles r/2 = h/8 or r = h/4.

       Thus $V = \pi h^3/48$ and

       $dV/dt = (\pi h^2/16)dh/dt$. Since

       $dV/dt = 0.5$ ft³/min and h = 5, we

       have $0.5 = (25\pi/16)(dh/dt)$ or $dh/dt = 8/25\pi$ ft/min.

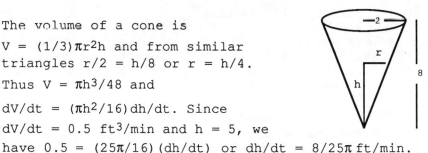

19.   Let x be the horizontal distance from the boat to the dock.
       Then the length r of the rope being wound is given by
       $r^2 = x^2 + 36$ so $r(dr/dt) = x(dx/dt)$. When x = 9,
       $r = \sqrt{117}$, thus $\sqrt{117}(-4) = 9(dx/dt)$ or $dx/dt = -4\sqrt{117}/9$ ft/min.

21.   Let x be the depth of the water in the trough. Then by
       similar triangles the cross sectional area of water is $x^2/2$,
       since the trough has equal height and base. Thus
       $V = 8(x^2/2) = 4x^2$ so that $dV/dt = 8x(dx/dt)$. We are given
       $dV/dt = 3$ft³/min, so $dx/dt = 3/8x = 1/4$ ft/min when x = 1.5.

Section 4.5, Page 219

1.   Using Eq.(4) we have $x_{n+1} = x_n - [(x_n)^2 + 2x_n - 4]/(2x_n + 2)$.
      Thus for $x_0 = 1$ we find $x_1 = 1.25$, $x_2 = 1.236111$,
      $x_3 = 1.2360680$ and $x_4 = 1.2360680$.

3.   We have $x_{n+1} = x_n - [(x_n)^4 - 2]/[4(x_n)^3]$ by Eq.(4). Thus, for
      $x_0 = 1$, we find $x_1 = 1.25$, $x_2 = 1.1935$, $x_3 = 1.18923$,
      $x_4 = 1.189207$, and $x_5 = 1.1892071$.

5.   By Eq.(4) we have $x_{n+1} = x_n - [(x_n)^3 + 5(x_n)^2 - 4]/[3(x_n)^2 + 10x_n]$.
      Thus, for $x_0 = 1$, we find $x_1 = 0.8461539$, $x_2 = 0.8286499$, and
      $x_3 = x_4 = 0.8284271$.

7.   We have $x_{n+1} = x_n - (x_n \sin x_n - \cos x_n)/(2\sin x_n + x_n \cos x_n)$.
      Thus, for $x_0 = 1$, $x_1 = 0.8645362$, $x_2 = 0.8603391$, and
      $x_3 = x_4 = 0.8603336$.

9.   We have $x_{n+1} = x_n - (x_n \cos x_n + 2\sin x_n)/(3\cos x_n - x_n \sin x_n)$.
      For $x_0 = 2$, we find $x_1 = 2.22855$, $x_2 = 2.266208$ and
      $x_n = 2.288929$ for $n \geq 13$.

11.   We have $x_{n+1} = x_n - [(2x_n)^{1/2} - 1/x_n - 1]/[1/(2x_n)^{1/2} + 1/x^2]$.
      Choosing $x_0 = 1$, we find $x_1 = 1.343146$, $x_2 = 1.433773$,
      $x_3 = 1.437559$, and $x_4 = x_5 = 1.4375649$

13.   Since $f(x) = x^2 - a$, we have $x_{n+1} = x_n - (f(x_n)/f'(x_n)) =$
      $x_n - [(x_n)^2 - a]/2x_n = (x_n + a/x_n)/2$ for $n \geq 0$.

15.   Using the result in Prob.13 we have $x_0 = 7$,
      $x_1 = (7 + 53/7)/2 = 7.285714$, $x_2 = (x_1 + 53/x_1)/2 = 7.280112$,
      and $x_3 = (x_2 + 53/x_2)/2 = 7.2801099$.

17.   Using the result of Prob.14 we have $p = 3$, $x_0 = 2$ and thus
      $x_1 = (4 + 5/4)/3 = 1.75$, $x_2 = [2x_1 + 5/(x_1)^2]/3 = 1.710884$,
      $x_3 = [2x_2 + 5/(x_2)^2]/3 = 1.7099764$, $x_4 = [2x_3 + 5/(x_3)^2]/3 = 1.7099759$.

19.   Setting $p = 5$ and $x_0 = 2$ in the result of Prob.14 we get
      $x_1 = [4x_0 + 50/(x_0)^4]/5 = 2.225$. In a similar fashion we then
      get $x_2 = 2.188019$, $x_3 = 2.1867257$ and $x_4 = 2.1867241$.

21.   Using $f(x) = a-(1/x)$ we find $f'(x) = 1/x^2$ so Eq.(4) becomes
       $x_{n+1} = x_n - (a - 1/x_n)/[1/(x_n)^2] = x_n - [a(x_n)^2 - x_n] = (2 - ax_n)x_n$,
       and thus $x_{n+1} \to 1/a$ as $n \to \infty$.

23a.   Using x_{n+1} from Prob.5 and $x_0 = 0.5$ we get $x_n \cong 0.8284271$ for
       $n \geq 4$.
   b. With $x_0 = 0.1$ we get $x_n \cong 0.8284271$ for $n \geq 7$.
   c. With $x_0 = -0.1$ we get

       $x_n \cong -4.8284271$ for $n \geq 6$.
   d. With $x_0 = -0.5$ we get

       $x_n \cong -1.0000000$ for $n \geq 4$.

       The behavior of the root selection
       is explained by the fact that the
       slope at $x = 0$ is 0. Thus the
       tangent line for points near $(0,-4)$
       intersects the x axis at widely
       separated points, thereby locating
       the three different roots by changing
       the initial choice for x_0 only slightly.

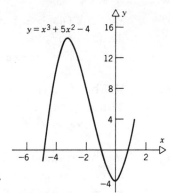

Chapter 4 Review, Page 220

1.   f has at most one real root provided $f'(x) \neq 0$. Since
       $f'(x) = 3ax^2 + 2bx + 1$, this is equivalent to $\sqrt{4b^2 - 12a}$ being
       imaginary, or $b^2 < 3a$.

3.   From the mean value theorem, with $f(x) = \sin^2 x$,
       $$\frac{\sin^2 x_2 - \sin^2 x_1}{x_2 - x_1} = \sin 2c \text{ for c in } (x_1, x_2). \quad \text{Thus}$$
       $$\frac{|\sin^2 x_2 - \sin^2 x_1|}{|x_2 - x_1|} = |\sin 2c| \leq 1 \text{ which gives the desired}$$
       inequality.

5.   As in Prob.3, with $f(x) = \tan x$, for any (x_1, x_2) in $(-\pi/2, \pi/2)$,
       $$\frac{|\tan x_2 - \tan x_1|}{|x_2 - x_1|} = \sec^2 c \geq 1. \quad \text{Thus the desired inequality}$$
       is false.

7. $f'(x) = -4\sin x \cos x$, and, $g'(x) = 2\tan x \sec^2 x$. Thus, since
        $f'(x) \neq g'(x)$, f and g are not antiderivatives of the same
        function.

9.   Since $f'(x) = -2\sin x \cos x$ and $g'(x) = 2\tan x \sec^2 x$,
        $f'(x) \neq g'(x)$, so that f and g are not antiderivatives of
        the same function.

11.   f is continuous on $[0,1]$ and $f'(x) = 1/2\sqrt{x+1}$ on $(0,1)$ so
        the mean value theorem applies and $f(1)-f(0) = f'(c)$ or
        $\sqrt{2} - 1 = 1/2\sqrt{c+1}$. Solving gives $c = (2\sqrt{2} - 1)/4$.

13.   Since f is continuous on $[-\pi/4, \pi/4]$ and differentiable on
        $(-\pi/4, \pi/4)$, there is a c in $(-\pi/4, \pi/4)$ such that
        $$\frac{f(\pi/4) - f(-\pi/4)}{\pi/2} = \sec^2 c + 1 = \tan^2 c + 2 \text{ or } \tan^2 c = (4/\pi)-1$$
        or $\tan c = \pm\sqrt{(4/\pi)-1}$.

15. $f'(x)$ does not exist at $x = 0$, so the mean value theorem
        does not apply.

17.   Since f is a polynomial, it is continuous on $[u,v]$ and
        differentiable on (u,v) for any finite u,v. Thus the mean
        value theorem applies and there is a c in (u,v) satisfying
        the conclusion of the theorem in each case.

19.   It is known that $\sin x$ satisfies all the hypothesis of
        Prob.20 of Section 4.1. Thus, with $a = -\pi$, $b = \pi$,
        $f'(x) = \cos x$, $f''(x) = -\sin x$, the equation becomes
        $0 = 0 - 2\pi - (1/2)\sin c (4\pi^2)$ or $\sin c = -1/\pi$.

21.   Since f and g are polynomials, they satisfy the hypothesis
        of Prob.21 of Section 4.1. Further, $f'(x) = 1+x$ and
        $g'(x) = 1+x+x^2$. With $a = -1$, $b = 1$, the equation becomes
        $(3/4) = (1+c)/(1+c+c^2)$, which gives $c = (1\pm\sqrt{13})/6$, both of
        which lie in the interval $(-1,1)$.

23a. $f(x) = \begin{cases} -\sin x & -\pi/2 \leq x < 0 \\ \sin x & 0 \leq x \leq \pi/2 \end{cases}$,   which has critical points at
        the endpoints $x = -\pi/2$, $x = \pi/2$ and at $x = 0$, where f' does
        not exist.

23b.   f is decreasing on $[-\pi/2,0]$ as $f'(x) = -\cos x < 0$ on
        $(-\pi/2,0)$. Similarly, since $f'(x) = \cos x > 0$ for x in
        $(0,\pi/2)$, f is increasing on $[0,\pi/2]$.
  c.   Since $f''(x) = \sin x < 0$ for x in $(-\pi/2,0)$ and
        $f''(x) = -\sin x < 0$ for x in $(0,\pi/2)$, f is concave down for x
        in $[-\pi/2,\pi/2]$.
  d.   The local maximum at the endpoints are both global,
        $f(-\pi/2) = f(\pi/2) = 1$, while the global minimum is $f(0) = 0$.

25a.   The critical points are the endpoints, $x = -1, 1$ and $x = 0$,
        where $f'(x)$ does not exist.
  b.   For x in $(-1,0)$, $f'(x) = 3x^2 > 0$, so f increases for $[-1,0]$
        and for x in $(0,1]$, $f'(x) = 1/2\sqrt{x} > 0$, so f increases for
        $[0,1]$. Thus, f increases for all x in $[-1,1]$.
  c.   As $f''(x) = 6x < 0$ for x in $(-1,0)$, f is concave down for
        $[-1,0]$. As $f''(x) = -1/4x^{3/2} < 0$ for x in $(0,1)$, f is concave
        down for $[0,1]$ as well.
  d.   f has a global minimum of $f(-1) = -1$ and a global maximum of
        $f(1) = 1$.

27a.   The only critical point is where $f'(x) = x/(1-x^2)^{3/2} = 0$, or
        $x = 0$.
  b.   With $f'(x) = \begin{cases} -1/(x+1)^2 & -\infty < x < -1 \\ x/(1-x^2)^{3/2} & -1 < x < 1 \\ 1/(x-1)^2 & 1 < x < \infty \end{cases}$, we find that f

        decreases for $-\infty < x < -1$ and $-1 < x \le 0$ and increases for
        $0 \le x < 1$ and $1 < x < \infty$.
  c.   With $f''(x) = \begin{cases} 2/(x+1)^3 & -\infty < x < -1 \\ (1+2x^2)/(1-x^2)^{5/2} & -1 < x < 1 \\ -2/(x-1)^3 & 1 < x < \infty \end{cases}$, we have f

        concave down for $-\infty < x < -1$ and $1 < x < \infty$, and concave up
        for $-1 < x < 1$.

  d.   f has no global maximum or minimum since there is no upper
        or lower bound as $x \to 1$. $(0,1)$ is a local minimum.

29a.   The distance from $(0,0)$ to $(x,1/x)$ is $D(x) = \sqrt{x^2+(1/x)^2}$ so
        $D'(x) = [2x-2/x^3][x^2+(1/x)^2]^{-1/2} = 0$ when $2x^4-2 = 0$ or $x = 1$,
        since $x > 0$. Thus $(1,1)$ is the closest to $(0,0)$.
  b.   With the distance from $(1,1)$ to $(x,x^2 + 1/2)$ being
        $D(x) = (x^4-2x+5/4)^{1/2}$, $D'(x) = (4x^3-2)(x^4-2x+5/4)^{-1/2}(1/2)$.
        Thus $D'(x) = 0$ when $x = (1/2)^{1/3}$, so that the nearest point
        is $((1/2)^{1/3}, (1/4)^{1/3} + 1/2)$.

29c.   Examination of the graph shows that $(\pi/2,1)$ is the point
       nearest to $(\pi/2,2)$. Also since
       $D(x) = [(x-\pi/2)^2 + (\sin x-2)^2]^{1/2}$, we find
       $D'(x) = [(x-\pi/2)+(\sin x-2)\cos x][D(x)]^{-1} = 0$ when $x = \pi/2$.

  d.   If $b > 0$, by symmetry, there are two points on $y = |x|$
       closest to $(0,b)$. We will find the point for $x > 0$, so
       $y = x$. Thus $D(x) = [(x^2+(x-b)^2]^{1/2}$ so $D'(x) = (2x-b)/D(x) = 0$
       for $x = b/2$. Thus $(b/2,b/2)$ is the point on the graph
       closest to $(0,b)$. By symmetry $(-b/2,b/2)$ is the second
       solution. For $b < 0$, $(0,0)$ is the closest, point by
       observing the graph.

31.   Letting the triangle have sides s,s, and $2b$, and height h,
       we have $s^2 = h^2+b^2$ and $A = bh$. Thus $s = [(A/b)^2+b^2]^{1/2}$. The
       perimeter is given by $P = 2s+2b$ or
       $P(b) = 2[(A/b)^2+b^2]^{1/2} + 2b$, so
       $P'(b) = [b^4-a^2+b^2(A^2+b^4)^{1/2}]/(A^2+b^4)^{1/2} = 0$ when $b^4 = A^2/3$ or
       $b = \sqrt{A}/(3)^{1/4}$. Thus the base has length $2\sqrt{A}/(3)^{1/4}$.
       Finally, from the expression for s above, $s = 2\sqrt{A}/(3)^{1/4}$
       also. Thus, the triangle is equilateral.

33.   Letting $x_0 = 0$, then $x = [(-g/2)t + v_0]t$, which is 0 for
       $t = 0$ and $t = 2v_0/g = 40/10 = 4$ sec. The maximum height,
       from Prob.32, is $x(2) = 400/20 = 20$m. Thus the total
       distance traveled is 40 meters.

35a. $P'(r) = (3\pi/R\sqrt{g})r^{1/2}$.
  b.   From $v = R\sqrt{g/r}$, $r = R^2g/v^2$, so $P(v) = 2\pi R^2g/v^3$.
       Thus $P'(v) = -6\pi R^2g/v^3$.
  c.   From part(b), $a_r = (R/r)^2g = v^4/R^2g$. Thus $a'_r(v) = 4v^3/R^2g$.

37.   With the perimeter $p = 2h+2b$, the area to be minimized is
       $A = bh - \pi h^2/64 = ph/2 - h^2(1+\pi/64)$. Since p is fixed,
       $A'(h) = p/2 - 2h(1+\pi/64) = 0$ for $h = p/(4+\pi/16)$.
       Substituting into the expression for the perimeter and
       solving for b gives $b = p[1+(\pi/32)]/[4+(\pi/16)]$.

39.   With $R^2 = x^2+y^2$, $0 = 2x(dx/dt) + 2y(dy/dt)$. Since
       $(x,y) = (a,b)$ and $dx/dt = -2$, we solve and find
       $dy/dt = 2a/b$. Similarly, $0 = 2x + 2y(dy/dx)$ or $dy/dx = -x/y$.

41a. $V = 4\pi r^3/3, S = 4\pi r^2$, $dA/dS = (dV/dr)(dr/dS)$ and $dV/dr = 4\pi r^2$.
       To find dr/dS, differentiate $S = 4\pi r^2$ with respect to S,
       $1 = 8\pi r(dr/dS)$, or $dr/dS = 1/(8\pi r)$. Thus $dA/dS = r/2$.

41b. $V = x^3$ and $S = 6x^2$ so that
       $dV/dS = (dV/dx)(dx/dS) = (3x^2)(1/12x) = x/4$.

43.   The area of the base of dimensions x and 2x is $2x^2$. Thus
       $S = 108 = 4x^2+6xy$ and $V = 2x^2y$, where y is the height of the
       box. Differentiating, $V' = 4xy+2x^2y'$. From the surface
       area, $0 = 8x+6y+6xy'$ or $y' = -(4x+3y)/3x$. Substituting into
       V' gives $V' = 4xy-(2x/3)(4x+3y) = 2x(3y-4x)/3 = 0$ when
       $x = 3y/4$. From the surface area, we then find $y = 4$ ft.

45.   We have $P(x) = Rx-h = 2500x - 35x^2 - 15,000$, so
       $P'(x) = 2500-70x = 0$ when $x = 250/7$ acres, which is a
       maximum since $P'' < 0$.

47a. $dy/dx = (dy/dt)/(dx/dt) = bx/ay$.
   b. $d^2y/dx^2 = (aby-abxy')/ay^2 = [b(y-bx^2/ay)]/ay^2 = b(ay^2-bx^2)/ay^2$
              $= 0$ for $ay^2 - bx^2 = 0$.
   c. If x and y satisfy $ay^2 - bx^2 = 0$,
       then $ay(dy/dt) - bx(dx/dt) = 0$. With $dx/dt = ay$ and
       $dy/dt = bx$, we have $abxy - abxy = 0$. Thus, for x and y
       satisfying $ay^2 - bx^2 = 0$, then x and y satisfy the
       differential equations as well.

CHAPTER 5

<u>Section 5.1, Page 237</u>

1.   Since $(-x)y = x(-y) = -xy$ and $(-x)(-y) = xy$, the graph of $xy = 4$
     is symmetric to the origin but not to either of the axis.

3.   Since $(-x-1)^2 = (x+1)^2$ and $(-y+2)^2 = (y-2)^2$, the graph is
     not symmetric to either axis or the origin.

5.   Since $\sin(-x) = -\sin x$ we have $y = (-x)\sin(-x) = x\sin x$ and
     thus the graph is symmetric to the y-axis.

7.   Since $\cos(-y) = \cos y$ we have $\cos(-y)\sin x = \cos y(\sin x)$ and
     thus the graph is symmetric to the x-axis.

9.   Since $|-y| = |y|$ and $|-x| = |x|$, the graph is symmetric to
     both axis and the origin.

11. f is unbounded at $x = -2$ and thus $x = -2$ is a vertical
     asymptote. Since $\lim_{x \to \pm\infty} f(x) = 1$ we have that $y = 1$ is a

     horizontal asymptote.

13. The denominator is never zero for real x and thus there is
     no vertical asymptote. Since $\lim_{x \to \pm\infty} f(x) = 1$, $y = 1$ is a
     horizontal asymptote.

15. As $x \to -1^+$, y^2 becomes unbounded and thus $x = -1$ is a

     vertical asymptote. As $x \to +\infty$, $y^2 \to 1$ and thus $y = -1$ and
     $y = 1$ are horizontal asymptotes.

17. Note that $f(x) = x/2 - 1/2x$. Thus $x = 0$ is a vertical
     asymptote and there is no horizontal asymptote since $f(x)$ is
     unbounded as $x \to \pm\infty$. $y = x/2$ is an asymptote since
     $$\lim_{x \to \pm\infty}\left[\frac{x}{2} - f(x)\right] = \lim_{x \to \pm\infty}\left[\frac{x}{2} - \left(\frac{x}{2} - \frac{1}{2x}\right)\right] = 0.$$

19. There is a vertical asymptote at $x = 0$ since y is unbounded
     there and there is no horizontal asymptote since y is
     unbounded as $x \to \pm\infty$. There are no other asymptotes since y
     is close to x^2, which is not a straight line, for $x \to \pm\infty$.

21.   Since $y^2 = (x^2-4)/2$ we see that there are no vertical or
      horizontal asymptotes. To determine other possible
      asymptotes we see that for large x we can neglect the four
      in the numerator to obtain $y^2 \cong x^2/2$. To prove that $y = x/\sqrt{2}$
      is an asymptote we have

$$\lim_{x \to \pm\infty}\left[\frac{x}{\sqrt{2}} - \sqrt{\frac{x^2-4}{2}}\right] = \lim_{x \to \pm\infty} \frac{\frac{x^2}{2} - \frac{(x^2-4)}{2}}{\frac{x}{\sqrt{2}} + \sqrt{\frac{x^2-4}{2}}} = \lim_{x \to \pm\infty} \frac{2}{\frac{x}{\sqrt{2}} + \sqrt{\frac{x^2-4}{2}}} = 0.$$

      A similar calculation shows that $-x/\sqrt{2}$ is also an asymptote.

23.   We have $f' = 3x^2 + 6x - 4$ and $f'' = 6x + 6 = 6(x+1)$. Thus
      for $x \leq -1$ the curve is concave down and for $x \geq -1$ the
      curve is concave up. Since f and f' are continuous at
      $x = -1$, f has a tangent line there and thus $x = -1$ is an
      inflection point since it changes concavity there.

25.   Since $f'' = 12(x-2)^2$, f is concave up for all x and there are
      no inflection points.

27.   Since $f'' = 2/(x+2)^3$, f is concave down for $x < -2$ and
      concave up for $x > -2$. However, $x = -2$ is not an inflection
      point since f is not continuous there.

29.   Since $f' = (3x+2)/2\sqrt{x}$ and $f'' = (3x-2)/4x^{3/2}$ we find f
      concave down for $0 \leq x \leq 2/3$ and concave up for $x \geq 2/3$.
      $x = 2/3$ is an inflection point since f and f' are continuous
      there and the concavity changes.

31.   Since $f' = 2\sin x\cos x = \sin 2x$ and $f'' = 2\cos 2x$ we find f
      concave up for $0 \leq x \leq \pi/4$, $3\pi/4 \leq x \leq 5\pi/4$ and
      $7\pi/4 \leq x \leq 2\pi$ and concave down for $\pi/4 \leq x \leq 3\pi/4$ and
      $5\pi/4 \leq x \leq 7\pi/4$. There are inflection points at $x = \pi/4$,
      $3\pi/4$, $5\pi/4$ and $7\pi/4$ since f and f' are continuous at these
      points and the concavity changes.

33. Since y is a polynomial there
    are no excluded intervals and
    the y intercept is y = 7.   There
    are no asymptotes and no symmetry
    to either axis or the origin.
    Since y' = 3(x+1)(x-3), y is

    increasing for x ≤ -1 and for

    x ≥ 3 and decreasing for

    -1 ≤ x ≤ 3.   Since y" = 6(x-1), y

    is concave down for x ≤ 1 and concave

    up for x ≥ 1.   Thus the graph is as shown.

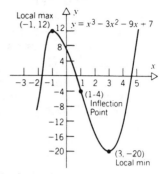

35. Since y is always non-negative,
    the graph must lie in the
    upper half plane.   The x

    intercepts are at x = ± 1
    and the y intercept is at
    y = 1.   Since
    $[(-x)-1]^2[(-x)+1]^2 =$
    $(x+1)^2(1-x)^2 = (x+1)^2(x-1)^2$
    the graph is symmetric about
    the y axis and there are no
    asymptotes because y is a polynomial.   y' = 4x(x-1)(x+1) and

    thus y is increasing on -1 ≤ x ≤ 0 and x ≥ 1 and decreasing

    on x ≤ -1 and 0 ≤ x ≤ 1.   Finally, y" = $4(3x^2-1)$ and thus y

    is concave up for x ≤ $-1/\sqrt{3}$ and x ≥ $1/\sqrt{3}$ and concave down
    for $-1/\sqrt{3}$ ≤ x ≤ $1/\sqrt{3}$.   Hence the graph is as shown.

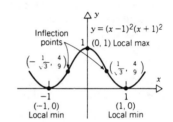

37. Since the left side of the
    equation has only positive terms
    they must each be less than or
    equal to four.   Thus the graph
    lies in the region -2 ≤ x ≤ 2
    and -1 ≤ y ≤ 1.   The graph is
    symmetric to both axis and to the
    origin and there are no asumptotes.
    The x intercepts are ± 2 and the y
    intercepts are ±1.   Using implicit differentiation we find
    dy/dx = -x/4y and thus y is increasing in the 2nd and 4th
    quadrants and decreasing in the 1st and 3rd quadrants.   Note

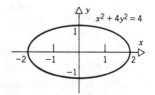

that the tangent line will be horizontal at the y intercepts
(x=0) and vertical at the x intercepts (y=0). Finally
$d^2y/dx^2 = -1/4y^3$ and thus the graph is concave down for
y > 0 and concave up for y < 0.

39. There is a vertical asymptote at
     x = -2 and a horizontal asymptote
     at y = 0. There is a y intercept
     at y = 1/2, no x intercept, and
     there is no symmetry or no
     restricted intervals. Since
     $y' = -1/(x+2)^2$ the graph is always
     decreasing and since $y'' = 2/(x+2)^2$,
     the graph is concave up for x > -2
     and concave down for x < -2.

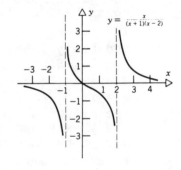

41. There are no restricted intervals,
     no symmetry and (0,0) is the only
     intercept. There are vertical
     asymptotes at x = -1 and x = 2 and
     a horizontal asymptote at y = 0.
     Since $y' = -(x^2+2)/(x+1)^2(x-2)^2$ the
     graph is decreasing for all x.
     Finally $y'' = 2(x^3+6x-2)/(x+1)^3(x-2)^3$
     and thus there is an inflection
     point somewhere between x = 0 and
     x = 1 [without too much work it can
     be deduced that the inflection point
     lies in the interval (.25,.50)].
     Thus the graph is concave down for x < -1 and for the
     interval from the inflection point to x = 2, and the graph
     is concave up for the interval x = -1 to the inflection
     point and for x > 2.

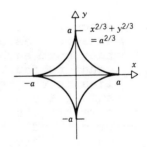

43. Just as in Prob.37, the graph must
     lie in the region -a ≤ x ≤ a and
     -a ≤ y ≤ a, where the end points
     are also intercepts, the graph
     is symmetric to both axis, and
     there are no asymptotes. Implicit
     differentiation yields $y' = -y^{1/3}/x^{1/3}$
     and thus the graph is decreasing in
     the 1st and 3rd quadrants and in-
     creasing in the 2nd and 4th quadrants.
     In this case, the graph is tangent to the x axis and y axis
     respectively at the intercepts. Finally,
     $y'' = a^{2/3}/(3y^{1/3} \cdot x^{4/3})$ and thus the graph is concave down for
     y < 0 and concave up for y > 0.

45.   Solving for y we obtain y =
        2 + 1/(x+2) and thus there are no
        restricted intervals, there is no
        symmetry, there is a y intercept
        at y = 2.5, there is a vertical
        asymptote at x= -2 and a horizontal
        asymptote at y = 2. Since
        y' = -1/(x+2)² the graph is always
        decreasing and since y" = 2/(x+2)³

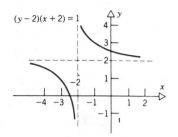

        the graph is concave down for x < 2 and concave up for x > 2.

47.   There are no restricted intervals,
        no asymptotes and no symmetry about
        either axis or the origin. Setting
        x = 0 we find y = 1 is an intercept
        and setting y = 0 we find x
        intercepts at all points such that
        tanx = -1.   Thus x = -π/4, 3π/4,
        7π/4, etc. are intercepts. Since
        y' = cosx - sinx, we find
        horizontal tangents at points where
        tanx = 1, or x = π/4, 5π/4, etc.
        Finally y" = -sinx - cosx and thus

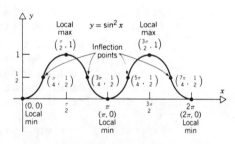

        there are inflection points at each of the x intercept
        points, which means that the concavity also changes at these
        points. Note that the equation can be written as
        y = √2 sin(x+π/4) and thus the same graph can be obtained by
        using knowledge of the sine function.

49.   Note that y is restricted to
        the interval 0 ≤ y ≤ 1, that
        the graph is symmetric to the
        y-axis, and there are no
        asymptotes. Setting y = 0, we
        find the x-axis intercepts
        to be all points where sinx=0,
        or x = 0, ±π, ±2π... . Since
        y' = 2sinx cosx we see that there

        are horizontal tangents at points where sinx = 0 or cosx = 0,
        which yield x = 0, ±π/2, ±π, ±3π/2 ... . The inflection
        points were found in Prob.31 and thus the graph is as shown.

51.   53.

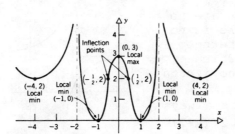   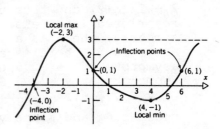

55a.   Since f and f' are continuous and since f" = 2(3ax+b) we
         see that f has exactly one inflection point at x = -b/3a.

  b.   Following the hint we have
         $f(x) = a(x-r_1)(x-r_2)(x-r_3) = a[x^3 + (b/a)x^2 + (c/a)x + d/a]$.
         Expanding the product of three terms we have:
         $$a(x - r_1)(x - r_2)(x - r_3) = a[x^2 - (r_1 + r_2)x + r_1 r_2](x - r_3)$$
         $$= a[x^3 - (r_1 + r_2 + r_3)x^2 + (r_1 r_2 + r_1 r_3 + r_2 r_3)x - r_1 r_2 r_3].$$

         Equating the coefficients of the x^2 terms then yields
         $b/a = -(r_1+r_2+r_3)$ and hence the inflection point
         $(-b/3a) = (r_1+r_2+r_3)/3$, which is the arithmetic mean of the
         three zeros of f.

Section 5.2, Page 244

1.   Putting the parabola in the
         form of Eq.(4) we find
         $x^2 = -(1/4)y$ and thus the vertex
         is at (0,0), p = |-1/4|/4 = 1/16
         and the parabola opens down.
         Thus the focus is at (0,-1/16)
         and the directrix is y = 1/16.

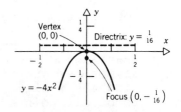

3.   Rewriting the equation we get
         $x^2 = (1/2)(y-1)$ and thus the
         vertex is at (0,1), the focus
         is at x = 0 and y = 1+1/8 = 9/8,
         and the directrix is y = 7/8.

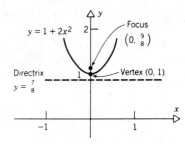

5.   Completing the square we have
      $2(y^2-2y+1) = -3x-8+2$, or
      $(y-1)^2 = (-3/2)(x+2)$. Thus from
      Eq.(8) the parabola opens left and
      $p = 3/8$. Hence the vertex is at
      $(-2,1)$, the focus is at $(-19/8,1)$,
      and the directrix is $x = -13/8$.

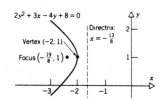

7.   Completing the square we have
      $4[x^2-(3/2)x+9/16] = 3y-9+9/4$, or
      $(x-3/4)^2 = (3/4)(y-9/4)$. Thus the
      vertex is at $(3/4,9/4)$, the focus is
      at $(3/4,39/16)$ and the directrix is
      $y = 33/16$.

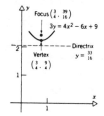

9.   The parabola is of the form of Eq.(4) and thus
      $(x-2)^2 = \alpha(y-1)$ since the vertex is at $(2,1)$. Setting
      $x = -1$ and $y = -1$ gives $9 = \alpha(-2)$ or $\alpha = -9/2$ and thus
      $(x-2)^2 = (-9/2)(y-1)$, or $9(y-1) = -2(x-2)^2$.

11.   Since the directrix is horizontal and below the vertex the
      parabola opens up and is of the form of Eq.(4) and thus
      $(x-1)^2 = \alpha(y-1)$, $\alpha > 0$. We know $p = |\alpha|/4$ and hence
      $\alpha = 4(3) = 12$ and therefore $12(y-1) = (x-1)^2$.

13.   Since the two points lying on the parabola are symmetric to
      the line $x = 2$, the parabola is of the form of Eq.(4). Thus
      $(x-2)^2 = \alpha(y+1)$, where we find $\alpha = 4/3$ by using $x = 4$ and
      $y = 2$. Hence $4(y+1) = 3(x-2)^2$.

15.   Since the axis is parallel to the x-axis, the parabola is of
      the form of Eq.(8) and hence $(y-3)^2 = \alpha(x+2)$. Also the
      distance from the focus to directrix is 2p and thus
      $p = 2 = |\alpha|/4$. We choose $\alpha = +8$ since the parabola opens to
      the right and therefore $8(x+2) = (y-3)^2$.

17.   Let (x,y) be any point on the parabola, which must open to
      the right. Thus $x+3$ is the distance from the parabola to
      the line $x = -3$ and $\sqrt{(x-2)^2+y^2}$ is the distance to $(2,0)$.
      Equating these two and squaring yields $(x+3)^2 = (x-2)^2+y^2$ or
      $x^2 + 6x + 9 = x^2 -4x + 4 + y^2$ and hence $y^2 = 10x + 5$.

19.   The curve must lie to the right of the line $x = -3$.   Thus
      $x+3$ is the distance from the curve to the line and

      $x+3 = 2 + \sqrt{(x+3)^2 + (y-2)^2}$, or $(x+1)^2 = (x+3)^2 + (y-2)^2$.
      Simplifying yields $8(x-1) = (y-2)^2$.

21a. Referring to Fig.5.2.9, we see that ψ is the exterior angle
      to the triangle which has θ and ϕ as interior angles.   Thus
      $\psi = \theta + \phi$, or $\theta = \psi - \phi$.

  b. Differentiation of $y^2 = 4px$ using the chain rule yields
      $2y\,dy/dx = 4p$, or $y' = 2p/y$.   Thus $\tan\phi = 2p/y_0$.

  c. From Fig.5.2.9 we have $\tan\psi = \dfrac{y_0}{-(p-x_0)} = \dfrac{y_0}{x_0-p}$.

  d. From the given identity we have

      $$\tan\theta = \frac{\dfrac{y_0}{x_0-p} - \dfrac{2p}{y_0}}{1 + \dfrac{y_0}{x_0-p}\cdot\dfrac{2p}{y_0}} = \frac{y_0^2 - 2px_0 + 2p^2}{y_0(x_0+p)}\ .$$ Since $y_0^2 = 4px_0$, the

      numerator becomes $2px_0 + 2p^2 = 2p(x_0+p)$ and thus
      $\tan\theta = 2p/y_0 = \tan\phi$.

23.   By adding $D^2/4A$ and $E^2/4C$ to both sides we have
      $A[x^2 + (D/A)x + D^2/4A^2] + C[y^2 + (E/C)y + E^2/4C^2] = F + D^2/4A + E^2/4C$.
      Thus $A(x + D/2A)^2 + C(y + E/2C)^2 = F + D^2/4A + E^2/4C$.   Hence
      $u = x-h$ if $h = -D/2A$ and $v = y-k$ if $k = -E/2C$.   The other
      constants are then $\alpha = A$, $\gamma = C$ and $\kappa = F + D^2/4A + E^2/4C$.

25a. Since $OR = u\cos\theta$ and $QR = v\sin\theta$ we have $x = u\cos\theta - v\sin\theta$.
      Likewise $QS = u\sin\theta$ and $SP = v\cos\theta$ and thus $y = u\sin\theta + v\cos\theta$.

  b. From the figure shown we have
      $u = NR' + R'P = N'M + R'P =$
      $y\sin\theta + x\cos\theta$ and
      $v = ON' - NN' = y\cos\theta - x\sin\theta$.

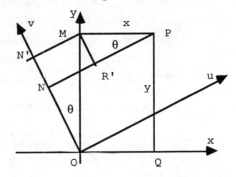

27a.   Letting $x = u\cos\theta - v\sin\theta$ and $y = u\sin\theta + v\cos\theta$ we have
$(u^2\cos^2\theta - 2uv\cos\theta + v^2\sin^2\theta) -$
$2\sqrt{3}(u^2\cos\theta\sin\theta + uv\cos^2\theta - uv\sin^2\theta - v^2\cos\theta\sin\theta) +$
$3(u^2\sin^2\theta + 2uv\sin\theta\cos\theta + v^2\cos^2\theta) -$
$2\sqrt{3}(u\cos\theta - v\sin\theta) - 2(u\sin\theta + v\cos\theta) + 8 = 0$. Collecting like
terms in u and v yields $[\cos^2\theta - 2\sqrt{3}\cos\theta\sin\theta + 3\sin^2\theta]u^2 +$
$[4\sin\theta\cos\theta - 2\sqrt{3}(\cos^2\theta - \sin^2\theta)]uv + [\sin^2\theta + 2\sqrt{3}\cos\theta\sin\theta + 3\cos^2\theta]v^2$
$- 2[\sin\theta + \sqrt{3}\cos\theta]u - 2[\cos\theta - \sqrt{3}\sin\theta]v + 8 = 0$.

 b.   Setting the coefficient of the uv term equal to zero we
obtain $4\sin\theta\cos\theta = 2\sqrt{3}(\cos^2\theta - \sin^2\theta)$ or $2\sin2\theta = 2\sqrt{3}\cos2\theta$.
Thus $\tan2\theta = \sqrt{3}$ and $2\theta = \pi/3 \pm 2n\pi$ or $2\theta = 4\pi/3 \pm 2n\pi$, which
gives $\theta = \pi/6 \pm n\pi$ or $\theta = 2\pi/3 \pm n\pi$.

 c.   Setting $\theta = \pi/6$ into the last
equation of part (a) we obtain
$0u^2 + 0uv + (16/4)v^2 - 2(2)u + 0v + 8 = 0$.
Thus $v^2 = u - 2$.

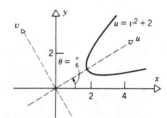

29.   For $x^2 = \alpha y$ the focus is at
$(0, \alpha/4)$. Thus the latus rectum
intersects the parabola at
$x^2 = \alpha(\alpha/4)$, or $x = \pm \alpha/2$ and
hence its length is $2(\alpha/2)$ or α.
Since the distance from the focus

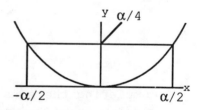

to the directrix is $2p = \alpha/2$, we conclude that the length of
the latus rectum is twice the distance from the focus to the
directrix.

Section 5.3, Page 253

1. Comparing to Eq.(5) we see that h = 0, k = 0, a^2 = 16, b^2 = 25, and thus $c^2 = b^2 - a^2$ = 9. Therefore the center is at (0,0), the foci are at (0,±3), the semimajor axis is b = 5, and the semiminor axis is a = 4.

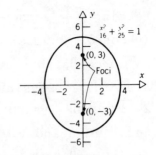

3. From Eq.(5) we have h = 1, k = -2. a^2 = 4, b^2 = 9 and thus c^2 = 9-4 = 5. Therefore the center is at (1,-2), the foci are at $(1,-2\pm\sqrt{5})$, the semimajor axis is 3 and the semiminor axis is 2.

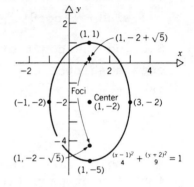

5. Completing the square in both x and y we have $2(x^2+2x+1)$ + $2(y^2-6y+9)$ = 5+2+18 = 25. Thus $(x+1)^2+(y-3)^2$ = 25/2, which is a circle with center at (-1,3) and radius $5/\sqrt{2}$.

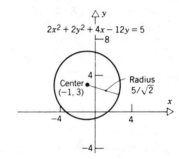

7. Completing the squares in both x and y we have (x^2-4x+4) + $4(y^2+4y+4)$ = -4+4+16 = 16 and thus $\dfrac{(x-2)^2}{16}+\dfrac{(y+2)}{4}=1$. Hence the ellipse has center is at (2,-2) with semimajor axis a = 4 and semiminor axis b = 2. Thus c^2 = 16-4 = 12. and therefore the foci are at $(2\pm2\sqrt{3},-2)$.

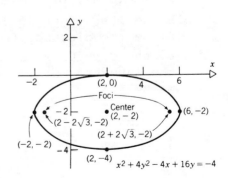

9.   The center is at (1,2) which is the midpoint between the
        foci. Since the distance between foci is 2c, we have 2c = 4,
        or c = 2. Finally, the distance from one foci to the vertex
        back to the second foci must be 2a and thus 2a = 8 or a = 4.
        Therefore $b^2 = a^2-c^2 = 16-4 = 12$ and thus the ellipse is

        given by $\dfrac{(x-1)^2}{16} + \dfrac{(y-2)^2}{12} = 1$.

11.   We have c = 3 (the distance from the center to the focus)
        and b = 5 (the distance from the center to the vertex).   The
        major axis must be vertical since the center and focus lie
        on a vertical line and thus $a^2 - b^2-c^2 = 16$.   Therefore the

        ellipse is given by $\dfrac{(x-1)^2}{16} + \dfrac{(y-2)^2}{25} = 1$.

13.   We have b = 3 (since the major axis must be horizontal) and
        2c+2 = 2a [since the distance from one focus to the vertex
        (which is 1) back to the second focus (which is 2c+1) must
        be 2a].   Thus a = c+1 and $a^2 = b^2+c^2$, or $(c+1)^2 = 9+c^2$.
        Solving we find c = 4, a = 5 and the center is at (−2+c,3)

        so that $\dfrac{(x-2)^2}{25} + \dfrac{(y-3)^2}{9} = 1$.

15.   We have $\sqrt{(x+2)^2 + y^2} + \sqrt{(x-2)^2 + y^2} = 10$, or

        $\sqrt{(x+2)^2 + y^2} = 10 - \sqrt{(x-2)^2 + y^2}$. Squaring both sides yields

        $x^2+4x+4+y^2 = 100 - 20\sqrt{(x-2)^2+y^2} + x^2-4x+4+y^2$, or

        $8x-100 = -20\sqrt{(x-2)^2+y^2}$. Squaring again yields
        $64x^2-1600x+10{,}000 = 400(x^2-4x+4y+y^2)$ or $336x^2+400y^2 = 8400$.
        Thus the ellipse is given by $x^2/25 + y^2/21 = 1$.

17.   We have $\sqrt{(x+1)^2 + (y-2)^2} + \sqrt{(x-3)^2+ (y+1)^2} = 8$, or

        $\sqrt{(x+1)^2 + (y-2)^2} = 8 - \sqrt{(x-3)^2 + (y+1)^2}$. Squaring and simplifying

        yields $8x-6y-69 = 16\sqrt{(x-3)^2 + (y+1)^2}$. Squaring again yields
        $(8x-6y)^2-138(8x-6y)+69^2 = 256(x^2-6x+9+y^2+2y+1)$, which
        simplifies to $192x^2+96xy+220y^2-432x-316y = 2201$.

19a.   Since a,b and c are always positve we have e > 0 in either
        case.   For b < a we have b/a < 1 and thus

$$e = \frac{\sqrt{a^2-b^2}}{a} = \sqrt{1-b^2/a^2} < 1 \quad \text{since } 1 - b^2/a^2 < 1.$$ Likewise for

a < b we have a/b < 1 and thus $e = \dfrac{\sqrt{b^2-a^2}}{b} = \sqrt{1-a^2/b^2} < 1.$

 b.   Suppose b > a, thus $e^2 = (b^2-a^2)/b^2$, or $a^2 = b^2(1-e^2)$.   Hence

the ellipse may be written as $\dfrac{(x-h)^2}{b^2(1-e^2)} + \dfrac{(y-k)^2}{b^2} = 1.$ If $e \to 0$,

then the ellipse becomes $(x-h)^2+(y-k)^2 = b^2$, which is a
circle.   Also $c^2 = b^2-a^2 = b^2e^2$ and thus $c \to 0$ as $e \to 0$ and
hence the foci become the center of the circle.   A similar
argument holds for a > b.

 c.   From part(b) we have $(x-h)^2+b^2(1-e^2)(y-k)^2 = b^2(1-e^2)$.   Thus
as $e \to 1$ the equation becomes $(x-h)^2 = 0$, or x = h.   Since
from part(b) we also have $c^2 = b^2e^2$, the foci approach the
vertex as $e \to 1$.   Thus as $e \to 1$ the ellipse approaches the
straight line segment joining the two foci.   A similar
argument holds for a > b.

21.   From Prob.1 we have c = 3 and b > a so that e = 3/5.

23.   From Prob.7 we have $c = 2\sqrt{3}$ and a>b so that $e = 2\sqrt{3}/4 = \sqrt{3}/2.$

25.   From the given foci we see that the center is at (1,3) and
        that 2c = 4, or c = 2.   Since the major axis is horizontal
        we have e = 2/a = 2/3 and thus a = 3 and $b^2 = a^2-c^2 = 5.$

Therefore $\dfrac{(x-1)^2}{9} + \dfrac{(y-2)^2}{5} = 1.$

27.   The distance from the center to the vertex is a and thus
        a = 4.   Since the vertex is at the end of the semimajor
        axis, we have e = c/a = c/4 = 1/2, or c = 2.   Finally,

$b^2 = a^2-c^2 = 12$ so that $\dfrac{(x+1)^2}{16} + \dfrac{(y+3)^2}{12} = 1.$

29.   Assume the equation of the ellipse is $\dfrac{x^2}{a^2} + \dfrac{y^2}{b^2} = 1$. Thus

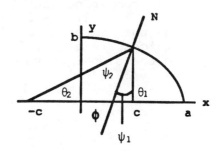

$\dfrac{dy}{dx} = \dfrac{-b^2}{a^2}\dfrac{x}{y}$ and hence the slope of the normal line N is given by $\tan\phi = a^2y_0/b^2x_0$. We

also have $\tan\theta_1 = \dfrac{y_0}{x_0 - c}$ and $\tan\theta_2 = \dfrac{y_0}{x_0 + c}$.   Now, using the

concept of exterior angles, we have $\theta_1 = \phi + \psi_1$ and $\phi = \theta_2 + \psi_2$

and thus $\psi_1 = \theta_1 - \phi$ and $\psi_2 = \phi - \theta_2$. Therefore

$$\tan\psi_1 = \frac{\tan\theta_1 - \tan\phi}{1 + \tan\theta_1\tan\phi} = \frac{y_0(b^2x_0 - a^2x_0 + a^2c)}{b^2x_0^2 - b^2x_0c + a^2y_0^2}$$

$$= \frac{y_0(x_0c^2 + a^2c)}{a^2b^2 + b^2cx_0} , \quad \begin{array}{l}\text{since } c^2 = a^2 - b^2 \text{ and} \\ \text{since } b^2x_0^2 + a^2y_0^2 = a^2b^2\end{array}$$

$$= \frac{cy_0}{b^2} \text{ and}$$

$$\tan\psi_2 = \frac{\tan\phi - \tan\theta_2}{1 + \tan\phi\tan\theta_2} = \frac{y_0(a^2x_0 + a^2c - b^2x_0)}{b^2x_0^2 + b^2x_0c + a^2y_0^2}$$

$$= \frac{y_0(a^2c + c^2x_0)}{a^2b^2 + b^2x_0c} = \frac{cy_0}{b^2}. \quad \text{Thus } \psi_1 = \psi_2.$$

Section 5.4, Page 260

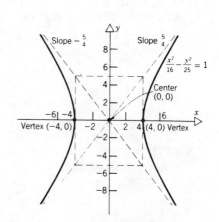

1.   Comparing the equation with Eq.(4) we find h = 0, k = 0, a = 4 and b = 5. Thus the center is at (0,0), the vertices are at (±4,0) and the slopes of the asymptotes are ±b/a = ±5/4.

3.   Again comparing with Eq.(4) we find
        h = 2, k = -2, a = 3 and b = 4.
        Thus the center is at (2,-2), the
        vertices are at (5,-2) and (-1,-2)
        and the slopes of the asymptotes
        are ±4/3.

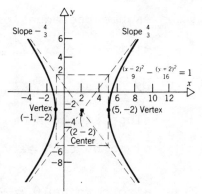

5.   Completing the squares we have
        $-4(x^2-2x+1)+(y^2+6y+9)=11-4+9 = 16$,
        or $-(x-1)^2/4 + (y+3)^2/16 = 1$. Thus
        we compare with Eq.(5) to find the
        center at (1,-3), the vertices at
        (1,1) and (1,-7), and the slopes
        of the asymptotes to be ±4/2 = ±2.

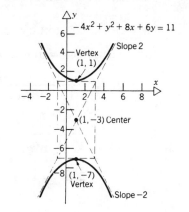

7.   Completing the squares we have
        $4(x^2-2x+1)-3(y^2-6y+9)=59+4-27 = 36$,
        or $(x-1)^2/9 - (y-3)^2/12 = 1$. Thus
        the center is at (1,3), the
        vertices are at (4,3) and (-2,3)
        and the slopes of the asymptotes
        are $\pm\sqrt{12}/3 = \pm 2\sqrt{3}/3$.

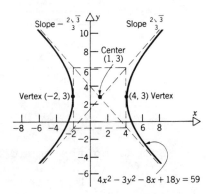

9.   The foci at (3,0) and (-3,0) imply that the center is at
        (0,0), that 2c = 6, or c = 3, and that the transverse axis
        is horizontal. The vertex at (2,0) tells us that a = 2 and
        thus $b^2 = c^2-a^2 = 5$. Hence $x^2/4 - y^2/5 = 1$.

11.   The given foci imply that the center is at $(-1,-1)$, that $c = 4$ and that the transverse axis is vertical. The slope of the asymptote is $1/2$ and thus $b/a = 1/2$ or $a = 2b$. Since $a^2+b^2 = c^2$ we then have $4b^2+b^2 = 16$ or $b^2 = 16/5$ and $a^2 = 64/5$. Hence $-(x+1)^2/(64/5) + (y+1)^2/(16/5) = 1$.

13.   The asymptotes intersect at $(-1,2)$, which is the center. Since one focus is at $(4,2)$ we see that $c = 5$ and that the transverse axis is horizontal. The slope of one asymptote is $b/a = 3/4$ and thus $a^2 + 9a^2/16 = 25$, or $a^2 = 16$ and $b^2 = 9$. Hence $(x+1)^2/16 - (y-2)^2/9 = 1$.

15.   We have $\sqrt{x^2+(y+2)^2} - \sqrt{x^2+(y-2)^2} = 2$, or

$\sqrt{x^2+(y+2)^2} = 2 + \sqrt{x^2+(y-2)^2}$. Squaring both sides yields

$x^2+y^2+4y = 4 + 4\sqrt{x^2+(y-2)^2} + x^2+y^2-4y+4$ or $2y-1 = \sqrt{x^2+(y-2)^2}$.
Squaring again and simplifying yields $-x^2/3 + y^2 = 1$.

17.   We have $\sqrt{(x+1)^2+(y-2)^2} - \sqrt{(x-3)^2+(y+1)^2} = 3$. Bringing the negative square root to the right side and squaring yields

$x^2+2x+1+y^2-4y+4 = 9 + 6\sqrt{(x-3)^2+(y+1)^2} + x^2-6x+9+y^2+2y+1$, which

simplifies to $4x-3y-7 = 3\sqrt{(x-3)^2+(y+1)^2}$. Squaring and simplifying again yields $7x^2-24xy-2x+24y-41 = 0$.

19a.   For Eq.(i) we have $e = \dfrac{\sqrt{a^2+b^2}}{a} = \sqrt{1+\dfrac{b^2}{a^2}} > 1$ and for

Eq.(ii) we have $e = \dfrac{\sqrt{a^2+b^2}}{b} = \sqrt{\dfrac{a^2}{b^2}+1} > 1$.

b.   For Eq.(i) we have $e = c/a$, or $a = c/e$. If $e \to 1$, then $a \to c$ for c fixed. Also, we have $a^2+b^2 = c^2$ and thus $b \to 0$. Hence the vertices approach the foci and the slope of the asymptote approaches zero. Therefore, as $e \to 1$ the hyperbola approaches the transverse axis with the segment joining the foci deleted. For $e \to \infty$ we have $a \to 0$ and $b \to c$. Therefore the vertex approaches the center and the slope of the asymptotes becomes infinite. Thus the hyperbola approaches the conjugate axis.

21.   We have $c^2 = 16+25 = 41$ and thus $e = \sqrt{41}/4$.

23.   From Prob.7 we have $a^2 = 9$, $b^2 = 12$ and a horizontal
       transverse axis. Thus $c^2 = 9+12 = 21$ and $e = \sqrt{21}/9 = \sqrt{7/3}$.

25.   The given foci tell us that the center is at $(1,2)$ and that
       $2c = 4$, or $c = 2$. Also $e = c/a = 3/2$ and thus $a = 4/3$.
       Finally $a^2+b^2 = c^2$ yields $b^2 = 20/9$ and thus
       $(x-1)^2/(16/9) - (y-2)^2/(20/9) = 1$.

27.   The asymptote may be written $v = 2u$, where $v = y+1$ and
       $u = x-3$. Thus the center is at $(3,-1)$ and $b/a = 2$. The
       focus at $(6,-1)$ yields $c = 3$ and $c^2 = a^2 + b^2$ gives $a^2 = 9/5$
       and $b^2 = 4a^2 = 36/5$. Thus $(x-3)^2/(9/5) - (y+1)^2/(36/5) = 1$.

29a.   From Prob.28 of Section 5.2 we find α,β,γ in terms of A,B,C

       and θ. Thus $\beta^2 = 4(A^2+C^2)\sin^2\theta\cos^2\theta - 8AC\sin^2\theta\cos^2\theta +$

       $4B(C-A)\sin\theta\cos\theta(\cos^2\theta-\sin^2\theta)-2B^2\sin^2\theta\cos^2\theta+B^2(\cos^4\theta+\sin^4\theta)$

       and $4\alpha\gamma = 4(A^2+C^2)\sin^2\theta\cos^2\theta - 4B^2\sin^2\theta\cos^2\theta -$

       $4AB\sin\theta\cos\theta(\cos^2\theta-\sin^2\theta) + 4AC(\sin^4\theta+\cos^4\theta) +$

       $4BC\sin\theta\cos\theta(\cos^2\theta-\sin^2\theta)$. Therefore

       $\beta^2 - 4\alpha\gamma = -8AC\sin^2\theta\cos^2\theta - 4AC(\sin^4\theta+\cos^4\theta) + 2B^2\sin^2\theta\cos^2\theta +$

       $B^2(\cos^4\theta+\sin^4\theta) = B^2(\cos^2\theta+\sin^2\theta)^2 - 4AC(\sin^2\theta+\cos^2\theta)^2 =$

       B^2-4AC since $\cos^2\theta+\sin^2\theta = 1$ for all θ.

   b. If $\beta = 0$ then $\alpha u^2 + \gamma v^2 + \delta u + \varepsilon v = \kappa$. Thus if $\alpha\gamma > 0$, α and
       γ have same sign and Eq.(ii) is an ellipse. If $\alpha\gamma = 0$, then
       either there is no u^2 or v^2 term and Eq.(ii) is a parabola
       (if both are zero then it is a degenerate parabola). If
       $\alpha\gamma < 0$ then α and γ differ in sign and Eq.(ii) is a
       hyperbola.

   c. Choose θ so that $\beta = 0$. Then $B^2-4AC = -4\alpha\gamma$ from part(a).

       Then $B^2-4AC < 0$ implies $\alpha\gamma > 0$ which means from part(b) that

       Eq.(i) is an ellipse. If $B^2-4AC = 0$ then $\alpha\gamma = 0$ and Eq.(i)

       is a parabola and if $B^2-4AC > 0$ then $\alpha\gamma < 0$ and Eq.(i) is a
       hyperbola, again using the results of part(b).

Section 5.5, Page 271

1.   Solving the second equation
      for t(t=y+2) and substituting
      into the first yields

      x = 3(y+2)-6 = 3y.   For t→2⁻

      we find x→0⁻ and y→0⁻ and
      thus x < 0 and y < 0 for all t.   Hence the graph is a ray in
      the third quadrant approaching (0,0) as t→2⁻.

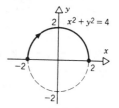

3.   Squaring each equation and
      adding we find $x^2+y^2 = 4$.   For
      the given t values we note that
      $0 \leq y \leq 2$ and $-2 \leq x \leq 2$.   Thus the
      graph is the upper half circle shown.

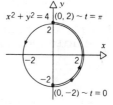

5.   Again squaring each equation and
      adding we find $x^2+y^2 = 4$.   In this
      case note that as t goes from 0 to
      $\pi/3$ that x goes from 0 to 2 to 0
      and that y goes from -2 to 0 to 2
      and thus the right half circle is
      described.   Similarly as t goes
      from $\pi/3$ to $2\pi/3$ the left half circle is described and as t
      goes from $2\pi/3$ to π the right half circle is described a
      second time.

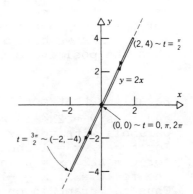

7.   Note for each t that y = 2x.
      However, the graph is only a
      segment of the straight line as
      $-2 \leq x \leq 2$ and $-4 \leq y \leq 4$.   Thus the
      graph starts at (0,0) goes to
      (2,4) then to (-2,-4) and back
      to (0,0) as t goes from 0 to 2π.

9.   Since x = sect and y/2 = tant
       and since $\sec^2 t - \tan^2 t = 1$, we

       have $x^2 - (y/2)^2 = 1$. For t→0+

       we have x→1+ and y→0+ and

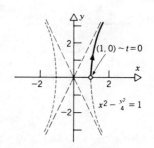

       as t→π/2 x→∞ and y→∞. Thus
       the graph is the portion of the
       hyperbola lying in the first
       quadrant asymptotic to the line y = 2x. The point (1,0) is
       not included.

11.   Note that $x^2 = 1/(1+t^2)$ and
       $y^2 = t^2/(1+t^2)$ and therefore
       $x^2 + y^2 = 1$. As t goes from
       0 to ∞, x goes from 1 to 0 and y
       goes from 0 to 1 and thus the
       graph is the quarter circle in the
       first quadrant. The point (1,0)
       is included, but (0,1) is not.

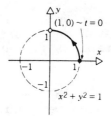

13.   From the given equations we have
       (x−1)/3 = sint and (y+2)/4 = cost.
       Thus $(x-1)^2/9 + (y+2)^2/16 = 1$, which
       is an ellipse with center at (1,−2),
       a horizontal semiminor axis of 3,
       and a vertical semimajor axis of 4.
       For t = 0 we have x = 1, y = 2 and
       for t = π/2 we have x = 4, y = −2 etc.
       Thus the positive direction is counterclockwise.

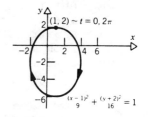

15.   Recalling that $\cos 2t = 2\cos^2 t - 1$
       and thus $y = 2x^2 - 1$, which is a
       parabola with vertex at (0,−1).
       For t = 0, x = 1 and y = 1;
       for t = π/2, x = 0 and y = −1;
       and for t = π, x = −1 and y = 1.
       Thus the graph is only the portion
       of the parabola shown, with the indicated positive direction.

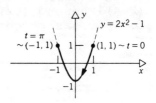

17.   From the given equations we find
$x^3 = t^6$ and $y^2 = t^6$ and thus $y^2 = x^3$.
For the given t values we have
$x \geq 0$ and $y \leq 0$. Also
$dy/dx = (3/2)(x^2/y) \leq 0$ and
$d^2y/dx^2 \leq 0$ and thus
the graph is always decreasing and concave down, with a
horizontal slope at $(0,0)$. The positive direction is as
shown since t starts at $-\infty$.

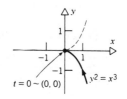

19.   Note that
$x^2+y^2 = (2+2t^2)/(1+t^2)^2 = 2/(1+t)^2 = x+y$.
Completing the squares we find
$(x-1/2)^2 + (y-1/2)^2 = 1/2$, and thus
the graph is a circle of radius
$1/\sqrt{2}$ with center at $(1/2,1/2)$.

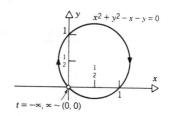

As $t \to -\infty$, $x \to 0^-$ and $y \to 0^+$ and as

$t \to \infty$, $x \to 0^+$ and $y \to 0^-$. Hence the positive direction is
counterclockwise with the origin excluded.

21.   From Eq.(15) we have $\dfrac{dy}{dx} = \dfrac{dy/dt}{dx/dt} = \dfrac{-3\cos t}{-2\sin t} = \dfrac{3}{2}$. For $t = \pi/4$,
we have $x = 2(\sqrt{2}/2) = \sqrt{2}$ and $y = -3(\sqrt{2}/2) = -3\sqrt{2}/2$ and thus
$(y + 3\sqrt{2}/2) = (3/2)(x - \sqrt{2})$ or $3x - 2y - 6\sqrt{2} = 0$.

23.   From Eq.(15) we have $\dfrac{dy}{dx} = \dfrac{2\sec 2t\,\tan 2t}{2\sec^2 2t} = \sin 2t$.

Thus $\dfrac{dy}{dx} = \sqrt{3}/2$, $x = \sqrt{3}$ and $y = 2$ at $t = \pi/6$ and therefore

$y-2 = (\sqrt{3}/2)(x-\sqrt{3})$ or $\sqrt{3}\,x - 2y + 1 = 0$.

25.   Again we have $\dfrac{dy}{dx} = \dfrac{1}{t^{-1/2}/(1/2)} = 2$ at $t = 1$. Evaluating x and

y at $t = 0$ we then have $y - (1+\sqrt{2}) = 2(x-3)$ or $2x - y - 5 + \sqrt{2} = 0$.

27.   From Eq.(15) we have $\dfrac{dy}{dx} = \dfrac{-8\sin 2t}{6\cos 2t} = -\dfrac{4}{3}\tan 2t$. In order to use

Eq.(16), we recall that $x = f(t)$ and $y = g(t)$ and thus

$$\frac{d^2y}{dx^2} = \frac{6\cos 2t\,(-16\cos 2t) - (-8\sin 2t)(-12\sin 2t)}{6^3\cos^3 2t}$$

$$= \frac{-96\,(\cos^2 2t + \sin^2 2t)}{216\cos^3 2t} = -\frac{4}{9}\sec^3 2t.$$

29.   As in Prob.27 we have $\dfrac{dy}{dx} = \dfrac{4\cos t}{-3\sin t} = -\dfrac{4}{3}\cot t$ and

$$\frac{d^2y}{dx^2} = \frac{-3\sin t\,(-4\sin t) - (4\cos t)(-3\cos t)}{-27\sin^3 t} = -\frac{4}{9}\csc^3 t.$$

31.   As in Prob.27 we have $\dfrac{dy}{dx} = \dfrac{2t-2}{3t^2} = \dfrac{2(t-1)}{3t^2}$ and

$$\frac{d^2y}{dx^2} = \frac{3t^2(2) - (2t-2)(6t)}{27t^6} = \frac{2}{9}\left(\frac{2-t}{t^5}\right).$$

33.   Note that $x+2y = t+5$, and hence squaring both sides yields
        $x^2+4xy+4y^2 = t^2+10t+25$
                    $= -y+12t+27$ (since $t^2 = -y+2t+2$)
                    $= -y+12x+24y-60+27$ (since $t = x+2y-5$),
        or $x^2+4xy+4y^2-12x-23y+33 = 0$. From Prob.29c, Section 5.4 we
        have $B^2-4AC = 0$ and hence this is a parabola.

35a.   For b < a we have from
        the top graph
        $x = a\theta - b\sin\theta$ and
        $y = a - b\cos\theta$.

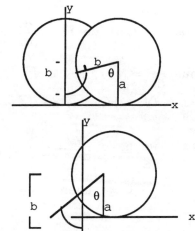

        For b > a
        we have from the second
        graph, $x = a\theta - b\sin\theta$ and
        $y = a - b\cos\theta$.

35b.

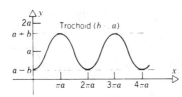

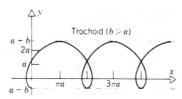

37a.   Substituting a = 4b into the equations of Prob.36 we get
        x = 3bcosθ + bcos 3θ = b(3cosθ + 4cos³θ − 3cosθ) and
        y = 3bsinθ − bsin3θ = b(3sinθ − 3sinθ + 4sin³θ).

        Thus x = 4bcos³θ = acos³θ and y = 4bsin³θ = asin³θ.

   c.   From part(a) we have   b.

        $(x/a)^{1/3} = \cos\theta$ and

        $(y/a)^{1/3} = \sin\theta$.   Therefore

        $(x/a)^{2/3} + (y/a)^{2/3} = 1$ or
        $x^{2/3} + y^{2/3} = a^{2/3}$.

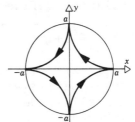

39.   We have 2a = 5b and thus from
       Prob.36 we have

       x = (3a/5)cosθ + (2a/5)cos(3θ/2),

       y = (3a/5)sinθ − (2a/5)sin(3θ/2).

       Note that for θ = 0 we have x = a,
       y = 0 and that x' = 0, y' = 0 for
       θ = 0°,144°,288°,432° and 576°.
       Thus there are cusps at each of
       these points on the circle of radius a.

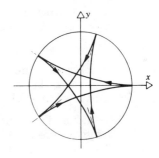

Chapter 5 Review, Page 272

1.   The relation possess origin symmetry as
       $(-y)^3 = [(-x)^2-1]/(-x)^3$ is equivalent to $y^3 = (x^2-1)/x^3$.

3.   The relation possesses x-axis symmetry,
       $\sin|-y| + \pi = \sin|y| + \pi$; y-axis symmetry, $\cos(-x) = \cos(x)$;
       and origin symmetry, as replacing (x,y) with (−x,−y) leaves
       the relation unchanged.

5.   The relation may be solved for x^2 and y^2 to get
       $x^2 = (y^2+2)/(1-y^2)$ and $y^2 = (x^2-2)/(x^2+1)$. Thus y = ±1 are
       horizontal asymptotes and there are no vertical asymptotes.

7.   The relation may be written as $16x^2 - 4 = 9y^2$ or
$y = \pm\sqrt{(16x^2 - 4)/9} = \pm(4x/3)\sqrt{1-(1/4x^2)}$. Since

$$\lim_{x \to \infty} [4x/3 - (4x/3)\sqrt{1-(1/4x^2)}] = \lim_{x \to \infty} [-4x/3 + (4x/3)\sqrt{1-(4/x^2)}]$$

$= 0$, $y = \pm 4x/3$ are skew asymptotes.

9.   The equation may be written as
$(y-1)^2 = x^2-x$ or $y = 1\pm\sqrt{x^2-x}$,
which has domain $x \le 0$, $x \ge 1$. This
relation defines two functions

$f(x) = 1+\sqrt{x^2-x}$, $g(x) = 1-\sqrt{x^2-x}$,

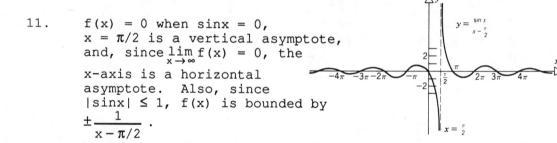

so $f'(x) = (2x-1)/2\sqrt{x^2-x}$ and
$f''(x) = -1/4(x^2-x)^{3/2}$. Thus $f(x) \ge 1$
for all x, decreases for $x < 0$ and
increases for $x > 1$, and is concave downward throughout its
domain. Similarly, $g(x)$ increases for $x < 0$, decreases for
$x > 1$ and is concave upward. Also, $y = -x + 3/2$ and
$y = x + 1/2$ are asymptotes.

11. $f(x) = 0$ when $\sin x = 0$,
$x = \pi/2$ is a vertical asymptote,
and, since $\lim_{x \to \infty} f(x) = 0$, the
x-axis is a horizontal
asymptote. Also, since
$|\sin x| \le 1$, $f(x)$ is bounded by
$\pm\dfrac{1}{x - \pi/2}$.

13.   By completing the square the
equation may be written as
$3(y+1) = (x-9)^2$, which is a parabola,
having vertex at $(9,-1)$, focus at
$(9,-1/4)$ and directrix $y = -7/4$ since
$p = 3/4$ is the focal length.

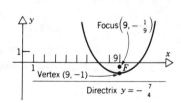

15.   Completing the square we have
       $x^2 - (y+2)^2/8 = 1$, which is a
       hyperbola with center $(0,-2)$,
       vertices $(\pm 1,-2)$, <u>foci</u> $(\pm 3,-2)$ and
       asymptotes $y = \pm 2\sqrt{2}\, x - 2$.

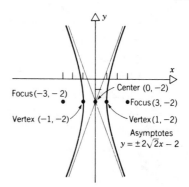

17.   Writing $(y-1/2)^2 - (x-4)^2/9 = 1$, the
       graph is a hyperbola, with center
       $(4,1/2)$, vertices $(4,3/2)$ and
       $(4,-1/2)$, foci $(4,1/2 \pm \sqrt{10}\,)$ and
       asymptotes $y = x/3 - 5/6$ and
       $y = -x/3 + 11/6$.

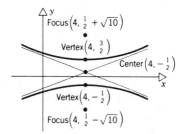

19.   Writing $(x+2)^2/4 - 2(y-3)^2/9 = 1$,
       the graph is a hyperbola, with
       center $(-2,3)$, vertices $(-4,3)$ and
       $(0,3)$, foci $(-2 \pm \sqrt{17/2}, 3)$, and
       asymptotes $y = \pm(3/2\sqrt{2})(x+2) + 3$.

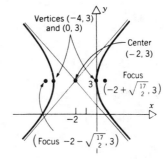

21.   We have $(x+5)^2/5 + (y-5)^2/4 = 1$,
       which is an ellipse with $a = \sqrt{5}$,
       $b = 2$ and $c = 1$, and with center
       at $(-5,5)$, major vertices $(-5 \pm \sqrt{5}, 5)$,
       minor vertices $(-5, 5 \pm 2)$, and
       foci $(-5 \pm 1, 5)$.

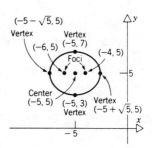

23.   We have $-(x-6)^2/36 + (y-2)^2 = 1$,
       which is seen to be a hyperbola with
       center at $(6,2)$. Since $a^2 = 36$, $b^2 = 1$,
       then $c^2 = 37$, the foci are at
       $(6,2\pm\sqrt{37})$, the vertices are $(6,8)$ and
       $(6,-4)$, and the asymptotes are
       $y-2 = \pm(x-6)/36$.

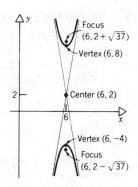

25.   From the given information, $p = 3$, the parabola opens down,
       and has vertex at $(1,-1)$. Thus the equation is
       $(x-1)^2 = -12(y+1)$.

27.   The center is the intersection of the transverse axis and
       the asymptote, $(2,1)$. Also $b = 5$, $a = 2$, so the equation is
       $(x-2)^2/4 - (y-1)^2/25 = 1$.

29.   From the information given, $c = 6$, so $e = c/a = 6$ gives
       $a = 1$ and $b^2 = 35$. Thus we have $(x+1)^2 - y^2/35 = 1$.

31.   The vertices are at $(-3,-1)$ and $(-3,7)$, with center at
       $(-3,3)$. Also $c = 5$, $b = 4$ and $a = 3$. Thus
       $(y-3)^2/16 - (x+3)^2/9 = 1$.

33.   The parabolas have foci at $(1,-3\pm3/4)$, that is, $p = 3/4$, so
       the equations are $3(y+3) = \pm(x-1)^2$.

35.   The centers are at $(3,1)$ and $(3,-5)$, as the vertex is on
       the major axis. Thus the equations are $(x-3)^2 + (y-1)^2/9 = 1$ or $(x-3)^2 + (y+5)^2/9 = 1$.

37.   The graph is a parabola with vertex $(-5/2,2)$ and $p = 3/2$.
       Its equation is $(y-2)^2 = 6(x+5/2)$.

39.   With $c = 2$ and $a = 1/2$, we have $b = \sqrt{15}/2$. The center is at
       $(4,1)$ so the equation is $(x-4)^2/(1/4) - (y-1)^2/(15/4) = 1$.

41.   From Section 5.2, Prob.28,

$\beta = -6\sin2\theta - 6\sqrt{3}\cos2\theta = 0$ for

$\theta = -\pi/6$. Thus $x = (\sqrt{3}u + v)/2$
and $y = (-u + \sqrt{3}v)/2$. Substituting
into the equation yields
$u^2 + v^2/4 = 1$, which is the ellipse
$x^2 + y^2/4 = 1$ rotated clockwise by
$\pi/6$ radians.

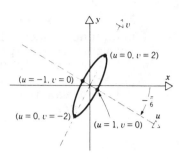

43.   The distance between the foci is $4|x_0|$, and with the
transverse diameter of 12, we need $4|x_0| > 12$ or $|x_0| > 3$.

45.   From $t = 5 - y$, we have
$y - 5 = -(x-3)/2$ for $-3 \le x \le 9$
and $8 \ge y \ge 2$.

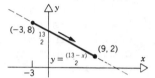

47.   Since $\sec t = 3(x-2)$ and
$\tan^2 t = \sec^2 t - 1$, we have
$y = 9(x-2)^2 - 2$ or $y+2 = 9(x-2)^2$. For
the given t values we obtain the
portion of the parabola above the
point $(7/3,-1)$.

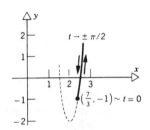

49.   Since $y = t^2$, $x = 1/\sqrt{1+y}$. Now
$y \ge 0$ so $0 < x \le 1$. The equation
may be written as $y = (1-x^2)/x^2$, so
there is a vertical asymptote as
x approaches 0.

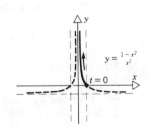

51.   From $x = t^3-1$, $t = (x+1)^{1/3}$ and so
$y = (x+1)^{2/3} + 2$. From the given t
values, the graph goes from
$(-1,2)$ to $(7,6)$.

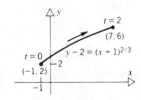

53.   As $t \to \pi^+$, $x \to +\infty$ and $y \to 0^+$,
      thus the x-axis is a
      horizontal asymptote. Also,
      $x \geq 1/\pi$ and $y \geq 0$. The only
      intercept is at $(1/\pi, 0)$ and
      y has a maximum value of 1
      at $t = \pi/2$.

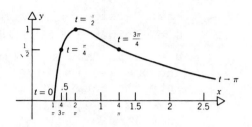

55.   The x-intercepts are at $t = 0, \pm\pi, \pm 2\pi$
      and thus are $(\pm 1, 0)$. Similarly,
      the y-intercepts are $(0, \pm\pi/2)$ and
      $(0, \pm 3\pi/2)$. Note that $|x| \leq 1$ and
      $|y| \leq 3\pi/2$, and the direction on
      the curves can be obtained by
      considering where the sine and
      cosine functions increase or
      decrease.

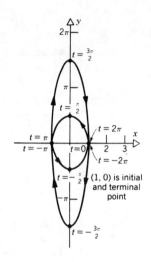

57. $dy/dx = -\pi\sin(\pi t)/(2\pi\sin\pi t \cos\pi t) = (-\sec\pi t)/2$ by Eq.(15)
      of Section 5.5. Then, by the first of Eqs.(16),

      $d^2y/dx^2 = (-\pi\sec\pi t \tan\pi t)/2(2\pi\sin\pi t \cos\pi t) = (-\sec^3\pi t)/4$. At
      $t = 1/2$, the curve goes through $(1,0)$ with slope undefined,
      so the tangent line has equation $x = 1$.

59.   Again, by Eqs.(15) and (16) of Section 5.5,
      $dy/dx = 2t/3t^2 = 2/3t$ and $d^2y/dx^2 = (-2/3t^2)/(3t^2) = -2/9t^4$.
      At $t = 1/2$, the curve passes through $(1/8, 5/4)$ with slope
      $4/3$. Thus the tangent line there has equation
      $y - 5/4 = (4/3)(x - 1/8)$.

CHAPTER 6

1. $\displaystyle\sum_{k=1}^{4}\frac{1}{k} = 1 + \frac{1}{2} + \frac{1}{3} + \frac{1}{4} = \frac{25}{12}$.

3. $\displaystyle\sum_{k=2}^{3} k^k = 2^2 + 3^3 = 4 + 27 = 31$.

5. $\displaystyle\sum_{k=1}^{n}\left(\frac{1}{k+1} - \frac{1}{k}\right) = \left(\frac{1}{2} - 1\right) + \left(\frac{1}{3} - \frac{1}{2}\right) +$

$\left(\frac{1}{4} - \frac{1}{3}\right) + \ldots + \left(\frac{1}{n} - \frac{1}{n-1}\right) + \left(\frac{1}{n+1} - \frac{1}{n}\right) = \frac{1}{n+1} - 1$.

7. $\displaystyle\sum_{i=1}^{1000} 1 = \underbrace{1 + 1 + \ldots + 1}_{\uparrow\ 1000\ \text{times}} = 1000\,(1) = 1000$.

9. $\displaystyle\sum_{k=1}^{50}\frac{1}{k(k+1)} = \sum_{k=1}^{50}\left(\frac{1}{k} - \frac{1}{k+1}\right) = -\left(\frac{1}{51} - 1\right) = \frac{50}{51}$ from Prob.5.

11. Letting $u_n = \sqrt{3n-2}$, then $u_{n+1} = \sqrt{3n+1}$, so by Eq. (12),

$\displaystyle\sum_{n=1}^{87}(\sqrt{3n+1} - \sqrt{3n-2}) = \sqrt{3(87)+1} - \sqrt{1} = \sqrt{262} - 1$.

13. Considering each sum separately we have

$\displaystyle\sum_{k=1}^{n} u_k = u_1 + u_2 + u_3 + \ldots + u_{n-1} + u_n,\quad \sum_{r=0}^{n-1} u_{r+1} = u_1 + u_2 + u_3 + \ldots + u_{n-1} + u_n,$

$\displaystyle\sum_{j=2}^{n+1} u_{j-1} = u_1 + u_2 + u_3 + \ldots + u_{n-1} + u_n,\quad \sum_{i=0}^{n-1} u_{n-i} = u_n + u_{n-1} + \ldots + u_3 + u_2 + u_1.$

Examining the right side of each expression verifies the equalities.

15.   By setting k = 1, k = 2 etc. in the general term we obtain

$$\sum_{k=1}^{n} (-1)^{k+1} x^k = x - x^2 + x^3 - \ldots + (-1)^{n+1} x^n.$$

17.   Note that each term involves (b-a)/n and for i = 0 and 1 the
        f term becomes f(a) and f[a + (b-a)/n] respectively. Thus
        the sum is:

$$\left(\frac{b-a}{n}\right)\left\{ f(a) + f\left[a + \frac{b-a}{n}\right] + f\left[a + \frac{2(b-a)}{n}\right] + \ldots + f\left[a + \frac{(n-1)(b-a)}{n}\right]\right\}.$$

19.   The terms alternate in sign so $(-1)^n$ must be in the general
        term and the terms look like 1/n so we have:

$$\frac{1}{2} - \frac{1}{3} + \frac{1}{4} - \ldots + \frac{1}{26} = \sum_{n=2}^{26} \frac{(-1)^n}{n}.$$

21.   This sum is similar to Prob.19 except that only odd terms
        appear so we must have (2k+1) rather than k. Thus

$$\frac{x^3}{3} - \frac{x^5}{5} + \frac{x^7}{7} + \ldots - \frac{x^{29}}{29} = \sum_{k=1}^{14} \frac{(-1)^{k+1} x^{2k+1}}{2k+1}.$$

23a.   If P(n) is the statement: 1 + 2 + 3 + ... + n = n(n+1)/2
        then P(1) is the equality: 1 = 1(1+1)/2 = 1, which is true.
   b.   Adding (k+1) to both sides of P(k) yields:

        1 + 2 + 3 + ... k + (k+1) = k(k+1)/2 + (k+1)

                                  = (k/2 + 1)(k+1)

                                  = (k+2)(k+1)/2.

        The last equality is P(k+1) and thus P(k+1) is true.

25.   Let P(n) be: $1^2 + 2^2 + \ldots + (n)^2 = n(n+1)(2n+1)/6$.
        i. P(1) is: $1^2 = 1(2)(3)/6 = 1$, so P(1) is true.
        ii. Adding $(n+1)^2$ to both sides of the equation for P(n):
        $1^2 + 2^2 + \ldots + n^2 + (n+1)^2 = n(n+1)(2n+1)/6 + (n+1)^2 =$
        $(n+1)[n(2n+1) + 6(n+1)]/6 = (n+1)[2n^2 + 7n + 6]/6 =$
        $(n+1)(n+2)(2n+3)/6$, which shows P(n+1) is true. Since P(1)
        is true and P(n+1) is true whenever P(n) is true, then P(n)
        is true for all integers n ≥ 1.

27.   Let P(n) be: $1(3)+3(5)+...+(2n-1)(2n+1) = n(4n^2+6n-1)/3$.
       i.   P(1): $1(3) = 1(4+6-1)/3 = 3$, so P(1) is true.
       ii. To show P(n+1) is true when P(n) is true we have P(n+1)
       is the statement:
       $$1(3)+...+(2n-1)(2n+1)+(2n+1)(2n+3) = (n+1)[4(n+1)^2+6(n+1)-1]$$
       $$= (n+1)(4n^2+14n+9)/3.$$
       Note that the left side is the left side of P(n) with
       $(2n+1)(2n+3)$ added. This suggests we add $(2n+1)(2n+3)$ to
       both sides of P(n) and simplify the right side to get
       agreement with P(n+1):
       $$1(3)+3(5)+...+(2n-1)(2n+1)+(2n+1)(2n+3)$$
       $$= n(4n^2+6n-1)/3+(2n+1)(2n+3) = (4n^3+18n^2+23n+9)/3$$
       $$= (n+1)(4n^2+14n+9)/3,$$ which verifies P(n+1). Therefore,
       since P(1) is true and P(n+1) is true whenever P(n) is true,
       P(n) is true for all integers $n \geq 1$.

29.   Let $s_n = 1+3+5+...+(2n-1)$. Then $s_1 = 1$, $s_2 = 4$, $s_3 = 9$, from
       which a good conjecture would be $s_n = n^2$. That is
       $1+3+5+...+(2n-1) = n^2$. Let P(n) be: $s_n = n^2$.
       i.   We have shown P(1) is true.
       ii. P(n+1) is: $1+3+5+...+(2n-1)+(2n+1) = (n+1)^2$. Thus,
       adding $(2n+1)$ to both sides of the equation P(n):
       $1+3+5+...+(2n-1)+(2n+1) = n^2+(2n+1) = n^2+2n+1 = (n+1)^2$ so
       that P(n+1) is true whenever P(n) is true. Since P(1) is
       also true, we know P(n) is true for all $n \geq 1$.

31. $$\sum_{i=1}^{n}(3-2i) = \sum_{i=1}^{n}3 - 2\sum_{i=1}^{n}i = 3n-2\left[\frac{1}{2}n(n+1)\right] = 2n - n^2.$$

33. $$\sum_{k=1}^{n}2k(1+k)^2 = \sum_{k=1}^{n}(2k+4k^2+2k^3) = 2\sum_{k=1}^{n}k + 4\sum_{k=1}^{n}k^2 + 2\sum_{k=1}^{n}k^3$$

       $$= (2)(1/2)(n)(n+1) + (4/6)(n)(n+1)(2n+1) + (1/2)n^2(n+1)^2$$
       $$= n(n+1)[1+(2/3)(2n+1)+(1/2)(n)(n+1)]$$
       $$= n(n+1)(3n^2+11n+10)/6.$$

35a.   Subdividing the interval $(0,b)$
       into n subintervals of length
       $h = b/n$ yields the partition
       points: $x_0, x_1, x_2, ..., x_{i-1}, x_i, ...x_n$
       whose coordinates are
       $0, h, 2h, ..., (i-1)h, ih, ...b$.
       Since $f(x) = x$, $f(x_i) = ih$. Also f

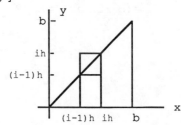

is increasing, so the height of the ith inscribed rectangle is $f(x_{i-1})$ while the height of the ith circumscribed rectangle will be $f(x_i)$. Thus the area a_i of the ith inscribed rectangle is $f(x_{i-1}) h = (i-1) h^2$ and the area of the ith circumscribed rectangle is $A_i = f(x_i) h = i h^2$.

Letting $s_n = \displaystyle\sum_{i=1}^{n} a_i$ and $\sigma_n = \displaystyle\sum_{i=1}^{n} A_i$ we have

$$s_n = \sum_{i=1}^{n} (i-1) h^2 = h^2 \sum_{i=1}^{n} (i-1) = h^2 \sum_{i=0}^{n-1} i = h^2 n (n-1)/2 \quad \text{from Prob. 23.}$$

Since $h = b/n$, $s_n = b^2 \left(\dfrac{n}{n}\right)\left(\dfrac{n-1}{n}\right)\left(\dfrac{1}{2}\right) = \dfrac{b^2}{2}\left(1-\dfrac{1}{n}\right)$. Similarly $\sigma_n = \dfrac{b^2}{2}\left(1+\dfrac{1}{n}\right)$.

35b. A bound for the error, from Eq, (33), is $|E_n| \le (\sigma_n - s_n)/2 = b^2/2n$.

c. The area is $b^2/2$ since $s_n \to \sigma_n \to b^2/2$ as $n \to \infty$.

37. For the triangular region of Prob. 35, with the starpoints as midpoints of the subintervals, $x_i^* = (x_{i-1} + x_i)/2 = \dfrac{(2i-1)h}{2}$, the height of the rectangles is $f(x_i^*) = (2i-1)h/2$ so that

$$A_i = [(2i-1) h/2] h. \text{ Thus } S_n = \sum_{i=1}^{n} A_i = \sum_{i=1}^{n} (h^2/2) (2i-1) = (h^2/2) \sum_{i=1}^{n} (2i-1)$$

or $S_n = (h^2/2) n^2$ from Prob. 29. Since $h = b/n$, $S_n = (b^2/2n^2) n^2 = b^2/2$. This is, in fact, the exact area.

39.   Using the identity sin(A ± B) = sinAcosB ± cosAsinB:
        sin(k+1/2)x - sin(k-1/2)x = sinkxcos(x/2) + coskxsin(x/2) -
        sinkxcos(x/2) + coskxsin(x/2) = 2sin(x/2)coskx. Thus

$$2\sin\left(\frac{x}{2}\right)\sum_{k=1}^{n}\cos kx = \sum_{k=1}^{n}2\sin\left(\frac{x}{2}\right)\cos kx = \sum_{k=1}^{n}\left[\sin\left(k+\frac{1}{2}\right)x - \sin\left(k-\frac{1}{2}\right)x\right]$$

        = sin(n+1/2)x - sin(x/2), by Eq.12 with $u_k = \sin(k-1/2)x$.
        Dividing by 2sin(x/2) gives

$$\sum_{k=1}^{n}\cos kx = \frac{\sin(n+1/2)x - \sin(x/2)}{2\sin(x/2)} \quad \text{for } x \neq 2n\pi.$$

Section 6.2, Page 295

1. $\Delta x_1 = 1/4$, $\Delta x_2 = 1/4$, $\Delta x_3 = 3/8$, $\Delta x_4 = 1/8$, $\Delta x_5 = \Delta x_6 = 1/2$.   Thus
        $\|\Delta\| = \max \Delta x_i = 1/2 = $ either Δx_5 or Δx_6.

3. $\Delta x_i = x_i - x_{i-1} = $ [a+(i/n)(b-a)] - [a+(i-1)/n)(b-a)] so
        $\Delta x_i = $ (b-a)/n for all i, $1 \leq i \leq n$.   Thus $\|\Delta\| = $(b-a)/n.

5. $x_1^* = 1 + (1/4)/2 = 9/8$, $f(x_1^*) = (9/8)/(1+9/8) = 9/17$, a1 =9/68,
        $x_2^* = 5/4 + (1/4)/2 = 11/8$, $f(x_2^*) = 11/19$, a2 = 11/76,
        $x_3^* = 3/2 + (1/2)/2 = 7/4$, $f(x_3^*) = 7/11$, a3 = 7/22, so the
        approximate value of the integral is a1 + a2 + a3 = 2115/3553.

7. $x_1^* = -1/4$, $\Delta x_1 = 1/2$, $f(x_1^*) = 1/4$ so a1 = 1/8 and
        $x_2^* = 1/2$, $\Delta x_2 = 1$, $f(x_2^*) = 1/2$ so a2 = 1/2 and the
        approximate value is 5/8.

9.   Since f(x) = 1/x is decreasing, the upper sums are computed
        at the left end, while the lower sums are computed at the
        right end of the intervals. Thus
        σ = (1)(1/4)+(4/5)(1/4)+(2/3)(1/4)+(4/7)(1/4) = 319/420 and
        s = (4/5)(1/4)+(2/3)(1/4)+(4/7)(1/4) + (1/2)(1/4) = 533/840.

11.   On the interval [-2,0], $1/(1+x^2)$ increases, so the lower
        sums are computed at the left endpoints, while the upper
        sums are computed at the right endpoints. Thus
        s = (1/5)(1/2)+(4/13)(1/2)+(1/2)(1/2)+(4/5)(1/2) = 47/52 and
        σ = (4/13)(1/2)+(1/2)(1/2)+(4/5)(1/2)+(1)(1/2) = 339/260.

13. On $0 \leq x \leq 1$ $f(x)$ is increasing and on $1 \leq x \leq 2$ $f(x)$ is
     constant. Note, $f(1) = 1$ is used for the upper sum for
     $1 \leq x \leq 3/2$. Thus $\sigma = (1/4 + 1 + 1 + 1/2)(1/2) = 11/8$ and
     $s = (0 + 1/4 + 1/2 + 1/2)(1/2) = 5/8$.

15. We let $\Delta x_i = 1/n$ for each i and thus $x_i = i/n$ for
     $i = 0, 1, \ldots n$. Since $\sqrt{x}$ is increasing on $(0,1)$ we evaluate
     it at the right end points for σ_n and at the left end
     points for s_n. Finally,

     $x_i^* = [(i-1)/n + i/n]/2 = (2i-1)/2n$. Thus we have

     $$s_n = (1/n)\sum_{i=1}^{n}\sqrt{(i-1)/n} \ , \quad \sigma_n = (1/n)\sum_{i=1}^{n}\sqrt{i/n} \ ,$$

     and $S_n = (1/n)\sum_{i=1}^{n}\sqrt{(2i-1)/2n}$. Using these sums for the

     various n values yields the results in the answer section
     of the text.

17. We let $\Delta x_i = 3/n$ for each i and thus $x_i = 3i/n$ for
     $i = 0, 1, \ldots n$. Since $\sqrt{1+x^2}$ is increasing on $(0,2)$ we
     evaluate it at the right end points for σ_n and at the left
     end points for s_n . Finally,
     $x_i^* = [3(i-1)/n + 3i/2]/2 = 3(2i-1)/2n$. Thus we have

     $$s_n = \frac{3}{n}\sum_{i=1}^{n}\left[1 + \frac{9(i-1)^2}{n^2}\right]^{1/2} = (3/n^2)\sum_{i=1}^{n}[n^2 + 9(i-1)^2]^{1/2} \ ,$$

     $$\sigma_n = (3/n^2)\sum_{i=1}^{n}[n^2 + 9i^2]^{1/2} \ , \quad \text{and } S_n = (3/2n^2)\sum_{i=1}^{n}[4n^2 + 9(2i-1)^2]^{1/2} .$$

     Using these expressions the corresponding values as given in
     the text are obtained.

19. We let $\Delta x_i = 2/n$ for each i and thus $x_i = 1+2i/n = (2i+n)/n$
     for $i = 0, 1, \ldots n$. Since $4 - x^2$ is decreasing on $(1,3)$, we
     evaluate it at the right end points for s_n and at the left
     end points for σ_n. Finally,
     $x_i^* = \{[1+2(i-1)/n]+[1+2i/n]\}/2 = 1 + (2i-1)/n$. Thus

$$S_n = (2/n)\sum_{i=1}^{n}[4 - (2i+n)^2/n^2] = (2/n^3)\sum_{i=1}^{n}[4n^2 - (2i+n)^2],$$

$$\sigma_n = (2/n^3)\sum_{i=1}^{n}[4n^2 - (2i-2+n)^2], \text{ and } S_n = (2/n^3)\sum_{i=1}^{n}[4n^2 - (2i-1+n)^2].$$

Using these expressions the corresponding values as given in the text are obtained.

21. The midpoint of the ith interval is $(2i-1)\pi/2n$ and thus

$$S_n = (\pi/n)\sum_{i=1}^{n}[(2i-1)\pi/2n]\cos[(2i-1)\pi/2n], \text{ which gives}$$

$S_{10} = -1.991704, S_{20} = -1.99794, S_{40} = -1.999486, S_{80} = -1.999872.$

23. Let $0 = t_0 < t_1 < \ldots < t_i < \ldots < t_n = T$ be a partition Δ of $[0,T]$. Choose a star point so that in each subinterval the velocity may be approximated by $v(t_i^*)$, a constant, and thus d_i, the distance traveled in the ith time interval, is approximated by $v(t_i^*)(t_i - t_{i-1})$. Thus the total distance d is approximated by

$$d \cong S(v,\Delta) = \sum_{i=1}^{n}v(t_i^*)\Delta t_i, \text{ where } \Delta t_i = t_i - t_{i-1}.$$

Thus, if v is integrable, $d = \lim_{\|\Delta\|\to 0} S(v,\Delta) = \int_0^T v(t)\,dt.$

25. Let $0 = x_0 < x_1 < \ldots < x_n = s$ be a partition Δ of $[0,s]$. Choose a star point so that in each subinterval the force may be approximated by $F(x_i^*)$, a constant, and W_i, the work done in the ith interval is approximated by $F(x_i^*)(x_i - x_{i-1})$. Thus the total work W is approximated by

$$W \cong S(F,\Delta) = \sum_{i=1}^{n}F(x_i^*)\Delta x_i, \text{ where } \Delta x_i = (x_i - x_{i-1}).$$

If F is integrable, then $W = \lim_{\|\Delta\|\to 0} S(F,\Delta) = \int_0^s F(x)\,dx.$

27a.   Since f is nondecreasing on [a,b] then in the ith
       subinterval x_i is used in evaluating the upper sum and x_{i-1}
       is used in evaluating the lower sum. Thus

$$\sigma(f,\Delta) = \sum_{i=1}^{n} f(x_i)\,\Delta x_i \text{ and } s(f,\Delta) = \sum_{i=1}^{n} f(x_{i-1})\,\Delta x_i .$$

   b.   Since $\Delta x_i \le \|\Delta\|$ for all i.

$$\sigma(f,\Delta) - s(f,\Delta) = \sum_{i=1}^{n}[f(x_i) - f(x_{i-1})]\,\Delta x_i \le \|\Delta\| \sum_{i=1}^{n}[f(x_i) - f(x_{i-1})] .$$

   c.   The sum in part (b) is telescoping and thus

$$\sum_{i=1}^{n}[f(x_i) - f(x_{i-1})] = f(x_n) - f(x_0) = f(b) - f(a), \text{ and by Theorem 2.3.4,}$$

$$\lim_{\|\Delta\|\to 0} [\sigma(f,\Delta) - s(f,\Delta)] = 0 .$$

   d.   With f nonincreasing, f is evaluated at the right endpoints
        for the lower sum and at the left endpoints for the upper
        sum. In this case we then have, as above,

$$\sigma(f,\Delta) - s(f,\Delta) = \sum_{i=1}^{n}[f(x_{i-1}) - f(x_i)]\Delta x_i \le \|\Delta\| \sum_{i=1}^{n}[f(x_{i-1}) - f(x_i)].$$

   The sum is again telescoping and thus
   $0 \le \sigma(f,\Delta) - s(f,\Delta) \le \|\Delta\|[f(a) - f(b)]$, so by Theorem 2.3.4,
   $$\lim_{\|\Delta\|\to 0} [\sigma(f,\Delta) - s(f,\Delta)] = 0 .$$

Section 6.3, Page 306

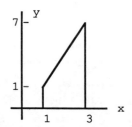

1.   The given integral represents the
       area of the trapezoid shown. The
       base is 2 and the average height is
       $(7+1)/2 = 4$ so the value of the
       integral is 8.

3. $\int_{-2}^{3} |x|\, dx = \int_{-2}^{0} -x\, dx + \int_{0}^{3} x\, dx.$ Each integral represents the area
       of a triangle and thus $\int_{-2}^{0} -x\, dx = (2)(2)/2 = 2$ and
       $\int_{0}^{3} x\, dx = (3)(3)/2 = 9/2$ so $\int_{-2}^{3} |x|\, dx = 13/2 .$

5. $\int_{-1}^{2} f(x)\,dx = \int_{-1}^{0} \sqrt{1-x^2}\,dx + \int_{0}^{1} 1\,dx + \int_{1}^{2} (2x-1)\,dx$. The first

integral is the area of a unit quarter circle in the second quadrant, the second is the area of a unit square in the first quadrant, and the third is the area of a trapezoid of base 1 and sides 1 and 3 in the first quadrant. Thus

$\int_{-1}^{1} f(x)\,dx = \pi/4 + 1 + (1/2)(1+3)(1) = 3 + \pi/4.$

7. $\int_{1}^{-1}\left(1-\sqrt{1-x^2}\right)dx = \int_{-1}^{1}\left(\sqrt{1-x^2}-1\right)dx = \int_{-1}^{1} \sqrt{1-x^2}\,dx - \int_{-1}^{1} dx$ which

represents the difference in the areas of a semicircle of radius 1 and a rectangle of sides 2 and 1.

Thus $\int_{1}^{-1}\left(1-\sqrt{1-x^2}\right)dx = \pi/2 - 2.$

9. $\int_{0}^{3} (2 - |x-1|)\,dx = \int_{0}^{1} (x+1)\,dx + \int_{1}^{3} (3-x)\,dx$

$= (1/2)(1+2)1 + (1/2)(2)(2) = 7/2,$ since the first integral is the area of a trapezoid of base 1 and sides 1 and 2 while the second is the area of a triangle of base 2 and height 2.

11. $\int_{-3}^{6} (2-3x)\,dx = 2\int_{-3}^{6} dx - 3\int_{-3}^{6} x\,dx = 2[6-(-3)] - 3[6^2-(-3)^2]/2$

$= 18 - 3(27)/2 = 36/2 - 81/2 = -45/2.$

13. $\int_{0}^{1} (2x+1)^2\,dx = \int_{0}^{1} (4x^2 + 4x + 1)\,dx = 4\int_{0}^{1} x^2\,dx + 4\int_{0}^{1} x\,dx + \int_{0}^{1} dx$

$= 4(1/3 - 0) + 4(1/2 - 0) + (1-0) = 13/3.$

15. $\int_{3}^{1} (2x^2 + 3x)\,dx = -\int_{1}^{3} (2x^2 + 3x)\,dx = -2\int_{1}^{3} x^2\,dx - 3\int_{1}^{3} x\,dx$

$= -2(3^3-1^3)/3 - 3(3^2-1^2)/2 = -52/3 - 24/2 = -88/3.$

17.   Since f is bounded and integrable on [a,b], then by

Corollary 2 to Theorem 6.3.4, $\left| \int_a^b f(x)\,dx \right| \le \int_a^b |f(x)|\,dx.$

Further, by Corollary 1 to the same Theorem,

$\int_a^b |f(x)|\,dx \le \int_a^b M\,dx$ since $|f(x)| \le M$ on [a,b]. But since

$\int_a^b M\,dx = M(b-a)$ we have $\left| \int_a^b f(x)\,dx \right| \le \int_a^b |f(x)|\,dx \le M(b-a).$

19.   For $0 \le x \le 2$, $1 \le 1 + x^2 \le 5$ so $1/5 \le 1/(1+x^2) \le 1$. Thus by

Corollary 1 to Theorem 6.3.4 $\int_0^2 \frac{1}{5}\,dx \le \int_0^2 \frac{1}{1+x^2}\,dx \le \int_0^2 dx$ or

$\frac{2}{5} \le \int_0^2 \frac{1}{1+x^2}\,dx \le 2$ by Eq.(1).

21.   For $0 \le x \le 1$, $x^6 \le x^4$, thus by Corollary 6.3.1, $\int_0^1 x^6\,dx \le \int_0^1 x^4\,dx.$

23.   Letting $\Delta x_i = 1/n$ we have $x_i = i/n$ for $i = 0,1,\ldots,n$. Since
$f = (1-x^2)^{1/2}$ is decreasing on [0,1] we evaluate f at the
right end of each subinterval for the lower sum and at the
left end for the upper sum. Finally, the midpoint is given
by $(2i-1)/2n$ and thus

$$s_n = (1/n)\sum_{i=1}^{n}(1-i^2/n^2)^{1/2} = (1/n^2)\sum_{i=1}^{n}(n^2-i^2)^{1/2},$$

$$\sigma_n = (1/n^2)\sum_{i=1}^{n}[n^2-(i-1)^2]^{1/2} \text{ and } S_n = (1/2n^2)\sum_{i=1}^{n}[4n^2-(2i-1)^2]^{1/2}.$$

The values given in the answer section of the text are
obtained with these expressions.

25.   The average velocity is $\dfrac{1}{5-2}\int_2^5 (2+3t^2)\,dt = (2/3)\int_2^5 dt + \int_2^5 t^2\,dt =$

$(2/3)(5-2) + (5^3 - 2^3)/3 = 41$ ft/sec.

27.   From the Mean Value Theorem, Theorem 6.3.5,

$$f(c) = \frac{1}{b-a} \int_a^b f(x)\,dx \quad \text{for some } c \text{ in } [a,b]. \quad \text{Since } \int_a^b f(x)\,dx = 0,$$

and f is continuous in [a,b], we have f(c) = 0.

Section 6.4, Page 316

1. $\displaystyle\int_0^2 3x^4\,dx = 3\int_0^2 x^4\,dx = 3\left.\frac{x^5}{5}\right|_0^2 = 3(32)/5 = 96/5$.

3. $\displaystyle\int_0^{\pi/2} (2 - \cos x)\,dx = \int_0^{\pi/2} 2\,dx - \int_0^{\pi/2} \cos x\,dx = 2x\Big|_0^{\pi/2} - \sin x\,\Big|_0^{\pi/2} =$

$2\pi/2 - 1 = \pi - 1$.

5. $\displaystyle\int_{-1}^1 s(2+s^2)\,ds = \int_{-1}^1 (2s + s^3)\,ds = \int_{-1}^1 2s\,ds + \int_{-1}^1 s^3\,ds =$

$s^2\Big|_{-1}^1 + \dfrac{s^4}{4}\Big|_{-1}^1 = (1-1) + \left(\dfrac{1}{4} - \dfrac{1}{4}\right) = 0$.

7. $\displaystyle\int_0^{\pi/4} (3+6s+2\sin s)\,ds = \int_0^{\pi/4} 3\,ds + \int_0^{\pi/4} 6s\,ds + 2\int_0^{\pi/4} \sin s\,ds =$

$3s\Big|_0^{\pi/4} + 6s^2/2\Big|_0^{\pi/4} - 2\cos s\Big|_0^{\pi/4} = 3\pi/4 + 3\pi^2/16 - (\sqrt{2} - 2) =$

$2 - \sqrt{2} + (3\pi/4)(1+\pi/4)$.

9. $\displaystyle\int_{-\pi/2}^{\pi} [3\cos x - (1/3)\sin x]\,dx = 3\sin x\Big|_{-\pi/2}^{\pi} + (1/3)\cos x\Big|_{-\pi/2}^{\pi} =$

$3[\sin\pi - \sin(-\pi/2)] + (1/3)[\cos\pi - \cos(-\pi/2)] = 8/3$.

11. $\displaystyle\int_0^x (2t + t^{1/2})\,dt = t^2\Big|_0^x + \left(\frac{2}{3}\right)t^{3/2}\Big|_0^x = x^2 + \left(\frac{2}{3}\right)x^{3/2}$.

13. $\displaystyle\int_0^x (3-\cos t)\,dt = (3t - \sin t)\Big|_0^x = 3x - \sin x$.

15. $\displaystyle\int_0^{x^2} (1+2t)dt = (t+t^2)\Big|_0^{x^2} = x^2 + x^4 .$

17. $\displaystyle\int_{2x}^{x^2} (3-4t)\,dt = (3t-2t^2)\,\Big|_{2x}^{x^2} = (3x^2 - 2x^4) - (6x - 8x^2) = -6x+11x^2-2x^4 .$

19. $\displaystyle\int_0^2 f(x)\,dx = \int_0^1 f(x)\,dx + \int_1^2 f(x)\,dx = \int_0^1 (2+x)\,dx + \int_1^2 (2-x)\,dx$

$$= \left(2x + \frac{x^2}{2}\right)\Big|_0^1 + \left(2x - \frac{x^2}{2}\right)\Big|_1^2 = \left(2 + \frac{1}{2}\right) + \left[(4-2) - \left(2 - \frac{1}{2}\right)\right] = 3 .$$

21.   By Eq.(24) we have

$$\int_{-10}^{10} f(x)dx = \int_{-10}^{-9} xdx + \int_{-9}^{-8} xdx + \ldots + \int_{-1}^{0} xdx + \int_{0}^{1} xdx + \ldots + \int_{8}^{9} xdx + \int_{9}^{10} xdx =$$

(81/2 - 100/2)+(64/2 - 81/2)+...+(0 - 1/2)+(1/2 - 0)+...+

(81/2 -64/2)+(100/2 - 81/2) = 0.

23. $\displaystyle\int_{-1}^{1} |2x+1|\,dx = \int_{-1}^{-1/2} -(2x+1)\,dx + \int_{-1/2}^{1} (2x+1)\,dx$

$$= -(x^2 + x)\Big|_{-1}^{-1/2} + (x^2 + x)\Big|_{-1/2}^{1} = 5/2 .$$

Alternatively, consider the integral as the area below
y = |2x+1|, above the x-axis between -1 and 1 which consists
of two triangles and hence A = (1/2)(1)/2 + [(3/2)3]/2 = 5/2.

25. $\displaystyle\int_{-2}^{2} |1-x^2|\,dx = \int_{-2}^{-1} (x^2-1)dx + \int_{-1}^{+1} (1-x^2)\,dx + \int_{1}^{2} (x^2-1)\,dx$

$$= \left(\frac{x^3}{3} - x\right)\Big|_{-2}^{-1} + \left(x - \frac{1}{3}x^3\right)\Big|_{-1}^{1} + \left(\frac{x^3}{3} - x\right)\Big|_{1}^{2} = \frac{4}{3} + \frac{4}{3} + \frac{4}{3} = 4 .$$

27.   If p and q are positive integers then

$$\int_0^1 (x^{p/q} + x^{q/p})\,dx = \frac{x^{p/q+1}}{p/q+1} + \frac{x^{q/p+1}}{q/p+1}\,\Big|_0^1$$

$$= \frac{1}{(p+q)/q} + \frac{1}{(q+p)/p} = \frac{p}{p+q} + \frac{q}{p+q} = \frac{p+q}{p+q} = 1.$$

29a.   The average velocity is $\dfrac{1}{2\pi/\omega}\displaystyle\int_{0}^{2\pi/\omega}A\sin\omega t\,dt$ which is

$$-\frac{A}{(2\pi/\omega)\omega}\cos\omega t\,\Big|_{0}^{2\pi/\omega} = -\frac{A}{2\pi}(\cos 2\pi - \cos 0) = 0.$$

  b.   Recall the trigonometric idientity: $\sin^2 B = (1-\cos 2B)/2$.
       Since the period is $2\pi/\omega$, the average velocity is

$$\frac{1}{2\pi/\omega}\int_{0}^{2\pi/\omega}A\sin^2\omega t\,dt = \frac{A\omega}{4\pi}\int_{0}^{2\pi/\omega}(1-\cos 2\omega t)\,dt$$

$$= \frac{A\omega}{4\pi}\left[t - \frac{1}{2\omega}\sin 2\omega t\right]\Big|_{0}^{2\pi/\omega} = \frac{A\omega}{4\pi}\frac{2\pi}{\omega} = \frac{A}{2}.$$

31a.   b.

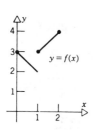

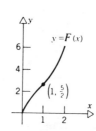

  b.   For $0 \le x < 1$, $F(x) = \displaystyle\int_{0}^{x}(3-t)\,dt = (3t - t^2/2)\Big|_{0}^{x} = 3x - x^2/2$

       while for $1 \le x < 2$ $F(x) =$

$$\int_{0}^{1}(3-t)dt + \int_{1}^{x}(2+t)dt = 5/2 + (2t + t^2/2)\Big|_{1}^{x} = 5/2 + (2x + x^2/2 - 5/2).$$

$$\text{Thus } F(x) = \begin{cases} 3x - (x^2/2) & 0 \le x < 1 \\ 2x + (x^2/2) & 1 \le x \le 2 \end{cases}.$$

  c.   From the graphs we see that f is discontinuous at x = 1,
       while F is continuous at x = 1, but has a discontinuous
       derivative at x = 1.

33.   From the Fundamental Theorem of Calculus, if

$$F(x) = \int_{0}^{x}\frac{\sin 2t}{1+t^2}\,dt, \text{ then } F'(x) = \frac{\sin 2x}{1+x^2}.$$

35.   If $F(x) = 2x + \int_0^x \frac{\sin 2t}{1+t^2}\,dt$,   then $F(0) = 0$,

$F'(x) = 2 + \frac{\sin 2x}{1+x^2}$, $F'(0) = 2$,

$F''(x) = \frac{2(1+x^2)\cos 2x - 2x \sin 2x}{(1+x^2)^2}$,   and $F''(0) = 2$.

37. $F(x) = \int_{1-x^2}^{1+x^2} \frac{\sin 2t}{1+t^2}\,dt = \int_{1-x^2}^{1} \frac{\sin 2t}{1+t^2}\,dt + \int_{1}^{1+x^2} \frac{\sin 2t}{1+t^2}\,dt$

or $F(x) = -\int_{1}^{1-x^2} \frac{\sin 2t}{1+t^2}\,dt + \int_{1}^{1+x^2} \frac{\sin 2t}{1+t^2}\,dt$.   Let $f_1(u)$ and

$f_2(v)$ be the first and second integrals respectively, with

$u = 1-x^2$, $v = 1+x^2$.   By the chain rule,

$\frac{d}{dx}[f_1(u)] = \frac{d}{du}[f_1(u)]\,\frac{du}{dx} = -(-2x)\frac{\sin 2u}{1+u^2} = 2x\frac{\sin 2(1-x^2)}{1+(1-x^2)^2}$ and

$\frac{d}{dx}[f_2(v)] = \frac{d}{dv}[f_2(v)]\frac{dv}{dx} = 2x\frac{\sin 2v}{1+v^2} = 2x\frac{\sin 2(1+x^2)}{1+(1+x^2)^2}$.

Thus $F'(x) = \frac{2x \sin 2(1-x^2)}{1+(1-x^2)^2} + \frac{2x \sin 2(1+x^2)}{1+(1+x^2)^2}$.

39. $F(x) = \int_0^{1+x^2} t\,f(t)\,dt = \int_0^u t\,f(t)\,dt$.   So

$F'(x) = F'(u)u'(x) = uf(u)\,du/dx = (1+x^2)\,f(1+x^2)(2x) = 2x(1+x^2)\,f(1+x^2)$.

41.   Let $g(t) = t\int_1^t f(s)\,ds$ then $F(x) = \int_0^x g(t)\,dt$ so that

(a) $F'(x) = g(x) = x\int_1^x f(s)\,ds$

(b) $F'(1) = \int_1^1 f(s)\,ds = 0$

41.   (c) $F''(x) = \int_1^x f(s)\,ds + xf(x)$

        (d) $F''(1) = \int_1^1 f(s)\,ds + f(1) = f(1)$.

43.   If $F(x) = \int_0^{x^2} f(t)\,dt = \sqrt{1+x^2}$ then $F'(x) = 2xf(x^2) = \dfrac{2x}{2\sqrt{1+x^2}}$

        so $f(x^2) = \dfrac{1}{2\sqrt{1+x^2}}$ or $f(x) = \dfrac{1}{2\sqrt{1+x}}$; $f(\pi/2) = \sqrt{2/(2+\pi)}\,/\,2$.

Section 6.5, Page 324

1.   Let $u = 3+2x$, then $du = 2dx$ or $dx = du/2$.   Thus

$$\int \sqrt{3+2x}\,dx = \frac{1}{2}\int \sqrt{u}\,du = \frac{1}{2}\int u^{1/2}du = \frac{1}{2}\frac{u^{3/2}}{3/2} + c = (1/3)(3+2x)^{3/2} + c.$$

3.   Let $u = 9+x^2$, $du = 2xdx$ and $xdx = du/2$. Thus

$$\int x(9+x^2)^3 dx = \int u^3\,du/2 = u^4/8 + c = (1/8)(9+x^2)^4 + c.$$

5.   Let $u = 1+4t^2$ so $tdt = du/8$ and

$$\int \frac{t\,dt}{(1+4t^2)^3} = \frac{1}{8}\int u^{-3}du = \left(\frac{1}{8}\right)(u^{-2}/-2) + c = -(1/16)(1+4t^2)^{-2} + c.$$

7.   Let $u = 4-x^4$ so $x^3 dx = -du/4$ and

$$\int x^3(4-x^4)^3 dx = -\int u^3 du/4 = -(1/4)(u^4/4) + c = -(4-x^4)^4/16 + c.$$

9.   Let $u = 3(x-\pi/4)$ so $dx = du/3$ and

$$\int \cos 3(x-\pi/4)\,dx = \int \cos u\,du/3 = (1/3)\sin u + c$$

$$= (1/3)\sin 3(x-\pi/4) + c.$$

11.   Let $u = \pi\theta$ then $d\theta = du/\pi$ and

$$\int \tan\pi\theta\sec\pi\theta \; d\theta = \int \tan u \sec u \; du/\pi = (1/\pi)\sec u + c$$
$$= (1/\pi)\sec\pi\theta + c.$$

13.   Let $u = \cos 2x$ so $\sin 2x \, dx = -du/2$ and

$$\int \cos^3 2x \sin 2x \, dx = -(1/2)\int u^3 \, du = -(1/2)u^4/4 + c = -(1/8)\cos^4 2x + c.$$

15.   Let $u = 1 + \sqrt{x}$. Thus $du = dx/2\sqrt{x}$ so $dx/\sqrt{x} = 2du$ and

$$\int \frac{dx}{\sqrt{x}(1+\sqrt{x})^2} = \int \frac{2du}{u^2} = \int 2u^{-2} du = -2u^{-1} + c = -2(1+\sqrt{x})^{-1} + c.$$

17.   Let $u = 2 + x^2$ so $x \, dx = du/2$ and $x^2 = u - 2$. Thus

$$\int x^5 (2+x^2)^{1/2} dx = \int x^2 \cdot x^2 (2+x^2)^{1/2}(x) \, dx = \int (u-2)(u-2)u^{1/2} \; du/2 =$$
$$\int (u^2-4u+4)u^{1/2}du/2 = (1/2)\int (u^{5/2} - 4u^{3/2} + 4u^{1/2}) \, du =$$
$$\left(\frac{1}{2}\right)\left(\frac{u^{7/2}}{(7/2)} - \frac{4u^{5/2}}{(5/2)} + \frac{4u^{3/2}}{3/2}\right) + c =$$
$$(1/7)(2+x^2)^{7/2} - (4/5)(2+x^2)^{5/2} + (4/3)(2+x^2)^{3/2} + c.$$

19.   Let $u = a^2 + x^2$ then $x \, dx = du/2$ and

$$\int x(a^2+x^2)^r dx = \int u^r du/2 = u^{r+1}/2(r+1) + c = (a^2+x^2)^{r+1}/2(r+1) + c.$$

21.   Compute $\int \sqrt{4+3x} \, dx$ first by letting $u = 4+3x$, $dx = du/3$, so

$$\int \sqrt{4+3x} \, dx = \left(\frac{1}{3}\right)\int \sqrt{u} \, du = 2u^{3/2}/9 + c = (2/9)(4+3x)^{3/2} + c.$$

Thus $\displaystyle\int_0^4 \sqrt{4+3x} \, dx = \left(\frac{2}{9}\right)(4+3x)^{3/2} \Big|_0^4 = (2/9)[16^{3/2} - 4^{3/2}] = 112/9.$

Note that the constant of integration, c, will subtract out
and thus is ommitted from the last computation.

23.　　Let $u = 16 - x^2$ so　$du = -2xdx$ or $xdx = -du/2$ and

$$\int x\sqrt{16-x^2}\, dx = -\left(\frac{1}{2}\right)\int u^{1/2}\, du = -(1/3)u^{3/2} = -(1/3)(16-x^2)^{3/2}\, .$$

Thus $\int_0^4 x\sqrt{16-x^2}\, dx = -(1/3)(16-x^2)^{3/2}\Big|_0^4 = -(1/3)[0-64] = 64/3.$

25.　　Let $u = \cot\theta$ so $-du = \csc^2\theta\, d\theta$　and

$$\int \cot\theta\,\csc^2\theta\, d\theta = -\int u\, du = (-1/2)u^2 + c = (-1/2)\cot^2\theta + c\, .$$

Hence $\int_{\pi/4}^{\pi/2} \cot\theta\,\csc^2\theta\, d\theta = (-1/2)\cot^2\theta\,\Big|_{\pi/4}^{\pi/2} = 1/2.$

27.　　Let $u = 4 + s^{1/2}$ then $s = 1 \Rightarrow u = 5$, $s = 4 \Rightarrow u = 6$ and
　　　$2du = ds/\sqrt{s}$.　Thus

$$\int_1^4 \frac{ds}{\sqrt{s}(4+\sqrt{s})^3} = 2\int_5^6 u^{-3}\, du = -u^{-2}\Big|_5^6 = -1/36 + 1/25 \cong 0.0122.$$

29.　　Let $u = a^2 + x^2$ so $xdx = du/2$, $x = 0 \Rightarrow u = a^2$, $x = a \Rightarrow u = a^2$. Thus

$$\int_0^a x(a^2 + x^2)^{-1/2}\, dx = (1/2)\int_{a^2}^{2a^2} u^{-1/2}\, du = u^{1/2}\Big|_{a^2}^{2a^2}$$

$$= \sqrt{2a^2} - \sqrt{a^2} = (\sqrt{2} - 1)|a|\, .$$

31a.　　If $x = a\sin\theta$　then $dx = a\cos\theta d\theta$ and for $0 \leq x \leq a$　we have
　　　$0 \leq \theta \leq \pi/2$.　Thus

$$\int_0^a \sqrt{a^2 - x^2}\, dx = \int_0^{\pi/2} \sqrt{a^2 - a^2\sin^2\theta}\ a\cos\theta d\theta =$$

$$a^2\int_0^{\pi/2} \sqrt{\cos^2\theta}\ \cos\theta\, d\theta = a^2\int_0^{\pi/2} \cos^2\theta\, d\theta.\quad \text{(Note for } 0 \leq \theta \leq \pi/2,$$

$\cos\theta \geq 0$ so $\sqrt{\cos^2\theta} = |\cos\theta| = \cos\theta).$

　b.　　Since $\cos^2\theta = (1+\cos2\theta)/2$, $\int_0^a \sqrt{a^2 - x^2}\, dx = \frac{a^2}{2}\int_0^{\pi/2} (1+\cos2\theta)\, d\theta.$

31c.   Now let $t = 2\theta$ so $d\theta = dt/2$ and for $0 \le \theta \le \pi/2$, $0 \le t \le \pi$.

Thus $\int_0^a \sqrt{a^2-x^2}\,dx = \frac{a^2}{4}\int_0^\pi (1+\cos t)\,dt = \frac{a^2}{4}(t+\sin t)\Big|_0^\pi = a^2\pi/4$.

33.   Let $x = a\sin\theta$, then $dx = a\cos\theta\,d\theta$, $a^2 - x^2 = a^2\cos^2\theta$, and for

$0 \le x \le a/2$, $0 \le \theta \le \pi/6$. Thus $\int_0^{a/2} \dfrac{dx}{(a^2-x^2)^{3/2}} =$

$\int_0^{\pi/6} \dfrac{a\cos\theta}{a^3\cos^3\theta}\,d\theta = \dfrac{1}{a^2}\int_0^{\pi/6} \sec^2\theta\,d\theta = (1/a^2)\tan\theta\,\Big|_0^{\pi/6} = \dfrac{1}{\sqrt{3}a^2}$.

35a.   We have

$$\int_{-a}^a f(x)\,dx = \int_{-a}^0 f(x)\,dx + \int_0^a f(x)\,dx = -\int_0^{-a} f(x)\,dx + \int_0^a f(x)\,dx.$$

In the first integral, let $x = -t$ so $dx = -dt$ and for $-a \le x \le 0$ we see that $0 \le t \le a$. Thus

$$\int_{-a}^a f(x)\,dx = -\int_0^a -f(-t)\,dt + \int_0^a f(x)\,dx = \int_0^a f(-t)\,dt + \int_0^a f(x)\,dx.$$

For an even function $f(-t) = f(t)$ and hence

$$\int_{-a}^a f(x)\,dx = \int_0^a f(t)\,dt + \int_0^a f(x)\,dx = 2\int_0^a f(x)\,dx.$$

b.   The development here is exactly the same until the last step, where now, f being odd, we have $f(-t) = -f(t)$ so

$$\int_{-a}^a f(x)\,dx = -\int_0^a f(t)\,dt + \int_0^a f(x)\,dx = 0.$$

Section 6.6, Page 334

1.   Let $h = (1-0)/n$ and thus $x_i = i/n$ and $y_i = f(x_i)$,

where $f(x) = x^{1/2}$. With these relationships Eqs.(5) and (16) yield the values appearing in the solutions section of the text. The values for the trapezoidal rule agree with those of $(s_n + \sigma_n)/2$ of Prob.15, Section 6.2.

The actual value is $\int_0^1 x^{1/2}\,dx = (2/3)x^{3/2}\Big|_0^1 = (2/3)$.

3.   Let $h = 2/n$ and thus $x_i = 2i/n$ and $y_i = 1/(1+x_i^2)$. Eqs.(5)
      and (16) then yield the values appearing in the solution
      section of the text. The trapezoidal approximation agrees
      with the average of the upper and lower sums of Prob.16,
      Section 6.2.   It can be shown that S_n, for $n \geq 20$, is the
      exact value to six decimal places.

5.   Let $h = [2 - (1/2)]/n = 3/2n$. Then $x_i = 1/2 + 3i/2n$ and
      $y_i = x_i - 1/x_i$, which when used in Eqs.(5) and (16) yield the
      results in the solution section. Again the trapezoidal rule
      agrees with the average of the upper and lower sums of
      Prob.20 in Section 6.2.

7. $h = \pi/2n$ so $x_i = \pi i/2n$ and $y_i = (x_i)^{1/2} \sin 2x_i$, which yield,
      by Eqs.(5) and (16), the results given in the solutions
      section. These results can be compared to those of Prob.22
      of Section 6.2.

9.   With $h = 1/n$, $x_i = i/n$ and $y_i = (x_i)^2 \sin x_i$, Eqs.(5) and (16)
      give the results in the solution section of the text.

11.   With $h = \pi/n$, we find $x_i = i\pi/n$ and $y_i = [\sin(x_i/2)]^{1/2}$. The
      results in the solution section of the text are then found
      using Eqs.(5) and (16).

13.   Letting $h = 2/n$, $x_i = 2i/n$ and $y_i = (1 + 2x_i)^{1/2}$ the desired
      results are found as in the above problems.

15.   With $h = \pi/n$, $\theta_i = \pi i/n$ and $y_i = \cos(\theta_i - 0.5\sin\theta_i)$ we
      obtain the desired results as in the above problems.

17a. From Eq.(9), the magnitude of the error is given by:
      $$|E_n| = |f''(\zeta)| (b-a)^3/12n^2 \leq M_2 (b-a)^3/12n^2.$$

      Since $f(x) = x - 1/x$, then $f''(x) = -2/x^3$. On the interval
      $1/2 \leq x \leq 2$, $|f''(x)| \leq 16$ and $b-a = 3/2$ so that
      $|E_n| \leq 16(3/2)^3/12n^2 < 10^{-4}$ or $|E_n| \leq 9/2n^2 < 10^{-4}$ when

      $n > \dfrac{300}{\sqrt{2}} \cong 212.132$. Thus for $n \geq 213$ the error in the

      trapezoidal approximation is less than 10^{-4}.

17b.   From Eq.(19), the magnitude of the error is bounded by:
$|E_n| \leq M_4(b-a)^5/180n^4$ where $|f^{iv}(x)| \leq M_4$ for all x in
[1/2,2]. Since $f^{iv}(x) = -24/x^5$, $M_4 = (3)2^8$ so that

$$|E_n| \leq \frac{3 \cdot 2^8(3/2)^5}{5 \cdot 3^2 2^2 \, n^4} = \frac{2 \cdot 3^4}{5n^4} \leq \frac{1}{10^4} \text{ for } n \geq 30(2/5)^{1/4}. \text{ Since}$$

$30(2/5)^{1/4} \cong 23.85$ we have that for $n \geq 24$ Simpson's
approximation will have an error bounded by 10^{-4}. Note that
in Prob.5, for $n > 40$ the change takes place in the sixth
decimal place.

19a.   We have $f''(x) = -(1+2x)^{-3/2}$ so $|f''(x)| \leq 1$ for all x in
[0,2]. Thus $M_2 = 1$ and
$$|E_n| \leq (b-a)^3/12n^2 = 2^3/12n^2 = 2/3n^2 < 10^{-4} \text{ for}$$

$n^2 \geq \dfrac{2 \times 10^4}{3}$ or $n \geq 100\sqrt{2/3} \cong 81.65$. Thus $|E_n| \leq 10^{-4}$ for $n \geq 82$.

  b.   Since $f^{iv}(x) = -5(3)(1+2x)^{-7/2}$ we have $|f^{iv}(x)| \leq 15$ for all x
in [0,2] so that $M_4 = 15$ and $|E_n| \leq 3 \cdot 5 \cdot 2^5/2^2 \cdot 3^2 \cdot 5n^4 = 2^3/3n^4 < 10^{-4}$
for $n^4 \geq (8/3)10^4 \cong 12.78$ or $n \geq 14$ since n must be even.

21.   Note that the integrand, $p(x)-1$, is simply the decimal part
of $p(x)$. Now choose $h = 1/12$ and let $y_i = p_i - 1$ and thus

W/BLpa $\cong (h/3)(y_0 + 4y_1 + 2y_2 + \cdots + 2y_{10} + 4y_{11} + y_{12})$
   $= (1/36)[0+4(.035)+2(.072)+ \ldots +2(.411)+4(.355)+0]$
   $= (1/36)(8.106000) = .2252$. Thus W = .2252 BLpa.

23.   The trapezoidal rule yields:
$T_n = (h/2)[y_0 + 2y_1 + 2y_2 + \cdots + 2y_{n-1} + y_n]$
   $= (h/2)[f(x_0)+2f(x_1)+2f(x_2)+\cdots+2f(x_{n-1})+f(x_n)]$
   $= h[f(x_1)+f(x_2)+\cdots+f(x_{n-1})] + (h/2)[f(x_0)+f(x_n)]$. Now add
$hf(x_n)$ to the first term and subtract the same from the
second term to obtain:
$T_n = h[f(x_1)+f(x_2)+\cdots+f(x_{n-1})+f(x_n)] + (h/2)[f(x_0)-f(x_n)]$

$$= \sum_{i=1}^{n} f(x_i)\Delta x_i + \frac{1}{2}[f(a) - f(b)]h, \text{ where } \Delta x_i = h, a = x_0, \text{ and } b = x_n.$$

25a.   With $f(x) = \sin x$ we have $f'(x) = \cos x$, $a = 0$, $b = \pi/2$,
$f'(b)-f'(a) = f'(\pi/2)-f'(0) = -1$, $b-a = \pi/2$ so $E_n \cong \pi^2/48n^2$.
The values for the bound from Table 6.5 and the estimated
error are:

n	Bound in Table 6.5	Error Estimate
10	0.0032298	0.0020562
20	0.0008075	0.0005410
40	0.0002019	0.0001285
80	0.0000505	0.0000321

Also, for all but $n = 10$, the approximate error agrees with
the actual error shown in the first column of Table 6.5.

b.   We have $f'(x) = -2x/(2+x^2)^2$, $b-a = 2$, $f'(2)-f'(0) = -1/9$ and
thus $E_n \cong 4/(108n^2)$. For E_n to be "less" than 5×10^{-4}, we
must choose n^2 so that $1/(27n^2) \le 5 \times 10^{-4}$ or
$n^2 \ge 10^4/(5)(27)$ or $n \ge 9$. For n to be even, choose
$n \ge 10$, a vast improvement over the 58 found in Ex.3.

Chapter 6 Review, Page 336

1. $\displaystyle\sum_{j=1}^{5} \frac{1}{j^2+1} = \frac{1}{1+1} + \frac{1}{4+1} + \frac{1}{9+1} + \frac{1}{16+1} + \frac{1}{25+1} = \frac{1983}{2210}$.

3. $\displaystyle\sum_{k=1}^{10} \frac{1}{(k+2)(k+3)} = \sum_{k=1}^{10}\left(\frac{1}{k+2} - \frac{1}{k+3}\right) = 1/3 - 1/13 = 10/39$,
since the sum is telescoping, as in Prob.5 Section 6.1.

5. $\displaystyle\sum_{k=0}^{52} (\sqrt{2k+1} - \sqrt{2k+3}) = -(\sqrt{2(52)+3} - \sqrt{1}) = 1-\sqrt{107}$ by

Eq.(12), Section 6.1.

7. $\displaystyle\sum_{n=1}^{32} (2n^2-5n-3) = 2\sum_{n=1}^{32} n^2 - 5\sum_{n=1}^{32} n - \sum_{n=1}^{32} 3$

$$= 2\frac{(32)(33)(65)}{6} - 5\frac{32(33)}{2} - 3(32) = 20,144.$$

9.   The sum is $\displaystyle\sum_{n=1}^{9} \sin[(n+2)x + (2n-1)]$, since the coefficients
     of x increase by ones and the numbers added are all odd.

11. We can write the sum as $\displaystyle\sum_{n=1}^{\infty} (-1)^{n-1}\frac{2^{n-1}}{nx^2} = \sum_{n=1}^{\infty}\frac{(-2)^{n-1}}{nx^2}$, since
     the terms alternate in sign.

13. Since (x^2+3) is increasing on $[1,6]$,

     $\sigma = (1.5^2+3)(.5) + (2^2+3)(.5) + (3^2+3)(1) + (4.5^2+3)(1.5)$

     $\quad + (4.75^2+3)(.25) + (5.5^2+3)(.75) + (6^2+3)(.5) = 103.828125$,

     $s = (1^2+3)(.5) + (1.5^2+3)(.5) + (2^2+3)(1) + (3^2+3)(1.5)$

     $\quad + (4.5^2+3)(.25) + (4.75^2+3)(.75) + (5.5^2+3)(.5) = 71.234375$,

     $S = (1.25^2+3)(.5) + (1.75^2+3)(.5) + (2.5^2+3)(1) + (3.75^2+3)(1.5)$

     $\quad + (4.625^2+3)(.25) + (5.125^2+37(.75) + (5.75^2+3)(.5) = 86.234375$.

15. Since x^3 increases on $[-3,3]$ and $\Delta x_i = 1$ for all i,

     $\sigma = (-2)^3 + (-1)^3 + (0)^3 + (1)^3 + (2)^3 + (3)^3 = 27$,

     $s = (-3)^3 + (-2)^3 + (-1)^3 + (0)^3 + (1)^3 + (2)^3 = -27$, and

     $S = (-2.5)^3 + (-1.5)^3 + (.5)^3 + (.5)^3 + (1.5)^3 + (2.5)^3 = 0$.

17. Since the function is not increasing or decreasing over the
     entire interval $[-2,5]$, we must find all relative maxima and
     minima for the computation of σ and s. By the techniques
     developed in Chapter 4 , we find the maximum on $(-2,5)$ is at
     $1-\sqrt{3}$, while the minimum is at $1+\sqrt{3}$, the former on the
     interval $[-1,0]$, the latter on $[2,3]$. Thus,
     $\sigma = 10+6\sqrt{3} +8+0+(-8)+0+28 = 38+6\sqrt{3} \cong 48.39230$,
     $s = 0+8+0+(-8)+(-6\sqrt{3})+(-10)+0 = -10-6\sqrt{3} \cong -20.39230$, and
     $S = 6.875+10.125+4.375-4.375-10.125-6.875+11.375 = 11.375$.

19. $\displaystyle I = \int_0^2 dx + 4\int_0^2 x\,dx + 4\int_0^2 x^2\,dx - \int_0^2 \sqrt{4-x^2}\,dx$

     $= 2 + 8 + 32/3 - \pi = 62/3 - \pi$,   by Exs.1,2,3 and 6 of
     Section 6.3 and Theorem 6.3.2.

21. $\displaystyle I = \int_0^5 \sqrt{25-x^2}\,dx + 9\int_5^{10} x^2\,dx - 12\int_5^{10} x\,dx + 4\int_5^{10} dx$

     $= 25\pi/4 + 2195$, as in Prob.20.

23. Since $\sqrt{13} \le \sqrt{x^2+9} \le \sqrt{109}$ for $2 \le x \le 10$, then

$\dfrac{1}{\sqrt{109}} \le \dfrac{1}{\sqrt{x^2+9}} \le \dfrac{1}{\sqrt{13}}$. Corollary 1 of Theorem 6.3.4,

then gives $\dfrac{8}{\sqrt{109}} = \displaystyle\int_2^{10} \dfrac{dx}{\sqrt{109}} \le \int_2^{10} \dfrac{dx}{\sqrt{x^2+9}} \le \int_2^{10} \dfrac{dx}{\sqrt{13}} = \dfrac{8}{\sqrt{13}}$.

25. avg $= [1/(5\pi/2 - 2\pi)] \displaystyle\int_{2\pi}^{5\pi/2} \sin 2x\, dx = -\left. \dfrac{\cos 2x}{\pi}\right|_{2\pi}^{5\pi/2} = 2/\pi$,

by Theorem 6.3.5 and Eq.(20) of Section 6.4.

27. avg $= (1/4)\displaystyle\int_0^4 \sqrt{16-x^2}\, dx = (1/4)(16\pi/4) = \pi$, as the integral

represents the area of a quarter circle of radius of 4.

29. $I = \displaystyle\int_{-5}^0 (x^2+2x)\, dx + \int_0^5 \sqrt{50-2x^2}\, dx + \int_5^{10} x\, dx$

$= (x^3/3 + x^2)\Big|_{-5}^0 + \sqrt{2}(25\pi/4) + x^2/2 \Big|_5^{10}$

$= 125/3 - 25 + 25\pi\sqrt{2}/4 + 50 - 25/2 = 325/6 + 25\pi\sqrt{2}/4$.

31. Using the identity $\sin^2 A = (1-\cos 2A)/2$,

$I = (1/2)\displaystyle\int_{\pi/2}^{7\pi/12} (1-\cos 6x)\, dx = (1/2)[x - (1/6)\sin 6x]\Big|_{\pi/2}^{7\pi/12}$

$= (1/2)[(\pi/12) - (1/6)(-1)] = (\pi+2)/24$.

33. $I = (1/2)\displaystyle\int_a^x [1-\cos(2t+\pi/2)]\, dt = (1/2)\int_a^x (1+\sin 2t)\, dt$

$= (1/2)[t-(1/2)\cos 2t]\Big|_a^x = (x-a)/2 + (\cos 2a - \cos 2x)/2$.

35. Let $u = \sin(3t^2)$, so $t\cos(3t^2)\, dt = du/6$ and

$\int t\sin^2(3t^2)\cos(3t^2)\, dt = (1/6)\int u^2\, du = (1/18)u^3$.

Thus $I = (1/18)\sin^3(3t^2)\Big|_{-\sqrt{\pi}/2}^{2\sqrt{\pi}/3}$

$= (1/18)[(-\sqrt{3}/2)^3 - (\sqrt{2}/2)^3] = -(3\sqrt{3}+2\sqrt{2})/144$.

37. Since $\sin(A+\pi) = -\sin A$, let $u = t^{1/2}$ so $0 \le t \le \pi^2/4$ becomes

$0 \le u \le \pi/2$ and $t^{-1/2}dt = 2du$. Thus $I = -2\displaystyle\int_0^{\pi/2} \sin u\, du = -2$.

39. $u = 1-4t$, $-1/2 \le t \le 0$ becomes $3 \ge u \ge 1$, and $dt = (-1/4)du$.

Thus $I = (-1/4)\displaystyle\int_3^1 u^{-1/2}\,du = (1/4)\int_1^3 u^{-1/2}\,du = (\sqrt{3} - 1)/2$.

41. Expanding the integrand gives $I = \displaystyle\int_0^1 (t^2 + 2t^{7/2} + t^5)\,dt = 17/18$.

43. Noting that $(1-t^6) = (1-t^3)(1+t^3) = (1-t^{3/2})(1+t^{3/2})(1+t^3)$,

we have $I = \displaystyle\int_0^1 (1-t^{3/2})(1+t^3)\,dt = \int_0^1 (1-t^{3/2}+t^3-t^{9/2})\,dt = 147/220$.

45. Let $u = 1+t$ so $I = \displaystyle\int_0^4 (u-1)^2 u^{1/2}\,du = \int_0^4 (u^{5/2} - 2u^{3/2} + u^{1/2})\,du$

$$= (2/7)u^{7/2} - (4/5)u^{5/2} + (2/3)u^{3/2}\Big|_0^4 = 1712/105.$$

47. If $u = x^3/2 + 1$, then $2x^2\,dx = (4/3)du$ and

$$I = (4/3)\int_{a^3/2+1}^{b^3/2+1} u^{1/2}\,du = (8/9)[(b^3/2 + 1)^{3/2} - (a^3/2 + 1)^{3/2}].$$

49. Let $u = x^2 + 2\pi x + \pi^2$, so $(x+\pi)\,dx = du/2$ and

$$I = (1/2)\int_{\pi^2}^{9\pi^2/4} \cos u\,du = [\sin(9\pi^2/4) - \sin(\pi^2)]/2.$$

51. From Eq.(9), Section 6.6, with $f''(x) = -4x^2\sin x^2 + 2\cos x^2$,
$|E_n| \le (4\pi^2+2)\pi^3/12n^2 \le 10^{-3}$ for $n \ge 328$. For Simpson's rule,
and $f^{(iv)}(x) = 16x^4\sin x^2 - 12\sin x^2 - 48x^2\sin x^2$, we have that
$|f^{(iv)}(x)| \le 16\pi^4+48\pi^2+12$ for $0 \le x \le \pi$. Thus, by Eq.(18) of
Section 6.6, $|E_n| \le \dfrac{(16\pi^4 + 48\pi^2 + 12)\pi^5}{180n^4} \le 10^{-3}$ for $n \ge 44$.

53.   We have $h = 2\pi/n$, so $x_i = -\pi + (2\pi i/n)$ and $y_i = f(x_i)$ where
        $f(x) = x^2 \sin^2 x$. Eqs. (5) and (16) of Section 6.6 then give
        the values found in the answer section of the text.

55.   With $h = 10/n$, we have $x_i = 10i/n$ and
        $y_i = f(x_i) = (x_i)^{3/2} \cos^2 x_i$. The values for the trapezoidal
        and Simpson's rules, as found by Eqs. (5) and (16)
        Section 6.6, are given in the text.

CHAPTER 7

Section 7.1, Page 348

1. The curve $y = 2x + x^2/2$ is a parabola with vertex at $(-2,-2)$
    and opening up. Thus the desired area is given by
    $$\int_0^2 (2x+x^2/2)\,dx = (x^2+x^3/6)\Big|_0^2 = 16/3.$$

3. Since $y = \sin x$ is positive in $(0,\pi)$ and intersects the x
    axis at $x = 0$ and $x = \pi$, the desired area is given by
    $$\int_0^\pi \sin x\,dx = -\cos x\Big|_0^\pi = 2.$$

5. The two curves intersect when $-3 = 1-x^2$ or $x = \pm 2$. Since
    $y = 1-x^2$ lies above $y = -3$ for $(-2,2)$, the desired area is
    then
    $$\int_{-2}^2 [(1-x^2)-(-3)]\,dx = \int_{-2}^2 (4-x^2)\,dx = 32/3.$$

7. The desired area is shown, where the
    point of intersection in the second
    quadrant is given by $-x = 2-x^2$, or
    $x = -1$. Thus the area is
    $$\int_{-1}^0 [(2-x^2)-(-x)]\,dx + \int_0^{\sqrt{2}} (2-x^2)\,dx =$$
    $$(2x+x^2/2 - x^3/3)\Big|_{-1}^0 + (2x-x^3/3)\Big|_0^{\sqrt{2}} = (7+8\sqrt{2})/6.$$

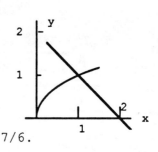

9. The two curves intersect when $\sqrt{x} = 2-x$,
    or $x = 1$. Thus the area is given by
    $$\int_0^1 \sqrt{x}\,dx + \int_1^2 (2-x)\,dx \text{ or by } \int_0^1 [(2-y)-y^2]\,dy.$$
    The latter integral is $(2y-y^2/2-y^3/3)\Big|_0^1 = 7/6$.

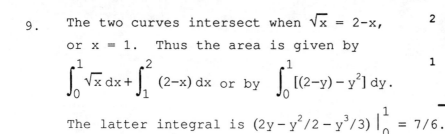

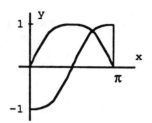

11. Since $\sin(x-\pi/2) = -\cos x$ we find that the two curves intersect at $x = 3\pi/4$ and thus the desired area is

$$\int_0^{3\pi/4} [\sin x - (-\cos x)]\,dx + \int_{3\pi/4}^{\pi} (-\cos x - \sin x)\,dx =$$

$$(-\cos x + \sin x)\,\Big|_0^{3\pi/4} + (-\sin x + \cos x)\,\Big|_{3\pi/4}^{\pi} = 2\sqrt{2}.$$

13. The two curves intersect at $x = 1$ and thus the desired area

is $\displaystyle\int_1^3 (x - 1/x^2)\,dx = (x^2/2 + 1/x)\,\Big|_1^3 = 10/3.$

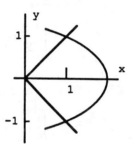

15. The curves intersect in the first quadrant at $y = 2-y^2$, or $y = 1$ and thus

the area is given by $\displaystyle 2\int_0^1 [(2-y^2) - y]\,dy$,

where symmetry has been used. This

integral has the value $2\,(2y - y^3/3 - y^2/2)\,\Big|_0^1 = 7/3.$

17. The two curves intersect at $x = \pi/4$ and $x = 5\pi/4$, and thus the area is given by

$$A = \int_0^{\pi/4} (\cos x - \sin x)\,dx + \int_{\pi/4}^{5\pi/4} (\sin x - \cos x)\,dx + \int_{5\pi/4}^{\pi} (\cos x - \sin x)\,dx$$

$$= 4\sqrt{2}.$$

19. Using elements parallel to the y axis and symmetry we find

$$A = 2\int_0^1 \sqrt{x}\,dx = (4/3)\,x^{3/2}\,\Big|_0^1 = 4/3.$$

Using elements parallel to the x axis and symmetry we find

$$A = 2\int_0^1 (1 - y^2)\,dy = 2\,(y - y^3/3)\,\Big|_0^1 = 4/3.$$

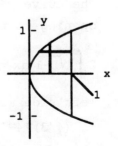

21.   The two curves intersect when $x^4 = 2-x^4$, or $x = 1$. Thus

$$A = \int_0^1 [(2-x^4) - x^4]\,dx = (2x - 2x^5/5)\Big|_0^1 = 8/5 \quad \text{and}$$

$$A = \int_0^1 y^{1/4}\,dy + \int_1^2 (2-y)^{1/4}\,dy = (4/5)y^{5/4}\Big|_0^1 - (4/5)(2-y)^{5/4}\Big|_1^2 = 8/5.$$

Using symmetry on the latter approach also yields

$$A = 2\int_0^1 y^{1/4}\,dy = 8/5.$$

23.   The two curves intersect at $2-x = x^3$, which has a root at
$x = 1$.   Thus $A = \int_0^1 [(2-x) - x^3]\,dx = 5/4$ and

$$A = \int_0^1 y^{1/3}\,dy + \int_1^2 (2-y)\,dy = 5/4.$$

25.   The two curves intersect when $\sqrt{4-x^2} = \sqrt{3}\,x$, or $x = 1$.   Thus

$$A = \int_0^1 \left(\sqrt{4-x^2} - \sqrt{3}\,x\right)dx.$$

27.   The two circles intersect at $(0,a)$ and $(a,0)$.   Thus

$$A = \int_0^a \left[\sqrt{a^2-x^2} - \left(a - \sqrt{a^2 - (x-a)^2}\right)\right]dx =$$

$$\int_0^a \left[\sqrt{a^2-x^2} - \left(a - \sqrt{2ax - x^2}\right)\right]dx.$$

29.   The curve $y^2 = x-1$ is a parabola with vertex at $(1,0)$ and
opening to the right.   Thus we must have

$$2\int_1^a \sqrt{x-1}\,dx = 2\int_a^5 \sqrt{x-1}\,dx, \quad \text{using symmetry.}$$ The value of the

left side is $(4/3)(a-1)^{3/2}$ and the value of the right side
is $(4/3)[4^{3/2} - (a-1)^{3/2}]$.   Equating these two and solving
for a yields $a = 4^{2/3} + 1 \cong 3.52$.

Section 7.2, Page 359

1.   Using Eq. (3) with $f = 2x^{1/4}$ we have

$$V = \pi \int_0^4 (2x^{1/4})^2 dx = 4\pi \int_0^4 x^{1/2} dx = (8/3) \pi x^{3/2} \Big|_0^4 = 64\pi/3.$$

3.   Using horizontal slices, we see that
     the radius of a slice is $2-x = 2 - \sqrt{y}$.
     Thus we have $V =$

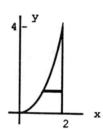

$$\pi \int_0^4 (2 - \sqrt{y})^2 dy = \pi (4y - 8y^{3/2}/3 + y^2/2) \Big|_0^4 = 8\pi/3.$$

5. $$V = \pi \int_0^{\pi/2} \cos^2 x \, dx = (\pi/2) \int_0^{\pi/2} (1 + \cos 2x) \, dx =$$

$$(\pi/2) [x + (\sin 2x)/2] \Big|_0^{\pi/2} = \pi^2/4.$$

7.   Using horizontal slices, we see
     that $A(y) = \pi (r_0{}^2 - r_i{}^2)$, where
     $r_0 = 4$ and $r_i = x = y^2$.
     Thus we have

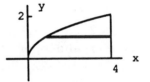

$$V = \pi \int_0^2 (4^2 - y^4) \, dy = \pi (16y - y^5/5) \Big|_0^2 = 128\pi/5.$$ We may also use the

method of shells in which case Eq. (8) holds and

$$V = 2\pi \int_0^4 x \sqrt{x} \, dx.$$

9.   Using vertical slices we have $A(x) = \pi(x^2 - x^4)$, so

$$V = \pi \int_0^1 (x^2 - x^4) \, dx = 2\pi/15.$$

11. We have $r_0 = 1 - x_0 = 1 - y$ and $r_i = 1 - x_i = 1 - \sqrt{y}$ and thus

$$V = \pi \int_0^1 [(1 - y)^2 - (1 - \sqrt{y})^2] \, dy = \pi \int_0^1 (y^2 + 2\sqrt{y} - 3y) \, dy = \pi/6.$$

13. $V = \pi \int_0^{\pi/6} (2/\cos 2x)^2 \, dx = 4\pi \int_0^{\pi/6} \sec^2 2x \, dx = 4\pi (\tan 2x)/2 \Big|_0^{\pi/6} = 2\pi\sqrt{3}$.

15.   We have $r_0 = 6-x_0 = 6-y^2$ and
      $r_i = 6-x_i = 6-4 = 2$.   Thus

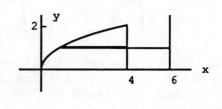

$$V = \pi \int_0^2 [(6-y^2)^2 - 4] \, dy$$

$$= \pi \int_0^2 (y^4 - 12y^2 + 32) \, dy = 192\pi/5 .$$

17.   Using Eq. (8) we have $V = 2\pi \int_0^2 x(x^2) \, dx = 8\pi$.

19.   The height of the cylindrical shell is $2x - x^2$ and thus
      Eq. (8) yields $V = 2\pi \int_0^2 x(2x - x^2) \, dx = 2\pi \int_0^2 (2x^2 - x^3) \, dx = 8\pi/3$.

21. $y = x^2 - 2x$ is a parabola with vertex at
      $(1,-1)$ and opening up. It intersects $y = 3x$
      at $3x = x^2 - 2x$, or $x = 0, 5$.   Using
      cylindrical shells we have

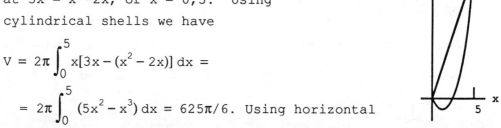

$$V = 2\pi \int_0^5 x[3x - (x^2 - 2x)] \, dx =$$

$$= 2\pi \int_0^5 (5x^2 - x^3) \, dx = 625\pi/6. \text{ Using horizontal}$$

slices, for $-1<y<0$, we have $r_0 = 1 + \sqrt{y+1}$ and
$r_i = 1 - \sqrt{y+1}$ and for $0<y<15$, we have $r_0 = 1 + \sqrt{y+1}$ and $r_i = y/3$.
Thus $V = \pi \int_{-1}^0 [(1 + \sqrt{y+1})^2 - (1 - \sqrt{y+1})^2] \, dy + \pi \int_0^{15} [(1 + \sqrt{y+1})^2 - y^2/9] \, dy$.

23.   Using cylindrical shells we have
      $$V = 2\pi \int_0^4 x[3x - x^2 - (-x)] \, dx = 2\pi \int_0^4 (4x^2 - x^3) \, dx = 128\pi/3 .$$

25. Since the base of the equilateral triangle, which forms the
     cross sectional area, is $2\sqrt{a^2-x^2}$ the height is $\sqrt{3}\sqrt{a^2-x^2}$
     and thus $A(x) = (1/2)\left(2\sqrt{a^2-x^2}\right)\left(\sqrt{3}\sqrt{a^2-x^2}\right) = \sqrt{3}(a^2-x^2)$. Thus
     Eq.(4) yields $V = 2\sqrt{3}\int_0^a (a^2-x^2)\,dx = 4\sqrt{3}\,a^3/3$.

27. The hemispherical bowl is obtained by revolving the quarter
     circle $y = -\sqrt{a^2-x^2}$, $0 \le x \le a$, about the y-axis. Note that
     the water will be between $y = -a$ and $y = -a+h$. Using
     horizontal slices, we then find the volume, v, of water to be
     $$v = \pi\int_{-a}^{-a+h} (a^2-y^2)\,dy = \pi\,(a^2 y - y^3/3)\,\Big|_{-a}^{-a+h} =$$
     $$\pi[a^2(h-a)-(h-a)^3/3 + 2a^3/3] = \pi h^2(a-h/3).$$

29. As seen in Fig.7.2.13, cutting the wedge with planes
     perpendicular to the x-axis will yield triangular
     crossectional areas of base y and height h, where
     $\tan\theta = h/y$. Thus the crossectional area is
     $hy/2 = y^2\tan\theta/2$, where $x^2+y^2 = a^2$. Therefore
     $$V = (1/2)\int_{-a}^a (a^2 - x^2)\tan\theta\,dx$$
     $$= (1/2)\tan\theta\,(a^2 x - x^3/3)\,\Big|_{-a}^a = (2/3)a^3\tan\theta.$$

31. From Fig.7.2.15, the volume is generated by revolving the
     area bounded by $x = 0$, $x = h$, $y = \left(\dfrac{r_2-r_1}{h}\right)x + r_1$ and $y = 0$
     about the x-axis. Setting $(r_2 - r_1)/h = \alpha$ we have
     $$V = \pi\int_0^h (\alpha x + r_1)^2\,dx =$$
     $$\pi\,(\alpha^2 x^3/3 + 2\alpha r_1 x^2/2 + r_1^2 x)\,\Big|_0^h = \pi h\,(2\alpha^2 h^2 + 6\alpha r_1 h + 6r_1^2)/6 =$$
     $(B_0+4M+B_1)h/6$, where $B_0 = \pi r_1^2$, $B_1 = \pi r_2^2$, and $M = \pi(r_1+r_2)^2/4$.

Section 7.3, Page 369

1. We have $y' = 3$ and thus Eq.(6) yields

$$L = \int_0^3 \sqrt{1+3^2} \, dx = 3\sqrt{10} \, .$$

3. We have $y' = 3x^{1/2}$ and thus

$$L = \int_1^4 \sqrt{1+9x} \, dx = (1/9)\int_{10}^{37} u^{1/2} \, du$$

$$= (2/27)u^{3/2}\Big|_{10}^{37} = (2/27)(37^{3/2} - 10^{3/2}).$$

5. We have $x = (2/3)y^{3/2}$ and thus $dx/dy = y^{1/2}$ and hence Eq.(7)

yields $L = \int_0^{3^{2/3}} \sqrt{1+y} \, dy = (2/3)(1+y)^{3/2}\Big|_0^{3^{2/3}} = 2[(1+3^{2/3})^{3/2} - 1]/3.$

7. We have $x' = at\cos t$ and $y' = at\sin t$ and thus Eq.(11)

yields $L = \int_0^{2\pi} \sqrt{a^2 t^2 \cos^2 t + a^2 t^2 \sin^2 t} \, dt = a\int_0^{2\pi} t \, dt = 2a\pi^2.$

9. We have $y' = (1/2)x^2 - 1/2x^2 = (x^4 - 1)/2x^2$. Thus
    $1+(y')^2 = 1+(x^8 - 2x^4 + 1)/4x^4 = (x^8 + 2x^4 + 1)/4x^4 = (x^4 + 1)^2/4x^4$.

    Therefore $L = \int_2^4 \dfrac{x^4 + 1}{2x^2} \, dx = \dfrac{1}{2}\left[\dfrac{x^3}{3} - \dfrac{1}{x}\right]\Big|_2^4 = \dfrac{227}{24}.$

11. Let $x = a\cos^3\theta$ and $y = a\sin^3\theta$, then $x^{2/3} + y^{2/3} = a^{2/3}$,
    $x' = -3a\sin\theta\cos^2\theta$, $y' = 3a\cos\theta\sin^2\theta$, and
    $(x')^2 + (y')^2 = 9a^2\sin^2\theta\cos^2\theta$. Thus

$$L = \int_0^{2\pi} |\sin\theta\cos\theta| \, d\theta = (3a/2)\int_0^{2\pi} |\sin 2\theta| \, d\theta$$

$$= 6a\int_0^{\pi/2} \sin 2\theta \, d\theta = 6a.$$

13.   For $1 \leq x \leq 3$ we find $y' = (1/6)x^{-1/2}(x-3) + (1/3)x^{1/2} = (x-1)/2x^{1/2}$

and therefore $1 + (y')^2 = 1 + \dfrac{(x-1)^2}{4x} = \dfrac{4x + x^2 - 2x + 1}{4x} = \dfrac{(x+1)^2}{4x}$.

Thus $L = \left(\dfrac{1}{2}\right)\displaystyle\int_1^3 (x^{1/2} + x^{-1/2})\,dx = (6\sqrt{3} - 4)/3$.

15a. Let $x = a\sin t$ and $y = b\cos t$. Then $x' = a\cos t$, $y' = -b\sin t$

and $L = 4\displaystyle\int_0^{\pi/2} \sqrt{a^2\cos^2 t + b^2\sin^2 t}\,dt$, using the symmetry of the

ellipse. Setting $\cos^2 t = 1 - \sin^2 t$ and $e^2 = (a^2-b^2)/a^2$ we

obtain $L = 4a\displaystyle\int_0^{\pi/2} \sqrt{1 - e^2\sin^2 t}\,dt$.

b.   Letting $u = e^2\sin^2 t$ in the given formula results in

$\sqrt{1 - e^2\sin^2 t} = 1 - (1/2)e^2\sin^2 t = 1 - (1/4)e^2(1-\cos 2t)$.

Integration then yields

$L = 4a[(1-e^2/4)t + (1/8)\sin 2t]\Big|_0^{\pi/2} = 2\pi a(1-e^2/4)$. Thus

$L/4a \cong 1.56687$, 1.54625, 1.47262, and 1.31947 when $e = .1$,
$.25$, $.5$ and $.8$ respectively.

17. $y' = 2x$ and thus $L = \displaystyle\int_0^2 \sqrt{1 + 4x^2}\,dx = 4.64678$, for $n = 20$.

19.   We have $y' = 2$ and thus Eq.(24) yields

$S = 2\pi\displaystyle\int_0^4 2x\sqrt{1+4}\,dx = 32\sqrt{5}\,\pi$.

21.   We have $y' = -x/\sqrt{r^2-x^2}$ and thus

$S = 2\pi\displaystyle\int_a^b \sqrt{r^2-x^2}\sqrt{1+x^2/(r^2-x^2)}\,dx = 2\pi r\displaystyle\int_a^b dx = 2\pi r(b-a)$.

23.   We have $y' = (x^4-1)/2x^2$ and hence $1+(y')^2 = (x^4+1)/2x^2$.   Thus

$S = 2\pi\displaystyle\int_1^4 \left(\dfrac{x^3}{6} + \dfrac{1}{2x}\right)\left(\dfrac{x^4+1}{2x^2}\right)dx = \dfrac{\pi}{6}\displaystyle\int_1^4 \dfrac{x^8 + x^4 + 3}{x^3}\,dx$

$= \dfrac{\pi}{6}\left(\dfrac{x^6}{6} + 2x^2 - \dfrac{3}{2}\dfrac{1}{x^2}\right)\Big|_1^4 = \dfrac{7615\pi}{64}$.

25.   We have $y' = 1/2\sqrt{x+1}$ so $S = 2\pi \int_0^3 \sqrt{x+1} \sqrt{1 + \dfrac{1}{4(x+1)}} \, dx =$

$\pi \int_0^3 \sqrt{4x+5} \, dx = (\pi/4) \int_5^{17} \sqrt{u} \, du = (\pi/6)(17^{3/2} - 5^{3/2}).$

27.   We have $y' = -2x$ so $S = 2\pi \int_0^2 (4 - x^2) \sqrt{1 + 4x^2} \, dx = 63.56045$
using Simpson's rule.

29.   If we let $u = x$ and $v = y - A = f(u) - A = h(u)$, then the (u, v)
coordinate system has its origin at $(0, A)$ in the (x, y)
coordinate system and the line $y = A$ becomes the u axis.
Thus in the u, v coordinate system Eq.(24) becomes

$S = 2\pi \int_a^b h(u) \sqrt{1 + [h'(u)]^2} \, du = 2\pi \int_a^b [f(u) - A]\sqrt{1 + [f'(u)]^2} \, du,$

which is the desired result with u replaced by x.

31.   For a given partition the length
of the horizontal line at y_i is x_i

and thus $S \cong \pi \sum_{i=1}^{n} (x_{i-1} + x_i) l_i$, where

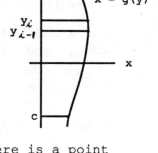

$l_i = \sqrt{\Delta x_i^2 + \Delta y_i^2}$.   Since $g'(y)$ is
continuous, then the Mean Value Theorem
will yield $\Delta x_i = g'(y_i^*) \Delta y_i$, $y_i^* \in (y_{i-1}, y_i)$,
which is the counterpart here to Eq.(21).
Also, since $g(y)$ is continuous, we know there is a point
$y_i^+ \in [y_{i-1}, y_i]$ such that $g(y_i^+) = [g(y_{i-1}) + g(y_i)]/2 = (x_{i-1} + x_i)/2$, by
the Intermediate Value Theorem.   Substituting l_i, Δx_i and $g(y_i^+)$

into S yields $S \cong 2\pi \sum_{i=1}^{n} g(y_i^+) \sqrt{[g'(y_i^*)]^2 + 1} \, \Delta y_i$.   As discussed in

the text, this term approaches a definite integral as $\|\Delta\| \to 0$

and thus $S = 2\pi \int_c^d g(y) \sqrt{[g'(y)]^2 + 1} \, dy$.   For $y = x^2/4$ we have
$x = 2\sqrt{y}$ and $x' = 1/\sqrt{y}$ and thus

$S = 2\pi \int_1^4 2\sqrt{y} \sqrt{1 + 1/y} \, dy = 4\pi \int_1^4 \sqrt{y+1} \, dy = (8\pi/3)(5^{3/2} - 2^{3/2}).$

33. We have x' = cost, y' = sint and thus the result of Prob.32
     gives

$$S = 2\pi \int_0^{\pi/2} (1 - \cos t)\sqrt{\cos^2 t + \sin^2 t}\, dt \ = 2\pi \int_0^{\pi/2} (1 - \cos t)\, dt \ = \pi(\pi-2).$$

35a. $V_b \ = \ \pi \int_1^b x^{-4/3}\, dx \ = \ 3(1 - b^{-1/3})\pi \ \to \ 3\pi$ as $b \to \infty$.

   b. We have $y' \ = \ -(2/3)x^{-5/3}$ and thus

   $$S_b \ = \ 2\pi \int_1^b x^{-2/3}\sqrt{1+(4/9)x^{-10/3}}\, dx.$$

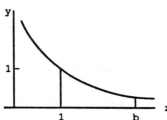

   c. Since $1 + (4/9)x^{-10/3} > 1$ for $x > 1$

   we have $S_b \ > \ 2\pi \int_1^b x^{-2/3} dx \ = \ 6\pi(b^{1/3} - 1) \to \infty$ as $b \to \infty$.

Section 7.4, Page 378

1a. From Eq.(5), the displacement of the particle is given by

$$\int_0^{2\pi} (1 - \sin t)\, dt \ = \ (t + \cos t)\Big|_0^{2\pi} \ = \ 2\pi \text{ ft.}$$

  b. Since v≥0 for 0≤t≤2π, we have the distance traveled is the
      same as the displacement.

3a. The displacement is given by

$$\int_0^2 (t^2 + t - 2)\, dt \ = \ (t^3/3 + t^2/2 - 2t)\Big|_0^2 \ = \ 2/3 \text{ ft.}$$

  b. The distance traveled is given by

$$\int_0^2 |t^2 + t - 2|\, dt \ = \ \int_0^1 (2 - t - t^2)\, dt \ + \ \int_1^2 (t^2 + 2t - 2)\, dt \ =$$

$$(2 - t^2/2 - t^3/3)\Big|_0^1 \ + \ (t^3/3 + t^2/2 - 2t)\Big|_1^2 \ = \ 3 \text{ ft.}$$

5a. Displacement $= \ \int_0^3 t\sqrt{1 + t^2}\, dt \ = \ (1/2)\int_1^{10} \sqrt{u}\, du$, where $u = 1+t^2$.

     Thus displacement $= (1/3)u^{3/2}\Big|_1^{10} \ = \ (10^{3/2} - 1)/3$ ft.

  b. Distance equals displacement since $v \geq 0$.

7a. Displacement $= \displaystyle\int_1^2 (t - 2/t^2)dt = (t^2/2 + 2/t)\Big|_1^2 = 1/2$ ft.

 b. Since $v = 0$ when $t^3 - 2 = 0$, we have distance s given by

$$\int_1^{2^{1/3}} (2/t^2 - t)dt + \int_{2^{1/3}}^2 (t - 2/t^2)dt$$

$$= -(2/t + t^2/2)\Big|_1^{2^{1/3}} + (t^2/2 + 2/t)\Big|_{2^{1/3}}^2 = 11/2 - 6(2)^{-1/3} \text{ ft.}$$

9. Since $F = kx$, we have $16 = k(1)$ and thus $k = 16$ lbs/in.

Hence by Eq.(7), $W = \displaystyle\int_0^3 16x\,dx = 8x^2\Big|_0^3 = 72$ in-lb. $= 6$ ft-lb.

Note that x is measured from the natural length of the spring
so that when $x = 0$ the spring has length 12 and when $x = 3$
the spring has length 15.

11. A cross section of the length is
shown, with x measured down from the
water surface. From Ex.(3) we know
that the minimum amount of work will
be done when the water is lifted from
the surface. Thus we have

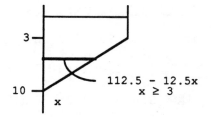

$$W = w\int_0^3 25(75)(x+1)\,dx + w\int_3^9 25(112.5 - 12.5x)(x+1)\,dx$$

$$= 1875\,w\,(x^2/2 + x)\Big|_0^3 + 25w\,(112.5x + 50x^2 - 12.5x^3/3)\Big|_3^9$$

$$= (14{,}062.5 + 33{,}750)w = 47{,}812.5w \text{ ft-lb.}$$

13. Measuring x from the original position of the electron being
moved, we have $F = k/(5-x)^2$, $0 \le x \le 5$. Thus

$$W = \int_0^3 \frac{k\,dx}{(5-x)^2} = \frac{k}{5-x}\Big|_0^3 = 3k/10.$$ If r is measured from the fixed

electron, then $F = k/r^2$ and thus

$$W = \int_5^2 (k/r^2)(-dr) = k/r\Big|_5^2 = 3k/10.$$ The $(-dr)$ is needed since

the motion is in the negative r direction.

15.   Following the second approach of Prob.13 we have

$$W = \int_{R}^{0} (mgr/R)(-dr) = -mgr^2/2R \bigg|_{R}^{0} = mgR/2.$$

17.   If p is a function of h, then partition the interval $[h_1, h_2]$
      and pick starpoints h_i^*.   Then the work done in the interval
      Δh_i is approximated by $p(h_i^*) A \Delta h_i$ and thus $W \cong \sum_{i=1}^{n} p(h_i^*)A\Delta h_i$. As

      $\|\Delta\| \to 0$ we then obtain $W = \int_{h_1}^{h_2} p(h) A\,dh.$ If p is a function of

      V then partition the interval $[V_1, V_2]$ and pick starpoints V_i^*
      with $V = Ah$.   Then the work done in the interval ΔV_i is

      approximated by $p(V_i)A\Delta h_i = p(V_i)\Delta V_i$ and thus $W \cong \sum_{i=1}^{n} p(V_i^*)\Delta V_i$. As

      $\|\Delta\| \to 0$ we then obtain $W = \int_{V_1}^{V_2} p(V)dV.$

19.   The generic element at y_i has volume
      $\pi x_i^2 \, \Delta y_i$ and must be lifted a
      distance $(10+y_i)$, where y_i is measured
      positively down from the center of
      the sphere and $x^2+y^2 = 100$.   Thus

      $$W = 60\pi \int_{0}^{10} (100 - y^2)(10 + y)\,dy$$

      $$= 60\pi(1000 + 50y^2 - 10y^3/3 - y^4/4) \bigg|_{0}^{10} = 55\pi \times 10^4 \text{ ft-lb.}$$

21.   The height of the gate is $\sqrt{5}$ and
      if x is measured down from the top
      we have $\dfrac{1}{\sqrt{5}-x} = \dfrac{4}{\sqrt{5}}$ or $1 = 4(\sqrt{5}-x)/\sqrt{5}.$
      Eq.(12) then yields

      $$F = (4w/\sqrt{5})\int_{0}^{\sqrt{5}} x(\sqrt{5} - x)\,dx = (4w/\sqrt{5})(\sqrt{5}\,x^2/2 - x^3/3)\bigg|_{0}^{\sqrt{5}} = 10w/3.$$

23. The generic element at y_i is
     $(3-y_i)$ ft below the top and
     has cross sectional area
     $2x_i\Delta y_i = 16\sqrt{1-y_i^2/9}\,\Delta y_i$ since
     the equation of the ellipse is
     $x^2/64 + y^2/9 = 1$.   Thus the force on

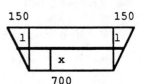

     the generic element is $16w(3-y)\sqrt{1-y^2/9}$ and hence the total

     force is $F = 16w\int_{-3}^{3}(3-y)\sqrt{1-y^2/9}\,dy$.

25a. If x is measured up from the bottom
      then a generic element is $(80-x)$
      feet below the surface and has cross
      sectional area $(700 + 2l_i)\Delta x_i$, where
      l_i is given by $l_i/x = 150/80$.   Thus

$$F = w\int_{0}^{80}(80-x)(700+15x/4)\,dx$$

$$= w(56{,}000x - 200x^2 - 5x^3/4)\Big|_{0}^{80} = (2.560\text{x}10^6)w \text{ lb}.$$

  b. In this case the generic element is only $(60-x)$ below the
      surface and thus

$$F = w\int_{0}^{60}(60-x)(700+15x/4)\,dx$$

$$= w(42{,}000x - 475x^2/2 - 5x^3/4)\Big|_{0}^{60} = (1.395\text{x}10^6)w \text{ lb}.$$

Section 7.5, Page 386

1a. This problem was discussed in the text with the volume of
     blood flow per unit time, Q, being given by Eq.(6).   Thus

$$Q = \frac{\pi(66.5\text{x}10^3 \text{ dyne/cm}^2)(30\text{x}10^{-4}\text{ cm})^4}{8(.04 \text{ dyne-sec/cm}^2)(.5 \text{ cm})} = 1.06\text{x}10^{-4} \text{ cm}^3/\text{sec}.$$

  b. Using Eq.(6) we have $R = \dfrac{\Delta p}{\pi(\Delta p)a^4/8\mu l} = \dfrac{8\mu l}{\pi a^4}$.   Since μ has

      dimensions dyne-sec/cm^2, R will have the units stated.

1c. Since $6000 \text{ cm}^3/\text{min.} = 100 \text{ cm}^3/\text{sec}$, we have
$R = (133 \times 10^3 \text{ dyne/cm}^2)/(100 \text{ cm}^3/\text{sec}) = 1330 \text{ dyne-sec/cm}^5$.
Using the formula for R from part (b) we have
$1330 = 8(.04)(200)/\pi a^4$ since 2 meters = 200 cm. Solving for
a yields a = .352 cm.

d. $R = 8\mu l/\pi a^4 = 8(.04)(.5)/\pi(3 \times 10^{-3})^4 = .6288 \times 10^9 \text{ dyne}$
sec/cm^5 is the resistance in one arteriole. The total
resistance is $R_t = \Delta p/Q = 65(1.33 \times 10^3)/100 = 864.5 \text{ dyne}$
sec/cm^5. For N equal resistors in parallel $\dfrac{1}{R_t} = \dfrac{1}{R} + \cdots + \dfrac{1}{R} = \dfrac{N}{R}$

and thus $N = R/R_t = \dfrac{6.288 \times 10^8}{8.645 \times 10^2} = 7.3 \times 10^5$.

3.   Let l_a be the length of the blood vessel of radius a and l_b be
the length of the blood vessel of radius b. Then the total
resistance from O to Q is $R = (8\mu/\pi)(l_a/a^4 + l_b/b^4)$. If l_1 is
the distance from O to P and l_2 is the distance from P to Q,
then $l_a = l_1 - l_b \cos\theta$ and $l_b = l_2/\sin\theta$. Substituting these values
into R we obtain $R = (8\mu/\pi)[l_1/a^4 - (l_2/a^4)\cot\theta + (l_2/b^4)\csc\theta]$
and thus $dR/d\theta = (8\mu/\pi)[(l_2/a^4)\csc^2\theta - (l_2/b^4)\cot\theta\csc\theta]$.
Setting this equal to zero yields $\cos\theta_{min} = (b/a)^4$.

5a. Total time $= \displaystyle\int_1^{401} 2x^{-1/2}dx = 4x^{1/2}\Big|_1^{401} = 76.10 \text{ hours}$.
Total cost = $(5)(76.10) = \$380.50$.

b.   Total time $= \displaystyle\int_1^{401} (2.5)x^{-1/3}dx = 3.75x^{2/3}\Big|_1^{401} = 200.17 \text{ hours}$.
Total cost = $(200.17)(4) = \$800.68$.

c.   For the faster interviewer to receive the equivalent pay as
the slower he should be paid at the rate of
$(800.68)/(76.10) = \$10.52$ per hour.

7a. Following the discussion on Reliability Theory we find

$$F(t) = \begin{cases} 0 & t \le 0 \\ \int_0^t k\,ds = kt, & 0 \le t \le 2. \\ 2k & t > 2 \end{cases}$$

Since $\lim\limits_{t \to \infty} F(t) = 1 = 2k$, we must have $k = 1/2$.

b. $F(t) = t/2$ for $0 \le t \le 2$, with $F(t) = 1$ for $t > 2$, is the failure distribution function.

c. By Eq.(14), the expected time of failure is

$$\tau = \int_0^2 t f(t)\,dt = \int_0^2 (t/2)\,dt = 1.$$

9a. A must satisfy $\int_0^2 f(t)\,dt = 1$ or

$$\int_0^1 At\,dt + \int_1^2 A(2-t)\,dt = A/2 + A/2 = 1 \text{ for } A = 1.$$

b. $F(t) = \begin{cases} \int_0^t s\,ds = t^2/2 & 0 \le t \le 1 \\ 1/2 + \int_1^t (2-s)\,dx = 2t - (t^2/2) - 1 & 1 \le t \le 2 \end{cases}$

and $F(t) = 1$ for $t > 2$.

c. $\tau = \int_0^2 t f(t)\,dt = \int_0^1 t^2\,dt + \int_1^2 t(2-t)\,dt = 1.$

11. f(T) consists of two parts: (a) the amount of the original population remaining at time T; and (b) the amount added to the population between the initial time and time T.

a. The amount of the original population, f(0), remaining at time T is f(0)s(T).

11   b. The population added due to renewal for $0 \leq t \leq T$ may be
      approximated by partitioning the interval $[0,T]$ with
      $0 = t_0 < t_1 < \ldots < t_{i-1} < t_i < \ldots < t_n = T$.   On the interval $[t_{i-1}, t_i]$
      there are approximately $r(t_i^*) \Delta t_i$, $t_{i-1} \leq t_i^* \leq t_i$, new members added
      to the population. Of these $s(T-t_i^*) r(t_i^*) \Delta t_i$ survive to time
      T, since they will survive from t_i^* to T, or for a duration of
      $T - t_i^*$. Summing over all subintervals we find approximately

$$\sum_{i=1}^{n} s(T - t_i^*) \, r(t_i^*) \, \Delta t_i \text{ new members. Letting } \Delta t_i \to 0, \text{ we have}$$

$$\int_0^T s(T-t) \, r(t) \, dt \text{ new members.}$$

      Thus $f(T) = f(0) s(T) + \displaystyle\int_0^T s(T-t) r(t) \, dt$.

13.   From Prob.(11), $f(T) = 10^4/(1+T^2)^2 + \displaystyle\int_0^T \frac{30(T-t)}{[1 + (T-t)^2]^2} \, dt$

      or $f(T) = 10^4 (1+T^2)^{-2} - 15(1+T^2)^{-1} + 15$.

15a. P.S. $= \displaystyle\int_0^{15} [18 - (3+x)] \, dx = \int_0^{15} (15-x) \, dx = 112.50$.

  b. P.S. $= \displaystyle\int_0^{5} [8 - (3+x)] \, dx = \int_0^{5} (5-x) \, dx = 12.50$.

Chapter 7 Review, Page 389

1. $A = \displaystyle\int_{-1}^{3} (1 + x^2) \, dx = (x + x^3/3) \big|_{-1}^{3} = 40/3$.

3.   The area is that of two regions. The first is the right
      semicircle (8π) and the second is the area of the region
      bounded on the left by $16 + 8x = y^2$ and on the right by the
      y-axis. Thus $A = 8\pi + \displaystyle\int_{-4}^{4} (-y^2/8 + 2) \, dy = 8\pi + 32/3$.

5. $f(x) = (x+1)(x-1)(x-2)$, so $f(x) \geq 0$ for $-1 \leq x \leq 1$ and
     $f(x) \leq 0$ for $1 \leq x \leq 2$. Thus

$$A = \int_{-1}^{1} (x^3-2x^2-x+2)\,dx + \int_{1}^{2} [-(x^3-2x^2-x+2)]\,dx = 8/3 + 5/12 = 37/12.$$

7. $A = \int_{-2}^{1} [(-x+2) - x^3]\,dx = (2x - x^2/2 - x^4/4) \Big|_{-2}^{1} = 45/4.$

9a. $f(x)$ and $g(x)$ intersect at $(1,1)$ and $(-2,4)$. Thus

$$A = \int_{-2}^{1} [(2-x)-x^2]\,dx = (2x - x^2/2 - x^3/3)\Big|_{-2}^{1} = 9/2.$$

 b. $A = \int_{0}^{1} [y^{1/2} - (-y^{1/2})]\,dy + \int_{1}^{4} [(2-y)-(-y^{1/2})]\,dy = 9/2.$

11a. f and g intersect at $(\sqrt{2}, 2\sqrt{2})$, $(0,0)$, and $(-\sqrt{2}, -2\sqrt{2})$. Using

     symmetry then yields $A = 2\int_{0}^{\sqrt{2}} (2x - x^3)\,dx = 2.$

 b. $A = 2\int_{0}^{2\sqrt{2}} (y^{1/3} - y/2)\,dy = 2.$

13a. The curves intersect at $(-1,1)$ and $(2,4)$ and $|2+x| = 2+x$, for

     $x \geq -2$. Thus $A = \int_{-1}^{2} [(2+x) - x^2]\,dx = 9/2.$

 b. $A = 2\int_{0}^{1} \sqrt{y}\,dy + \int_{1}^{4} [\sqrt{y} - (y-2)]\,dy = 9/2.$

15a. $V = \pi\int_{0}^{4} x\,dx = 8\pi.$ b. $V = 2\pi\int_{0}^{2} y(4 - y^2)\,dy = 8\pi.$

17a. $V = \pi\int_{0}^{2} (y^2 - 4)^2\,dy = \pi\int_{0}^{2} (16 - 8y^2 + y^4)\,dy = 256\pi/15.$

 b. $V = 2\pi\int_{0}^{4} (4-x)\sqrt{x}\,dx = 256\pi/15.$

19a. $V = \pi\int_{0}^{4} [(\sqrt{x}+1)^2 - (1)^2]\,dx = \pi\int_{0}^{4} (x + 2\sqrt{x})\,dx = 56\pi/3.$

 b. $V = 2\pi\int_{0}^{2} (4-y^2)(1+y)\,dy = 56\pi/3.$

21a. By symmetry, $V = 2\{\pi\int_0^1 (3^2 - [3 - (2-2x)]^2)\,dx\}$

$$= 2\pi\int_0^1 (8 - 4x - 4x^2)\,dx = 28\pi/3.$$

  b. By symmetry, $V = 2\{2\pi\int_0^2 (3-y)[(2-y)/2]\,dy\}$

$$= 2\pi\int_0^2 (6 - 5y + y^2)\,dy = 28\pi/3.$$

23b. Needed intersection points are $(0,6), (1/2,6), (1,4)$ and $(3,0)$.

  Thus $V = \pi\int_0^4 [\sqrt{4-y}+1]^2 dy + \pi\int_4^6 [(8-y)/4]^2 dy.$

  c. $V = 2\pi\int_0^{1/2} 6x\,dx + 2\pi\int_{1/2}^1 x(8-4x)\,dx + 2\pi\int_1^3 x[4-(1-x)^2]\,dx.$

23a. 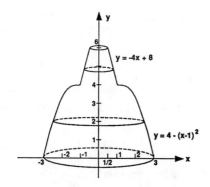   25a.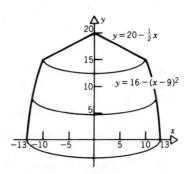

25b. The curves intersect at $(10,15)$. Thus we have
$$V = \pi\int_0^{15} (9 + \sqrt{16-y})^2\,dy + \pi\int_{15}^{20} (40 - 2y)^2\,dy.$$
  c. $V = 2\pi\int_0^{10} x(20 - x/2)\,dx + 2\pi\int_{10}^{13} x[16 - (x-9)^2]\,dx.$

27. $x' = -2\cos 2t$, $y' = -2\sin t$, so $(x')^2 + (y')^2 = 4$ and
$$L = \int_0^\pi 2\,dt = 2\pi.$$

29. Using $x = (2/5)^{1/2} y^{3/2}$ gives $x' = (2/5)^{1/2}(3/2)y^{1/2}$ so

$$L = \int_0^1 \sqrt{1+(9/10)\,y}\ dy = (20/27)(1+9y/10)^{3/2}\Big|_0^1$$

$$= (20/27)[(19/10)^{3/2} - 1].$$

31. $$S = 2\pi \int_1^4 (x^3/6)\sqrt{1+(x^2/2)^2}\ dx = (\pi/6)\int_1^4 x^3\sqrt{4+x^4}\ dx$$

$$= (\pi/36)(4+x^4)^{3/2}\Big|_1^4 = (\pi/36)[(260)^{3/2} - (5)^{3/2}].$$

33. Since $y' = 3$, $S = 2\pi \int_0^{10} (3x+1)\sqrt{10}\ dx = 320\sqrt{10}\,\pi.$

35a. From Eq.(5a), Section 7.4, the displacement of the particle

is $\int_0^\pi (\cos 2t + 1/2)\,dt = (1/2)\sin 2t + (1/2)t\Big|_0^\pi = \pi/2$ ft.

  b. Since $v \geq 0$ for $0 \leq t \leq \pi/3$ and $2\pi/3 \leq t \leq \pi$ and $v \leq 0$ for $\pi/3 \leq t \leq 2\pi/3$, the total distance traveled is given by

$$\int_0^{\pi/3} (\cos 2t + 1/2)\,dt - \int_{\pi/3}^{2\pi/3} (\cos 2t + 1/2)\,dt + \int_{2\pi/3}^{\pi} (\cos 2t + 1/2)\,dt$$

$$= (\pi/6 + \sqrt{3})\ \text{ft}.$$

37a. The displacement is $\int_1^9 2t^{-1/2}\,dt = 8$ ft.

  b. As $v \geq 0$ for all t in $1 \leq t \leq 9$, the distance traveled equals the displacement, or 8 ft.

39. Since $F = kx$, we have $12 = k(1)$ or $k = 12$ lbs/in. Thus, by

Eq.(7) Section 7.4, $W = \int_0^6 12x\,dx = 216$ in-lb.

41. The water is raised from the position y to 5 and the cross sectional area is $\pi x^2 = 4y$. Thus

$$W = 250\pi \int_2^4 (5 - y)y\,dy = 8500\pi/3\ \text{ft-lb}.$$

43. We measure x positive down from the top of the pond and use Eq.(12) of Section 7.4. A representative strip at depth x will have length $l(x) = 3 + (x-10)/2$ and thus

$$F = 62.5 \int_{10}^{12} x[3+(x-10)/2]\,dx = 14,500/3\ \text{lb}.$$

CHAPTER 8

Section 8.1, Page 399

1. $y = f(x) = x-3$ has domain $[-2,4]$ and
    range $[-5,1]$. Solving for x yields
    $x = y+3 = f^{-1}(y)$. Thus $f^{-1}(x) = x+3$
    has domain $[-5,1]$ and range $[-2,4]$.
    $f[f^{-1}(y)] = f(y+3) = (y+3) - 3 = y$.
    $f^{-1}[f(x)] = f^{-1}[x-3] = (x-3) + 3 = x$.

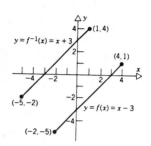

3. $y = f(x) = 5-2x$ with domain $(1,\infty)$
    has range $(-\infty,3)$. Solving for x
    gives $x = f^{-1}(y) = (5-y)/2$ which
    has domain $(-\infty,3)$ and range $(1,\infty)$.
    $f[f^{-1}(y)] = f[(5-y)/2] = 5-2(5-y)/2 = y$,
    $f^{-1}[f(x)] = f^{-1}[5-2x] = [5-(5-2x)]/2 = x$.

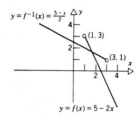

5. $y = f(x) = x^2-2x+3$ has $f'(x) = 2(x-1)$
    which is negative for x in $(-4,1)$,
    thus f is decreasing for x
    in $(-4,1]$ and possesses an inverse.
    With domain $(-4,1]$, $f(x)$ has range
    $[2,27)$. From $x^2-2x+(3-y) = 0$, and
    using the quadratic formula
    $x = f^{-1}(y) = 1-\sqrt{y-2}$, with domain
    $[2,27)$ and range $(-4,1]$.
    $f[f^{-1}(y)] = f[1 - \sqrt{y-2}] = (1 - \sqrt{y-2})^2 - 2(1-\sqrt{y-2}) + 3 = y$

    and $f^{-1}[f(x)] = 1 - \sqrt{x^2-2x+3-2} = x$.

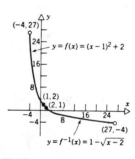

7. $y = \dfrac{x-1}{x+2}$ with domain $(0,3]$ has range
    $\left(-\dfrac{1}{2}, \dfrac{2}{5}\right]$. Solving for x gives
    $f^{-1}(y) = (1+2y)/(1-y)$ or
    $y = f^{-1}(x) = (1+2x)/(1-x)$ with domain $\left(-\dfrac{1}{2}, \dfrac{2}{5}\right]$ and range $(0,3]$.
    $f[f^{-1}(y)] = \dfrac{[(1+2y)/(1-y) - 1]}{[(1+2y)/(1-y) + 2]} = \dfrac{3y}{3} = y$
    and $f^{-1}[f(x)] = \dfrac{[1+2(x-1)/(x+2)]}{[1-(x-1)/(x+2)]} = \dfrac{3x}{3} = x$.

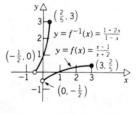

9. $y = \sqrt{x+2}$ with domain $[2,7]$ has range
      $[2,3]$. $f^{-1}(y) = y^2-2$ has domain $[2,3]$
      and range $[2,7]$.
      $f^{-1}[f(x)] = (\sqrt{x+2})^2 - 2 = x$ and
      $f[f^{-1}(y)] = \sqrt{(y^2-2)+2} = y$.

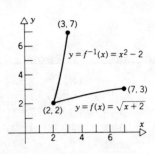

11.   For $y = x-3$, $dx/dy = 1/(dy/dx)$ by Eq.(18). Thus since
      $dy/dx = 1$, we have $dx/dy = 1$. Note also that since $x = y+3$,
      $dx/dy = 1$.

13.   For $y = 5-2x$, $dx/dy = 1/(dy/dx)$. Since $dy/dx = -2$,
      $dx/dy = -1/2$. Note that $x = -(1/2)(y-5)$ so $dx/dy = -1/2$.

15.   For $y = x^2-2x+3$, $dx/dy = 1/(dy/dx) = 1/(2x-2)$ since
      $dy/dx = 2x-2$.

17.   For $y = (x-1)/(x+2)$, $dx/dy = (x+2)^2/3$ since $dy/dx = 3/(x+2)^2$.

19.   For $y = \sqrt{x+2}$, $dx/dy = 2\sqrt{x+2} = 2y$ since $dy/dx = 1/2\sqrt{x+2}$. Note, from
      Prob.9, $x = y^2-2$ so $dx/dy = 2y$.

21.   If $y = f(x) = x^3 - 12x + 4$ then $f'(x) = 3x^2 - 12$ or
      $f'(x) = 3(x+2)(x-2) < 0$ for x in $(-2,2)$. Thus by Theorem
      8.1.1, f has an inverse in $(-2,2)$, with derivative
      $dx/dy = 1/3(x+2)(x-2)$.

23.   For $y = f(x) = \sqrt{4-x^2}$, $f'(x) = -x/\sqrt{4-x^2} < 0$ for x in
      $(0,2)$. Thus f has an inverse whose derivative is
      $dx/dy = -\sqrt{4-x^2}/x = -y/\sqrt{4-y^2}$ since $x = \sqrt{4-y^2}$.

25.   For $y = \cos x = f(x)$, $f(\pi/2) = f(3\pi/2) = 0$, thus f does not
      have an inverse.

27a. For $y = f(x) = \sin x$, $f'(x) = \cos x > 0$ for x in $(0,\pi/2)$. Thus,
      $y = \sin x$ has an inverse for x in $(0,\pi/2)$ by Theorem 8.1.1.
      and $dx/dy = \dfrac{1}{\cos x} = \dfrac{1}{\sqrt{1-\sin^2 x}} = \dfrac{1}{\sqrt{1-y^2}}$.

   b. Since $\sin(\pi/4) = \sin(3\pi/4)$, $y = \sin x$ does not possess an
      inverse for x in $(0,\pi)$.

27c. For x in $(\pi/2, 3\pi/2)$, $f'(x) = \cos x < 0$, so by Theorem 8.1.1
$f(x) = \sin x$ has an inverse. Again $dx/dy = 1/\cos x$. Now,
however, since $\cos x < 0$, we use $\cos x = -\sqrt{1-\sin^2 x} = -\sqrt{1-y^2}$.
Therefore, for x in $(\pi/2, 3\pi/2)$, $dx/dy = -1/\sqrt{1-y^2}$.

29. For $y = f(x) = 5-2x$, $f^{-1}(y) = x = (y-5)/2$ so $dx/dy = 1/2$ and
$d^2x/dy^2 = 0$.

31. For $y = (x-1)/(x+2)$, $x = (1+2y)/(1-y)$ from Prob.7. Thus
$dx/dy = 3/(1-y)^2 = 3(1-y)^{-2}$ and $d^2x/dy^2 = 6/(1-y)^3 = 2(x+2)^3/9$.

33. For $y = \sqrt{x+2}$, $x = y^2-2$, $dx/dy = 2y$, $d^2x/dy^2 = 2$. Alternatively,
from Prob.28, $dx/dy = -y''/(y')^3$ where $y' = (1/2)(x+2)^{-1/2}$,
$y'' = -(1/4)(x+2)^{-3/2}$ and thus
$d^2x/dy^2 = [(1/4)(x+2)^{-3/2}]/[(1/8)(x+2)^{-3/2}] = 2$.

35. For $y = \sqrt{4-x^2}$, $x = \sqrt{4-y^2}$ since x is positive. Thus
$dx/dy = -y(4-y^2)^{-1/2}$ and
$d^2x/dy^2 = -(4-y^2)^{-1/2} - y^2(4-y^2)^{-3/2} = -4/(4-y^2)^{3/2} = -4/x^3$.

37a. If $y = [f(x)]^2$ then $y' = 2ff'$ and hence y has an inverse
since y satisfies the conditions of Theorem 8.1.1 with the
given conditions on f, which imply that f' is either positive
or negative on [0,1], and f is always positive on [0,1].
  b. No; for example, $y = 2x-1$ has domain [0,1], range [-1,1], and
an inverse, but $y^2 = (2x-1)^2 = g(x)$ does not possess an
inverse, since $g(1/4) = g(3/4) = 1/4$.
  c. Yes; as in part(a).

39. If $y = f(x) = (x+a)/(x+b)$ then $f^{-1}(y) = x = (a-by)/(y-1)$ or
$y = f^{-1}(x) = (a-bx)/(x-1)$. Thus for $f^{-1}(x) = f(x)$ we must
have $(a-bx)/(x-1) = (x+a)/(x+b)$ and hence $b = -1$, $a = 1$.

41a. Since $f'(x) = g(x)$ by the Fundamental Theorem of Calculus, f
will have an inverse as long as g is integrable and $g(x) > 0$
or $g(x) < 0$ for all x in [a,b].
  b. Since $dx/dy = 1/(dy/dx)$, dx/dy will exist if $dy/dx = g(x)$ is
continuous and non-zero.
  c. $dx/dy = 1/g(x)$ from part(b).

43. By Theorem 8.1.1, we need $f'(x) > 0$ or $f'(x) < 0$. That is
     $f'(x) \neq 0$. Now $f'(x) = 3ax^2+2bx+c$, and thus $f'(x) = 0$ when
     $$x = \frac{-2b\pm\sqrt{4b^2-12ac}}{6a}.$$ For $f'(x) \neq 0$, the discriminant must be
     negative so we need $4b^2 - 12ac < 0$.

Section 8.2, Page 409

1.   From Eq.(22) $\dfrac{d}{dx}[\ln|x+4|] = \dfrac{1}{x+4}$.

3.   From Eq.(22) $\dfrac{d}{dx}[\ln(x^4+4)] = \dfrac{4x^3}{x^4+1}$.

5.   Using Eq.(22) and the product rule
     $$\frac{d}{dx}[x\ln(x^2+9)+3/x] = \ln(x^2+9)+\frac{2x^2}{x^2+9} - 3/x^2.$$

7. $f(x) = \ln 2 + \ln|x| - \ln(x^2+1)$ so $\dfrac{d}{dx}[f(x)] = \dfrac{1}{x}-\dfrac{2x}{x^2+1} = \dfrac{1-x^2}{x(x^2+1)}$.

9. $\dfrac{d}{dx}[\ln(1+\sqrt{x})] = \dfrac{1/2\sqrt{x}}{1+\sqrt{x}} = \dfrac{1}{2(\sqrt{x}+x)}$.

11. $\dfrac{d}{dx}[x\ln|x| - x] = \ln|x| + 1 - 1 = 1x|x|$.
     (Note: This means that $\int \ln|x|dx = x\ln|x| - x + c$.)

13. Using Eq.(22) with $u(x) = \ln x$, $\dfrac{d}{dx}[\ln|\ln x|] = \dfrac{1}{x \ln x}$.

15. $\dfrac{d}{dx}[\sin(\ln|x|)] = \cos(\ln|x|)\dfrac{d}{dx}[\ln|x|] = (1/x)\cos(\ln|x|)$.

17. Let $u = x-3$, so $\int \dfrac{dx}{x-3} = \int \dfrac{du}{u} = \ln|u| + c = \ln|x-3| + c$.

19. Let $u = x^2-3x+7$, so $du = (2x-3)dx$ and thus
     $$\int\frac{(2x-3)dx}{x^2-3x+7} = \int\frac{du}{u} = \ln|u| + c = \ln|x^2-3x+7| + c.$$

21. $\int\dfrac{dx}{\sqrt{x}(1+\sqrt{x})} = 2\int\dfrac{dx}{2\sqrt{x}(1+\sqrt{x})} = 2\ln(1+\sqrt{x})+c$ from Prob. 9.

23. Let $u = \sin 2x$, then $du = 2\cos 2x\,dx$ and
$$\int \cot 2x\,dx = \int \frac{\cos 2x\,dx}{\sin 2x} = \frac{1}{2}\int \frac{du}{u} = (1/2)\ln|u|+c = (1/2)\ln|\sin 2x|+c.$$

25. $\displaystyle\int_2^6 \frac{dx}{x-1} = \ln|x-1|\Big|_2^6 = \ln 5 - \ln 1 = \ln 5.$

27. $\displaystyle\int_2^4 \frac{4x-6}{x^2-3x+4}dx = 2\int_2^4 \frac{2x-3}{x^2-3x+4}dx = 2\ln|x^2-3x+4|\,\Big|_2^4 = 2(\ln 8-\ln 2) = 2\ln 4.$

29. $\displaystyle\int_{-1}^1 \frac{dx}{2x-5} = \frac{1}{2}\ln|2x-5|\,\Big|_{-1}^1 = (\ln 3 - \ln 7)/2 = -(1/2)\ln(7/3).$

31. $\displaystyle\int_{\pi/6}^{\pi/3} \cot x\,dx = \int_{\pi/6}^{\pi/3} \frac{\cos x}{\sin x}dx = \ln|\sin x|\,\Big|_{\pi/6}^{\pi/3}$
$$= \ln(\sqrt{3}/2) - \ln(1/2) = \ln\sqrt{3} = (\ln 3)/2.$$

33. $y' = 1/x \Rightarrow y'' = -1/x^2 \Rightarrow y''' = (-1)^2\, 2/x^3 \Rightarrow$
$y^{iv} = (-1)^3\, 3!/x^4$ so $y^{(n)} = (-1)^{n-1}(n-1)!/x^n.$

35. $y' = -1/(1-x),\ y'' = -1/(1-x)^2,\ y''' = -2/(1-x)^3,$
$y^{iv} = 3!/(1-x)^4,$ so $y^{(n)} = -(n-1)!/(1-x)^n.$

37. For $f(x) = x - \ln x$ on $(0,2]$, the critical points are the
endpoint of the domain, $x = 2$, and the points when $f'(x) = 0$.
Now $f'(x) = 1 - 1/x = 0$ for $x = 1$ and $f''(x) = 1/x^2$ so
$f''(1) = 1 > 0$. Thus f has a relative minimum at $x = 1$.
Since $f''(x) > 0$ for all x in $(0,2]$, f has an absolute minimum
at $x = 1$. Since $\lim\limits_{x\to 0+} f(x) = +\infty$, f has no maximum.

39. If $f(x) = \ln(\sin x)$, then $f'(x) = \cot x$, for all x in $(0,\pi)$.
Also, $\lim\limits_{x\to 0+} f(x) = \lim\limits_{x\to\pi-} f(x) = -\infty$, so f has no minimum in $(0,\pi)$.
A maximum is possible only when $f'(x) = 0 = \cot x$, so
$x = \pi/2$. Since $f''(x) = -\csc^2 x < 0$ for all x in $(0,\pi)$, $f(x)$
has a maximum of $f(\pi/2) = \ln[\sin(\pi/2)] = 0.$

41. The volume of the solid of revolution is given by:
$$V = \int_a^b \pi y^2\,dx = \int_0^8 \frac{\pi\,dx}{1+x} = \pi\ln(1+x)\Big|_0^8 = \pi\ln 9.$$

43. Notice that $1/(x+2)$ in $[-1,2]$ is the negative of $1/(x+2)$ in
$[-6,-3]$ and thus $\displaystyle\int_{-6}^{-3} \frac{dx}{x+2} = -\int_{-1}^2 \frac{dx}{x+2} = -\ln|x+2|\,\Big|_{-1}^2 = -\ln 4.$

45. $\lim\limits_{x\to 0}\dfrac{\ln(1+x)}{x} = \lim\limits_{x\to 0}\dfrac{\ln(1+x)-\ln 1}{x} = f'(1)$ where $f(v) = \ln v$. Thus

$f'(v) = 1/v$ so $f'(1) = 1$. Therefore, $\lim\limits_{x\to 0}\dfrac{\ln(1+x)}{x} = f'(1) = 1$.

47. If $y = g_1(x)g_2(x)\ldots g_n(x)$, then
$\ln|y| = \ln|g_1(x)g_2(x)\ldots g_n(x)|$
$= \ln|g_1(x)| + \ln|g_2(x)| + \ldots + \ln|g_n(x)|$. Differentiating
both sides gives:

$$\frac{1}{y}\frac{dy}{dx} = \frac{g_1'(x)}{g_1(x)} + \frac{g_2'(x)}{g_2(x)} + \ldots + \frac{g_n'(x)}{g_n(x)} = \sum_{k=1}^{n}\frac{g_k'(x)}{g_k(x)} \quad \text{so that}$$

$$dy/dx = g_1(x)g_2(x)\cdots g_n(x)\sum_{k=1}^{n}\frac{g_k'(x)}{g_k(x)}\,.$$

49. $\ln|f(x)| = \ln|x| + \ln|x^2+x+1| + \ln|\sin 2x|$, so
$\dfrac{f'(x)}{f(x)} = \dfrac{1}{x} + \dfrac{2x+1}{x^2+x+1} + 2\cot 2x$, or

$$f'(x) = x(x^2+x+1)\sin 2x\left[\frac{1}{x} + \frac{2x+1}{x^2+x+1} + 2\cot 2x\right].$$

51a. For $x>1$, $x\geq\sqrt{x}$ so that $\dfrac{1}{x}\leq\dfrac{1}{\sqrt{x}}$. Thus, by Corollary 1 of

Sect.6.3, $\displaystyle\int_{1}^{x}\frac{dt}{t}\leq\int_{1}^{x}\frac{dt}{\sqrt{t}}$ for $x>1$.

 b. Integrating, we have $\ln x \leq 2(\sqrt{x}-1)$. Since
$2(\sqrt{x}-1) = 2x(1/\sqrt{x} - 1/x)$, $0 \leq \ln x \leq 2x(1/\sqrt{x} - 1/x)$ so
$0 \leq \ln x/x \leq 2(1/\sqrt{x} - 1/x)$. Since $\lim\limits_{x\to\infty}(2/\sqrt{x} - 2/x) = 0$ and
$0 \leq \lim\limits_{x\to\infty}(\ln x/x) \leq \lim\limits_{x\to\infty}(2/\sqrt{x} - 2/x)$ we have $\lim\limits_{x\to\infty}(\ln x/x) = 0$.

 c. In $\ln x/x$, let $x = 1/y$ so $\ln x/x = y\ln(1/y) = -y\ln y$. Thus
$0 = \lim\limits_{x\to\infty}(\ln x/x) = \lim\limits_{y\to 0^+} y\ln y$. Therefore, $\lim\limits_{y\to 0^+} x\ln x = 0$.

51d. Letting

$$x = u^{\in}, \quad \lim_{x \to \infty} \ln x / x = \lim_{x \to \infty} \frac{\ln u^{\in}}{u^{\in}}. \text{ Thus, } 0 = \lim_{x \to \infty} (\ln x / x) = \in \lim_{u \to \infty} \frac{\ln u}{u^{\in}},$$

or, $\lim_{u \to \infty} \dfrac{\ln u}{u^{\in}} = 0$. For the second limit replace u with 1/y to

obtain $0 = \lim_{u \to \infty} \dfrac{\ln u}{u^{\in}} = \lim_{y \to 0^{+}} y^{\in} \ln(1/y) = -\lim_{y \to 0^{+}} y^{\in} \ln y.$

Section 8.3, Page 419

1.   Referring to Eq. (25) with $u(x) = 3x$, $u'(x) = 3$ then
$\dfrac{d}{dx}(e^{3x}) = e^{3x}(3) = 3e^{3x}$.

3.   Using the product and chain rules we have
$f'(x) = e^{-x} - xe^{-x} - 2\sin 2x$.

5.   Using Eq. (25), with $u = \sqrt{1-x^2}$, $u' = -x/\sqrt{1-x^2}$ so that
$f'(x) = \left(-x \exp\sqrt{1-x^2}\right)/\sqrt{1-x^2}$.

7. $f'(x) = (e^x + e^{-x})/2$ since if $u(x) = -x$, $u'(x) = -1$.

9.   From Eq. (22) of Section 8.2, with $u(x) = 1+e^{-x}$, $u'(x) = -e^{-x}$
we have $f'(x) = \dfrac{-e^{-x}}{1+e^{-x}}$.

11. Referring to Eq. (25) with $u(x) = 1/x$, $u'(x) = -x^{-2}$,
$f'(x) = -x^{-2}e^{1/x}$.

13. Using the product rule and Eq. (25), with $u(x) = 2/x$,
$u'(x) = -2x^{-2}$, we obtain
$f'(x) = (1/2)e^{2/x} - x^{-1}e^{2/x} = [(1/2) - (1/x)]e^{2/x}$.

15. Using the product rule:
$f'(x) = 2e^{-x}\cos 2x - e^{-x}\sin 2x = (2\cos 2x - \sin 2x)e^{-x}$.

17. With $u = 2x$, $du = 2dx$, or $dx = du/2$, we have
$\int e^{2x}dx = \frac{1}{2}\int e^u du = (1/2)e^u + c = (1/2)e^{2x} + c$.

19. With $u = -x^3$, $du = -3x^2 dx$, so $x^2 dx = -du/3$, and
$\int x^2 e^{-x^3} dx = -(1/3)\int e^u du = (1/3)e^u + c = -(1/3)e^{-x^3} + c$.

21. With $u = 1+e^x$, $du = e^x dx$, and
$\int e^x (1+e^x)^3 dx = \int u^3 du = u^4/4 + c = (1+e^x)^4/4 + c$.

23. With $u = e^{-x}$, $du = -e^{-x} dx$, and $e^{-x} dx = -du$, so
$\int e^{-x}\cos(e^{-x}) dx = -\int \cos u\, du = -\sin u + c = -\sin(e^{-x}) + c$.

25. Note that $\dfrac{e^{2x}-1}{e^{2x}+1} = \dfrac{e^x(e^x-e^{-x})}{e^x(e^x+e^{-x})} = \dfrac{e^x-e^{-x}}{e^x+e^{-x}}$, thus with $u = e^x + e^{-x}$,

$\displaystyle \int \frac{e^{2x}-1}{e^{2x}+1} dx = \int \frac{e^x-e^{-x}}{e^x+e^{-x}} dx = \int \frac{du}{u} = \ln(e^x + e^{-x}) + c$.

27. $\displaystyle \int_0^3 e^{3x} dx = e^{3x}/3 \Big|_0^3 = (e^9 - 1)/3$.

29. Compute $\int xe^{-2x^2} dx$ using $u = -2x^2$ so $\int xe^{-2x^2} dx = -e^{-2x^2}/4 + c$,

and thus, $\displaystyle \int_0^1 xe^{-2x^2} dx = -e^{-2x^2}/4 \Big|_0^1 = -(e^{-2}-1)/4 = (1-e^{-2})/4$.

31. With $u = e^{2x} + 4$, $\int e^{2x}(e^{2x}+4)^2 dx = \frac{1}{2}\int u^2 du = u^3/6 = (e^{2x}+4)^3/6$.

Thus, $\displaystyle \int_{\ln 2}^{\ln 4} e^{2x}(e^{2x}+4)^2 dx = (e^{2x}+4)^3/6 \Big|_{\ln 2}^{\ln 4} = (20^3-8^3)/6 = 1248$.

33. $\displaystyle \int_0^{\pi/2} \sin x\, e^{\cos x} dx = -e^{\cos x} \Big|_0^{\pi/2} = e-1$.

35. Since $f(-x) = f(x)$, the graph
possesses y-axis symmetry. Also,
$f(x) > 0$ for all x,

$f'(x) = -2xe^{-x^2} = 0$ for $x = 0$, and

$f''(x) = (4x^2-2)e^{-x^2}$ and thus
$f''(0) = -2$. By the 2nd derivative
test, f has a maximum of $f(0) = 1$ at $x = 0$.   f also has

inflection points at $x = \pm\sqrt{2}/2$. Finally, $\displaystyle \lim_{x\to\infty} e^{-x^2} = 0$.

37. Since $f(-x) = -f(x)$ f has origin
     symmetry. Also,
     $\lim\limits_{x\to\infty} f(x) = \infty$ and $\lim\limits_{x\to-\infty} f(x) = -\infty$.

     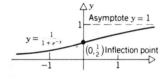

     $f'(x) = (e^x + e^{-x})/2 > 0$,
     so f is monotone increasing and
     $f''(x) = (e^x - e^{-x})/2$.
     Thus $f''(x) < 0$ for $x<0$, $f''(x) > 0$
     for $x>0$, so f is concave downward
     for $x<0$ and concave upward for $x>0$, with $x = 0$ a point of
     inflection.

39. $f(x) = (1+e^{-x})^{-1}$ so
     $\lim\limits_{x\to\infty} f(x) = 1$, $\lim\limits_{x\to-\infty} f(x) = 0$, $f(0) = 1/2$.

     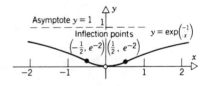

     Also, $f'(x) = e^{-x}(1+e^{-x})^{-2} > 0$ for
     all x, so f is monotone increasing,

     and $f''(x) = \dfrac{1-e^x}{e^{2x}(1+e^{-x})^3} > 0$ for $x<0$ and $f''(x) < 0$ for $x>0$, giving

     upward concavity for $x<0$ and downward concavity for $x>0$.

41. $f(x) = \exp[-1/|x|] = f(-x)$, so the
     function has y-axis symmetry.
     For $x < 0$, $f(x) = \exp[1/x]$,
     $f'(x) = -e^{1/x}/x^2$, and
     $f''(x) = (2x+1)e^{1/x}/x^4$, and thus f
     is always decreasing and is concave
     down for $x < -1/2$ and concave up for $-1/2 < x < 0$. For
     $x > 0$, we have the reflection across the y-axis.

43. $A = \displaystyle\int_0^{\ln 10} e^{-x}dx = -e^{-x}\Big|_0^{\ln 10} = 1-e^{-\ln 10} = 1-1/10 = 9/10.$

45. For $s(t) = e^{-kt}$, $r(t) = R$, then $s(t-\tau) = e^{-k(t-\tau)}$ and

     $f(t) = f(0)e^{-kt} + R\displaystyle\int_0^t e^{-k(t-\tau)}d\tau = f(0)e^{-kt} + (R/k)e^{k(t-\tau)}\Big|_{\tau=0}^{\tau=t}$. So

     $f(t) = f(0)e^{-kt} + (R/k)[1-e^{-kt}]$. and $L = \lim\limits_{t\to\infty} f(t) = R/k.$

47. If $t = e^{\alpha x}$, then $x = (\ln t)/\alpha$. Now, as $x\to\infty$, $t\to\infty$; thus
     $\lim\limits_{x\to\infty} x/e^{\alpha x} = \lim\limits_{t\to\infty}[(\ln t)/\alpha t] = 0$ from Prob.51(b) of Section 8.2.

49a. If $k = \dfrac{1}{h}$, then $h = 1/k$ so as $k \to \infty$, we have $h \to 0$. Using this

substitution we have $\displaystyle\lim_{k\to\infty}\left(1+\frac{1}{k}\right)^k = \lim_{h\to 0}(1+h)^{1/h}$.

b. $\exp\left[\dfrac{\ln(1+h)}{h}\right] = \exp[\ln(1+h)^{1/h}] = (1+h)^{1/h}$ by Definition 8.3.2.

Thus, $\displaystyle\lim_{h\to 0}(1+h)^{1/h} = \lim_{h\to 0}\exp\left[\frac{\ln(1+h)}{h}\right]$.

c. The limit may be taken inside the exponential function as the exponential is a continuous function.

d. Letting $f(x) = \ln x$, we have $f(1) = 0$ and $f'(1) = 1$. Now by definition, $f'(1) = \displaystyle\lim_{h\to 0}\frac{f(1+h)-f(1)}{h}$, and thus, $1 = \displaystyle\lim_{h\to 0}\frac{\ln(1+h)}{h}$.

51. Let $h = r/k$, $k = r/h$ so that

$$\lim_{k\to\infty}\left(1+\frac{r}{k}\right)^k = \lim_{h\to 0}(1+h)^{r/h} = \lim_{h\to 0}[(1+h)^{1/h}]^r.$$

Thus, $\displaystyle\lim_{k\to\infty}\left(1+\frac{r}{k}\right)^k = \left[\lim_{h\to 0}(1+h)^{1/h}\right]^r = e^r$ from Prob. 49.

53. Let $h = 1/k^2$ so $k = 1/\sqrt{-h}$. Then

$$\lim_{k\to\infty}(1-1/k^2)^k = \lim_{h\to 0^-}(1+h)^{1/\sqrt{-h}} = \lim_{h\to 0^-}\exp\left[\frac{\ln(1+h)}{\sqrt{-h}}\right] =$$

$$\exp\left[\lim_{h\to 0^-}\sqrt{-h}\,\frac{\ln(1+h)}{-h}\right] = \exp\left[\lim_{h\to 0^-}-\sqrt{-h}\,\lim_{h\to 0^-}\frac{\ln(1+h)}{h}\right] = \exp(0) = 1.$$

55. $\displaystyle\lim_{k\to\infty}\left(1+\frac{r}{k}\right)^{kt} = \lim_{k\to\infty}\left[\left(1+\frac{r}{k}\right)^k\right]^t = \left[\lim_{k\to\infty}\left(1+\frac{r}{k}\right)^k\right]^t = e^{rt}$ by Prob. 51.

57. Using Simpson's rule and letting $I = \displaystyle\int_1^b (1/t)\,dt$, we find for

b = 2.7, I ≅ .9932535; for b = 2.71, I ≅ .9969488; for
b = 2.72, I ≅ 1.000632; for b = 2.715, I ≅ .9987919; and for
b = 2.719, I ≅ 1.000265. Thus $2.715 < e < 2.719$.

59a. If $x = y = 0$, then the given equation becomes $f(0) = [f(0)]^2$ or $f(0)[f(0)-1] = 0$, so either $f(0) = 0$ or $f(0) = 1$.

b. If $f(0) = 0$, then $f(x) = f(x+0) = f(x)f(0) = 0$ for all x.

59c. If for some x, f(x) = 0, then for all y, f(x+y)=f(x)f(y) = 0,
     so that f(0) = f(x-x) = f(x)f(-x) = 0, contrary to the
     assumption.
  d. If f(a) ≠ 0 for some a ≠ 0, then f(a) = f(a+0) = f(a)f(0), so
     f(0) = f(a)/f(a) = 1, and thus, from part(c), f(x) ≠ 0 for
     all x.
  e. Suppose for some a, f(a) = 0 and for some x, f(x) ≠ 0. The
     latter implies, by Part(d), that f(y) ≠ 0 for all y, and in
     particular, f(a) ≠ 0, which contradicts the hypothesis.
     Thus, we can not have both a and x for which f(a) = 0,
     f(x) ≠ 0. Therefore, if f(a) = 0 for some a, then f(x) = 0
     for all x.
  f. Suppose for some a ≠ b, f(a) ≠ f(b). Then f(a)f(b) ≠ [f(a)]^2
     so that f(a)[f(b)-f(a)] ≠ 0. Thus, for this a, f(a) ≠ 0 and
     by part(d), f(0) = 1.

61a. $\dfrac{f(x+h)-f(x)}{h} = \dfrac{f(x)f(h)-f(x)}{h} = f(x)\dfrac{f(h)-1}{h}$.   But, by Prob.59(f), if f

     is non-constant, then f(0) = 1 so $\dfrac{f(x+h)-f(x)}{h} = f(x)\dfrac{f(h)-f(0)}{h}$.

  b. $f'(x) = \lim\limits_{h\to 0}\dfrac{f(x+h)-f(x)}{h} = f(x)\lim\limits_{h\to 0}\dfrac{f(h)-f(0)}{h}$ so

     f'(x) = f'(0)f(x), provided f'(0) exists.
  c. If f'(0) = 1, then f satisfies the initial value problem
     f'(x) = f(x), and f(0) = 1.

Section 8.4, Page 428

1.   According to Eq.(4), the solution of $\dfrac{dQ}{dt} = 2Q$ is Q(t) = ce^{2t}.

     At t = 0, Q(0) = 4 = c.   Thus, Q(t) = 4e^{2t}.

3.   Rewrite du/dt + 3u = 0 as du/dt = -3u, whose general
     solution, given by Eq.(4), is u(t) = ce^{-3t}.   With u(2) = -1,
     we have -1 = ce^{-6} or c = -e^6.   Thus u(t) = -e^{6-3t} = -e$^{-3(t-2)}$.

5.   To solve dQ/dt - 2Q = 7, we proceed as in Ex.3.   Multiply
     both sides by e^{-2t}:   e^{-2t}(dQ/dt-2Q) = 7e^{-2t}.   Rewriting the

     left side, the equation becomes $\dfrac{d}{dt}$(Qe^{-2t}) = 7e^{-2t}.

     Integrating both sides yields Qe^{-2t} = -(7/2)e^{-2t} + c or
     Q(t) = -7/2 + ce^{2t}. Thus Q(0) = -7/2 + c = 3 or c = 13/2 so
     that Q(t) = -7/2 + (13/2)e^{2t}.

7.   Following Ex.3, multiply the equation by $e^{-1/2t}$ to obtain
$\frac{d}{dt}(ue^{-t/2}) = 2t$ so $ue^{-t/2} = t^2+c$ and $u(t) = t^2e^{t/2} + ce^{t/2}$.
Letting $t = 0$ and applying the initial condition, $u(0) = -2$,
yields $-2 = c$. The particular solution then becomes
$u(t) = t^2e^{t/2} - 2e^{t/2}$.

9.   The general solution is $Q(t) = ce^{rt}$. Letting $t = t_0$ and
applying the initial condition $Q(t_0) = Q_0$ yields $Q_0 = ce^{rt_0}$
or $c = Q_0e^{-rt_0}$. Thus, the particular solution is
$Q(t) = Q_0e^{-rt_0}e^{rt} = Q_0e^{r(t-t_0)}$.

11.   For $dQ/dt = rQ$, $Q(0) = Q_0$, we have the solution $Q(t) = Q_0e^{rt}$.
At $t = 11.7$ days, the material loses one-third its mass, with
two-thirds remaining so $Q(11.7) = (2/3)Q_0 = Q_0e^{11.7r}$.
Solving for r gives $r = [\ln(2/3)]/11.7$ so that
$Q(t) = Q_0e^{\ln(2/3)t/11.7}$. To find the half life, we determine
τ so that $Q(\tau) = Q_0/2$. Thus $Q(\tau) = Q_0/2 = Q_0e^{(\tau/11.7)\ln(2/3)}$.
Taking the logarithm of both sides: $\ln(1/2) = (\tau/11.7)\ln(2/3)$
or $\tau = \dfrac{-11.7\ln2}{\ln(2/3)} \cong 20.0$ days.

13a. The rate of change of the mass of thorium-234 comes from the
radioactive decay, rQ, as well as the addition of mass per
day of 1 mg/day. Thus the model is $dQ/dt = rQ + 1$. (Note
that r was determined in Ex.2 to be -0.02828). Employing the
technique of Ex.4, we obtain $e^{-rt}[dQ/dt - Q] = e^{-rt}$ or,
$\frac{d}{dt}[Qe^{-rt}] = e^{-rt}$, or $Q(t) = -1/r + ce^{rt}$. Since $Q(0) = 100\,mg$,
we have $100 = -1/r + c$ so $(100 + 1/r) = c$ or
$Q(t) = 35.36 + 64.64e^{-0.02828t}$ mg.

  b. $\lim\limits_{t\to\infty} Q(t) = \lim\limits_{t\to\infty} [35.36+64.64e^{-0.02828t}] = 35.36mg$.

  c. We must determine τ so that $Q(\tau) = 35.86$. Thus,
$Q(\tau) = 35.86 = 35.36 + 64.64e^{-0.02828\tau}$ so
$0.5/(64.64) = e^{-0.02828\tau}$ or
$\tau = -\ln[0.5/64.64]/(.02828) = 171.9$ days.

15a. The mathematical model for the insect population is
$P'(t) = rP(t)$, $P(0) = 10^4$, $P(2) = 2\cdot10^4$ where time is measured
in weeks. Thus, $P(t) = 10^4e^{rt}$ and hence $P(2) = 2\cdot10^4 = 10^4e^{2r}$
giving $r = \ln2/2$, or $P(t)=10^4\exp[(\ln2)t/2]$.

b. $P(26) = 10^4 \exp[(13)\ln 2] = 81.92 \cdot 10^6$.

c. $\lim_{t \to \infty} P(t) = \infty$. This does not appear realisitc.

d. A limited food supply would affect the death rate, which in turn would affect the growth rate (= birthrate - death rate).

17a. To show $u(t) = T + ce^{kt}$ satisfies $u'(t) = k(u-T)$, we may either solve the differential equation as in Ex.4 or substitute the function into the equation. We will choose the latter, in which case $u'(t) = kce^{kt}$ and $u(t) - T = ce^{kt}$ so that $k(u-T) = kce^{kt} = u'(t)$. Thus, $u'(t) = k(u-T)$.

b. Let $t = t_o$ so $u(t_o) = u_o = T + ce^{kt_o}$ or $c = (u_o-T)e^{-kt_o}$ and $u(t) = T + (u_o-T)e^{k(t-t_o)}$.

c. Since $k < 0$, and for $t > t_o$, we have

$$\lim_{t \to \infty} u(t) = \lim_{t \to \infty}[T + (u_0 - T)e^{k(t-t_o)}] = T, \text{ which is the ambient}$$

temperature, as expected.

19a. Since there are no additions to the account following the initial deposit of S_o, the model is $S'(t) = rS$, $S(0) = S_o$, which has the solution $S(t) = S_o e^{rt}$. We need to find T so that $S(T) = 2S_o$. Therefore, $S(T) = 2S_o = S_o e^{rT}$ or $T = (\ln 2)/r$ or $r = (\ln 2)/T$.

b. If T is to be 8 then $r = (\ln 2)/8 = 8.67\%$.

21a. With withdrawals at the rate of \$1000/month, or \$12,000/yr., $S'(t) = rS - 12,000$, $S(0) = 1.2 \cdot 10^5$ whose solution is: $S(t) = (12,000)/r + (120,000 - 12,000/r)e^{rt}$. To find T, solve $S(T) = 0 = (12,000)/r + (120,000 - 12,000/r)e^{rT}$ or $T = (1/r)\ln[1/(1-10r)]$.

b. For $r = 7\%$, $T = 17.2$ years and for $r = 9\%$, $T = 25.58$ years.

Section 8.5, Page 436

1. $y = 5^x = e^{x\ln 5}$ so $y' = \ln 5 e^{x\ln 5} = 5^x \ln 5$.

3. $y = 10^{\ln x} = e^{(\ln x)\ln 10}$ so
    $y' = e^{(\ln x)\ln 10}(\ln 10/x) = 10^{\ln x}(\ln 10/x)$.

5. $y = x^{-2} + e^{-x\ln 2}$ so $y' = -2x^{-3} - e^{-x\ln 2}\ln 2 = -2x^{-3} - 2^{-x}\ln 2$.

7. $y = 4^{\log_2 x} = 2^{2\log_2 x} = 2^{\log_2 x^2} = x^2$. Thus, $y' = 2x$.

9.   Using Eq. (29) with $u = x^2+x+1$, $y' = \dfrac{2x+1}{(x^2+x+1)\ln10}$.

11. $y = \log_5 25x^2 = x^2\log_5 5^2 = 2x^2\log_5 5 = 2x^2$. Thus, $y' = 4x$.

13. $y = \log_{10}(x^2+1)^2 = 2\log_{10}(x^2+1)$. By Eq. (29), then,
$$y' = \frac{2(2x)}{(x^2+1)\ln10} = \frac{4x}{(x^2+1)\ln10}.$$

15. Using logarithmic differentiation,
$\ln y = \sqrt{x}\ln x$ so $y'/y = (\sqrt{x}/x)+(\ln x/2\sqrt{x})$, or
$y' = (x^{\sqrt{x}}/\sqrt{x})+x^{\sqrt{x}}\ln x/2\sqrt{x} = x^{\sqrt{x}}(2+\ln x)/2\sqrt{x}$.

17. Using logarithmic differentiation, $\ln y = (\sin x)\ln x$ $(x>0)$, so
$y'/y = (\cos x)\ln x + (\sin x)/x$ and $y'= x^{\sin x}[(\cos x)\ln x + x^{-1}\sin x]$.

19. Using logarithmic differentiation, $\ln y = (1+x^2)\ln x$ so
$y'/y = 2x\ln x + (1+x^2)/x$ and $y' = x^{1+x^2}(2x\ln x + x^{-1} + x)$.

21. From Eq. (17), $\int 3^x dx = (3^x/\ln3) + c$.
(Note: $\int 3^x dx = \dfrac{1}{\ln3}\int 3^x\ln3\,dx = \dfrac{1}{\ln3}\int d(3^x)$).

23. Let $u = 2x$ so $dx = (1/2)du$. Thus,
$$\int 4^{2x}dx = \frac{1}{2}\int 4^u du = \frac{4^u}{2\ln4}+c = (4^{2x}/\ln16)+c.$$

25. Let $u = \cos2x$ so $du = -2\sin2x\,dx = -4\sin x\cos x\,dx$. Thus,
$$\int \sin x\cos x10^{\cos2x}dx = -\frac{1}{4}\int 10^u du = -\frac{10^u}{4\ln10}+c = (-10^{\cos 2x}/4\ln10)+c.$$

27. $\displaystyle\int_1^4 2^x dx = 2^x/\ln2\Big|_1^4 = (2^4-2)/\ln2 = 14/\ln2$.

29. $\displaystyle\int_{-1}^1 10^{2x}dx = 10^{2x}/2\ln10\Big|_{-1}^1 = (10^2 - 10^{-2})/\ln100 = 9999/100\ln100$.

31. $\displaystyle\int_0^5 5^{x/5}\,dx = 5^{x/5}/[(1/5)\ln5]\Big|_0^5 = 5(5-1)/\ln5 = 20/\ln5$.

33a. $a^x a^y = (e^{x\ln a})(e^{y\ln a}) = e^{(x\ln a\, +\, y\ln a)}$ by Eq. (5) of
      Section 8.3. Thus, $a^x a^y = e^{(x+y)\ln a} = a^{x+y}$.

33b. $a^{-x} = e^{-x\ln a} = 1/e^{x\ln a}$ by Eq. (6) of Section 8.3.
      Thus, $a^{-x} = 1/a^x$.

  c. $(a^x)^r = (e^{x\ln a})^r = e^{rx\ln a}$ by Eq. (8) of Section 8.3.
      Thus, $(a^x)^r = a^{rx}$.

35.   We will start with the given equation and from it derive a
      true statement. Each step being reversible then yields the
      desired result. Taking exponentials of both sides of the
      given equation gives $e^{(\ln a)\log_a |x|} = e^{(\ln b)\log_b |x|}$, so
      $e^{\ln(a^{\log_a |x|})} = e^{\ln(b^{\log_b |x|})}$. Thus, $e^{\ln|x|} = e^{\ln|x|}$, which
      is known to be true.

37.   The area will be $A = \displaystyle\int_0^1 5^{-x}\,dx = -5^{-x}/\ln 5\,\Big|_0^1 = 4/5\ln 5$.

39.   If $x = 2^{y+3}$, then $-\infty < y < \infty$ is the domain and $0 < x < \infty$ is the
      range. Next, $\log_2 x = y+3$, so $y = -3 + \log_2 x = f^{-1}(x)$ is the
      inverse function which has domain $0 < x < \infty$ and range $-\infty < y < \infty$.

41.   If $x = \log_a \dfrac{y-1}{2}$; then $1 < y < \infty$ is the domain and $-\infty < x < \infty$ is the

      range. Thus, $a^x = a^{\log_a\left(\frac{y-1}{2}\right)} = (y-1)/2$, so $y = 2a^x + 1 = f^{-1}(x)$
      is the inverse with domain $-\infty < x < \infty$ and range $y > 1$.

43a. Let $f(x) = x + a^x$. Note, $f(-1) = -1 + 1/a < 0$ and
      $f(1) = 1 + a > 0$ for $a > 1$. Since f is continuous on $[-1,1]$,
      $f(t) = 0$ for at least one t in $[-1,1]$. Also,
      $f'(x) = 1 + a^x\ln a > 0$ for all x, thus f is monotone
      increasing so can have no more than one zero. Thus, $f(x)$ has
      a unique root.

  b. If $f(x) = x + \log_a x$ then $f(a^{-2}) = a^{-2} - 2 < 0$ and
      $f(a^2) = a^2 + 2 > 0$ so, since f is continuous for all $x > 0$,
      there is at least one t in $[a^{-2}, a^2]$ for which $f(t) = 0$.
      Also, since $f'(x) = 1 + 1/x\ln a > 0$ for $x > 0$ and $a > 1$, f is
      monotone increasing so can have no more than one root. Thus
      $x + \log_a x = 0$ has a unique solution. (Note, in both (a) and
      (b) we showed f had at least one root on a given interval and
      at most one root for all x, which guarantees f has exactly
      one root.)

Section 8.6, Page 448

1. $\arctan(1) = \pi/4$ since $\tan(\pi/4) = 1$ and $\pi/4 \in (-\pi/2, \pi/2)$.

3. $\arcsin(1/\sqrt{2}) = \pi/4$ since $\sin(\pi/4) = 1/\sqrt{2}$ and $\pi/4 \in [-\pi/2, \pi/2]$.

5. $\arcsin(-1) = -\pi/2$ since $\sin(-\pi/2) = -1$ and $-\pi/2 \in [-\pi/2, \pi/2]$.

7.   Since $\cos(\pi/3) = \sin(\pi/6)$,
      $\arcsin[\cos(\pi/3)] = \arcsin[\sin(\pi/6)] = \pi/6$.

9. $\cos[\arcsin(1/2)] = \cos(\pi/6) = \sqrt{3}/2$.

11.   Let $\theta = \arctan x$ so $\tan\theta = x$ and $\cos(\arctan x) = \cos\theta = 1/\sqrt{1+x^2}$.

13.   Let $\sin w = x$, $\sin v = y$ and thus $\cos w = \sqrt{1-x^2}$, $\cos v = \sqrt{1-y^2}$,
      and $\sin(w+v) = \sin w \cos v + \sin v \cos w = x\sqrt{1-y^2} + y\sqrt{1-x^2}$. Hence
      $\arcsin x + \arcsin y = w + v = \arcsin(x\sqrt{1-y^2} + y\sqrt{1-x^2})$.

15.   Let $\theta = \arctan x$ and $\beta = \arctan y$, then
      $\tan(\theta+\beta) = (x+y)/(1-xy)$, so
      $\arctan x + \arctan y = \arctan[\tan(\theta+\beta)] = \arctan[(x+y)/(1-xy)]$.

17.   Let $\theta = \arcsin(1/x)$ so $\sin\theta = 1/x$ and $\csc\theta = x$. Thus
      $\text{arccsc}\, x = \text{arccsc}(\csc\theta) = \theta = \arcsin(1/x)$.

19.   Referring to Eq.(10) with $u = x^3/3$, $u' = x^2$, we have
      $\frac{d}{dx}[\arcsin(x^3/3)] = x^2/\sqrt{1-(x^3/3)^2} = 3x^2/\sqrt{9-x^6}$.

21.   Combining the power rule with Eq.(32) we have
      $\frac{d}{dx}[\arctan(x/2)]^2 = 2\arctan(x/2)\dfrac{1/2}{1+(x/2)^2} = 4\arctan(x/2)/(4+x^2)$.

23.   Using the chain rule with $u = 1-x^2$ and Eq.(23) we have
      $\frac{d}{dx}[\arccos(1-x^2)] = -(-2x)/\sqrt{1-(1-x^2)^2} = 2x/\sqrt{2x^2-x^4}$.

25.   From Eq.(44) we have $\dfrac{d}{dx}[\text{arcsec}(x^2)] = \dfrac{2x}{x^2\sqrt{x^4-1}} = \dfrac{2}{x\sqrt{x^4-1}}$.

27. $f'(x) = \left(\sqrt{1-x^2}\right)/\sqrt{1-x^2} - (x\arcsin x)/\sqrt{1-x^2}$

$\quad = 1 - (x\arcsin x)/\sqrt{1-x^2}.$

29. Using Eq. (32) with $u = 2\tan x$, $\quad f'(x) = \dfrac{2\sec^2 x}{1+4\tan^2 x}.$

31. Let $u = 2x/\sqrt{1-4x^2}$ then $u' = 2/(1-4x^2)3/2$, so by Eq. (32)

$\quad f'(x) = \dfrac{2/(1-4x^2)3/2}{1+\left(2x/\sqrt{1-4x^2}\right)^2} = 2/\sqrt{1-4x^2}.$ Alternatively, if

$\quad \theta = \arctan(2x/\sqrt{1-4x^2})$, then $\theta = \arcsin(2x)$ also, so

$\quad d\theta/dx = f'(x) = \dfrac{2}{\sqrt{1-4x^2}}.$

33. $f'(x) = 2/[(\arctan 2x)(1+4x^2)].$

35. Referring to Eq. (16) with $a = 3$, $\displaystyle\int\dfrac{dx}{\sqrt{9-x^2}} = \arcsin(x/3) + c.$

37. $\displaystyle\int\dfrac{dx}{\sqrt{1-4x^2}} = \int\dfrac{dx}{\sqrt{1-(2x)^2}} = \frac{1}{2}\int\dfrac{du}{\sqrt{1-u^2}}$ with $u = 2x$.   Thus,

$\displaystyle\int\dfrac{dx}{\sqrt{1-4x^2}} = \frac{1}{2}\int\dfrac{du}{\sqrt{1-u^2}} = \frac{1}{2}\arcsin(u)+c = \frac{1}{2}\arcsin(2x)+c.$

39. $\displaystyle\int\dfrac{dx}{4+9x^2} = \frac{1}{4}\int\dfrac{dx}{1+(3x/2)^2}. = (1/6)\arctan(3x/2) + c.$

41. $\displaystyle\int\dfrac{dx}{|x|\sqrt{x^2-4}} = \int\dfrac{dx}{2|x|\sqrt{\left(\frac{x}{2}\right)^2-1}} = \frac{1}{4}\int\dfrac{dx}{\left|\frac{x}{2}\right|\sqrt{\left(\frac{x}{2}\right)^2-1}} = \frac{1}{2}\int\dfrac{du}{|u|\sqrt{u^2-1}},$

$\quad = \frac{1}{2}\operatorname{arcsec}(u)+c = \frac{1}{2}\operatorname{arcsec}(x/2)+c.$

43. $\displaystyle\int\dfrac{dx}{\sqrt{1-(x+1)^2}} = \int\dfrac{du}{\sqrt{1-u^2}}, = \arcsin(u) + c = \arcsin(x+1) + c.$

45. $\int \dfrac{(\arcsin x)^2}{\sqrt{1-x^2}} dx = \int u^2\, du = u^3/3 + c = (\arcsin x)^3/3 + c.$

47. Let $x = \cos y$, $0 \le y \le \pi$, then $\sin y = \sqrt{1-x^2}$ and implicit differentation yields $1 = -(\sin y)dy/dx$ or
$$dy/dx = -1/\sin y = -1/\sqrt{1-x^2}.$$

49. Let $x = \sec y$, then $\tan y = \sqrt{x^2-1}$ and $1 = (\sec y \tan y)y'$ or
$y' = 1/(\sec y \tan y) = 1/|x|\sqrt{x^2-1}$, where $|x|$ is needed since
$y' > 0$, according to Fig.8.6.6b.

51a. Referring to Eq.(10) with $u = \cos x$, we have
$$f'(x) = \dfrac{-\sin x}{\sqrt{1-\cos^2 x}} = -\dfrac{\sin x}{\sqrt{\sin^2 x}} = -\dfrac{\sin x}{|\sin x|}.$$

b.

c. f is continuous as shown by:

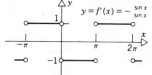

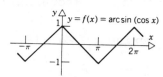

53a. $A(\alpha) = \displaystyle\int_0^{\alpha} \dfrac{dx}{\sqrt{1-x^2}} = \arcsin x \Big|_0^{\alpha} = \arcsin \alpha.$

b. $\displaystyle\lim_{\alpha\to 1-}\int_0^{\alpha} \dfrac{dx}{\sqrt{1-x^2}} = \lim_{\alpha\to 1-} \arcsin \alpha = \pi/2.$

55. $\alpha = \arctan(a/x)$, $\beta = \arctan(b/x)$,
$A = \beta - \alpha = \arctan(b/x) - \arctan(a/x)$,
and $\dfrac{dA}{dx} = \dfrac{-b/x^2}{1+b^2/x^2} - \dfrac{-a/x^2}{1+a^2/x^2}.$

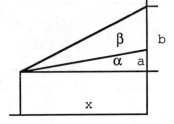

Thus $\dfrac{dA}{dx} = -\dfrac{b(x^2+a^2)-a(x^2+b^2)}{(x^2+b^2)(x^2+a^2)} = 0$

when $(b-a)x^2 - ab(b-a) = 0$ so $x = \sqrt{ab}$ or $x = -\sqrt{ab}$. Now dA/dx
is positive for $-\sqrt{ab}<x<\sqrt{ab}$ and negative for $x>\sqrt{ab}$. Thus
$x = \sqrt{ab}$ maximizes the angle.

Section 8.7, Page 458

1. $\sinh x \cosh y + \cosh x \sinh y = (e^x - e^{-x})(e^y + e^{-y})/4 + (e^x + e^{-x})(e^y - e^{-y})/4$

$= (e^{x+y} + e^{x-y} - e^{-(x-y)} - e^{-(x+y)} + e^{x+y} - e^{x-y} + e^{y-x} - e^{-(x+y)})/4$

$= (e^{x+y} + e^{-(x+y)})/2 = \sinh(x+y)$. From this result we have
$\sinh(x-y) = \sinh[x+(-y)] = \sinh x \cosh(-y) + \cosh x \sinh(-y)$

$= \sinh x \cosh y - \cosh x \sinh y$ by Eqs. (5) and (6).

3. $\sinh^2(x/2) = (e^{x/2} - e^{-x/2})^2/4 = (e^x + e^{-x})/4 - 2/4$

$= (\cosh x)/2 - 1/2 = (\cosh x - 1)/2$. Thus,
for $x < 0$, $\sinh(x/2) = -\sqrt{(\cosh x - 1)/2}$ and,
for $x > 0$, $\sinh(x/2) = \sqrt{(\cosh x - 1)/2}$.

5. Following the discussion from Eq. (47) and following:
If $y = \text{arccosh} x$ then $\cosh y = x$ for $x \geq 1$ so $x = (e^y + e^{-y})/2$,
and $e^y - 2x + e^{-y} = 0$ or $e^{2y} - 2xe^y + 1 = 0$. Now,
from the quadratic formula, $e^y = \left(2x \pm \sqrt{4x^2 - 4}\right)/2 = x \pm \sqrt{x^2 - 1}$ or
$y = \ln(x \pm \sqrt{x^2 - 1})$. Choose $x + \sqrt{x^2 - 1}$ so that $y \to \infty$ as $x \to \infty$.

7. Using Eq. (22) with $u(x) = 3x$: $\dfrac{d}{dx}[\sinh 3x] = 3\cosh 3x$.

9. Using the product rule: $\dfrac{d}{dx}[x\cosh x] = \cosh x + x\sinh x$.

11. Using the product rule:
$\dfrac{d}{dx}[\sinh x \exp(\cosh x)] = \sinh^2 x \exp(\cosh x) + \cosh x \exp(\cosh x)$

$= (\sinh^2 x + \cosh x)\exp(\cosh x)$.

13. Differentiate as a quotient:
$$f'(x) = \frac{(4 + \sinh^2 x)\sinh x - 2\sinh x \cosh^2 x}{(4 + \sinh^2 x)^2}$$

$$= \frac{4\sinh x + 2\sinh^3 x - 2\sinh x \cosh^2 x - \sinh^3 x}{(4 + \sinh^2 x)^2}$$

$$= \frac{4\sinh x + 2\sinh x(\sinh^2 x - \cosh^2 x) - \sinh^3 x}{(4 + \sinh^2 x)^2}$$

$$= (2\sinh x - \sinh^3 x)/(4 + \sinh^2 x)^2, \quad \text{by Eq. (15)}.$$

15. Using Eq.(35) with $u(x) = x/2$:
     $f'(x) = 1/2\sqrt{1+(x/2)^2} = (4+x^2)^{-1/2}$.

17. Using Eq.(25) with $u(x) = 2x$:
     $\int \sinh 2x\,dx = (1/2)\int 2\sinh 2x\,dx = (1/2)\int \sinh u\,du$
     $\qquad\qquad = (1/2)\cosh u + c = (1/2)\cosh(2x) + c$.

19. Let $u = \sinh x$, then
     $\int \sinh^2 x \cosh x\,dx = \int u^2 du = (1/3)u^3 + c = (1/3)\sinh^3 x + c$.

21. Using Eq.(27), $\int \coth 2x\,dx = (1/2)\int 2\dfrac{\cosh 2x}{\sinh 2x}dx = (1/2)\int \dfrac{\cosh u}{\sinh u}du =$
     $(1/2)\ln|\sinh u| + c. = (1/2)\ln|\sinh 2x| + c$.

23. Using Eq.(45), $\displaystyle\int \frac{dx}{\sqrt{4x^2 - 1}} = (1/2)\int \frac{2dx}{\sqrt{(2x)^2 - 1}} = (1/2)\int \frac{du}{\sqrt{u^2 - 1}} =$

     $(1/2)\text{arccosh}\,u + c = (1/2)\text{arccosh}\,2x + c,\ x > 1/2$.

25. Using Eq.(39) with $a = 4$, $\displaystyle\int \frac{dx}{\sqrt{x^2+16}} = \text{arcsinh}(x/4) + c$.

27. $\displaystyle\int_{\ln 2}^{\ln 3} \cosh x\,dx = \sinh x\Big|_{\ln 2}^{\ln 3} = \sinh(\ln 3) - \sinh(\ln 2)$.   In general,

     $\sinh(\ln a) = \dfrac{e^{\ln a} - e^{-\ln a}}{2} = \dfrac{a^2 - 1}{2a}$ $(a>0)$.   Thus,

     $\displaystyle\int_{\ln 2}^{\ln 3} \cosh x\,dx = (9-1)/6 - (4-1)/4 = 8/6 - 3/4 = 7/12$.

29. Following Prob.(21), $\displaystyle\int_{\ln 8}^{\ln 64} \coth\frac{x}{3}\,dx = 3\ln\left|\sinh\frac{x}{3}\right|\,\Big|_{3\ln 2}^{3\ln 4} =$

     $3[\ln(\sinh \ln 4) - \ln(\sinh \ln 2)] = 3[\ln(15/8) - \ln(3/4)] = 3\ln(5/2)$.

31. $\displaystyle\int_1^2 \frac{dx}{\sqrt{9x^2 - 1}} = \frac{1}{3}\int_1^2 \frac{3dx}{\sqrt{(3x)^2 - 1}} = (1/3)\text{arccosh}\,3x\Big|_1^2$

     $\qquad\qquad\qquad = (\text{arccosh}\,6 - \text{arccosh}\,3)/3$.

33. $\dfrac{d}{dx}(\tanh x) = \dfrac{d}{dx}\left(\dfrac{\sinh x}{\cosh x}\right) = \dfrac{\cosh^2 x - \sinh^2 x}{\cosh^2 x} = 1/\cosh^2 x = \operatorname{sech}^2 x.$

35. $\dfrac{d}{dx}(\operatorname{sech} x) = \dfrac{d}{dx}(\cosh x)^{-1} = -(\cosh x)^{-2}\sinh x$

$\qquad\qquad = -(1/\cosh x)(\sinh x/\cosh x) = -\operatorname{sech} x \tanh x.$

37a. y = arctanh x is defined as the inverse of the function
      y = tanh x. Thus, the domain and range of arctanh x are the
      range and domain of tanh x, respectively. Recall that
      tanh x = (sinh x)/(cosh x) is defined
      and continuous for all x, thus, it
      has domain $-\infty < x < \infty$. Further, from

      Prob. 33, $\dfrac{d}{dx}(\tanh x) = \operatorname{sech}^2 x > 0$ for

      all x, so tanh x is monotone
      increasing. Finally, since
      $\displaystyle\lim_{x \to -\infty} \tanh x = -1$ and $\displaystyle\lim_{x \to \infty} \tanh x = 1,$

      we have that tanh x has range $-1 < y < 1$.
      Therefore the domain of

      y = arctanh x is $-1 < x < 1$ and the range is $-\infty < y < \infty$. The graph
      of y = arctanh x is the reflection of the graph of y = tanh x
      across the line y = x. To compute the derivative of
      y = arctanh x, we differentiate x = tanh y implicitly getting
      $1 = y'\operatorname{sech}^2 y$ or $y' = \cosh^2 y$. Note that

      $1 - \tanh^2 x = \dfrac{\cosh^2 x - \sinh^2 x}{\cosh^2 x} = \operatorname{sech}^2 x.$ Thus

      $y' = \cosh^2 y = 1/\operatorname{sech}^2 y = 1/(1 - \tanh^2 y) = 1/(1 - x^2),$ for $|x| < 1.$

  b. $\dfrac{d}{dx}\left[\dfrac{1}{2}\ln\dfrac{|1+x|}{|1-x|}\right] = \dfrac{d}{dx}\left[\dfrac{1}{2}(\ln|1+x| - \ln|1-x|)\right] = \dfrac{1}{2}\left[\dfrac{1}{1+x} + \dfrac{1}{1-x}\right] = \dfrac{1}{1-x^2}.$

  c. Since $\dfrac{d}{dx}[\operatorname{arctanh} x] = \dfrac{d}{dx}\left[\dfrac{1}{2}\ln\dfrac{|1+x|}{|1-x|}\right],$ we have that

      $\operatorname{arctanh} x - \dfrac{1}{2}\ln\dfrac{|1+x|}{|1-x|} = c$ for some constant c and all $|x| < 1.$

      But $\operatorname{arctanh} 0 - (1/2)\ln\dfrac{|1|}{|1|} = 0,$ so $\operatorname{arctanh} x = \dfrac{1}{2}\ln\left(\dfrac{1+x}{1-x}\right),$ $|x| < 1.$

39a. With $x = \cosh t$, $y = \sinh t$, $x^2-y^2 = \cosh^2 t-\sinh^2 t = 1$, so that the point (x,y) is on the hyperbola. Further, for $-\infty < t < \infty$, $-\infty < \sinh t < \infty$ and $1 \le \cosh t < \infty$, so the entire right branch is traced.

b. From Fig.8.7.9(b), $A(t) = \text{Area}(OQP) = \text{Area}(OPR) - \text{Area}(QPR)$. But, $\text{Area}(OPR) = xy/2 = (1/2)\cosh t \sinh t$ and

$$\text{Area}(QPR) = \int_{x=Q}^{x=R} y\, dx = \int_1^{\cosh t} \sqrt{x^2-1}\, dx. \quad \text{Thus,}$$

$$A(t) = (1/2)\cosh t \sinh t - \int_1^{\cosh t} \sqrt{x^2-1}\, dx.$$

c. $A'(t) = (1/2)(\cosh^2 t + \sinh^2 t) - \sqrt{\cosh^2 t-1}\,\sinh t =$

$(1/2)(\cosh^2 t+\sinh^2 t)-\sinh^2 t = (1/2)(\cosh^2 t-\sinh^2 t) = 1/2.$

d. Since $A'(t) = 1/2$, $A(t) = t/2 + c$. But $A(0) = 0 = c$ so that $A(t) = t/2$ or $t = 2A(t)$.

e. Following the procedure in parts (b),(c),and(d), $A(t) = \text{Area}(OPQ) = \text{Area}(OPR) + \text{Area}(RPQ)$. Where $\text{Area}(OPR) = xy/2 = (1/2)\cos t \sin t$ and

$$\text{Area}(RPQ) = \int_{\cos t}^1 \sqrt{1-x^2}\, dx \quad |x|<1, \text{ so that}$$

$$A(t) = (1/2)\cos t \sin t - \int_1^{\cos t} \sqrt{1-x^2}\, dx. \quad \text{Next,}$$

$A'(t) = (1/2)(\cos^2 t - \sin^2 t) + \sin t \sqrt{1-\cos^2 t}$

$= (1/2)(\cos^2 t + \sin^2 t) = 1/2$ which yields $A(t) = t/2$ or $t = 2A(t)$ as in part (d).

f. The parameter t for the circle may also be interpreted as $\angle POQ$, and since the arclength $s = rt$ and $r = 1$, t also represents arclength. For the hyperbola, however, t no longer represents $\angle POQ$ [Fig.8.7.9(b)]. Also, for the hyperbola, the arclength from $t = 1$ to $t = \alpha$ $(\alpha>1)$ is given by $s(\alpha) = \int_1^\alpha \sqrt{\cosh^2 t+\sinh^2 t}\, dt > \int_1^\alpha \sqrt{\cosh^2 t}\, dt$. Thus

$s(\alpha) > \int_1^\alpha \cosh t\, dt = \sinh \alpha > \alpha$ for $\alpha>1$ and hence $s(\alpha) \ne \alpha$, which indicates that the parameter is not the arclength.

Chapter 8 Review, Page 459

1. By Theorem 8.1.1, since $f'(x) = 1/2f(x)$, $d[f^{-1}(y)]/dy = 2y$ so
    $(f^{-1})'(x) = 2x$ and $f^{-1}(x) = x^2+c$. Also, as $f(0) = 0$,
    $f^{-1}(0) = 0$, so $c = 0$, $f^{-1}(x) = x^2$, and $f(x) = \sqrt{x}$.

3. $d[f^{-1}(y)]/dy = 1/(1+y^2)$ so $(f^{-1})'(x) = 1/(1+x^2)$ and
    $f^{-1}(x) = \arctan x + c$. Since $f(0) = 0$, $f^{-1}(0) = 0$ and $c = 0$.
    Hence $f^{-1}(x) = \arctan x$ and $f(x) = \tan x$.

5. Since $y' = 2x > 0$ for $x > 0$, $f(x) = x^2$ has an inverse for
    $x \geq 0$, and $f^{-1}(x) = \sqrt{x}$.

7. With $f'(x) = 2ax+b$, $f^{-1}(x)$ exists for $x \geq -b/2a$. Now solve
    $ax^2 + bx + (c-y) = 0$ for x to obtain $x = (1/2a)[-b+\sqrt{b^2 - 4a(c-y)}]$
    since $x \geq -b/2a$. Thus $f^{-1}(x) = (1/2a)[-b + \sqrt{b^2 - 4a(c - x)}]$.

9. Since $f(-1) = 1 = f(1)$, f has no inverse for $-\infty < x < \infty$.

11. Let $y = (e^x+e^{-x})/(e^x-e^{-x}) = (e^{2x}+1)/(e^{2x}-1)$. Thus
    $e^{2x} = (y+1)/(y-1)$ or $x = (1/2)\ln[(y+1)/(y-1)]$, and hence
    $f^{-1}(x) = (1/2)\ln[(x+1)/(x-1)]$ for $x < -1$ or $x > 1$.

13. We have $\ln y = x \ln x \ln 2$.   Thus $y'/y = \ln x \ln 2 + (x\ln 2)/x$ or
    $y' = \ln 2(\ln x + 1)y = 2^{x\ln x}\ln 2(\ln x + 1)$.

15. $y' = -(e^{-x})/\sqrt{1-(e^{-x})^2} = -e^{-x}/\sqrt{1-e^{-2x}}$.

17. Since $y = x\log_a e$, $y' = \log_a e = 1/\ln a$.

19. $y' = [1/\sqrt{(\ln \sinh x)^2 + 1}](1/\sinh x)(\cosh x)$
    $= \coth x/\sqrt{(\ln \sinh x)^2 + 1}$.

21. $y' = \cosh x/(1+\sinh^2 x) = 1/\cosh x$.

23. Let $(ax+b)^2 = u$, then $(ax+b)dx = du/2a$ and
    $I = (1/2a)\int e^u du = (1/2a)e^u + c = (1/2a)e^{(ax+b)^2} + c$.

25. Let $u = \tan x$, then $du = \sec^2 x\, dx$.   Recall that $1+\tan^2 x = \sec^2 x$,
    so $I = \int 2^u du = 2^u/\ln 2 + c = (2^{\tan x})/\ln 2 + c$.

27. With $u = \sin x$, $I = \int \pi^u du = \pi^u / \ln \pi + c = (\pi^{\sin x}) / \ln \pi + c$.

29. With $u = \sin(\pi/x)$, $du = -(\pi/x^2)\cos(\pi/x)\,dx$ and

$$I = \frac{-1}{\pi} \int_0^1 2^u du = \left. \frac{-2^u}{\pi \ln 2} \right|_0^1 = -1/\pi \ln 2.$$

31. $u = (x-4)/2$, then $I = \dfrac{1}{2} \int \dfrac{du}{|u|\sqrt{u^2-1}} = (1/2)\operatorname{arcsec}[(x-4)/2] + c$.

33. With $u = \sqrt{x}$, we have $du = (1/2\sqrt{x})\,dx$ and

$$I = 2\int \frac{du}{1+u^2} = 2\arctan u + c = 2\arctan\sqrt{x} + c.$$

35. Let $u = \ln x$, then as $x \to 1$, $u \to 0$, so the required limit is
$$\lim_{u \to 0} \frac{\sin u}{u} = 1.$$

37. Let $u = 1/t$ so as $t \to 0+$, $u \to \infty$, and the limit becomes
$$\lim_{u \to \infty} \frac{u^m}{e^u} = 0, \text{ by Prob.48 Section 8.3.}$$

39. With $f(x) = \arcsin x$, then $f'(0) = 1 = \lim\limits_{x \to 0} \dfrac{\arcsin x}{x}$.

41a. For $f(t) = 1/(1+t^2)$, Theorem 6.3.5 gives
$$\frac{1}{x} \int_0^x \frac{dt}{1+t^2} = \frac{1}{1+c^2}, \text{ where } 0 \le c \le x. \text{ Since } \frac{1}{1+x^2} \le \frac{1}{1+c^2} \le 1$$

for $0 \le c < x$, we have $\dfrac{1}{1+x^2} \le \dfrac{1}{x} \int_0^x \dfrac{dt}{1+t^2} = \dfrac{1}{x}\arctan x \le 1$.

Thus $\dfrac{x}{1+x^2} \le \arctan x \le x$.

b. If $x<0$, then at the last step, multiplying by x reverses the
sense of all inequalities yielding $x \le \arctan x \le x/(1+x^2)$.

43. The equation may be written as $(\ln x)(\ln x - 2) = 0$, so $x = 1$
and $x = e^2$ are the solutions.

45.   Using Newton's method, $x_{n+1} = x_n - \dfrac{\sinh^2 x_n - 2\sqrt{\cosh^2 x_n - 1} + 1}{\sinh 2x_n \left(1 - 1/\sqrt{\cosh^2 x_n - 1}\right)}$.

With $x_0 = 1$, we find $x_1 = .9432$, $x_2 = .9130$, ... $x_{11} \cong .8814$.
Using the identity $\cosh^2 x - 1 = \sinh^2 x$ gives the equation
$\sinh^2 x - 2\sinh x + 1 = (\sinh x - 1)^2 = 0$ or $\sinh x = 1$ so that
$x = \operatorname{arcsinh} 1 = .8814$. With $x_0 = 0$, $\cosh^2 x_0 - 1 = 0$, which
results in division by zero. Also, the only root of the
equation is a double root.

47.   From Prob. 35 of Section 8.5, $\log_a |x| = (\ln b/\ln a) \log_b |x|$. Let
$a = 10$, $b = e = x$, so we have $\log_{10} e = (\ln e/\ln 10) \ln e = 1/\ln 10$.

49.   With $f'(x) = 2 \ln x\, e^{(\ln x)^2}/x$ and
$f''(x) = (2/x^2) e^{(\ln x)^2} [2(\ln x)^2 - \ln x + 1]$
we have $f'(x) < 0$ for $0 < x < 1$ and
$f'(x) > 0$ for $x > 1$. Also $f''(x) > 0$
for all x. Thus f is decreasing for
$0 < x \le 1$, increasing for $x \ge 1$,
concave upward for all $x > 0$, and
has a global minimum at $x = 1$ of 1.

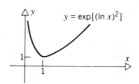

51.   As $f'(x) = \cos x\, e^{\sin x}$ and
$f''(x) = (\cos^2 x - \sin x) e^{\sin x}$, f has
critical points at 0, $\pi/2$, $3\pi/2$ and 2π.
Since $f''(\pi/2) = -e$ and $f''(3\pi/2) = 1/e$,
f has a relative maximum of e at
$x = \pi/2$ and a relative minimum of $1/e$
at $x = 3\pi/2$. Since $f(0) = 1 = f(2\pi)$,
$x = \pi/2$ gives a global maximum while $x = 3\pi/2$ gives a global
minimum. f increases for $0 \le x \le \pi/2$ and $3\pi/2 \le x \le 2\pi$, and
decreases otherwise.   There is a point of inflection when
$x = \alpha$, where $\cos^2 \alpha - \sin \alpha = 1 - \sin \alpha - \sin^2 \alpha = 0$, or
$\alpha = \arcsin[(-1+\sqrt{5})/2]$, as well as when $x = \pi - \alpha$.

53.   Note that $f(x) = e^x = f'(x) = f''(x)$.
Thus f has no critical points,
increases for all x, and is
concave up everywhere.

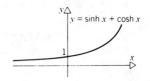

CHAPTER 9

Section 9.1, Page 466

1. $du = 8xdx \Rightarrow xdx = du/8$ and thus
$$\int \frac{3xdx}{1+4x^2} = (3/8)\int \frac{du}{u} = (3/8)\ln|u| + c = (3/8)\ln(1+4x^2) + c.$$

3. $du = e^x dx$ and thus
$$\int \frac{e^x\,dx}{\sqrt{1+e^x}} = \int \frac{du}{\sqrt{1+u}} = 2(1+u)^{1/2} + c = 2\sqrt{1+e^x} + c.$$

5. $du = dx/x$ and thus $\int \frac{(\ln x)^3\,dx}{x} = \int u^3 du = u^4/4 + c = (\ln x)^4/4 + c.$

7. $\int \cot x\,dx = \int \frac{\cos x\,dx}{\sin x}$. Now let $u = \sin x$, then $du = \cos x\,dx$ and
$$\int \cot x\,dx = \int \frac{du}{u} = \ln|u| + c = \ln|\sin x| + c.$$

9. Let $u = \sin x$, then $du = \cos x\,dx$ and thus
$$\int \frac{2\cos x\,dx}{1 + \sin^2 x} = \int \frac{2du}{1+u^2} = 2\arctan u + c = 2\arctan(\sin x) + c.$$

11. Let $u = e^{2x}$ then $du = 2e^{2x}dx$ and thus
$$\int \frac{e^{2x}\,dx}{\sqrt{1-e^{4x}}} = (1/2)\int \frac{du}{\sqrt{1-u^2}} = (1/2)\arcsin u + c$$
$$= (1/2)\arcsin(e^{2x}) + c, \quad x < 0.$$

13. Let $u = 1+x$, then $x = u-1$ and $dx = du$. Thus
$$\int_1^3 \frac{xdx}{1+x} = \int_2^4 \frac{(u-1)du}{u} = \int_2^4 (1 - 1/u)du = (u - \ln|u|)\Big|_2^4$$
$$= (4-\ln 4) - (2-\ln 2) = 2 - \ln 2 \cong 1.3069.$$

15. Let $u = \sinh(x/2)$, then $du = (1/2)\cosh(x/2)\,dx$ and thus
$$\int \cosh(x/2)\sinh^4(x/2)\,dx = 2\int u^4\,du = (2/5)u^5 + c$$
$$= (2/5)\sinh^5(x/2) + c.$$

17.   Let $u = \ln x$, then $du = dx/x$ and thus
$$\int \frac{\cos(\ln x)}{x}\, dx = \int \cos u\, du = \sin u + c = \sin(\ln x) + c.$$

19.   Let $u = 7 + 2x^3$, then $du = 6x^2\, dx$ and thus
$$\int x^2 (7+2x^3)^{1/3}\, dx = (1/6)\int u^{1/3}\, du = (1/6)u^{4/3}/(4/3) + c$$
$$= (1/8)(7 + 2x^3)^{4/3} + c.$$

21.   Let $u = \ln x$, then $du = dx/x$ and thus
$$\int_2^e \frac{dx}{x\ln x} = \int_{\ln 2}^1 \frac{du}{u} = \ln|u|\Big|_{\ln 2}^1 = -\ln(\ln 2) \cong .3665.$$

23.   Let $u = \cosh 2x$, then $du = (1/2)\sinh 2x\, dx$ and thus
$$\int \tanh 2x\, dx = \int (\sinh 2x/\cosh 2x)\, dx = (1/2)\int du/u$$
$$= (1/2)\ln|u| + c = (1/2)\ln(\cosh 2x) + c.$$

25.   Let $u = \sqrt{x^2+4}$, then $x^2 = u^2 - 4$ and $du = \dfrac{x\, dx}{\sqrt{x^2+4}}$ or $x\, dx = u\, du$.

Thus $\displaystyle\int_0^2 x^3\sqrt{x^2+4}\, dx = \int_2^{2\sqrt{2}} (u^2-4)\, u\, (u\, du) = \int_2^{2\sqrt{2}} (u^4 - 4u^2)\, du =$

$u^5/5 - (4/3)u^3\Big|_2^{2\sqrt{2}} = (2^7\sqrt{2}/5 - 4 \cdot 2^4\sqrt{2}/3) - (2^5/5 - 4 \cdot 2^3/3)$

$= (32/15)[(12\sqrt{2} - 10\sqrt{2}) - (3 - 5)] = 64(\sqrt{2} + 1)/15 \cong 10.30.$

27. $\displaystyle\int \frac{dx}{\sqrt{4x - x^2}} = \int \frac{dx}{\sqrt{4 - (x^2 - 4x + 4)}} = \frac{1}{2}\int \frac{dx}{\sqrt{1 - [(x-2)/2]^2}}.$

Thus let $u = (x-2)/2$, so $du = dx/2$ and
$$\int \frac{dx}{\sqrt{4x - x^2}} = \int \frac{du}{\sqrt{1-u^2}} = \arcsin u + c = \arcsin[(x-2)/2] + c.$$

29. $\displaystyle\int \frac{(4x + 18)\, dx}{x^2 + 4x + 29} = \int \frac{(4x + 8)\, dx}{x^2 + 4x + 29} + \int \frac{10\, dx}{(x + 2)^2 + 25}.$ Let $u = x^2 + 4x + 29$

in the first integral, so $du = (2x+4)\, dx$, and in the second

let $w = x+2$, so $dw = du$. Thus $\displaystyle\int \frac{(4x + 18)\, dx}{x^2 + 4x + 29} = 2\int \frac{du}{u} + 10\int \frac{dw}{w^2 + 25} =$

$2\ln(x^2 + 4x + 29) + 2\arctan[(x+2)/5] + c.$

31. $$\int \frac{dx}{\sqrt{4x^2+16x+17}} = \int \frac{dx}{\sqrt{4(x^2+4x+4)+1}} = \int \frac{dx}{\sqrt{[2(x+2)]^2+1}}.$$

Thus let $u = 2(x+2)$ so $du = 2dx$ and $\int \dfrac{dx}{\sqrt{4x^2+16x+17}} =$

$(1/2)\int \dfrac{du}{\sqrt{u^2+1}} = (1/2)\,\text{arcsinh}\,u + c = (1/2)\,\text{arcsinh}\,(2x+4) + c.$

33. Since $2x^2+8x+26 = 2[(x^2+4x+4)+9] = (18)\{[(x+2)/3]^2+1\}$
we let $u = (x+2)/3$ and hence $du = dx/3$. Thus

$$\int_1^4 \frac{3\,dx}{2x^2+8x+26} = (1/6)\int_1^4 \frac{dx}{[(x+2)/3]^2+1} = (1/2)\int_1^2 \frac{du}{u^2+1} =$$

$(1/2)\,\text{arctan}\,u \Big|_1^2 = (1/2)(\text{arctan}\,2 - \text{arctan}\,1) \cong .1609.$

35. If $\phi = \pi - \theta$ then $d\phi = -d\theta$ and we have

$$\int_0^\pi \phi f(\sin\phi)\,d\phi = \int_\pi^0 (\pi-\theta)f[\sin(\pi-\theta)](-d\theta) = \int_0^\pi (\pi-\theta)f(\sin\theta)\,d\theta,$$

since $\sin(\pi-\theta) = \sin\theta$. Writing the last integral as the difference of two integrals we then have

$$\int_0^\pi \phi f(\sin\phi)\,d\phi = \pi\int_0^\pi f(\sin\theta)\,d\theta - \int_0^\pi \theta f(\sin\theta)\,d\theta.$$

$$= \pi\int_0^\pi f(\sin\phi)\,d\phi - \int_0^\pi \phi f(\sin\phi)\,d\phi \quad \text{where we have let}$$

$\theta = \phi$ and $d\theta = d\phi$ on the right. By adding the last integral to

both sides we then have $2\displaystyle\int_0^\pi \phi f(\sin\phi)\,d\phi = \pi\int_0^\pi f(\sin\theta)\,d\theta$ which

gives the desired result when both sides are divided by 2.

Section 9.2, Page 472

1. $u = x$, $dv = \cos x\,dx \Rightarrow du = dx$, $v = \sin x$. Thus

$$\int x\cos x\,dx = x\sin x - \int \sin x\,dx = x\sin x + \cos x + c.$$

3.　$u = x^2$, $dv = \sin\pi x\, dx \Rightarrow du = 2x\,dx$, $v = -(1/\pi)\cos\pi x$. Thus

$$\int x^2\sin\pi x\, dx = -(1/\pi)x^2\cos\pi x + (2/\pi)\int x\cos\pi x\, dx. \text{ Now let}$$

$u = x$, $dv = \cos\pi x\, dx \Rightarrow du = dx$, $v = (1/\pi)\sin\pi x$. Thus

$$\int x^2\sin\pi x\, dx = -(1/\pi)x^2\cos\pi x + (2/\pi^2)(x\sin\pi x - \int \sin\pi x\, dx)$$

$$= -(1/\pi)x^2\cos\pi x + (2/\pi^2)x\sin\pi x + (2/\pi^3)\cos\pi x + c.$$

5.　$u = x$, $dv = \sin\pi x\, dx \Rightarrow du = dx$, $v = -(1/\pi)\cos\pi x$. Thus

$$\int_0^{\frac{1}{2}} x\sin\pi x\, dx = [-(1/\pi)x\cos\pi x]\Big|_0^{\frac{1}{2}} + (1/\pi)\int_0^{\frac{1}{2}}\cos\pi x\, dx$$

$$= -0 + 0 + (1/\pi^2)\sin\pi x\Big|_0^{1/2} = 1/\pi^2 \cong .1013.$$

7.　$u = \sin 2x$, $dv = e^x dx \Rightarrow du = 2\cos 2x\, dx$, $v = e^x$. Thus

$$\int_0^{\frac{\pi}{2}} e^x\sin 2x\, dx = e^x\sin 2x\Big|_0^{\frac{\pi}{2}} - 2\int_0^{\frac{\pi}{2}} e^x\cos 2x\, dx. \text{ Now let}$$

$u = \cos 2x$, $dv = e^x\, dx \Rightarrow du = -2\sin 2x\, dx$, $v = e^x$. Thus

$$\int_0^{\frac{\pi}{2}} e^x\sin 2x\, dx = (0-0) - 2(e^x\cos 2x\Big|_0^{\frac{\pi}{2}} + 2\int_0^{\frac{\pi}{2}} e^x\sin 2x\, dx)$$

$$= -2(-e^{\pi/2} - 1) - 4\int_0^{\frac{\pi}{2}} e^x\sin 2x\, dx. \text{ Adding the last integral to}$$

both sides and then dividing by 5 yields

$$\int_0^{\frac{\pi}{2}} e^x\sin 2x\, dx = (2/5)(e^{\pi/2} + 1) \cong 2.3242.$$

9.　$u = x^2$, $dv = e^{-x} dx \Rightarrow du = 2x\, dx$, $v = -e^{-x}$. Thus

$$\int_0^2 x^2 e^{-x}\, dx = -x^2 e^{-x}\Big|_0^2 + 2\int_0^2 xe^{-x}\, dx = -4e^{-2} + 2\int_0^2 xe^{-x}\, dx. \text{ Now let}$$

$u = x$, $dv = e^{-x}\, dx \Rightarrow du = dx$, $v = -e^{-x}$. Thus

$$\int_0^2 x^2 e^{-x}\, dx = -4e^{-2} + 2(-xe^{-x}\Big|_0^2 + \int_0^2 e^{-x}\, dx)$$

$$= -4e^{-2} - 4e^{-2} - 2e^{-x}\Big|_0^2 = 2 - 10e^{-2} \cong .6466.$$

11. $u = \arctan x$, $dv = dx \Rightarrow du = dx/(1+x^2)$, $v = x$. Thus
$$\int \arctan x \, dx = x\arctan x - \int \frac{x \, dx}{1+x^2} = x\arctan x - (1/2)\ln(1+x^2) + c.$$

13. $u = \sin(\ln x)$, $dv = dx \Rightarrow du = \cos(\ln x)dx/x$, $v = x$. Thus
$$\int \sin(\ln x) \, dx = x\sin(\ln x) - \int \cos(\ln x) \, dx. \text{ Now let}$$

$u = \cos(\ln x)$, $dv = dx \Rightarrow du = -\sin(\ln x)dx/x$, $v = x$. Thus
$$\int \sin(\ln x) \, dx = x\sin(\ln x) - [x\cos(\ln x) + \int \sin(\ln x) \, dx].$$

Adding the last integral to both sides and dividing by 2
yields $\int \sin(\ln x) \, dx = [x\sin(\ln x) - x\cos(\ln x)]/2 + c.$

15. $u = x^2$. $dv = dx/\sqrt{1-x} \Rightarrow u = 2xdx$, $v = -2(1-x)^{1/2}$. Thus
$$\int \frac{x^2 \, dx}{\sqrt{1-x}} = -2x^2(1-x)^{1/2} + 4\int x(1-x)^{1/2} \, dx. \text{ Now}$$

$u = x$, $dv = (1-x)^{1/2} \Rightarrow du = dx$, $v = -(2/3)(1+x)^{3/2}$. Thus
$$\int \frac{x^2 \, dx}{\sqrt{1-x}} = -2x^2(1-x)^{1/2} + 4[-(2/3)x(1-x)^{3/2} + (2/3)\int (1-x)^{3/2} \, dx]$$
$$= -2x^2(1-x)^{1/2} - (8/3)x(1-x)^{3/2} - (16/15)(1-x)^{5/2} + c.$$

17. $u = x$, $dv = 2^x dx = e^{x\ln 2} \, dx \Rightarrow du = dx$, $v = 2^x/\ln 2$. Thus
$$\int x2^x \, dx = x2^x/\ln 2 - (1/\ln 2)\int 2^x \, dx$$
$$= x2^x/\ln 2 - 2^x/(\ln 2)^2 + c = 2^x(x\ln 2 - 1)/(\ln 2)^2 + c.$$

19. $u = \ln x$, $dv = \sqrt{x} \, dx$, $\Rightarrow du = dx/x$, $v = 2\sqrt{x}$. Thus
$$\int_1^3 \frac{\ln x}{\sqrt{x}} \, dx = 2\sqrt{x}\ln x \Big|_1^3 - 2\int_1^3 \frac{dx}{\sqrt{x}}$$
$$= 2\sqrt{3}\ln 3 - 4\sqrt{x}\Big|_1^3 = 2\sqrt{3}\ln 3 - 4\sqrt{3} + 4 \cong .8775.$$

21. Let $u = 9+x^2$ so $du = 2xdx$ and using the result of Prob.10,
$$\int_0^1 x\ln(9+x^2) \, dx = (1/2)\int_9^{10} \ln u \, du = (1/2)(u\ln u - u)\Big|_9^{10}$$
$$= (10\ln 10 - 9\ln 9 - 1)/2.$$

23. $u = x^3$, $dv = x^2(x^3-1)^{1/2} dx \Rightarrow du = 3x^2 dx$, $v = (2/9)(x^3-1)^{3/2}$.

Thus $\int x^5(x^3-1)^{1/2} dx = (2/9)x^3(x^3-1)^{3/2} - (2/3)\int x^2(x^3-1)^{3/2} dx$

$$= (2/9)x^3(x^3-1)^{3/2} - (4/45)(x^3-1)^{5/2} + c.$$

25. $u = e^{ax}$, $dv = \sin bx\, dx \Rightarrow du = ae^{ax} dx$, $v = -(1/b)\cos bx$. Thus

$\int e^{ax}\sin bx\, dx = -(1/b)e^{ax}\cos bx + (a/b)\int e^{ax}\cos bx\, dx$. Now let

$u = e^{ax}$, $dv = \cos bx\, dx \Rightarrow du = ae^{ax} dx$, $v = (1/b)\sin bx$. Thus

$\int e^{ax}\sin bx\, dx = -(1/b)e^{ax}\cos bx + (a/b^2)e^{ax}\sin bx - (a^2/b^2)\int e^{ax}\sin bx\, dx$

or $(1 + a^2/b^2)\int e^{ax}\sin bx\, dx = (ae^{ax}\sin bx - be^{ax}\cos bx)/b^2$. Divide

by $(a^2 + b^2)/b^2$ and then add c to get the desired result.

27. $u = \ln x$, $dv = x^p dx \Rightarrow du = dx/x$, $v = x^{p+1}/(p+1)$. Thus

$$\int x^p\ln x\, dx = \frac{x^{p+1}\ln x}{p+1} - \int\frac{x^p dx}{p+1} = \frac{x^{p+1}\ln x}{p+1} - \frac{x^{p+1}}{(p+1)^2} + c.$$

29. $u = x^m$, $dv = \sin ax\, dx \Rightarrow du = mx^{m-1} dx$, $v = -(1/a)\cos ax$. Thus

$\int x^m\sin ax\, dx = -(1/a)x^m\cos ax + (m/a)\int x^{m-1}\cos ax\, dx$.

31. $u = x^n$, $dv = e^{ax} dx \Rightarrow du = nx^{n-1} dx$, $v = (1/a)e^{ax}$. Thus

$\int x^n e^{ax} dx = (x^n/a)e^{ax} - (n/a)\int x^{n-1}e^{ax} dx$.

33. $u = (\ln x)^n$, $dv = dx \Rightarrow du = (n/x)(\ln x)^{n-1} dx$, $v = x$. Thus

$\int(\ln x)^n dx = x(\ln x)^n - n\int(\ln x)^{n-1} dx$.

35. $\int x^2\sin 3x\, dx = -(1/3)x^2\cos 3x + (2/3)\int x\cos 3x\, dx$; (#29, m=2, a=3)

$= -(1/3)x^2\cos 3x + (2/3)[(1/3)x\sin 3x - (1/3)\int\sin 3x\, dx]$; (#30, m=1)

$= -(1/3)x^2\cos 3x + (2/9)x\sin 3x + (2/27)\cos 3x + c.$

37. $\displaystyle\int_0^\pi x^3 \sin x\, dx = -x^3\cos x\Big|_0^\pi + 3\int_0^\pi x^2\cos x\, dx;$ (#29; m=3, a=1)

$\qquad\qquad = \pi^3 + 3[x^2\sin x\Big|_0^\pi - 2\int_0^\pi x\sin x\, dx];$ (#30; m=2, a=1)

$\qquad\qquad = \pi^3 - 6[-x\cos x\Big|_0^\pi + \int_0^\pi \cos x\, dx];$ (#29; m=1, a=1)

$\qquad\qquad = \pi^3 - 6\pi - 6\sin x\Big|_0^\pi = \pi^3 - 6\pi.$

39. $\displaystyle\int_0^{\pi/2} \sin^{2m}x\, dx = \frac{-\cos x\,\sin^{2m-1}x}{2m}\Big|_0^{\pi/2} + \frac{2m-1}{2m}\int_0^{\pi/2}\sin^{2m-2}x\, dx$

$\qquad\qquad = \frac{2m-1}{2m}\left[\frac{-\cos x\,\sin^{2m-3}x}{2m-2}\Big|_0^{\pi/2} + \frac{2m-3}{2m-2}\int_0^{\pi/2}\sin^{2m-4}x\, dx\right]$

$\qquad\qquad \vdots$

$\qquad\qquad = \frac{2m-1}{2m}\cdot\frac{2m-3}{2m-2}\cdot\cdots\cdot\frac{3}{4}\int_0^{\pi/2}\sin^2 x\, dx$

$\qquad\qquad = \frac{2m-1}{2m}\cdot\frac{2m-3}{2m-2}\cdot\cdots\cdot\frac{3}{4}\left[\frac{-\cos x\,\sin x}{2}\Big|_0^{\pi/2} + \frac{1}{2}\int_0^{\pi/2}dx\right]$

$\qquad\qquad = \frac{2m-1}{2m}\cdot\frac{2m-3}{2m-2}\cdot\cdots\cdot\frac{3}{4}\cdot\frac{1}{2}\cdot\frac{\pi}{2}.$

41a. $\displaystyle\int \ln(x+1)\, dx = x\ln(x+1) - \int\frac{x\, dx}{x+1} = x\ln(x+1) - \int\frac{(u-1)du}{u}$

$\qquad\qquad = x\ln(x+1) - u + \ln u + c = (x+1)\ln(x+1) - (x+1) + c$

$\qquad\qquad = (x+1)\ln(x+1) - x + (c-1).$

b. $\displaystyle\int \ln(x+1)\, dx = (x+1)\ln(x+1) - \int dx = (x+1)\ln(x+1) - x + c_1.$

By choosing $c_1 = c-1$, the answers are the same.

43a. $u = f'(t),\ dv = dt \Rightarrow du = f''(t)dt,\ v = t-b.$ Thus

$\qquad f(b) - f(a) = (t-b)f'(t)\Big|_a^b - \int_a^b f''(t)(t-b)\, dt$

$\qquad\qquad = f'(a)(b-a) - \int_a^b f''(t)(t-b)\, dt.$

43b. $u = f''(t)$, $dv = (t-b)dt \Rightarrow du = f'''(t)dt$, $v = (t-b)^2/2$. Thus

$$f(b) - f(a) = f'(a)(b-a) - f''(t)(t-b)^2/2 \Big|_a^b + \int_a^b [f'''(t)(t-b)^2/2]dt$$

$$= f'(a)(b-a) + f''(a)(b-a)^2/2 + \int_a^b (1/2)f'''(t)(t-b)^2 dt.$$

45. Differentiating and using the Fundamental Theorem of Calculus
     we have
$$e^{2x}(2\sin 3x + 6\cos 3x) = 2Ae^{2x}\sin 3x + 3Ae^{2x}\cos 3x + 2Be^{2x}\cos 3x - 3Be^{2x}\sin 3x$$
$$= (2A-3B)e^{2x}\sin 3x + (3A+2B)e^{2x}\cos 3x. \quad \text{Thus}$$
$2A-3B = 2$ and $3A+2B = 6$, which yield $A = 22/13$ and $B = 6/13$.

Section 9.3, Page 479

1.   By Eq.(7), $\displaystyle\int_0^{\pi/4} \cos^2\theta d\theta = (1/2)\int_0^{\pi/4} (1 + \cos 2\theta)d\theta$

$$= (1/2)[\theta + (1/2)\sin 2\theta]\Big|_0^{\pi/4} = (\pi + 2)/8.$$

3. $\displaystyle\int \sin^3 x \cos^2 x\, dx = \int \sin x(1 - \cos^2 x)\cos^2 x\, dx$

$$= \int (\cos^2 x - \cos^4 x)\sin x\, dx = -(1/3)\cos^3 x + (1/5)\cos^5 x + c.$$

5. $\displaystyle\int \cos^{1/3} x \sin^3 x\, dx = \int \cos^{1/3} x(1 - \cos^2 x)\sin x\, dx$

$$= \int (\cos^{1/3} x - \cos^{7/3} x)\sin x\, dx = -(3/4)\cos^{4/3} x + (3/10)\cos^{10/3} x + c.$$

7. $\displaystyle\int \cos^4 x \sin^2 x\, dx = (1/8)\int (1 + \cos 2x)^2(1 - \cos 2x)dx$

$$= (1/8)\int (1 + \cos 2x)(1 - \cos^2 2x)dx$$

$$= (1/8)\int (\sin^2 2x + \sin^2 2x \cos 2x)dx$$

$$= (1/8)\int [(1 - \cos 4x)/2 + \sin^2 2x \cos 2x]dx$$

$$= (1/16)x - (1/64)\sin 4x + (1/48)\sin^3 2x + c.$$

9. $\int_{0}^{\pi/2} \cos^{1/2}x \sin^3 x\, dx = \int_{0}^{\pi/2} \cos^{1/2}x(1 - \cos^2 x)\sin x\, dx$

$= \int_{0}^{\pi/2} (\cos^{1/2}x - \cos^{5/2}x)\sin x\, dx$

$= [-(2/3)\cos^{3/2}x + (2/7)\cos^{7/2}x]\Big|_{0}^{\pi/2} = 0 - [-2/3 + 2/7] = 8/21.$

11. $\int_{\pi/6}^{\pi/2} \sin^{-1/2}x \cos x\, dx = 2\sin^{1/2}x\Big|_{\pi/6}^{\pi/2} = 2(1 - 1/\sqrt{2}) = 2 - \sqrt{2}.$

13. $\int_{0}^{\pi} \sin^4 x\, dx = (1/4)\int_{0}^{\pi} (1 - \cos 2x)^2 dx$

$= (1/4)\int_{0}^{\pi} (1 - 2\cos 2x + \cos^2 2x)\, dx$

$= (1/4)\int_{0}^{\pi} [1 - 2\cos 2x + (1/2)(1 + \cos 4x)]\, dx$

$= (1/4)[(3/2)x - \sin 2x + (1/8)\sin 4x]\Big|_{0}^{\pi} = 3\pi/8.$

15. $\int \sin^{-2}x \cos^3 x\, dx = \int \sin^{-2}x(1 - \sin^2 x)\cos x\, dx$

$= \int (\sin^{-2}x \cos x - \cos x)\, dx = -\csc x - \sin x + c.$

17   Using Eq. (12b), $\int \sin 3x \sin x\, dx = (1/2)\int (\cos 2x - \cos 4x)\, dx$

$= (1/4)\sin 2x - (1/8)\sin 4x + c.$

19. $\int (\cos 4x - \cos 2x)^2\, dx = \int (\cos^2 4x - 2\cos 4x \cos 2x + \cos^2 2x)\, dx$

$= \int [(1/2)(1 + \cos 8x) - (\cos 2x + \cos 6x) + (1/2)(1 + \cos 4x)]\, dx$

$= x - (1/2)\sin 2x + (1/8)\sin 4x - (1/6)\sin 6x + (1/16)\sin 8x + c.$

21. $\int \cos 3x \sin 2x\, dx = (1/2)\int [\sin(-x) + \sin 5x]\, dx$

$= (1/2)\cos x - (1/10)\cos 5x + c.$

23. $\displaystyle\int \sin 2x \sin 4x \sin 6x\, dx = (1/2)\int (\cos 2x - \cos 6x)\sin 6x\, dx$

$$= (1/4)\int [(\sin 4x + \sin 8x) - \sin 12x]\, dx$$

$$= -(1/16)\cos 4x - (1/32)\cos 8x + (1/48)\cos 12x + c.$$

25. $\displaystyle\int x^3 (1 - x^2)^n\, dx = \int \sin^3 u (1 - \sin^2 u)^n \cos u\, du$

$$= \int (\cos^{2n+1} u - \cos^{2n+3} u)\sin u\, du$$

$$= -[1/(2n+2)]\cos^{2n+2} u + [1/(2n+4)]\cos^{2n+4} u + c$$

$$= -(1 - x^2)^{n+1}/(2n+2) + (1 - x^2)^{n+2}/(2n+4) + c.$$

27. $\displaystyle V = \pi\int_0^{\pi/4} (\cos^2 x - \sin^2 x)\, dx = \pi\int_0^{\pi/4} \cos 2x\, dx$

$$= (\pi/2)\sin 2x\Big|_0^{\pi/4} = \pi/2.$$

29. $\displaystyle V = \pi\int_a^b y^2\, dx = \pi\int_0^{2\pi} a^2(1 - \cos\theta)^2 [a(1 - \cos\theta)\, d\theta]$

$$= \pi a^3\int_0^{2\pi} (1 - \cos\theta)^3\, d\theta = \pi a^3\int_0^{2\pi} (1 - 3\cos\theta + 3\cos^2\theta - \cos^3\theta)$$

$$= \pi a^3\int_0^{2\pi} [1 - 3\cos\theta + (3/2)(1 + \cos 2\theta) - (1 - \sin^2\theta)\cos\theta]\, d\theta$$

$$= \pi a^3[(5/2)\theta - 4\sin\theta + (3/4)\sin 2\theta + (1/3)\sin^3\theta]\Big|_0^{2\pi} = 5\pi a^3.$$

31. $\displaystyle S = 2\pi\int_0^{2\pi} y(\theta)\sqrt{[x'(\theta)]^2 + [y'(\theta)]^2}\, d\theta$

$$= 2\pi\int_0^{2\pi} a(1 - \cos\theta)\sqrt{a^2(1 - \cos\theta)^2 + a^2\sin^2\theta}\, d\theta$$

$$= 2\pi a^2\int_0^{2\pi} (1 - \cos\theta)\sqrt{2 - 2\cos\theta}\, d\theta = 2\sqrt{2}\pi a^2\int_0^{2\pi} (1 - \cos\theta)^{3/2}\, d\theta$$

$$= 8\pi a^2\int_0^{2\pi} \sin^3(\theta/2)\, d\theta = 8\pi a^2\int_0^{2\pi} [1 - \cos^2(\theta/2)]\sin(\theta/2)\, d\theta$$

$$= 8\pi a^2[-2\cos(\theta/2) + (2/3)\cos^3(\theta/2)]\Big|_0^{2\pi} = 64\pi a^2/3.$$

33a. Using the results of Prob.32, we have, for $m \neq n$

$$\int_{-\pi}^{\pi} \cos nx \cos mx \, dx = \left[\frac{\sin(n-m)x}{2(n-m)} + \frac{\sin(n+m)x}{2(n+m)} \right] \Big|_{-\pi}^{\pi} = 0 \text{ and for } m = n,$$

$$\int_{-\pi}^{\pi} \cos^2 nx \, dx = (1/2) \int_{-\pi}^{\pi} (1+\cos 2nx) \, dx = [x/2 + (1/4n)\sin 2nx] \Big|_{-\pi}^{\pi} = \pi.$$

b. $m \neq n \Rightarrow \int_{-\pi}^{\pi} \cos nx \sin mx \, dx = \left[\frac{\cos(m-n)x}{2(m-n)} - \frac{\cos(m+n)x}{2(m+n)} \right] \Big|_{-\pi}^{\pi} = 0,$

since cosine is an even function. For $m = n$,

$$\int_{-\pi}^{\pi} \cos nx \sin nx \, dx = (1/2n)\sin^2 nx \Big|_{-\pi}^{\pi} = 0.$$

c. $m \neq n \Rightarrow \int_{-\pi}^{\pi} \sin nx \sin mx \, dx = \left[\frac{\sin(n-m)x}{2(n-m)} - \frac{\sin(n+m)x}{2(n+m)} \right] \Big|_{-\pi}^{\pi} = 0,$ and

$$\int_{-\pi}^{\pi} \sin^2 nx \, dx = (1/2) \int_{-\pi}^{\pi} (1-\cos 2nx) \, dx = [x/2 - (1/2n)\sin 2nx] \Big|_{-\pi}^{\pi} = \pi.$$

Section 9.4, Page 487

1.   By Eq. (6), $\int \sec ax \, dx = (1/a) \int \sec u \, du = (1/a) \ln|\sec ax + \tan ax| + c.$

3.   Let $u = \tan x$, then $\int \tan^2 x \sec^2 x \, dx = \int u^2 \, du = (1/3) \tan^3 x + c.$

5. $\int \tan^4 x \, dx = \int \tan^2 x (\sec^2 x - 1) \, dx = \int \tan^2 x \sec^2 x \, dx - \int \tan^2 x \, dx$

$$= (1/3) \tan^3 x - \int (\sec^2 x - 1) \, dx$$

$$= (1/3) \tan^3 x - \tan x + x + c, \text{ as in Ex.1.}$$

7.   As in Ex.4, $\int \tan^3 2x \sec^3 2x \, dx = \int \tan^2 2x \sec^2 2x (\sec 2x \tan 2x) \, dx$

$$= (1/2) \int (\sec^4 2x - \sec^2 2x)(2\sec 2x \tan 2x) \, dx$$

$$= (1/10) \sec^5 2x - (1/6) \sec^3 2x + c.$$

9. $\int \tan x \sec^5 x \, dx = \int \sec^4 x (\sec x \tan x) \, dx = (1/5) \sec^5 x + c.$

11. As in Ex.4, $\int \cot^5 x \csc^3 x \, dx = \int \cot^4 x \csc^2 x (\csc x \cot x) \, dx$

$$= \int (\csc^2 x - 1)^2 \csc^2 x (\csc x \cot x) \, dx$$

$$= -\int (\csc^6 x - 2\csc^4 x + \csc^2 x)(-\csc x \cot x) \, dx$$

$$= -(1/7)\csc^7 x + (2/5)\csc^5 x - (1/3)\csc^3 x + c.$$

13. $\int \cot^2 x \csc x \, dx = \int (\csc^3 x - \csc x) \, dx$

$$= -(1/2)\csc x \cot x + (1/2)\int \csc x \, dx - \int \csc x \, dx$$

$$= -(1/2)\csc x \cot x - (1/2)\ln|\csc x - \cot x| + c.$$

Eq.(10), with n = 3, and Eq.(9) have been used.

15. $\int \tan^{(1/2)} x \sec^4 x \, dx = \int \tan^{(1/2)} x (\tan^2 x + 1) \sec^2 x \, dx$

$$= (2/7)\tan^{7/2} x + (2/3)\tan^{3/2} x + c.$$

17. $\int \tan^3 x \sec^{1/2} x \, dx = \int (\sec^2 x - 1) \sec^{-1/2} x (\sec x \tan x) \, dx$

$$= (2/5)\sec^{5/2} x - 2\sec^{1/2} x + c.$$

19. Since $1 - \sin^2 x = (1 + \sin x)(1 - \sin x) = \cos^2 x$, divide both sides by $\cos x (1 + \sin x)$ to obtain the desired result. Now $(1 - \sin x)/\cos x = \sec x - \tan x$ and thus $\sec x = \tan x + \cos x/(1 + \sin x)$.

So $\int \sec x \, dx = \int \dfrac{\sin x}{\cos x} \, dx + \int \dfrac{\cos x \, dx}{1 + \sin x} = -\ln|\cos x| + \ln|1 + \sin x| + c$

$$= \ln\left|\dfrac{1 + \sin x}{\cos x}\right| + c = \ln|\sec x + \tan x| + c.$$

21. $\int \cot^n x \, dx = \int \cot^{n-2} x (\csc^2 x - 1) \, dx = \int \cot^{n-2} x \csc^2 x \, dx - \int \cot^{n-2} x \, dx$

$$= -[1/(n-1)]\cot^{n-1} x - \int \cot^{n-2} x \, dx.$$

23. $\int \cot^3 x \, dx = -(1/2)\cot^2 x - \int \cot x \, dx = -(1/2)\cot^2 x - \int \dfrac{\cos x \, dx}{\sin x}$

$$= -(1/2)\cot^2 x - \ln|\sin x| + c, \text{ by setting } n = 3.$$

25. n = 5, $\int \cot^5 x \, dx = -(1/4)\cot^4 x - \int \cot^3 x \, dx$ (and from Prob.23)

$$= -(1/4)\cot^4 x + (1/2)\cot^2 x + \ln|\sin x| + c.$$

27. $\int \csc^3 x\, dx = -(1/2)\csc x \cot x + (1/2)\int \csc x\, dx$ (for n = 3)

$\qquad\qquad = -(1/2)\csc x \cot x + (1/2)\ln|\csc x - \cot x| + c.$

29. $\int \csc^5 x\, dx = -(1/4)\csc^3 x \cot x + (3/4)\int \csc^3 x\, dx$ (for n = 5)

$\qquad\quad = -(1/4)\csc^3 x \cot x - (3/8)\csc x \cot x + (3/8)\ln|\csc x - \cot x| + c,$
from Prob.27.

31. $\displaystyle\int \frac{d\theta}{1-\cos\theta} = \int \frac{d\theta}{2\sin^2(\theta/2)} = (1/2)\int \csc^2(\theta/2)\, d\theta = -\cot(\theta/2) + c.$

Section 9.5, Page 492

1. $x = \sin\theta \Rightarrow dx = \cos\theta\, d\theta$ and $1-x^2 = \cos^2\theta$. Thus
$\displaystyle\int \sqrt{1-x^2}\, dx = \int \cos^2\theta\, d\theta = (1/2)\int (1 + \cos 2\theta)\, d\theta$

$\qquad\qquad = \theta/2 + (1/4)\sin 2\theta + c = \theta/2 + (1/2)\sin\theta \cos\theta + c$

$\qquad\qquad = (1/2)(\arcsin x + x\sqrt{1-x^2}) + c.$

3. $x = \sqrt{5/2}\tan\theta \Rightarrow dx = \sqrt{5/2}\sec^2\theta\, d\theta$ and $2x^2 + 5 = 5\sec^2\theta$. Thus
$\displaystyle\int \frac{dx}{(2x^2+5)^{3/2}} = \int \frac{\sqrt{5/2}\sec^2\theta\, d\theta}{5^{3/2}\sec^3\theta} = (1/5\sqrt{2})\int \cos\theta\, d\theta$

$\qquad\quad = (1/5\sqrt{2})\sin\theta + c = (1/5\sqrt{2})x/\sqrt{x^2+5/2} + c = x/5\sqrt{2x^2+5} + c.$

5. $-x = u \Rightarrow \displaystyle\int \frac{dx}{(x^2-4)^{3/2}} = -\int \frac{du}{(u^2-4)^{3/2}}.$ Now

$u = 2\sec\theta \Rightarrow du = 2\sec\theta \tan\theta\, d\theta$ and $u^2 - 4 = 4\tan^2\theta$. Thus
$\displaystyle\int \frac{dx}{(x^2-4)^{3/2}} = -\int \frac{2\sec\theta \tan\theta\, d\theta}{8\tan^3\theta} = -(1/4)\int \frac{\cos\theta}{\sin^2\theta}\, d\theta$

$\qquad\qquad = 1/4\sin\theta + c = u/4\sqrt{u^2-4} + c = -x/4\sqrt{x^2-4} + c.$

7. $x = 2\sin\theta \Rightarrow \int \dfrac{\sqrt{4-x^2}}{x}\, dx = \int \dfrac{(2\cos\theta)\,2\cos\theta\, d\theta}{2\sin\theta}$

$$= 2\int (\csc\theta - \sin\theta)\, d\theta = 2\ln|\csc\theta - \cot\theta| + 2\cos\theta + c$$

$$= 2\ln|(2 - \sqrt{4-x^2})/x| + \sqrt{4-x^2} + c.$$

9. $\int \dfrac{3x-1}{\sqrt{x^2-9}}\, dx = \int \dfrac{9\sec\theta - 1}{3\tan\theta}\, 3\sec\theta\tan\theta\, d\theta \quad (x = 3\sec\theta)$

$$= 9\int \sec^2\theta\, d\theta - \int \sec\theta\, d\theta = 9\tan\theta - \ln|\sec\theta + \tan\theta| + c_1$$

$$= 3\sqrt{x^2-9} - \ln|(x/3) + \sqrt{x^2-9}/3| + c_1$$

$$= 3\sqrt{x^2-9} - \ln|x + \sqrt{x^2-9}| + c.$$

11. $x = \sqrt{3}\tan\theta \Rightarrow \int \dfrac{dx}{x\sqrt{3+x^2}} = \int \dfrac{\sqrt{3}\sec^2\theta\, d\theta}{\sqrt{3}\tan\theta\,\sqrt{3}\sec\theta} = (1/\sqrt{3})\int \csc\theta\, d\theta$

$$= (1/\sqrt{3})\ln|\csc\theta - \cot\theta| + c = (1/\sqrt{3})\ln|(\sqrt{3+x^2} - \sqrt{3})/x| + c.$$

13. $x = \sec\theta \Rightarrow \int \dfrac{x^2 dx}{\sqrt{x^2-1}} = \int \dfrac{(\sec^2\theta)\sec\theta\tan\theta\, d\theta}{\tan\theta}$

$$= \int \sec^3\theta\, d\theta = (1/2)\sec\theta\tan\theta + (1/2)\int \sec\theta\, d\theta$$

$$= (1/2)\sec\theta\tan\theta + (1/2)\ln|\sec\theta + \tan\theta| + c$$

$$= (1/2)x\sqrt{x^2-1} + (1/2)\ln|x + \sqrt{x^2-1}| + c. \quad \text{Thus}$$

$\displaystyle\int_2^4 \dfrac{x^2 dx}{\sqrt{x^2-1}} = 2\sqrt{15} + (1/2)\ln|4 + \sqrt{15}| - [\sqrt{3} + (1/2)\ln|2 + \sqrt{3}|]$

$$= 2\sqrt{15} - \sqrt{3} + (1/2)\ln[(4 + \sqrt{15})/(2 + \sqrt{3})] \cong 6.3872.$$

15. $x = -u \Rightarrow \int_{-5}^{-3} \frac{dx}{(x^2-1)^{3/2}} = \int_{3}^{5} \frac{du}{(u^2-1)^{3/2}}$. $u = \sec\theta \Rightarrow$

$\int \frac{du}{(u^2-1)^{3/2}} = \int \frac{\sec\theta \tan\theta \, d\theta}{\tan^3\theta} = \int \frac{\cos\theta \, d\theta}{\sin^2\theta} = -1/\sin\theta + c$

$= -u/\sqrt{u^2-1} + c$. Thus $\int_{3}^{5} \frac{du}{(u^2-1)^{3/2}} = -5/\sqrt{24} + 3/\sqrt{8}$ and hence

$\int_{-5}^{-3} \frac{dx}{(x^2-1)^{3/2}} = 3\sqrt{2}/4 - 5\sqrt{6}/12 \cong .0400$.

17. $\int \frac{\sqrt{x^2-3}}{x} \, dx = \int \frac{(\sqrt{3}\tan\theta)\sqrt{3}\sec\theta \tan\theta \, d\theta}{\sqrt{3}\sec\theta}$ $(x = \sqrt{3}\sec\theta)$

$= \sqrt{3} \int \tan^2\theta \, d\theta = \sqrt{3} \int (\sec^2\theta - 1) \, d\theta = \sqrt{3}(\tan\theta - \theta) + c$

$= \sqrt{x^2-3} - \sqrt{3} \operatorname{arcsec}(x/\sqrt{3}) + c$.

19. $\int \frac{dx}{\sqrt{4x^2+9}} = \int \frac{(3/2)\sec^2\theta \, d\theta}{3\sec\theta}$ $[x = (3/2)\tan\theta]$

$= (1/2)\ln|\sec\theta + \tan\theta| + c = (1/2)\ln\left|\frac{\sqrt{4x^2+9}+2x}{3}\right| + c$.

Thus $\int_{-2}^{2} \frac{dx}{\sqrt{4x^2+9}} = (1/2)\ln[(5+4)/3] - (1/2)\ln[(5-4)/3] = \ln 3$.

21. $\int \sqrt{a^2-x^2} \, dx = \int (a\cos\theta) a\cos\theta \, d\theta$ $(x = a\sin\theta)$

$= (a^2/2) \int (1 + \cos 2\theta) \, d\theta = (a^2/2)(\theta + \sin\theta \cos\theta) + c$

$= (a^2/2)[\arcsin(x/a) + (x/a)(\sqrt{a^2-x^2}/a)] + c$

$= (x/2)\sqrt{a^2-x^2} + (a^2/2)\arcsin(x/a) + c$.

23. $\displaystyle\int\frac{x^2 dx}{\sqrt{a^2-x^2}} = \int\frac{(a^2\sin^2\theta)a\cos\theta\,d\theta}{a\cos\theta}$ $(x = a\sin\theta)$

$\displaystyle = (a^2/2)\int(1-\cos 2\theta)\,d\theta = (a^2/2)(\theta - \sin\theta\cos\theta) + c$

$\displaystyle = (a^2/2)[\arcsin(x/a) - (x/a)(\sqrt{a^2-x^2}/a)] + c$

$\displaystyle = -(x/2)\sqrt{a^2-x^2} + (a^2/2)\arcsin(x/a) + c.$

25. $\displaystyle\int\sqrt{a^2+x^2}\,dx = \int(a\sec\theta)a\sec^2\theta\,d\theta = a^2\int\sec^3\theta\,d\theta$ $(x = a\tan\theta)$

$\displaystyle = (a^2/2)[\sec\theta\tan\theta + \int\sec\theta\,d\theta]$

$\displaystyle = (a^2/2)[\sec\theta\tan\theta + \ln|\sec\theta + \tan\theta|] + c_1$

$\displaystyle = \frac{a^2}{2}\left[\frac{\sqrt{a^2+x^2}}{a}\frac{x}{a} + \ln\left|\frac{\sqrt{a^2+x^2}}{a} + \frac{x}{a}\right|\right] + c_1$

$\displaystyle = (x/2)\sqrt{a^2+x^2} + (a^2/2)\ln|\sqrt{a^2+x^2} + x| + c.$

27. $\displaystyle\int\frac{dx}{(a^2+x^2)^{3/2}} = \int\frac{a\sec^2\theta\,d\theta}{a^3\sec^3\theta} = (1/a^2)\int\cos\theta\,d\theta$ $(x = a\tan\theta)$

$\displaystyle = (1/a^2)\sin\theta + c = (1/a^2)(x/\sqrt{x^2+a^2}) + c.$

29. $\displaystyle\int\frac{dx}{x\sqrt{a^2+x^2}} = \int\frac{a\sec^2\theta\,d\theta}{a\tan\theta\,a\sec\theta} = (1/a)\int\csc\theta\,d\theta$ $(x = a\tan\theta)$

$\displaystyle = (1/a)\ln|\csc\theta - \cot\theta| + c = (1/a)\ln\left|\frac{\sqrt{a^2+x^2}}{x} - \frac{a}{x}\right| + c.$

31. $\displaystyle\int\sqrt{x^2-a^2}\,dx = \int(a\tan\theta)a\sec\theta\tan\theta\,d\theta$ $(x = a\sec\theta)$

$\displaystyle = a^2\int(\sec^3\theta - \sec\theta)\,d\theta = (a^2/2)(\sec^2\theta\sin\theta - \ln|\sec\theta + \tan\theta|) + c_1$

$\displaystyle = \frac{a^2}{2}\left(\frac{x^2}{a^2}\frac{\sqrt{x^2-a^2}}{x} - \ln\left|\frac{x}{a} + \frac{\sqrt{x^2-a^2}}{a}\right|\right) + c_1$

$\displaystyle = (x/2)\sqrt{x^2-a^2} - (a^2/2)\ln|x + \sqrt{x^2-a^2}| + c.$

33. Solving for y and using symmetry we have

$$A = 4\int_0^a b\sqrt{1 - x^2/a^2}\,dx = (4b/a)\int_0^a \sqrt{a^2 - x^2}\,dx.$$ Letting $x = a\sin\theta$,

we have $A = (4b/a)\int_0^{\pi/2} (a\cos\theta)a\cos\theta\,d\theta = 2ab\int_0^{\pi/2} (1 + \cos2\theta)\,d\theta$

$$= 2ab[\theta + (1/2)\sin2\theta]\Big|_0^{\pi/2} = \pi ab.$$

Section 9.6, Page 500

1. $\dfrac{1}{x(x+1)} = \dfrac{A}{x} + \dfrac{B}{x+1}$ or $1 = A(x+1) + Bx$ or $1 = (A+B)x + A$. Thus

$A + B = 0$ and $A = 1$ and hence $B = -1$. Therefore

$$\int \frac{dx}{x(x+1)} = \int \frac{dx}{x} - \int \frac{dx}{x+1} = \ln|x| - \ln|x+1| + c$$

$$= \ln|x/(x+1)| + c.$$

3. $\dfrac{-4}{x^2 - 4} = \dfrac{A}{x+2} + \dfrac{B}{x-2}$ or $-4 = A(x-2) + B(x+2)$ or

$-4 = (A+B)x + (2B-2A)$. Thus $A + B = 0$ and $2B - 2A = -4$ and

hence $A = 1$, $B = -1$, and therefore

$$\int \frac{-4dx}{x^2 - 4} = \int \frac{dx}{x+2} - \int \frac{dx}{x-2} = \ln|x+2| - \ln|x-2| + c$$

$$= \ln|(x+2)/(x-2)| + c.$$

5. $\dfrac{x^2 + 4x + 5}{(x+1)(x+2)(x+3)} = \dfrac{A}{x+1} + \dfrac{B}{x+2} + \dfrac{C}{x+3}$ so

$x^2 + 4x + 5 = A(x+2)(x+3) + B(x+1)(x+3) + C(x+1)(x+2)$. Let $x = -1$,
then $1 - 4 + 5 = A(1)(2)$ or $A = 1$; $x = -2$, then $4 - 8 + 5 = B(-1)1$ or
$B = -1$; and $x = -3$, then $9 - 12 + 5 = C(-2)(-1)$ or $C = 1$. Thus

$$\int \frac{(x^2 + 4x + 5)\,dx}{(x+1)(x+2)(x+3)} = \int \frac{dx}{x+1} - \int \frac{dx}{x+2} + \int \frac{dx}{x+3}$$

$$= \ln|x+1| - \ln|x+2| + \ln|x+3| + c = \ln|(x+1)(x+3)/(x+2)| + c.$$

7. $\dfrac{x}{(x+1)^2(x-1)^2} = \dfrac{A}{x+1} + \dfrac{B}{(x+1)^2} + \dfrac{C}{x-1} + \dfrac{D}{(x-1)^2}$ or

$x = A(x+1)(x-1)^2 + B(x-1)^2 + C(x+1)^2(x-1) + D(x+1)^2$.
Let $x = 1$, then $1 = 4D$ or $D = 1/4$; $x = -1$, then $-1 = 4B$
or $B = -1/4$; $x = 0$, then $0 = A - 1/4 + C + 1/4$ or $A + C = 0$;
$x = 2$, then $2 = 3A - 1/4 + 9C + 9/4$ or $3A + 9C = 0$.
The last two equations yield $A = C = 0$ and thus

$$\int \frac{x\,dx}{(x+1)^2(x-1)^2} = -\int \frac{dx}{4(x+1)^2} + \int \frac{dx}{4(x-1)^2} = \frac{1}{4(x+1)} - \frac{1}{4(x-1)} + c$$

$$= \frac{-1}{2(x^2-1)} + c.$$

9. $\dfrac{x^3+3x^2-x+3}{x(x^2+1)} = 1 + \dfrac{3x^2-2x+3}{x(x^2+1)}$. Now $\dfrac{3x^2-2x+3}{x(x^2+1)} = \dfrac{A}{x} + \dfrac{Bx+C}{x^2+1}$

or $3x^2 - 2x + 3 = A(x^2 + 1) + (Bx+C)x = (A+B)x^2 + Cx + A$, and

therefore $A+B = 3$, $C = -2$ and $A = 3$, so that $B = 0$ and thus

$$\int \frac{x^3+3x^2-x+3}{x(x^2+1)}\,dx = \int dx + 3\int \frac{dx}{x} - 2\int \frac{dx}{x^2+1} = x + 3\ln|x| - 2\arctan x + c.$$

11. $\dfrac{2x^2+4}{x(x^2+2x+2)} = \dfrac{A}{x} + \dfrac{Bx+C}{x^2+2x+2}$ so that

$2x^2 + 4 = A(x^2+2x+2) + x(Bx+C) = (A+B)x^2 + (2A+C)x + 2A$.

Thus $A+B = 2$, $2A+C = 0$ and $2A = 4$, which yield $A = 2$, $B = 0$

and $C = -4$. Hence

$$\int \frac{(2x^2+4)dx}{x(x^2+2x+2)} = 2\int \frac{dx}{x} - 4\int \frac{dx}{x^2+2x+2} = 2\ln|x| - 4\int \frac{dx}{(x+1)^2+1}$$

$$= 2\ln|x| - 4\arctan(x+1) + c.$$

13. $\dfrac{1}{x^4-1} = \dfrac{Ax+B}{x^2+1} + \dfrac{C}{x+1} + \dfrac{D}{x-1}$ or

$1 = (Ax+B)(x^2-1) + C(x^2+1)(x-1) + D(x^2+1)(x+1)$. Let $x = 1$,

then $1 = 4D$ or $D = 1/4$; $x = -1$, then $1 = -4C$ or $C = -1/4$; and

$x = 0$ then $1 = -B + 1/4 + 1/4$ or $B = -1/2$. To find A we note

that the coefficient of x^3 on the right side is $A + C + D$,

which must be zero, and hence $A = -C - D = 0$. Thus

$$\int \frac{dx}{x^4-1} = -\frac{1}{2}\int \frac{dx}{x^2+1} - \frac{1}{4}\int \frac{dx}{x+1} + \frac{1}{4}\int \frac{dx}{x-1}$$

$$= (1/4)\ln|(x-1)/(x+1)| - (1/2)\arctan x + c.$$

15 $\dfrac{1}{(x-1)^2(x+2)^2} = \dfrac{A}{x-1} + \dfrac{B}{(x-1)^2} + \dfrac{C}{x+2} + \dfrac{D}{(x+2)^2}$ so that

$1 = A(x-1)(x+2)^2 + B(x+2)^2 + C(x+2)(x-1)^2 + D(x-1)^2$.

Let $x = 1$, then $1 = 9B$ or $B = 1/9$; $x = -2$ then $1 = 9D$

or $D = 1/9$; $x = 0$, then $1 = -4A + 4/9 + 2C + 1/9$

or $-4A + 2C = 4/9$; and $x = 2$, then $1 = 16A + 16/9 + 4C + 1/9$

or $16A + 4C = -8/9$. Solving the last two equations for A and

C we find A = -2/27 and C = 2/27, and thus

$$\int \frac{dx}{(x-1)^2(x+2)^2} = \frac{-2}{27}\int \frac{dx}{x-1} + \frac{1}{9}\int \frac{dx}{(x-1)^2} + \frac{2}{27}\int \frac{dx}{(x+2)^2} + \frac{1}{9}\int \frac{dx}{(x+2)^2}$$

$$= (2/27)\ln|(x+2)/(x-1)| - (1/9)(x-1)^{-1} - (1/9)(x+2)^{-1} + c.$$

17. $\dfrac{1}{(x+b)(x+d)} = \dfrac{A}{x+b} + \dfrac{B}{x+d}$ or 1 = A(x+d) + B(x+b). x = -d gives

B = 1/(b-d) and x = -b gives A = 1/(d-b). Thus

$$\int \frac{dx}{(x+b)(x+d)} = \frac{1}{d-b}\int \frac{dx}{x+b} + \frac{1}{b-d}\int \frac{dx}{x+d}$$

$$= \frac{1}{d-b}\ln|x+b| + \frac{1}{b-d}\ln|x+d| + c = \frac{1}{d-b}\ln\left|\frac{x+b}{x+d}\right| + c.$$

19. $\dfrac{1}{a^2-x^2} = \dfrac{A}{a-x} + \dfrac{B}{a+x} = \dfrac{1}{2a}\left[\dfrac{1}{a-x} + \dfrac{1}{a+x}\right]$ by the same steps as in

Prob.17. Thus $\displaystyle\int \frac{dx}{a^2-x^2} = (1/2a)\int \frac{dx}{a-x} + (1/2a)\int \frac{dx}{a+x}$

$$= -(1/2a)\ln|a-x| + (1/2a)\ln|a+x| + c$$
$$= (1/2a)\ln|(a+x)/(a-x)| + c.$$

21. $\displaystyle\int \frac{x\,dx}{(a^2-x^2)} = -(1/2)\int \frac{du}{u} = -(1/2)\ln|a^2-x^2| + c.$

23. $\displaystyle\int \frac{x^2\,dx}{a^2-x^2} = -\int dx + a^2\int \frac{dx}{a^2-x^2} = -x + (a/2)\ln|(a+x)/(a-x)| + c,$

using the results of Prob.21.

25. $\dfrac{1}{x(x^2+a^2)} = \dfrac{A}{x} + \dfrac{Bx+C}{x^2+a^2}$ or

1 = A(x^2 + a^2) + x(Bx + C) = (A + B)x^2 + Cx + Aa^2. Hence
A+B = 0, C = 0, and Aa^2 = 1 which yield A = 1/a^2, B = -1/a^2
and C = 0. Thus

$$\int \frac{dx}{x(x^2+a^2)} = (1/a^2)\int \frac{dx}{x} - (1/a^2)\int \frac{x\,dx}{x^2+a^2}$$

$$= (1/a^2)\ln|x| - (1/2a^2)\ln(x^2+a^2) + c = (1/2a^2)\ln[x^2/(x^2+a^2)] + c.$$

27. $\dfrac{1}{x^3(x^2+a^2)} = \dfrac{A}{x} + \dfrac{B}{x^2} + \dfrac{C}{x^3} + \dfrac{Dx+E}{x^2+a^2}$ and therefore

1 = Ax^2(x^2 + a^2) + Bx(x^2 + a^2) + C(x^2 + a^2) + (Dx + E)x^3, or
1 = (A+D)x^4 + (B+E)x^3 + (Aa^2 + C)x^2 + Ba^2x + Ca^2. Hence
A+D = 0, B+E = 0, Aa^2 + C = 0, Ba^2 = 0 and Ca^2 = 1, which

yield $A = -1/a^4$, $B = 0$, $C = 1/a^2$, $D = 1/a^4$ and $E = 0$. Thus

$$\int \frac{dx}{x^3(x^2+a^2)} = (-1/a^4)\int \frac{dx}{x} + (1/a^2)\int \frac{dx}{x^3} + (1/a^4)\int \frac{x\,dx}{x^2+a^2}$$

$$= -(1/a^4)\ln|x| - (1/2a^2)x^{-2} + (1/2a^4)\ln(x^2+a^2) + c$$

$$= -(1/2a^2)x^{-2} - (1/2a^4)\ln[x^2/(x^2+a^2)] + c.$$

29. $\int \frac{x^2\,dx}{x^2+a^2} = \int dx - a^2\int \frac{dx}{x^2+a^2} = x - a\arctan(x/a) + c.$

31. $\int \frac{x^4\,dx}{x^2+a^2} = \int x^2 dx - a^2\int dx + a^4\int \frac{dx}{x^2+a^2}$

$$= x^3/3 - a^2 x + a^3\arctan(x/a) + c.$$

Section 9.7, Page 505

1. Using Eq. (4) with $n = 1$ and $k^2 = 9$ we have

$$\int \frac{dx}{(x^2+9)^2} = \frac{x}{18(x^2+9)} + \frac{1}{18}\int \frac{dx}{x^2+9} = \frac{x}{18(x^2+9)} + \frac{1}{54}\arctan(x/3) + c.$$

3. $\int \frac{dx}{(x^2+4)^3} = \frac{x}{16(x^2+4)^2} + \frac{3}{16}\int \frac{dx}{(x^2+4)^2}$ [by Eq. (4), $n=2$, $k^2=4$]

$$= \frac{x}{16(x^2+4)^2} + \frac{3}{16}\left[\frac{x}{8(x^2+4)} + \frac{1}{8}\int \frac{dx}{x^2+4} \right]$$

$$= \frac{x}{16(x^2+4)^2} + \frac{3x}{128(x^2+4)} + \frac{3}{256}\arctan(x/2) + c.$$

5. $\int \frac{dx}{(x^2-4x+8)^2} = \int \frac{dx}{[(x-2)^2+4]^2} = \int \frac{ds}{(s^2+4)^2}$ (for $s = x-2$)

$$= \frac{s}{8(s^2+4)} + \frac{1}{8}\int \frac{ds}{s^2+4} = \frac{s}{8(s^2+4)} + (1/16)\arctan(s/2) + c$$

$$= \frac{x-2}{8(x^2-4x+8)} + (1/16)\arctan[(x-2)/2] + c.$$

7. $\displaystyle\int \frac{dx}{(x^2 - 2x + 2)^2} = \int \frac{dx}{[(x-1)^2 + 1]^2} = \int \frac{ds}{(s^2 + 1)^2}$ $(s = x-1)$

$\displaystyle = \frac{s}{2(s^2 + 1)} + (1/2)\arctan s + c$

$\displaystyle = \frac{x - 1}{2(x^2 - 2x + 2)} + (1/2)\arctan(x-1) + c.$

9. $\displaystyle\int \frac{dx}{(2x^2 + 6)^3} = \frac{1}{8}\int \frac{dx}{(x^2 + 3)^3} = \frac{1}{8}\left[\frac{x}{12(x^2 + 3)^2} + \frac{1}{4}\int \frac{dx}{(x^2 + 3)^2}\right]$

$\displaystyle = \frac{x}{96(x^2 + 3)^2} + \frac{1}{32}\left[\frac{x}{6(x^2 + 3)} + \frac{1}{6}\int \frac{dx}{x^2 + 3}\right]$

$\displaystyle = \frac{x}{96(x^2 + 3)^2} + \frac{x}{192(x^2 + 3)} + \frac{1}{192\sqrt{3}}\arctan(x/3) + c.$

11. $\displaystyle\frac{1}{x(x^2 + 4)^2} = \frac{A}{x} + \frac{Bx + C}{x^2 + 4} + \frac{Dx + E}{(x^2 + 4)^2}$ so that

$1 = A(x^2+4)^2 + (Bx+C)x(x^2+4) + (Dx+E)x$
$\quad = (A+B)x^4 + Cx^3 + (8A+4B+D)x^2 + (4C+E)x + 16A.$ Thus
$A+B = 0$, $C = 0$, $8A+4B+D = 0$, $4C+E = 0$, and $16A = 1$ and hence
$A = 1/16$, $B = -1/16$, $C = 0$, $D = -1/4$ and $E = 0$.   Therefore

$\displaystyle\int \frac{dx}{x(x^2 + 4)^2} = (1/16)\int \frac{dx}{x} - (1/16)\int \frac{x\,dx}{x^2 + 4} - (1/4)\int \frac{x\,dx}{(x^2 + 4)^2}$

$\displaystyle = (1/16)\ln|x| - (1/32)\ln(x^2+4) + (1/8)(x^2+4)^{-1} + c.$

13. $\displaystyle\frac{2x^2 + 5x + 5}{(x - 1)(x^3 - 1)} = \frac{A}{x - 1} + \frac{B}{(x - 1)^2} + \frac{Cx + D}{x^2 + x + 1}$ so that

$2x^2+5x+5 = A(x-1)(x^2+x+1) + B(x^2+x+1) + (Cx+D)(x-1)^2$
$\quad = (A+C)x^3 + (B+D-2C)x^2 + (B+C-2D)x + (-A+B+D).$ Thus
$A+C = 0$, $B+D-2C = 2$, $B+C-2D = 5$ and $-A+B+D = 5$ and hence
$A = -1$, $B = 4$, $C = 1$ and $D = 0$.   Therefore

$\displaystyle\int \frac{2x^2 + 5x + 5}{(x - 1)(x^3 - 1)}\,dx = -\int \frac{dx}{x - 1} + 4\int \frac{dx}{(x - 1)^2} + \int \frac{x\,dx}{x^2 + x + 1}$

$\displaystyle = -\ln|x-1| - 4(x-1)^{-1} + \int \frac{(s - 1/2)\,ds}{s^2 + 3/4}$,   where $s = x+1/2$. Now

$\displaystyle\int \frac{(s - 1/2)\,ds}{s^2 + 3/4} = \int \frac{s\,ds}{s^2 + 3/4} - (1/2)\int \frac{ds}{s^2 + 3/4}$

$\displaystyle = (1/2)\ln(s^2 + 3/4) - \frac{1}{2\sqrt{3/4}}\arctan(s/\sqrt{3/4}) + c$

$\displaystyle = (1/2)\ln(x^2+x+1) - (1/\sqrt{3})\arctan[2(x+1/2)/\sqrt{3}] + c,$ and thus

$$\int \frac{2x^2 + 5x + 5}{(x-1)(x^3-1)}\,dx = -\ln|x-1| - 4(x-1)^{-1} + (1/2)\ln(x^2+x+1)$$

$$- (1/\sqrt{3})\arctan[(2x+1)/\sqrt{3}] + c.$$

15. $\displaystyle \int \frac{dx}{(x^2+a^2)^2} = \frac{x}{2a^2(x^2+a^2)} + \frac{1}{2a^2}\int \frac{dx}{x^2+a^2}$ [by Eq. (4)]

$$= \frac{x}{2a^2(x^2+a^2)} + \frac{1}{2a^3}\arctan(x/a) + c.$$

17. $\displaystyle \int \frac{dx}{(x^2+a^2)^4} = \frac{x}{6a^2(x^2+a^2)^3} + \frac{5}{6a^2}\int \frac{dx}{(x^2+a^2)^3}$

$$= \frac{x}{6a^2(x^2+a^2)^3} + \frac{5}{6a^2}\left[\frac{x}{4a^2(x^2+a^2)^2} + \frac{3x}{8a^4(x^2+a^2)} + \frac{3}{8a^5}\arctan(x/a)\right] + c$$

$$= \frac{x}{6a^2(x^2+a^2)^3} + \frac{5x}{24a^4(x^2+a^2)^2} + \frac{5x}{16a^6(x^2+a^2)} + \frac{5}{16a^7}\arctan(x/a) + c,$$

where the results of Prob.16 have been used.

19. $\displaystyle \frac{1}{x(x^2+a^2)^3} = \frac{A}{x} + \frac{Bx+C}{x^2+a^2} + \frac{Dx+E}{(x^2+a^2)^2} + \frac{Fx+G}{(x^2+a^2)^3}$ so that

$$1 = A(x^2+a^2)^3 + x(Bx+C)(x^2+a^2)^2 + x(Dx+E)(x^2+a^2) + (Fx+G)x$$
$$= (A+B)x^6 + Cx^5 + (3a^2A+2a^2B+D)x^4 + (2a^2C+E)x^3$$
$$+ (3a^4A+a^4B+a^2D+F)x^2 + (a^4C+a^2E+G)x + a^6A. \quad \text{Hence}$$

$A+B = 0$, $C = 0$, $3a^2A+2a^2B+D = 0$, $2a^2C+E = 0$, $3a^4A+a^4B+a^2D+F = 0$, $a^4C+a^2E+G = 0$ and $a^6A = 1$, or $A = 1/a^6$, $B = -1/a^6$, $C = 0$, $D = -1/a^4$, $E = 0$, $F = -1/a^2$ and $G = 0$. Thus

$$\int \frac{dx}{x(x^2+a^2)^3} = \frac{1}{a^6}\int \frac{dx}{x} - \frac{1}{a^6}\int \frac{x\,dx}{x^2+a^2} - \frac{1}{a^4}\int \frac{x\,dx}{(x^2+a^2)^2} - \frac{1}{a^2}\int \frac{x\,dx}{(x^2+a^2)^3}$$

$$= (1/a^6)\ln|x| - (1/2a^6)\ln(x^2+a^2) + (1/2a^4)(x^2+a^2)^{-1}$$
$$+ (1/4a^2)(x^2+a^2)^{-2} + c$$
$$= (1/2a^6)\ln[x^2/(x^2+a^2)] + (1/2a^4)(x^2+a^2)^{-1} + (1/4a^2)(x^2+a^2)^{-2} + c.$$

21. $\displaystyle \frac{1}{x^2(x^2+a^2)^3} = \frac{A}{x} + \frac{B}{x^2} + \frac{Cx+D}{x^2+a^2} + \frac{Ex+F}{(x^2+a^2)^2} + \frac{Gx+H}{(x^2+a^2)^3}$ or

$$1 = Ax(x^2+a^2)^3 + B(x^2+a^2)^3 + x^2(Cx+D)(x^2+a^2)^2 + x^2(Ex+F)(x^2+a^2)$$
$$+ x^2(Gx+H). \quad \text{Expanding the right side and equating like}$$

coefficients, we get $A+C = 0$, $B+D = 0$, $3a^2A+2a^2C+E = 0$, $3a^2B+2a^2D+F = 0$, $3a^4A+a^4C+a^2E+G = 0$, $3a^4B+a^4D+a^2F+H = 0$, $a^6A = 0$, and $a^6B = 1$, which yield $A = 0$, $B = 1/a^6$, $C = 0$,

$D = -1/a^6$, $E = 0$, $F = -1/a^4$, $G = 0$, and $H = -1/a^2$. Thus

$$\int \frac{dx}{x^2(x^2+a^2)^3} = \frac{1}{a^6}\int \frac{dx}{x^2} - \frac{1}{a^6}\int \frac{dx}{x^2+a^2} - \frac{1}{a^4}\int \frac{dx}{(x^2+a^2)^2} - \frac{1}{a^2}\int \frac{dx}{(x^2+a^2)^3}$$

$$= -(1/a^6)x^{-1} - (1/a^7)\arctan(x/a) - \frac{1}{a^4}\left[\frac{x}{2a^2(x^2+a^2)} + \frac{1}{2a^3}\arctan(x/a)\right]$$

$$- \frac{1}{a^2}\left[\frac{x}{4a^2(x^2+a^2)^2} + \frac{3x}{8a^4(x^2+a^2)} + \frac{3}{8a^5}\arctan(x/a)\right] + c$$

$$= -\frac{1}{a^6 x} - \frac{x}{4a^4(x^2+a^2)^2} - \frac{7x}{8a^6(x^2+a^2)} - \frac{15}{8a^7}\arctan(x/a) + c, \text{ where}$$

the results of Probs. 15 and 16 were used.

23. Let $u = x^2+a^2$, then $\displaystyle\int \frac{x\,dx}{(x^2+a^2)^2} = \frac{1}{2}\int \frac{du}{u^2} = -(1/2)(x^2+a^2)^{-1} + c$.

25. $\displaystyle\frac{x^3}{(x^2+a^2)^2} = \frac{Ax+B}{x^2+a^2} + \frac{Cx+D}{(x^2+a^2)^2}$ so that

$x^3 = (Ax+B)(x^2+a^2) + Cx + D = Ax^3 + Bx^2 + (a^2A+C)x + (a^2B+D)$. Hence
$A = 1$, $B = 0$, $a^2A+C = 0$, and $a^2B+D = 0$, so $C = -a^2$ and $D = 0$,

and thus $\displaystyle\int \frac{x^3 dx}{(x^2+a^2)^2} = \int \frac{x\,dx}{x^2+a^2} - a^2\int \frac{x\,dx}{(x^2+a^2)^2}$

$$= (1/2)\ln(x^2+a^2) + (a^2/2)(x^2+a^2)^{-1} + c.$$

27. $x = a\tan u \Rightarrow x^2+a^2 = a^2\sec^2 u$ and $dx = a\sec^2 u\,du$. Thus

$$I_n = \int \frac{a\sec^2 u\,du}{a^{2n}\sec^{2n}u} = \frac{1}{a^{2n-1}}\int \frac{du}{\sec^{2(n-1)}u} = \frac{1}{a^{2n-1}}\int \cos^{2(n-1)}u\,du.$$ If $n = 2$,

then $I_2 = (1/a^3)\int\cos^2 u\,du = (1/2a^3)\int(1+\cos 2u)\,du$

$$= (1/2a^3)[u + (1/2)\sin 2u] + c$$

$$= (1/2a^3)u + (1/2a^3)\sin u\cos u + c$$

$$= (1/2a^3)\arctan(x/a) + (1/2a^2)x/(x^2+a^2) + c.$$

Chapter 9 Review, Page 506

1. With $u = \ln x$, $I = \int \dfrac{du}{1+u^2} = \arctan u + c = \arctan(\ln x) + c$.

3. With $u = \ln x$, $I = \int(u+1)\,du = (1/2)(u+1)^2 + c = (1/2)[\ln x + 1]^2 + c$.

5. $\displaystyle\int_{-\pi/4}^{\pi/4} \tan^5 x\, dx = \int_{-\pi/4}^{\pi/4} (\tan^3 x\, \sec^2 x - \tan x\, \sec^2 x + \tan x)\, dx$

$$= (1/4)\tan^4 x - (1/2)\tan^2 x - \ln|\cos x|\Big|_{-\pi/4}^{\pi/4} = 0.$$

7. With $x = \sin u$, $dx = \cos u\, du$ and the integral becomes

$I = \int \sin^4 u\, \cos^2 u\, du = (1/8)\int (1 - \cos 2u)^2(1 + \cos 2u)\, du$

$= (1/16)u - (1/64)\sin 4u - (1/48)\sin^3 2u + c$

$= (1/16)\arcsin x - (3/48)x\sqrt{1-x^2}\,(1 - 2x^2) - (8/48)x3(1-x^2)3/2 + c$

$= (1/16)\arcsin x + (1/6)x\sqrt{1-x^2}\,[x^2 - 3/4][x^2 + 1/2] + c.$

9. $\tan x^2 = \sin x^2/\cos x^2$, so $I = \displaystyle\int \frac{2x(\cos x^2 - \sin x^2)}{\sin x^2 + \cos x^2}\,dx$

$$= \ln|\sin x^2 + \cos x^2| + c.$$

11. Integrating by parts with $u = x^2$, $dv = x\sqrt{1+x^2}$,

$I = (x^2/3)(1+x^2)3/2 - (2/3)\int x(1+x^2)3/2\, dx$

$= (x^2/3)(1+x^2)3/2 - (2/15)(1+x^2)5/2 + c.$

13. $(x^3+x^2-7x-2)/x(x+3) = (x-2) - [(x+2)/x(x+3)]$ so
$I = \int [(x-2) - (x+2)/x(x+3)]\, dx$

$= (1/2)(x-2)2 - (2/3)\int dx/x - (1/3)\int dx/(x+3)$, by part. fract.

$= (1/2)(x-2)2 - (1/3)\ln|x^2(x+3)| + c.$

15. $\int \sin(x/2)\cos(\pi x/4)\, dx = (1/2)\int [\sin(\dfrac{x}{2} - \dfrac{\pi x}{4}) + \sin(\dfrac{x}{2} + \dfrac{\pi x}{4})]\, dx$

$= (-1/2)[(1/2-\pi/4)^{-1}\cos(x/2-\pi x/4) + (1/2+\pi/4)^{-1}\cos(x/2+\pi x/4)]+c$

$= [4/(\pi^2-4)][2\cos(x/2)\cos(\pi x/4) + \pi\sin(x/2)\sin(\pi x/4)] + c.$

17. With $u = \sec^3 x$, $du = 3\tan x\, \sec^3 x\, dx$ and
$I = (1/3)\int du/(1+u^2) = (1/3)\arctan u + c = (1/3)\arctan(\sec^3 x) + c.$

19. Using Eq.(12a),Sect.9.3, then (12b), then (12a) yields
$I = (1/8)\int (-3\sin x - \sin 3x + 3\sin 5x - \sin 9x)\, dx$

$= (3/8)\cos x + (1/24)\cos 3x - (3/40)\cos 5x + (1/72)\cos 9x + c.$

21. By division and partial fractions,
$(2x^4-3x^3-5x^2-4)/x^2(x^2-4) = 2 + 1/x^2 - 1/(x-2) - 2/(x+2)$ so that
$I = 2x - 1/x - \ln|x-2| - 2\ln|x+2| + c$
$= (2x^2-1)/x + \ln[|x-2|/(x^2-4)^2] + c.$

23. Using integration by parts with $u = x$ and $dv = xdx/\sqrt{4+x^2}$,
$$I = x(4+x^2)^{1/2}\Big|_0^{\pi/2} - \int_0^{\pi/2} (4+x^2)^{1/2}dx$$
$$= [x(4+x^2)^{1/2} - (x/2)(4+x^2)^{1/2} - 2\,\text{arcsinh}(x/2)]\Big|_0^{\pi/2}$$
$$= (\pi/4)\sqrt{4+(\pi^2/4)} - 2\,\text{arcsinh}(\pi/4).$$

25. Let $8u = x^{3/2}$, so
$$I = (1/12)\int \frac{du}{u\sqrt{1-u^2}} = (1/12)\ln|(1 - \sqrt{1-u^2})/u| + c$$
$$= (1/12)\ln|8x^{-3/2} - x^{-3/2}\sqrt{64-x^3}| + c.$$

27. $I = 2\displaystyle\int \frac{dx}{(2x+1)^2 + 1} = \arctan(2x+1) + c.$

29. $I = \displaystyle\int_0^{\pi/4} \frac{\sec^2 x}{1+\tan x}dx = \ln|1+\tan x|\,\Big|_0^{\pi/4} = \ln 2.$

31. $I = \int [\sec x + 1/(x+1)]dx = \ln|\sec x+\tan x| + \ln|x+1| + c$
$= \ln|(x+1)(\sec x+\tan x)| + c.$

33. $I = \int \sin x(\sec^2 x - 1)dx = -\int \sin x(1-\cos^{-2} x)dx = \cos x + \sec x + c.$

35. $I = \displaystyle\int_1^2 \frac{dx}{2(x-3/2)^2+1/2} = 2\int_1^2 \frac{dx}{[2(x-3/2)]^2+1} = \arctan[2(x-3/2)]\Big|_1^2$
$= \arctan 1 - \arctan(-1) = \pi/2.$

37. Letting $w = x^2$ and then integrating by parts yields
$I = (1/2)\int w \sinh w\, dw = (1/2)[w \cosh w - \sinh w] + c$
$= (1/2)[x^2\cosh x^2 - \sinh x^2] + c.$

39. $u = x$, $dv = \sec^2 x \Rightarrow I = x\tan x - \int \tan x\, dx = x\tan x + \ln|\cos x| + c.$

41. $I = \int (\dfrac{1}{x} - \dfrac{1}{x+1} + \dfrac{1}{x+2})\, dx = \ln|x(x+2)/(x+1)| + c.$

43. $I = \displaystyle\int \dfrac{(x-2)^2\, dx}{\sqrt{(x-2)^2 - 3}} + 4\int \dfrac{(x-2)dx}{\sqrt{(x-2)^2 - 3}} + 4\int \dfrac{dx}{\sqrt{(x-2)^2 - 3}}$

$= \dfrac{3}{2}[\dfrac{(x-2)\sqrt{(x-2)^2-3}}{3} + \ln|(x-2) + \sqrt{(x-2)^2-3}|]$

$+ 4\sqrt{(x-2)^2-3} + 4\ln|(x-2) + \sqrt{(x-2)^2-3}| + c$

$= [(x-2)/2 + 4]\sqrt{x^2-4x+1} + (11/2)\ln|(x-2) + \sqrt{x^2-4x+1}| + c.$

Techniques of Sections 9.4 and 9.5 have been used here.

45.   With $u = \arctan x$, $I = \int \tan(au)\, du = (1/a)\ln|\sec(au)| + c$
$= (1/a)\ln|\sec(a \arctan x)| + c.$

47. $I = \int [\dfrac{2}{x+1} + \dfrac{8/3}{(x-1)} + \dfrac{3}{(x-1)^2} - (4/3)\dfrac{2x+2}{x^2+2}]\, dx$

$= 2\ln|x+1| + (8/3)\ln|x-1| - 3/(x-1)$

$- (4/3)\ln(x^2+2) - (4\sqrt{2}/3)\arctan(x/\sqrt{2}) + c.$

49.   With $u = e^x$, $I = \int (1-u^2)^{3/2} du/u = \int (\csc\theta - 2\sin\theta + \sin^3\theta)\, d\theta$
after substituting $\sin\theta$ for u.   Thus
$I = \int (\csc\theta - \sin\theta - \cos^2\theta \sin\theta)\, d\theta$

$= \ln|\csc\theta - \cot\theta| + \cos\theta + (1/3)\cos^3\theta + c$

$= \ln|(1 - \sqrt{1-u^2}/u| + \sqrt{1-u^2} + (1/3)(1-u^2)^{3/2} + c$

$= \ln[(1 - \sqrt{1-e^{2x}})e^{-x}] + (1-e^{2x})^{1/2} + (1/3)(1-e^{2x})^{3/2} + c.$

51.   Letting $u = x^{1/3}$ gives $I = \displaystyle\int_0^1 3u^2\sqrt{1+u^2}\, du$, which becomes

$I = 3\displaystyle\int_0^{\pi/4} \tan^2\theta \sec^3\theta\, d\theta = 3\int_0^{\pi/4} (\sec^5\theta - \sec^3\theta)\, d\theta$ with $u = \tan\theta$.
Thus by Eq.(10), Section 9.4,
$I = (3/4)[\sec^3\theta \tan\theta - (1/2)\sec\theta \tan\theta - (1/2)\ln|\sec\theta + \tan\theta|]_0^{\pi/4}$
$= 3\sqrt{8}/4 - 3\sqrt{2}/8 - (3/8)\ln(\sqrt{2} + 1) = (9\sqrt{2}/8) - (3/16)\ln(3+2\sqrt{2}).$

53. With $2\sin^2(\theta/2) = 1-\cos\theta$,

  $I = (1/2\sqrt{2})\int \csc^3(\theta/2)\,d\theta$

  $= (1/\sqrt{2})[-(1/2)\csc(\theta/2)\cot(\theta/2)$

  $+ (1/2)\ln|\csc(\theta/2) - \cot(\theta/2)|] + c$.

  by Eqs.(9) and (10) of Section 9.4.

55. Since the integrand is an odd function, $I = 0$.

57. With the suggested substitutions,

  $I = \int \dfrac{4u\,du}{(1-u)(1+u)^3} = \dfrac{1}{2}\int [\dfrac{1}{1+u} + \dfrac{1}{1-u}]\,du + \int \dfrac{du}{(1+u)^2} - 2\int \dfrac{du}{(1+u)^3}$

  $= (1/2)\ln|(1+u)/(1-u)| - 1/(1+u) + 1/(1+u)^2 + c$

  $= (1/2)\ln|[1+\tan(x/2)]/[1-\tan(x/2)]|$

  $- [\tan(x/2)]/[1 + \tan(x/2)]^2 + c$.

59. $I = \int \dfrac{2\,du}{(a-1)u^2 + (a+1)} = \dfrac{2}{a-1}\int \dfrac{du}{u^2 + \alpha^2}$, where $\alpha = \sqrt{(a+1)/(a-1)}$ for $a > 1$.

  So, for $a > 1$, $I = \dfrac{2}{(a-1)\alpha}\arctan(u/\alpha)$

  $= (2/\sqrt{a^2-1})\arctan[\sqrt{(a-1)/(a+1)}\tan(x/2)] + c$.

  If $0 < a < 1$, then $I = \dfrac{2}{a-1}\int \dfrac{du}{u^2-\alpha^2}$, where $\alpha = \sqrt{(1+a)/(1-a)}$, so

  $I = \dfrac{2}{a-1}\int [(\dfrac{1}{2\alpha})\dfrac{1}{u-\alpha} - (\dfrac{1}{2\alpha})\dfrac{1}{u+\alpha}]\,du$

  $= (-1/\sqrt{1-a^2})\{\ln|[\alpha - \tan(x/2)]/[\alpha + \tan(x/2)]|\} + c$.

CHAPTER 10

1.	The integrating factor is $\mu(t) = \exp(\int -2dt) = e^{-2t}$. Thus, $e^{-2t}y' - 2e^{-2t}y = t^2$, or $(e^{-2t}y)' = t^2$. Thus $e^{-2t}y = t^3/3 + c$ so that the general solution of the differential equation is $y = (1/3)t^3e^{2t} + ce^{2t}$.

3.	The integrating factor is $\mu(t) = \exp(\int 3dt) = e^{3t}$ which gives $y'e^{3t} + 3e^{3t}y = (e^{3t}y)' = te^{3t} + e^{2t}$. Integration by parts then gives
$e^{3t}y = \int te^{3t}dt + \int e^{2t}dt = (1/3)te^{3t} - (1/9)e^{3t} + (1/2)e^{2t} + c$
or $y = ce^{-3t} + (t/3) - (1/9) + (1/2)e^{-t}$.

5.	The integrating factor is $\mu(t) = \exp[\int(1/t)dt] = t$ giving $(ty)' = 3t\cos 2t$ or, with integration by parts,
$ty' = (3/2)t\sin 2t - (3/2)\int \sin 2t = (3/2)t\sin 2t + (3/4)\cos 2t + c$
so $y = (c/t) + (3/2)\sin 2t + (3/4t)\cos 2t$.

7.	The equation, in standard form, is $y' + (2/t)y = e^t/t$, which has the integrating factor $\mu(t) = \exp[\int(2/t)dt] = t^2$. Thus $t^2y' + 2ty = te^t$ or $(t^2y)' = te^t$, so integrating by parts yields $t^2y = te^t - e^t + c$ or $y = (c + te^t - e^t)/t^2$.

9.	With $\mu(t) = t^3$ as integrating factor, $(t^3y)' = \sin t$ so $t^3y = -\cos t + c$ and $y = (c - \cos t)/t^3$.

11.	With $\mu = e^t/t$ as integrating factor, $(ye^t/t)' = 1/t$ so $ye^t/t = \ln t + c$ and $y = cte^{-t} + te^{-t}\ln t$ for $t > 0$.

13.	$\mu(t) = e^{-t}$ is the integrating factor so, $(e^{-t}y)' = 2t$ and $e^{-t}y = t^2 + c$. Thus $y = ce^t + t^2e^t$. Substituting $t = 0$ we then have $y(0) = ce^0 + 0 = 1$ or $c = 1$ and $y = e^t + t^2e^t$.

15.	With $\mu(t) = t^2$ we find $t^2y' + 2ty = \cos t$ or $(t^2y)' = \cos t$, so that $y = (\sin t + c)/t^2$. To determine c, set $t = \pi$ and $y(\pi) = 0$ so $0 = (0+c)/\pi^2$ or $c = 0$. Thus $y = \sin t/t^2$.

17. In standard form the equation is $y' + [(2/(t+1)]y = t/(t+1)$
     which has $\mu(t) = (t+1)^2$, so $[(t+1)^2 y]' = t(t+1) = t^2+t$. Thus
     $(t+1)^2 y = (t^3/3) + (t^2/2) + c/6 = (2t^3+3t^2+c)/6$. Therefore
     $y = (2t^3+3t^2+c)/6(t+1)^2$ and hence $y(0) = c/6 = 3$ or $c = 18$
     and the solution of the initial value problem is thus
     $y = (2t^3+3t^2+18)/6(t+1)^2$.

19. Let $a > 0$, $\lambda > 0$ and b be real.
     Case (i): $a \neq \lambda$ so $\mu(t) = e^{at}$ is the integrating factor, which
     gives $(e^{at}y)' = be^{(a-\lambda)t}$. Thus $e^{at}y = [b/(a-\lambda)]e^{(a-\lambda)t} + c$ or
     solving for y gives $y = [b/(a-\lambda)]e^{-\lambda t} + ce^{-at}$ and thus
     $$\lim_{t\to\infty} y = \lim_{t\to\infty}\left[\frac{b}{a-\lambda}e^{-\lambda t} + ce^{-at}\right] = 0 \text{ for } a > 0, \lambda > 0.$$
     Case (ii): $a = \lambda$ so the integrating factor is $\mu(t) = e^{\lambda t}$
     which gives $(e^{\lambda t}y)' = b$ and $e^{\lambda t}y = bt + c$. Therefore
     $y = (bt+c)e^{-\lambda t}$ and again $\lim_{t\to\infty} y = \lim_{t\to\infty}(bt+c)e^{-\lambda t} = 0$, for $\lambda > 0$.

21. From the stated conditions and temperature values, the
     mathematical model for the problem is $d\theta/dt = -k(\theta-T)$,
     $\theta(0) = 28°C$, $\theta(1) = 22°C$, $T = 10°C$, when t is measured in
     hours. Writing the equation as $d\theta/(\theta-10) = -kdt$ and
     integrating, gives $\ln(\theta-10) = -kt + c$, $(\theta>10)$. Applying the
     initial condition $\theta(0) = 28$, gives $\ln 18 = c$ so
     $\ln(\theta-10) = -kt + \ln 18$. Now using the condition $\theta(1) = 22$, we
     find $\ln 12 = -k + \ln 18$, or $k = \ln 1.5$ and
     $\ln(\theta-10) = (-\ln 1.5)t + \ln 18$. To find t_1 such that $\theta(t_1) = 37$
     we need to solve $\ln 27 = -\ln(1.5)t_1 + \ln 18$ or $t_1 = -1$ hours.
     Thus, death occurred one hour before the body was discovered.

23. From the discussion and the conditions, the model for the
     problem is: $Q'(t) + (5/300)Q = .5$, $Q(0) = 20kg$. The
     equation has integrating factor $\mu(t) = e^{t/60}$ so
     $(e^{t/60}Q)' = .5e^{t/60}$ giving $e^{t/60}Q = 30e^{t/60} + c$ or
     $Q(t) = 30 + ce^{-t/60}$. Now $Q(0) = 30 + c = 20$ so $c = -10$ and
     thus $Q(t) = 30 - 10e^{-t/60}$.

25a. In this problem, $V = 50$, $r = 5$ gal/min, $q = \gamma$ lb/gal and
      $Q(0) = 0$, so the model becomes $Q' + (1/10)Q = 5\gamma$. $\mu(t) = e^{t/10}$
      so $(e^{t/10}Q)' = 5\gamma e^{t/10}$ and $e^{t/10}Q = 50\gamma e^{t/10} + c$ or
      $Q = 50\gamma + ce^{-t/10}$. With $Q(0) = 0$, we find $c = -50\gamma$ so that
      $Q(t) = 50\gamma - 50\gamma e^{-t/10}$.
  b. If $Q(60) = 5$, then $5 = \gamma(50-50e^{-60/10})$ or $\gamma = 1/10(1-e^{-6})$.

27.   If $v = y^{1-n}$ then $y = v^{1/(1-n)}$ and $y' = [1/(1-n)]v^{n/(1-n)}v'$ so
      the equation $y'+p(t)y = q(t)y^n$ becomes
      $[1/(1-n)]v^{n/(1-n)}v' + p(t)v^{1/(1-n)} = q(t)v^{n/(1-n)}$. Dividing by
      $v^{n/(1-n)}$ yields $[1/(1-n)]v' + p(t)v^{[1/(1-n) - n/(1-n)]} = q(t)$
      or, simplifying, $v' + (1-n)p(t)v = (1-n)q(t)$.

29.   Using the technique of Prob.27, with $n=3$, $v=y^{-2}$ or $y = 1/\sqrt{v}$,
      we have $v' + 2\epsilon v = 2\sigma$. Since $\mu(t) = e^{2\epsilon t}$, $(e^{2\epsilon t}v)' = 2\sigma e^{2\epsilon t}$
      and $e^{2\epsilon t}v = (\sigma/\epsilon)e^{2\epsilon t} + c$ or $v = \sigma/\epsilon + ce^{-2\epsilon t}$ and thus
      $y = (\sigma/\epsilon + ce^{-2\epsilon t})^{-1/2}$.

Section 10.2, Page 524

1.   The equation may be separated as $ydy - t^2 dt = 0$. Integrating
      gives $y^2/2 - t^3/3 = c$ or $3y^2 - 2t^3 = c$. (Note that if c is
      arbitrary then 6c is also, so we rename 6c as c which is
      common practice.)

3.   Separating the equation gives $[(1+2y^2)/2y]dy - \cos 2t\, dt = 0$,
      or $[(1/2y)+y]dy - \cos 2t\, dt = 0$. Integrating gives
      $(1/2)\ln|y| + y^2/2 - (1/2)\sin 2t = c$ or, multiplying by 2,
      $\ln|y| + y^2 - \sin 2t = c$ for $y \neq 0$. Note that $y(t) = 0$ also is
      a solution, as $dy/dt = 0$.

5.   The equation may be written as $ydy = [t^2/(1+t^3)]dt$.
      Integrating gives $y^2/2 - (1/3)\ln|1+t^3| = c$ or, multiplying by
      6, $3y^2 - 2\ln|1+t^3| = c$.

7.   Separating the variables gives $dy/\sqrt{1-y^2} = dt/t$, so
      $\arcsin y - \ln|t| = c$.

9. Separating yields $y\,dy - [2t/(1+t^2)]dt = 0$. Integrating gives
    the solution $y^2/2 - \ln(1+t^2) = c$ or $y^2 - 2\ln(1+t^2) = c$. To
    determine c, let $t=0$, $y=2$: $2^2 - 2\ln 1 = c$ or $4 = c$. Thus the
    solution to the initial value problem is $y^2 - 2\ln(1+t^2) = 4$
    or $y = \sqrt{2\ln(1+t^2)+4}$.

11. Separating: $dy/y^3 - t\,dt/\sqrt{1+t^2} = 0$. Integrating gives
    $-(1/2)y^{-2} - \sqrt{1+t^2} = c$. Letting $t=0, y=1$ we have $-1/2 - 1 = c$
    or $c = -3/2$ and thus $-(1/2)y^{-2} - \sqrt{1+t^2} = -3/2$ or
    $y^{-2} = 3 - 2\sqrt{1+t^2}$ so $y = [3 - 2\sqrt{1+t^2}]^{-1/2}$ is the solution of
    the initial value problem.

13. The equation separates as $(2y+2)dy = (2t-1)dt$. Integrating
    we find $y^2+2y = t^2-t+c$ or $y^2+2y-t^2+t = c$. Setting $y = -3$
    when $t=3$ yields $9-6-9+3 = c$, or $c = -3$. Thus $y^2+2y-t^2+t = -3$.
    To solve for y we rewrite the equation as $y^2+2y+1 = t^2-t-2$ or
    $(y+1)^2 = t^2-t-2$ or $y = -1 - \sqrt{t^2-t-2}$. (Note that we choose
    $-\sqrt{t^2-t-2}$ so that $y(3) < 0$).

15. Rewrite the equation as $y' = 2te^{3t}e^y$ to separate:
    $e^{-y}dy = 2te^{3t}dt$. Integrating gives
    $-e^{-y} = (2/3)te^{3t} - (2/9)e^{3t} + c$ or $9e^{-y} = c - 6te^{3t} + 2e^{3t}$.
    With $y(0) = 0$ we find c by $9 = c+2$ or $c = 7$. Thus
    $e^{-y} = [7-6te^{3t}+2e^{3t}]/9$ so $y = -\ln|(7-6te^{3t}+2e^{3t})/9|$.

17. $yy' = 3(2t+1)^2$ may be written $y\,dy = 3(2t+1)^2dt$ and integrated
    to give $y^2/2 = (2t+1)^3/2 + c$ or $y^2 = (2t+1)^3 + c$. Set $y = -1$
    when $t=0$ to obtain $1 = 1 + c$, or $c = 0$. Thus $y^2 = (2t+1)^3$ and
    $y = -\sqrt{(2t+1)^3}$, the negative being chosen so the solution
    satisfies the initial condition.

19a. For $y_0 = k/3$, Eq.(37) becomes $y = \dfrac{K/3}{1/3 + (2/3)e^{-rt}} = \dfrac{K}{1 + 2e^{-rt}}$.
    We now find τ so that $y(\tau) = 2y_0 = 2K/3$. That is
    $2K/3 = \dfrac{K}{1 + 2e^{-r\tau}}$ or $3/2 = 1+2e^{-r\tau}$ so that $e^{r\tau} = 4$. Thus
    $\tau = (1/r)\ln4$ years and if $r = .025$/year, $\tau = 40\ln4 \cong 55.452$.

19b. Setting $y_0 = \alpha K$, we find $y = \dfrac{\alpha K}{\alpha + (1 - \alpha)e^{-rt}}$. Now

$y(T) = \beta K = \dfrac{\alpha K}{\alpha + (1-\alpha)e^{-rT}}$. So $\alpha/\beta = \alpha + (1-\alpha)e^{-rT}$ or

$\dfrac{\alpha(1 - \beta)}{\beta(1 - \alpha)} = e^{-rT}$ and $T = \dfrac{1}{r}\ln\dfrac{\beta(1-\alpha)}{\alpha(1-\beta)}$. We have omitted the

absolute value as the fraction is positive for the given α
and β. If $r = .025$, $\alpha = .1$, $\beta = .9$ then $T = 40\ln 81 \cong 175.78$yrs.

Note that T may be written as $T = \dfrac{1}{r}[\ln\beta(1-\alpha) - \ln\alpha(1-\beta)]$.

Thus, as $\alpha \to 0$, or $\beta \to 1$, $\ln\alpha(1-\beta) \to -\infty$, and $T \to \infty$.

21. $\dfrac{dy}{dt} = \dfrac{at+b}{ct+d}$ may be written as $dy = \left(\dfrac{at+b}{ct+d}\right)dt$ so that

$y = \displaystyle\int \dfrac{at+b}{ct+d}\,dt$. To integrate the right side, divide to get:

$\dfrac{at+b}{ct+d} = \dfrac{a}{c} + \dfrac{(bc-ad)/c}{ct+d}$. Thus $y = \displaystyle\int\left[\dfrac{a}{c} + \dfrac{(bc-ad)/c}{ct+d}\right]dt$.

Therefore $y = \dfrac{a}{c}t + \dfrac{bc-ad}{c^2}\ln|ct+d| + k$.

23a. $\dfrac{dx}{dt} = -\beta x$ becomes $dx/x = -\beta dt$ or $\ln|x| = -\beta t + c$ so that

$x(t) = e^{-\beta t + c} = Ke^{-\beta t}$ $(K>0)$. Then $x(0) = x_0 = K$, gives

$x(t) = x_0 e^{-\beta t}$.

b. Substituting this value into $\dfrac{dy}{dt} = -\alpha xy$ yields $\dfrac{dy}{dt} = -\alpha x_0 e^{-\beta t}y$

or $dy/y = -\alpha x_0 e^{-\beta t}dt$. Thus $\ln|y| = (\alpha x_0/\beta)e^{-\beta t} + c$ so that

$y = ce^{(\alpha x_0/\beta)e^{-\beta t}}$ and $y(0) = y_0 = ce^{\alpha x_0/\beta}$. Hence $c = y_0 e^{-\alpha x_0/\beta}$

and $y(t) = y_0 e^{-\alpha x_0(1-e^{-\beta t})/\beta}$.

c. As $t \to \infty$, $e^{-\beta t} \to 0$ so $\lim\limits_{t\to\infty} y(t) = y_0 e^{-\alpha x_0/\beta}$. Note that for α

small, that is for a slowly spreading disease, the number of
well, but susceptible people, $y(\epsilon)$, remains close to y_0,

while for a rapidly spreading disease, α larger, $e^{-\alpha x_0/\beta}$ is
small so the limiting number of well people is smaller than
the original number, y_0. On the other hand, suppose β is

large, (then the disease carriers are being removed rapidly), $-\alpha x_0/\beta$ is small so $e^{-\alpha x_0/\beta}$ is near 1 and the limiting well population is close to the original well population. Similarly, if the carriers are removed slowly, β small, then $\alpha x_0/\beta$ is large so $e^{\alpha x_0/\beta}$ is large and $y_0 e^{-\alpha x_0/\beta}$ is small. That is, if carriers are removed slowly, then there will be few members of the population who are well.

25. The mathematical model for the reaction is assumed as $dx/dt = \alpha(p-x)(q-x)$ with $x(0) = 0$. The differential equation may be written as $\dfrac{dx}{(p-x)(q-x)} = \alpha\,dt$. Since

$$\frac{1}{(p-x)(q-x)} = \frac{1/(q-p)}{p-x} + \frac{1/(p-q)}{q-x} \text{ we have}$$

$$c_1 + \alpha t = \int \left[\frac{1}{(q-p)(p-x)} + \frac{1}{(p-q)(q-x)} \right] dx = \frac{1}{p-q} \ln\left|\frac{p-x}{q-x}\right|.$$

Therefore $\ln\left|\dfrac{p-x}{q-x}\right| = c_2 + \alpha t$ and $(p-x)/(q-x) = ce^{\alpha(p-q)t}$. Now

setting $x = 0$ when $t = 0$ we have that $c = p/q$, giving

$(p-x)/(q-x) = (p/q)e^{\alpha(p-q)t}$. To solve for x explicitly, set the right side equal to R, which makes the manipulations easier to perform. Thus $(p-x)/(q-x) = R$ so $p-x = Rq-Rx$ or

$$x = \frac{Rq-p}{R-1} = \frac{\left[(p/q)\,e^{\alpha(p-q)t}\right]q-p}{\left[(p/q)\,e^{\alpha(p-q)t}-1\right]} = pq[e^{\alpha(p-q)t}-1]/[pe^{\alpha(p-q)t}-q].$$

27a. Under the assumption $k = 0$, the equation becomes $dv/dt = -9.8$ from which $v(t) = -9.8t + c$. $v(0) = 20$ gives $c = 20$ so $v(t) = -9.8t + 20$. We will take $x = 0$ to be at ground level, so that $x(0) = 30$ ft. Since $v = dx/dt$, we have $dx/dt = -9.8t + 20$. Integration yields $x = -4.9t^2 + 20t + k$. Since $x(0) = 30$, we find $k = 30$. At this point we know that the position of the ball at time t is $x(t) = -4.9t^2 + 20t + 30$, with velocity $v(t) = -9.8t + 20$. At maximum height $v(t) = 0$. Thus $-9.8t + 20 = 0$, or $t = 2.04$. At this time the height is given by $x(2.04) = 50.4$ m.

b. The ball will hit the ground at time T such that $x(T) = 0$. That is $-4.9T^2 + 20T + 30 = 0$, $T = \dfrac{-20 - \sqrt{988}}{-9.8} = 5.25$ seconds.

29a. The equation is again $dv/dt = -g$ so $v(t) = -gt + c$. With
     $v(0) = v_0$, we find $c = v_0$. Thus $dx/dt = v(t) = -gt + v_0$,
     giving $x(t) = -(g/2)t^2 + v_0 t + k$, so $x(0) = 0$, gives $k = 0$.
     Thus the body is at $x(t) = -(g/2)t^2 + v_0 t$ at time t, and
     travels with velocity $v(t) = -(g/2)t + v_0$. At its maximum
     height the body has velocity $v(T) = 0$, so $v(T) = 0 = -gT + v_0$
     or $T = v_0/g$ and $x(T) = x(v_0/g) = v_0^2/2g$ is the maximum height
   b. From part (a) the maximum height is reached at $T = v_0/g$.
   c. The body returns to its starting point at time τ for which
     $x(\tau) = 0$. That is $-(g/2)\tau^2 + v_0\tau = 0$ or $\tau(v_0-g\tau/2) = 0$. Thus
     $\tau = 0$ (starting time) and $\tau = 2v_0/g$.

Section 10.3, Page 535

1a. From Eq. (3), $y(.1) \cong 1 + (.1)(2 - 1) = 1.1$,
    $y(.2) \cong 1.1 + (.1)(2.2 - 1) = 1.22$,
    $y(.3) \cong 1.22 + (.1)(2.44 - 1) = 1.364$, and
    $y(.4) \cong 1.364 + (.1)(2.728 - 1) = 1.5368$.
  b. As in part (a), $y(.05) \cong 1 + (.05)(2 - 1) = 1.05$, so
    $y(.1) \cong 1.05 + (.05)(2.1 - 1) = 1.105$. Similarly
    $y(.2) \cong 1.23205$, $y(.3) \cong 1.38579$, and $y(.4) \cong 1.57179$.
  c. By Eq. (8), $Y_1 = 1 + (.1)(2-1) = 1.1$, $f(.1,Y_1) = 2.2 - 1 = 1.2$,
    so $y(.1) \cong y(0) + (.05)\{[2y(0) - 1] + 1.2\} = 1.11$. Likewise
    $y(.2) \cong y(.1) + (.05)\{[2y(.1) - 1] + 1.464\} = 1.2442$.
    Similarly $y(.3) \cong 1.40792$ and $y(.4) \cong 1.60767$.
  d. From Eqs. (10), $k_{01} = 2y(0)-1 = 1$, $k_{02} = 2[y(0)+(.05)(1)]-1 = 1.1$,
    $k_{03} = 2[y(0)+(.05)(1.1)]-1 = 1.11$, $k_{04} = 2[1+(.1)(1.11)]-1 = 1.222$
    so by Eq. (9) $y(.1) \cong 1+(.1/6)(1 + 2.2 + 2.22 + 1.222) = 1.1107$.
    Similarly, $y(.2) \cong 1.24587$, $y(.3) \cong 1.41106$, $y(.4) \cong 1.61263$.
  e. Multiplying the given equation by the integrating factor
    e^{-2t}, integrating, and applying the initial condition yields
    $y = (1 + e^{2t})/2$ as the exact solution, from which we find
    $y(.1) = 1.11070$, $y(.2) = 1.24591$, $y(.3) = 1.41106$, and
    $y(.4) = 1.61277$.

3a. From Eq. (3), $y_{n+1} = y_n + h(t_n^2 + y_n^2)$. Thus
    $y(.1) \cong 1 + (.1)(0+1) = 1.1$ and
    $y(.2) \cong 1.1 + (.1)[(.1)^2 + (1.1)^2] = 1.222$. Similarly
    $y(.3) \cong 1.37533$ and $y(.4) = 1.57348$.

3b. From Eq. (8), $y_{n+1} = y_n + (h/2)[t_n^2 + y_n^2 + t_{n+1}^2 + (y_n + hy_n')^2]$.
     Thus $y(.1) \cong 1 + (.05)[0 + 1 + (.1)^2 + (1.1)^2] = 1.111$ and
     $y(.2) \cong 1.111 + (.05)[(.1)^2 + (1.111)^2 + (.2)^2 + (1.2354)^2] = 1.25153$.
     Similarly, $y(.3) \cong 1.43606$ and $y(.4) \cong 1.68801$.

 c. From Eqs. (10), $k_{01} = 1$, $k_{02} = (.1)^2 + (1.1)^2 = 1.22$,
     $k_{03} = 1.268884$, and $k_{04} = 1.611956$. Thus from Eq. (9),
     $y(.2) \cong 1 + (.2/6)(1 + 2.44 + 2.53777 + 1.61196) = 1.25299$.
     Similarly $y(.4) \cong 1.69592$.

5a. $y_{n+1} = y_n + h(t_n + y_n)^{1/2}$, so $y(1.1) \cong 3 + (.1)(1+3)^{1/2} = 3.2$
     and $y(1.2) \cong 3.2 + (.1)(1.1 + 3.2)^{1/2} = 3.40736$. Similarly,
     $y(1.3) \cong 3.62201$ and $y(1.4) \cong 3.84387$.

 b. $y_{n+1} = y_n + (h/2)\left(y_n' + \sqrt{t_{n+1} + y_n + h\sqrt{t_n + y_n}}\right)$, which gives
     $y(1.1) \cong 3 + (.05)[2 + (1.1 + 3.2)^{1/2}] = 3.20368$,
     $y(1.2) \cong 3.20368 + (.05)[2.05029 + (1.2 + 3.41133)^{1/2}] = 3.41478$,
     $y(1.3) \cong 3.63320$, and $y(1.4) \cong 3.85888$.

 c. For $t = 1$, $k_{01} = (1+3)^{1/2} = 2$, $k_{02} = (1.1 + 3.2)^{1/2} = 2.07364$,
     $k_{03} = (1.1 + 3.20736)^{1/2} = 2.07542$, and
     $k_{04} = (1.2 + 3.41508)^{1/2} = 2.14827$. Thus
     $y(1.2) \cong 3 + (.2/6)(2 + 4.1473 + 4.15084 + 2.14827) = 3.41488$.
     Likewise $y(1.4) \cong 3.85908$.

7a. $y(1.1) \cong y(1) + (.1)\{y^2(1) + 2(1)[y(1)]/(3+1)\} = 2.2$ since
     $y(1) = 1$. Likewise
     $y(1.2) \cong 2.2 + (.1)\{(2.2)^2 + 2(1.1)(2.2)/[3 + (1.1)^2]\} = 2.42993$,
     $y(1.3) \cong 2.69426$, and $y(1.4) \cong 2.99840$.

 b. $y_{n+1} = y_n + (h/2)[(y_n^2 + 2t_n y_n)/(3+t_n^2) + f(t_{n+1}, Y_{n+1})]$ where
     $Y_{n+1} = y_n + h(y_n^2 + 2t_n y_n)/(3+t_n^2)$. Thus
     $y(1.1) \cong 2 + (.05)(2 + 2.29929) = 2.21497$, $y(1.2) \cong 2.46469$,
     $y(1.3) \cong 2.75525$, and $y(1.4) \cong 3.09420$.

 c. For $t = 1$, $k_{01} = (4+4)/4 = 2$, $k_{02} = 2.29929$, $k_{03} = 2.34642$,
     and $k_{04} = 2.70802$. Thus Eq. (9) gives $y(1.2) \cong 2.46665$.
     Similarly, $y(1.4) \cong 3.09994$.

9a. We have $y_{n+1} = y_n + h(0.5 - t_n + 2y_n)$, so for $h = .025$ we get
     $y(1) \cong 7.53999$.
 b. Setting $h = .0125$ in part (a) then gives $y(1) \cong 7.70957$.

9c. $y_{n+1} = y_n + (h/2)[(0.5 - t_n + 2y_n) + (0.5 - t_{n+1} + 2Y_{n+1})]$
     where $Y_{n+1} = y_n + h[(0.5 - t_n + 2y_n)$. Setting $h = .05$ then
     gives $y(1) \cong 7.86623$.
  d. Setting $h = .025$ in part(c) then gives $y(1) \cong 7.88313$.
  e. Using Eqs.(9) and (10) with $h = .1$ gives $y(1) \cong 7.88889$.
  f. Using Eqs.(9) and (10) with $h = .05$ gives $y(1) \cong 7.88905$.

11a. $y_{n+1} = y_n + h(t_n + y_n)^{1/2}$ with $h = .025$ yields $y(2) \cong 5.35218$.
   b. Setting $h = .0125$ then gives $y(2) \cong 5.35712$.
   c. Using the relationships developed in Prob.5, with $h = .05$,
       give $y(2) \cong 5.36194$.
   d. Setting $h = .025$ then gives $y(2) \cong 5.36203$.
   e. As in Prob.5, with $h = .1$, $y(2) \cong 5.36206$.
   f. Letting $h = .05$ then gives $y(2) \cong 5.36206$.

13a. As in Prob.9, with $h = .025$, $y(1) \cong 7.88906$.
   b. With $h = .01$, $y(1) \cong 7.88906$.

15a. As in Prob.11(e), with $h = .025$, $y(2) \cong 5.36206$.
   b. With $h = .01$, $y(2) \cong 5.36206$.

17. From Eq.(8), if $y' = f(t)$, then $f(t_n, y_n) = f(t_n)$ and
     $f(t_{n+1}, y_n + hy_n') = f(t_{n+1})$. Thus $y_{n+1} = y_n + (h/2)[f(t_n) + f(t_{n+1})]$.

Chapter 10 Review, Page 536

1. The linear equation in standard form is
    $$y' + \left(\frac{e^x - e^{-x}}{e^x + e^{-x}}\right) y = \frac{e^x}{e^x + e^{-x}},$$ which has integrating factor
    $\mu = \exp[\int [(e^x - e^{-x})/(e^x + e^{-x})]dx = \exp[\ln(e^x + e^{-x})] = e^x + e^{-x}$. Hence
    $[(e^x + e^{-x})y]' = e^x$, which yields $y = (e^x + c)/(e^x + e^{-x})$. Since
    $y(0) = 1 = (1+c)/2$, we have $c = 1$ and $y = (e^x + 1)/(e^x + e^{-x})$.

3. $y' + (\sin x/\cos x)y = \sec^2 x$ has integrating factor
    $\mu = \exp(-\ln|\cos x|) = 1/\cos x$ for x near 0. Thus
    $(y/\cos x)' = \sec^3 x$, which yields
    $y(x) = (1/2)\tan x + (1/2)\cos x \ln|\sec x + \tan x| + c\cos x$. $y(0) = 1/2 = c$
    then gives $y = (1/2)[\tan x + \cos x + \cos x \ln|\sec x + \tan x|]$.

5.   Using Prob.27 Section 10.1, with $n = 2$, $v = 1/y$ and the
      equation becomes $v' + e^t v = -e^{-e^t}$. This equation has
      integrating factor $\mu = \exp[\int e^t dt] = e^{e^t}$ so $(e^{e^t}v)' = -1$. Hence
      $v = (-t+c)e^{-e^t}$. Since $v = 1/y$, we have $y = e^{e^t}/(c-t)$ and
      then $y(0) = 1$ gives $c = e$. Thus $y = e^{e^t}/(e-t)$.

7.   Separating variables gives $dy/y - dx/\sqrt{1-x^2} = 0$, so
      $\ln|y| - \arcsin x = c$ or $|y| = ke^{\arcsin x}$. With $y(0) = 0$ we find
      $k = 0$, so $y = 0$ is the solution.

9.   Separating variables gives $[y/(4+y^2)]dy - dx = 0$ so that
      $(1/2)\ln(4+y^2) - x = c$. Applying the initial condition yields
      $c = (1/2)\ln 5$, and thus $\ln(4+y^2) = 2x + \ln 5$. Solving for y we
      find $y^2 = 5e^{2x} - 4$ or $y = -\sqrt{5e^{2x} - 4}$, since $y(0) < 0$.

11a. Using Eq.(3), Section 10.3, we have
      $Y_{n+1} = y_n + h[y_n/t_n + \exp(y_n/t_n)]$. With $h = 0.01$ we find
      $y(2) \cong 2.321579$.

  b. From Eq.(8), Section 10.3, we find
      $Y_{n+1} = y_n + (h/2)[y_n/t_n + \exp(y_n/t_n) + Y_{n+1}/t_{n+1} + \exp(Y_{n+1}/t_{n+1})]$
      where $Y_{n+1} = y_n + h[y_n/t_n + \exp(y_n/t_n)]$. Repeated applications of
      this with $h = 0.025$ gives $y(2) \cong 2.360865$, while for $h = 0.01$
      we have $y(2) \cong 2.362464$.

  c. Using Eqs.(9) and (10) of Section 10.3 gives
      $Y_{n+1} = y_n + (h/6)(k_{n1} + 2k_{n2} + 2k_{n3} + k_{n4})$, where
      $k_{n1} = y_n/t_n + \exp(y_n/t_n)$,
      $k_{n2} = z_n + \exp(z_n)$ for $z_n = [y_n + hk_{n1}/2]/[t_n + h/2]$,
      $k_{n3} = w_n + \exp(w_n)$ for $w_n = [y_n + hk_{n2}/2]/[t_n + h/2]$, and
      $k_{n4} = u_n + \exp(u_n)$ for $u_n = [y_n + hk_{n3}]/[t_n + h]$. Using these
      gives $y(2) \cong 2.362773$, when $h = 0.05$, and $y(2) \cong 2.362775$
      when $h = 0.025$.

13a. $Y_{n+1} = y_n + h[y_n/t_n + (y_n/t_n)^{1/2}]$ gives $y(2) \cong 6.189172$.

  b. $Y_{n+1} = y_n + (h/2)[y_n/t_n + (y_n/t_n)^{1/2} + Y_{n+1}/t_{n+1} + (Y_{n+1}/t_{n+1})^{1/2}]$ where
      $Y_{n+1} = y_n + h[y_n/t_n + (y_n/t_n)^{1/2}]$. Thus $y(2) \cong 6.200323$ for
      $h = 0.025$ and $y(2) \cong 6.200680$ for $h = 0.01$.

13c. Using Eq. (9) of Section 10.3, with $k_{n1} = y_n/t_n + (y_n/t_n)^{1/2}$,
   $k_{n2} = z_n + z_n^{1/2}$ for $z_n = [y_n + hk_{n1}/2]/[t_n + h/2]$,
   $k_{n3} = w_n + w_n^{1/2}$ for $w_n = [y_n + hk_{n2}/2]/[t_n + h/2]$, and
   $k_{n4} = u_n + u_n^{1/2}$ for $u_n = [y_n + hk_{n3}]/[t_n + h]$ gives
   $y(2) \cong 6.200744$ for both $h = 0.05$ and 0.025.

15. From Eq. (9), Section 10.3, with $f(t_n, y_n) = 0.5 y_n (1 - y_n/10)$ and
   $h = 0.05$ we find $y(1) \cong 1.548281$, $y(2) \cong 2.319693$,
   $y(3) \cong 3.324278$, $y(4) \cong 4.508531$, $y(5) \cong 5.751209$, and
   $y(6) \cong 6.905679$.

17. The mathematical model for the problem is $P' = -kP$, $P(0) = P_0$,
   $P(13.20) = P_0/2$. The general solution of the differential
   equation is $P(t) = ce^{-kt}$. Since $P(0) = P_0 = c$, we have that
   $P(t) = P_0 e^{-kt}$. Now $P(13.20) = P_0/2 = P_0 e^{-13.2k}$, which gives
   $k = \ln2/13.2 \cong .0525$. Thus $P(t) = P_0 e^{-(\ln2/13.2)t}$. With
   $P_0 = 50$ mg, $P(10) = 50e^{-(\ln2)/1.32} \cong 29.6$ mg.

19. Letting $y' = u$ and $y'' = u'$, the equation becomes $u' + u = 0$,
   which has general solution $u = ce^{-x}$. With $y'(0) = u(0) = 1$, we
   have $c = 1$. Thus $y' = e^{-x}$ and $y = k - e^{-x}$, so $y(0) = 0 = k-1$
   gives $k = 1$. Hence $y = 1-e^{-x}$.

21. We have $uu' = 2$ or $\dfrac{d}{dx}(u^2/2) = 2$ so $u^2/2 = 2x + c$.
   Since $y'(0) = u(0) = 2$, then $2 = c$ and $u^2 = 4x+4 = 0$ or
   $y' = u = 2(x+1)^{1/2}$. This yields $y = (4/3)(x+1)^{3/2} + k$, so
   $y(0) = 1 = 4/3 + k$ gives $k = -1/3$ and $y = (4/3)(x+1)^{3/2} - 1/3$.

CHAPTER 11

Section 11.1, Page 547

1. $\lim\limits_{x\to 0}\dfrac{\sin 2x}{x}=\dfrac{0}{0}$ and thus we need l'Hospital's rule, Hence

$\lim\limits_{x\to 0}\dfrac{\sin 2x}{x}=\lim\limits_{x\to 0}\dfrac{2\cos 2x}{1}=2$.

3. Form of 0/0. Thus $\lim\limits_{x\to 0}\dfrac{1-\cos^2 x}{x^2}=\lim\limits_{x\to 0}\dfrac{-2\cos x\,(-\sin x)}{2x}=$

$\lim\limits_{x\to 0}\dfrac{\sin 2x}{2x}=\lim\limits_{x\to 0}\dfrac{2\cos 2x}{2}=1$.

5. $\lim\limits_{x\to 0}\dfrac{\sin^2 x}{1+\cos x}=\dfrac{\lim\limits_{x\to 0}\sin^2 x}{\lim\limits_{x\to 0}(1+\cos x)}=\dfrac{0}{1}=0$.

7. Form of $(0)(\infty)$. Thus

$\lim\limits_{x\to \pi/2}(x-\pi/2)\tan x=\lim\limits_{x\to \pi/2}\dfrac{x-\pi/2}{\cot x}=\lim\limits_{x\to \pi/2}\dfrac{1}{-\csc^2 x}=-1$.

9. $\lim\limits_{x\to 0^-}\dfrac{e^{-x}+1}{e^x-1}=\dfrac{\lim\limits_{x\to 0^-}(e^{-x}+1)}{\lim\limits_{x\to 0^-}(e^x+1)}=\dfrac{2}{0^-}=-\infty$.

11. Form of ∞/∞. Thus $\lim\limits_{x\to 0^+}\dfrac{\ln(1+1/x)}{1/\sqrt{x}}=\lim\limits_{x\to 0^+}\dfrac{\ln(x+1)-\ln x}{1/\sqrt{x}}=$

$\lim\limits_{x\to 0^+}\dfrac{1/(x+1)-1/x}{-1/2x^{3/2}}=\lim\limits_{x\to 0^+}-2\left(\dfrac{x^{3/2}}{x+1}-x^{1/2}\right)=0$.

13. $(0/0)$, thus $\lim\limits_{x\to 0^+}\dfrac{\ln(1+x^2)}{x^3}=\lim\limits_{x\to 0^+}\dfrac{2x/(1+x^2)}{3x^2}=\lim\limits_{x\to 0^+}\dfrac{2}{3x(1+x^2)}=\infty$.

15. Form of ∞/∞. Thus $\lim\limits_{x\to 0^+}\dfrac{\ln x}{x^{-q}}=\lim\limits_{x\to 0^+}\dfrac{1/x}{-q\,x^{-q-1}}=\lim\limits_{x\to 0^+}-(1/q)\,x^q=0$.

17. Form of ∞/∞. $\lim\limits_{x\to 0^+}\dfrac{e^{1/x}}{1/x}=\lim\limits_{x\to 0^+}\dfrac{(-1/x^2)e^{1/x}}{-1/x^2}=\lim\limits_{x\to 0^+}e^{1/x}=\infty$.

19.   Form of $(-\infty + \infty)$.   Now $\dfrac{1}{\ln x} + \dfrac{1}{\sqrt{1-x^2}} = \dfrac{1 + \ln x/\sqrt{1-x^2}}{\ln x}$,   where

$\ln x/\sqrt{1-x^2}$ has the form $0/0$ as $x \to 1^-$.   Thus

$$\lim_{x \to 1^-} \frac{\ln x}{\sqrt{1-x^2}} = \lim_{x \to 1^-} \frac{1/x}{-x(1-x^2)^{-1/2}} = \lim_{x \to 1^-} \frac{\sqrt{1-x^2}}{-x^2} = 0 \text{ and hence}$$

$$\lim_{x \to 1^-} \left(\frac{1}{\ln x} + \frac{1}{\sqrt{1-x^2}} \right) = \lim_{x \to 1^-} \frac{1 + \ln x/\sqrt{1-x^2}}{\ln x} = \frac{1+0}{0} = \infty.$$

21. $(\infty - \infty)$.   Thus $\lim\limits_{x \to \pi/2} (\sec x - \tan x) = \lim\limits_{x \to \pi/2} \dfrac{1 - \sin x}{\cos x} = \lim\limits_{x \to \pi/2} \dfrac{-\cos x}{-\sin x} = 0.$

23. $(0/0)$.   Thus $\lim\limits_{x \to 0} \dfrac{\sqrt{1+x^2} - \sqrt{1-x^2}}{x} = \lim\limits_{x \to 0} \dfrac{x(1+x^2)^{-1/2} + x(1-x^2)^{-1/2}}{1} = 0.$
This problem can also be done without L'Hospital's rule by
rationalizing the numerator.

25.   Form ∞/∞.   Thus $\lim\limits_{x \to 0} \dfrac{\cot a x}{\cot b x} = \lim\limits_{x \to 0} \dfrac{\tan b x}{\tan a x} = \lim\limits_{x \to 0} \dfrac{b \sec^2 b x}{a \sec^2 a x} = b/a.$

27. $0/0$, so $\lim\limits_{x \to 2} \dfrac{x - 2}{(x+6)^{1/3} - 2} = \lim\limits_{x \to 2} \dfrac{1}{(1/3)(x+6)^{-2/3}} = \lim\limits_{x \to 2} 3(x+6)^{2/3} = 12.$

29.   Form $0/0$.   Thus $\lim\limits_{x \to 0} \dfrac{\arctan 2x}{\arcsin x} = \lim\limits_{x \to 0} \dfrac{2/(1+4x^2)}{1/\sqrt{1-x^2}} = 2.$

31.   Form $0/0$.   Thus $\lim\limits_{x \to 0} \dfrac{\sin^2 x - \sin(x^2)}{x^4} = \lim\limits_{x \to 0} \dfrac{\sin 2x - 2x \cos(x^2)}{4x^3} =$

$$\lim_{x \to 0} \frac{2 \cos 2x - 2 \cos(x^2) + 4x^2 \sin(x^2)}{12x^2} =$$

$$\lim_{x \to 0} \left\{ \frac{\cos 2x - \cos(x)^2}{6x^2} + (1/3) \sin(x^2) \right\} =$$

$$\lim_{x \to 0} \left\{ \frac{-2 \sin 2x + 2x \sin(x^2)}{12x} \right\} + 0 =$$

$$\lim_{x \to 0} \left\{ \frac{-\sin 2x}{6x} + \frac{\sin(x^2)}{6x} \right\} = -1/3 + 0 = -1/3.$$

33. Form 0/0. Thus

$$\lim_{x\to 0}\frac{\arcsin 2x - 2\arcsin x}{x\sin^2 x} = \lim_{x\to 0}\frac{2(1-4x^2)^{-1/2} - 2(1-x^2)^{-1/2}}{\sin^2 x + x\sin 2x} =$$

$$\lim_{x\to 0}\frac{8x(1-4x^2)^{-3/2} - 2x(1-x^2)^{-3/2}}{2\sin 2x + 2x\cos 2x} =$$

$$\lim_{x\to 0}\frac{4(1-4x^2)^{-3/2} - (1-x^2)^{3/2}}{2(\sin 2x/2x) + \cos 2x} = \frac{4-1}{2+1} = 1.$$

35. Form $(0)(\infty)$. Thus $\lim_{x\to 0^+} x(\ln x)^2 = \lim_{x\to 0^+}\frac{(\ln x)^2}{1/x} = \lim_{x\to 0^+}\frac{2(\ln x)/x}{-1/x^2} =$

$$\lim_{x\to 0^+}\frac{2\ln x}{-1/x} = \lim_{x\to 0^+}\frac{2/x}{1/x^2} = \lim_{x\to 0^+} 2x = 0.$$

37. Form (0/0). Thus $\lim_{x\to 0}\dfrac{\left(\displaystyle\int_0^x \sin t\, dt\right)^2}{\displaystyle\int_0^x \sin t^2\, dt} = \lim_{x\to 0}\dfrac{2\sin x\displaystyle\int_0^x \sin t\, dt}{\sin x^2} =$

$$\lim_{x\to 0}\frac{2\cos x\displaystyle\int_0^x \sin t\, dt + 2\sin^2 x}{2x\cos x^2} =$$

$$\lim_{x\to 0}\frac{-2\sin x\displaystyle\int_0^x \sin t\, dt + 6\sin x\cos x}{2\cos x^2 - 4x^2\sin x^2} = 0.$$

39. We must have $\lim_{x\to 0}\cos ax - b = 0$ in order for the limit to exist. Thus $b = 1$ and L'Hospital's rule then applies to yield $\lim_{x\to 0}\dfrac{\cos ax - 1}{2x^2} = \lim_{x\to 0}\dfrac{-a\sin ax}{4x} = \lim_{x\to 0}\dfrac{-a^2\cos ax}{4}$. Hence we must have $-a^2/4 = -1$ or $a = \pm 2$.

41a. $a(\theta) = (1/2)\sin\theta(1-\cos\theta)$ and $b(\theta) = (1/2)(\theta - \sin\theta\cos\theta)$. Thus

$$\lim_{\theta\to 0}\frac{a(\theta)}{b(\theta)} = \lim_{\theta\to 0}\frac{\sin\theta(1-\cos\theta)}{\theta - \sin\theta\cos\theta} = \lim_{\theta\to 0}\frac{\cos\theta(1-\cos\theta) + \sin^2\theta}{1 - \cos^2\theta + \sin^2\theta} =$$

$$\lim_{\theta\to 0}\frac{\cos\theta - \cos 2\theta}{1 - \cos 2\theta} = \lim_{\theta\to 0}\frac{-\sin\theta + 2\sin 2\theta}{2\sin 2\theta} = \lim_{\theta\to 0}\frac{-\cos\theta + 4\cos 2\theta}{4\cos 2\theta} = \frac{3}{4}.$$

41b. $c(\theta) = (1/2)(\theta - \sin\theta)$ and thus $\lim\limits_{\theta \to 0} \dfrac{a(\theta)}{c(\theta)} = \lim\limits_{\theta \to 0} \dfrac{\sin\theta\,(1 - \cos\theta)}{\theta - \sin\theta} =$

$\lim\limits_{\theta \to 0} \dfrac{\cos\theta\,(1 - \cos\theta) + \sin^2\theta}{1 - \cos\theta} = \lim\limits_{\theta \to 0} [\cos\theta + (1 + \cos\theta)] = 3.$

43. Since $\lim\limits_{x \to a^+} f(x)/g(x)$ exists we have $\lim\limits_{x \to a^+} f(x)/g(x) = A$, where

we are to determine A. Now $\lim\limits_{x \to a^+} \dfrac{f(x)}{g(x)} = \lim\limits_{x \to a^+} \dfrac{1/g(x)}{1/f(x)} = \dfrac{0}{0}$ and thus

we may apply L'Hospital's rule for this form to obtain

$\lim\limits_{x \to a^+} \dfrac{f(x)}{g(x)} = \lim\limits_{x \to a^+} \dfrac{-g'(x)/g^2(x)}{-f'(x)/f^2(x)} = \lim\limits_{x \to a^+} \dfrac{f^2(x)}{g^2(x)} \lim\limits_{x \to a^+} \dfrac{g'(x)}{f'(x)} = \dfrac{A^2}{L}.$ This last

step can be done since each of the limits exist by hypothesis
(this is where the assumption that $\lim\limits_{x \to a^+} f(x)/g(x)$ exists is

used). Finally we have $A^2/L = A$, or $A = L$. Note also that
$L \neq 0, \infty$ in order to carry out this last step, which is also
a variance with the proof in the text.

<u>Section 11.2, Page 556</u>

1. (∞/∞), $\lim\limits_{x \to \infty} \dfrac{(\ln x)^n}{x} = \lim\limits_{x \to \infty} \dfrac{n(\ln x)^{n-1}(1/x)}{1} = \lim\limits_{x \to \infty} \dfrac{n(\ln x)^{n-1}}{x} = \cdots =$

$\lim\limits_{x \to \infty} \dfrac{n(n-1)\cdots(2)\ln x}{x} = \lim\limits_{x \to \infty} \dfrac{n!}{x} = 0.$

3. (∞/∞), $\lim\limits_{x \to \infty} \dfrac{x^n}{(1.01)^x} = \lim\limits_{x \to \infty} \dfrac{n\,x^{n-1}}{(1.01)^x[\ln 1.01)} = \cdots =$

$\lim\limits_{x \to \infty} \dfrac{n!}{(1.01)^x(\ln 1.01)^n} = 0.$

5. (∞/∞), $\lim\limits_{x \to \infty} \dfrac{\ln(1 + e^x)}{x} = \lim\limits_{x \to \infty} \dfrac{e^x/(1 + e^x)}{1} = \lim\limits_{x \to \infty} \dfrac{e^x}{1 + e^x} = \lim\limits_{x \to \infty} \dfrac{e^x}{e^x} = 1.$

7. $(0/0)$, $\lim\limits_{x \to \infty} \dfrac{\sin(e^{-x})}{\sin(1/x)} = \lim\limits_{x \to \infty} \dfrac{-e^{-x}\cos(e^{-x})}{-(1/x^2)\cos(1/x)} =$

$\left(\lim\limits_{x \to \infty} x^2 e^{-x}\right)\left(\lim\limits_{x \to \infty} \dfrac{\cos(e^{-x})}{\cos(1/x)}\right) = (0)(1) = 0,$ where the result of

Ex.(2) has been used.

9. (∞^0), $\lim\limits_{x\to\infty} \left(\dfrac{3^x+5^x}{2}\right)^{1/x} = \lim\limits_{x\to\infty} \exp\left[(1/x)\ln\left(\dfrac{3^x+5^x}{2}\right)\right]$. Now

$$\lim_{x\to\infty} \frac{\ln\left(\dfrac{3^x+5^x}{2}\right)}{x} = \lim_{x\to\infty} \frac{2}{3^x+5^x}\left(\frac{3^x\ln 3 + 5^x\ln 5}{2}\right) =$$

$$\lim_{x\to\infty} \frac{(3/5)^x\ln 3}{(3/5)^x+1} + \lim_{x\to\infty} \frac{\ln 5}{(3/5)^x+1} = 0 + \ln 5. \quad\text{Thus}$$

$$\lim_{x\to\infty}\left(\frac{3^x+5^x}{2}\right)^{1/x} = \exp(\ln 5) = 5.$$

11. (∞/∞), $\lim\limits_{x\to\infty} \dfrac{\sqrt{1+x^2}}{x} = \lim\limits_{x\to\infty} \dfrac{(1/2)(1+x^2)^{-1/2}2x}{1} = \lim\limits_{x\to\infty} \dfrac{x}{\sqrt{1+x^2}}$, which

also has the form ∞/∞. If L'Hospital's rule is applied again

we obtain $\lim\limits_{x\to\infty} \dfrac{x}{\sqrt{1+x^2}} = \lim\limits_{x\to\infty} \dfrac{\sqrt{1+x^2}}{x}$ and thus L'Hospital's rule

will not yield the limit. We may find the limit, though,

since $\dfrac{\sqrt{1+x^2}}{x} = \sqrt{(1/x^2)+1}$ and thus

$$\lim_{x\to\infty} \frac{\sqrt{1+x^2}}{x} = \lim_{x\to\infty} \sqrt{(1/x^2)+1} = 1.$$

13. $(0)(\infty)$, $\lim\limits_{x\to 0^+} x(\ln x)^n = \lim\limits_{x\to 0^+} \dfrac{(\ln x)^n}{1/x} = \lim\limits_{x\to 0^+} \dfrac{n(\ln x)^{n-1}(1/x)}{-1/x^2} =$

$$\lim_{x\to 0^+} \frac{n(\ln x)^{n-1}}{-1/x} = \cdots = \lim_{x\to 0^+} \frac{n!\ln x}{(-1)^{n-1}/x} = \lim_{x\to 0^+} \frac{n!}{(-1)^n/x} = 0.$$

15. (∞/∞), $\lim\limits_{x\to\infty} \dfrac{x^a}{b^x} = \lim\limits_{x\to\infty} \dfrac{ax^{a-1}}{b^x\ln b} = \cdots = \lim\limits_{x\to\infty} \dfrac{a(a-1)\ldots(a-n+1)}{b^x(x^{n-a})(\ln b)^n} = 0,$

where $0 \le n-a < 1$.

17. (1^∞), $\displaystyle\lim_{x\to\infty}(1+1/x^2)^x = \lim_{x\to\infty} \exp\left[\dfrac{\ln(1+1/x^2)}{1/x}\right]$. Now

$$\lim_{x\to\infty}\dfrac{\ln(1+1/x^2)}{1/x} = \lim_{x\to\infty}\dfrac{(-2/x^3)/(1+1/x^2)}{-1/x^2} = \lim_{x\to\infty}\dfrac{2x}{x^2+1} = 0$$

and thus $\displaystyle\lim_{x\to\infty}(1+1/x^2)^x = e^0 = 1$.

19. (1^∞), $\displaystyle\lim_{x\to\infty}(1-1/x)^{x^2} = \lim_{x\to\infty}\exp\left[\dfrac{\ln(1-1/x)}{1/x^2}\right]$. Now

$$\lim_{x\to\infty}\dfrac{\ln(1-1/x)}{1/x^2} = \lim_{x\to\infty}\dfrac{(1/x^2)/(1-1/x)}{-2/x^3} = \lim_{x\to\infty}\dfrac{-x}{2(1-1/x)} = -\infty$$

and thus $\displaystyle\lim_{x\to\infty}(1-1/x)^{x^2} = e^{-\infty} = 0$.

21. (∞^0), $\displaystyle\lim_{x\to\infty}(1+x^3)^{x^{-2}} = \lim_{x\to\infty}\exp[x^{-2}\ln(1+x^3)]$. Now

$$\lim_{x\to\infty}\dfrac{\ln(1+x^3)}{x^2} = \lim_{x\to\infty}\dfrac{3x^2/(1+x^3)}{2x} = \lim_{x\to\infty}\dfrac{(3/2)x}{1+x^3} = \lim_{x\to\infty}\dfrac{3/2}{3x^2} = 0 \text{ and}$$

thus $\displaystyle\lim_{x\to\infty}(1+x^3)^{x^{-2}} = e^0 = 1$.

23. (1^∞), $\displaystyle\lim_{x\to0}(\cos x)^{1/x^2} = \lim_{x\to0}\exp\left(\dfrac{\ln\cos x}{x^2}\right)$. Now $\displaystyle\lim_{x\to0}\dfrac{\ln\cos x}{x^2} =$

$$\lim_{x\to0}\dfrac{-\sin x/\cos x}{2x} = \left(\lim_{x\to0}\dfrac{\sin x}{x}\right)\left(\lim_{x\to0}\dfrac{-1}{2\cos x}\right) = -1/2 \text{ and thus}$$

$\displaystyle\lim_{x\to0}(\cos x)^{-1/x^2} = e^{-1/2} = 1/\sqrt{e}$.

25. (∞^0), $\displaystyle\lim_{x\to\infty}(1+e^x)^{x^{-q}} = \lim_{x\to\infty}\exp[x^{-q}\ln(1+e^x)]$. Now

$$\lim_{x\to\infty}\dfrac{\ln(1+e^x)}{x^q} = \lim_{x\to\infty}\dfrac{e^x/(1+e^x)}{qx^{q-1}} = \lim_{x\to\infty}\dfrac{1}{qx^{q-1}}\dfrac{1}{1+e^{-x}} = \begin{cases} \infty & 0 < q < 1 \\ 1 & q = 1 \\ 0 & q > 1 \end{cases}$$

and thus $\displaystyle\lim_{x\to\infty}(1+e^x)^{x^{-q}} = \begin{cases} \infty & 0 < q < 1 \\ e & q = 1 \\ 1 & q > 1 \end{cases}$.

27. (1^∞), $\lim\limits_{x\to\infty} (1+x^{-p})^{x^q} = \lim\limits_{x\to\infty} \exp\left[\dfrac{\ln(1+x^{-p})}{x^{-q}}\right]$. Now

$$\lim_{x\to\infty} \frac{\ln(1+x^{-p})}{x^{-q}} = \lim_{x\to\infty} \frac{-p\,x^{-p-1}/(1+x^{-p})}{-q\,x^{-q-1}} = \lim_{x\to\infty} \frac{p\,x^q}{q\,x^p}\,\frac{1}{1+x^{-p}} = \begin{cases} \infty & q>p \\ 1 & q=p \\ 0 & q<p \end{cases}$$

and thus $\lim\limits_{x\to\infty} (1+x^{-p})^{x^q} = \begin{cases} \infty & q>p \\ e & q=p \\ 1 & q<p \end{cases}$.

29. (0^0), $\lim\limits_{x\to\infty} [\sin(1/x)]^{1/x} = \lim\limits_{x\to\infty} \exp\left[\dfrac{\ln\sin(1/x)}{x}\right]$. Now

$$\lim_{x\to\infty} \frac{\ln\sin(1/x)}{x} = \lim_{x\to\infty} \frac{\cos(1/x)}{\sin(1/x)}(-1/x^2) = \lim_{w\to 0} \frac{-w^2\cos w}{\sin w} = 0 \text{ and thus}$$

$\lim\limits_{x\to\infty} [\sin(1/x)]^{1/x} = e^0 = 1.$

31. (1^∞), $\lim\limits_{x\to 1} x^{1/(1-x)} = \lim\limits_{x\to 1} \exp\left(\dfrac{\ln x}{1-x}\right)$. Now

$$\lim_{x\to 1} \frac{\ln x}{1-x} = \lim_{x\to 1} \frac{1/x}{-1} = -1 \text{ and thus } \lim_{x\to 1} x^{1/(1-x)} = e^{-1} = 1/e.$$

33. (1^∞), $\lim\limits_{x\to 0+} (1+x)^{1/x^2} = \lim\limits_{x\to 0+} \exp\left[\dfrac{\ln(1+x)}{x^2}\right]$. Now

$$\lim_{x\to 0+} \frac{\ln(1+x)}{x^2} = \lim_{x\to 0+} \frac{1/(1+x)}{2x} = \infty \text{ and thus } \lim_{x\to 0+} (1+x)^{1/x^2} = e^\infty = \infty.$$

35. (1^∞), $\lim\limits_{x\to 0+} (1-x)^{1/x} = \lim\limits_{x\to 0+} \exp\left[\dfrac{\ln(1-x)}{x}\right]$. Now

$$\lim_{x\to 0+} \frac{\ln(1-x)}{x} = \lim_{x\to 0+} \frac{-1}{1-x} = -1 \text{ and thus } \lim_{x\to 0+} (1-x)^{1/x} = e^{-1} = 1/e.$$

Section 11.3, Page 564

1. $\displaystyle\int_0^\infty \frac{x\,dx}{1+x^2} = \lim_{b\to\infty}\int_0^b \frac{x\,dx}{1+x^2} = \lim_{b\to\infty} (1/2)\ln(1+x^2)\,\Big|_0^b =$

$\lim\limits_{b\to\infty} (1/2)\ln(1+b^2) = \infty$ and thus the given integral diverges.

3. $\int_0^1 \dfrac{dx}{(1-x)^{1/3}} = \lim\limits_{c \to 1-} \int_0^c \dfrac{dx}{(1-x)^{1/3}} = \lim\limits_{c \to 1-} (-3/2)(1-x)^{2/3}\Big|_0^c =$

$\lim\limits_{c \to 1-} (3/2)[1-(1-c)^{2/3}] = 3/2$ and thus the given integral

converges.

5. $\int_0^9 \dfrac{dx}{\sqrt{9-x}} = \lim\limits_{c \to 9-} \int_0^c \dfrac{dx}{\sqrt{9-x}} = \lim\limits_{c \to 9-} -2(9-x)^{1/2}\Big|_0^c = \lim\limits_{c \to 9-} 2[3-(9-c)^{1/2}]$

$= 6$.   Integral converges.

7. $\int_{-\infty}^0 xe^{-x^2} dx = \lim\limits_{b \to -\infty} \int_b^0 xe^{-x^2} dx = \lim\limits_{b \to -\infty} (-1/2)e^{-x^2}\Big|_b^0 =$

$\lim\limits_{b \to -\infty} (1/2)e^{-b^2} - 1/2 = -1/2$.   Integral converges.

9. $\int_0^3 \dfrac{dx}{\sqrt{9-x^2}} = \lim\limits_{c \to 3-} \int_0^c \dfrac{dx}{\sqrt{9-x^2}} = \lim\limits_{c \to 3-} \arcsin(x/3)\Big|_0^c =$

$\lim\limits_{c \to 3-} \arcsin(c/3) = \pi/2$.   Integral converges.

11. $\int_2^\infty \dfrac{dx}{x \ln x} = \lim\limits_{b \to \infty} \int_2^b \dfrac{dx}{x \ln x} = \lim\limits_{b \to \infty} \ln(\ln x)\Big|_2^b =$

$\lim\limits_{b \to \infty} [\ln(\ln b) - \ln(\ln 2)] = \infty$.   Integral diverges.

13. $\int_0^\infty \dfrac{dx}{(x-2)^3} = \int_0^2 \dfrac{dx}{(x-2)^3} + \int_2^3 \dfrac{dx}{(x-2)^3} + \int_3^\infty \dfrac{dx}{(x-2)^3}$ if all three exist.

Now $\int_2^3 \dfrac{dx}{(x-2)^3} = \lim\limits_{c \to 2+} \int_c^3 \dfrac{dx}{(x-2)^3} = \lim\limits_{c \to 2+} \dfrac{-1}{2(x-2)^2}\Big|_c^3 =$

$\lim\limits_{c \to 2+}\left[-\dfrac{1}{8} + \dfrac{1}{2(c-2)^2} \right] = \infty$ and thus $\int_0^\infty \dfrac{dx}{(x-2)^3}$ diverges.

15 . $\int_0^\infty \dfrac{x \, dx}{(x^2+1)^{3/2}} = \lim\limits_{b \to \infty} \int_0^b \dfrac{x \, dx}{(x^2+1)^{3/2}} = \lim\limits_{b \to \infty} -(x^2+1)^{-1/2}\Big|_0^b =$

$\lim\limits_{b \to \infty}[1 - (b^2+1)^{-1/2}] = 1$.   Integral converges.

17. $\int_0^1 \dfrac{e^x dx}{1-e^x} = \lim\limits_{c \to 0+} \int_c^1 \dfrac{e^x dx}{1-e^x} = \lim\limits_{c \to 0+} (-\ln|1-e^x|)\Big|_c^1 =$

$\lim\limits_{c \to 0+} (\ln|1-e^c| - \ln|1-e|) = \infty$.   Integral diverges.

19. $\int_{-\infty}^{\infty} \dfrac{dx}{1+|x|} = \lim\limits_{a\to-\infty} \int_{a}^{0} \dfrac{dx}{1-x} + \lim\limits_{b\to\infty} \int_{0}^{b} \dfrac{dx}{1+x}$, provided both integrals

exist. Now $\lim\limits_{b\to\infty} \int_{0}^{b} \dfrac{dx}{1+x} = \lim\limits_{b\to\infty} \ln(1+x)\Big|_{0}^{b} = \lim\limits_{b\to\infty} \ln(1+b) = \infty$ and

thus the original integral diverges.

21. $\int_{0}^{1} \dfrac{\ln x}{x}\, dx = \lim\limits_{c\to0+} \int_{c}^{1} \dfrac{\ln x}{x}\, dx = \lim\limits_{c\to0+} (1/2)(\ln x)^2 \Big|_{c}^{1} =$

$\lim\limits_{c\to0+} -(1/2)(\ln c)^2 = -\infty$. Integral diverges.

23. $\int_{-1}^{1} \dfrac{dx}{x^2-1} = \lim\limits_{a\to-1+} \int_{a}^{0} \dfrac{dx}{x^2-1} + \lim\limits_{b\to1-} \int_{0}^{b} \dfrac{dx}{x^2-1}$ provided both integrals

exist. Now $\lim\limits_{b\to1-} \int_{0}^{b} \dfrac{dx}{x^2-1} = \lim\limits_{b\to1-} \int_{0}^{b} \left[\dfrac{1/2}{x+1} - \dfrac{1/2}{x-1} \right] dx =$

$\lim\limits_{b\to1-} (1/2)\ln|(b+1)/(b-1)| = \infty$ and thus $\int_{-1}^{1} \dfrac{dx}{x^2-1}$ diverges.

25. $\int_{a}^{\infty} u(x)v'(x)\, dx = \lim\limits_{b\to\infty} \int_{a}^{b} u(x)v'(x)\, dx$

$= \lim\limits_{b\to\infty} [u(x)v(x) \Big|_{a}^{b} - \int_{a}^{b} u'(x)v(x)\, dx]$

$= \lim\limits_{b\to\infty} [u(b)v(b) - u(a)v(a) - \int_{a}^{b} u'(x)v(x)\, dx]$

$= \lim\limits_{b\to\infty} u(b)v(b) - u(a)v(a) - \int_{0}^{\infty} u'(x)v(x)\, dx.$

27. $\int_{0}^{\infty} x^n e^{-x} dx = \lim\limits_{b\to\infty} \int_{0}^{b} x^n e^{-x} dx = \lim\limits_{b\to\infty} [-x^n e^{-x}\Big|_{0}^{b} + n\int_{0}^{b} x^{n-1} e^{-x} dx] =$

$\lim\limits_{b\to\infty} [-b^n e^{-b} + n\int_{0}^{b} x^{n-1} e^{-x} dx] = n\int_{0}^{\infty} x^{n-1} e^{-x} dx = \ldots =$

$n! \int_{0}^{\infty} e^{-x} dx = n!$ by Ex.(1).

29. $\int_0^\infty e^{-ax}\sin bx\, dx = \lim_{c\to\infty}\int_0^c e^{-ax}\sin bx\, dx$

$= \lim_{c\to\infty}\left[(-1/b)e^{-ax}\cos bx\,\Big|_0^c - (a/b)\int_0^c e^{-ax}\cos bx\, dx\right]$

$= 1/b - (a/b)\lim_{c\to\infty}\left[(1/b)e^{-ax}\sin bx\,\Big|_0^c + (a/b)\int_0^c e^{-ax}\sin bx\, dx\right]$

$= (1/b) - (a^2/b^2)\int_0^\infty e^{-ax}\sin bx\, dx.$ Solving for the

desired integral we get $\int_0^\infty e^{-ax}\sin bx\, dx = b/(a^2 + b^2).$

31. $\int_1^\infty\left(\dfrac{x}{x^2+1} - \dfrac{\alpha}{2x+3}\right)dx = \lim_{b\to\infty}[(1/2)\ln(x^2+1) - (\alpha/2)\ln(2x+3)]\,\Big|_1^b =$

$(1/2)\lim_{b\to\infty}\{\ln[(b^2+1)/(2b+3)^\alpha] - \ln(2/5^\alpha)\}.$ This last limit

will be finite only if $\lim_{b\to\infty}\dfrac{b^2+1}{(2b+3)^\alpha}$ is finite and not zero.

Therefore we must have $\alpha > 0$ and then

$\lim_{b\to\infty}\dfrac{b^2+1}{(2b+3)^\alpha} = \lim_{b\to\infty}\dfrac{2b}{2\alpha(2b+3)^{\alpha-1}} = \lim_{b\to\infty}\dfrac{2}{4\alpha(\alpha-1)(2b+3)^{\alpha-2}},$ which is

finite and not zero only for $\alpha = 2.$ In this case

$\lim_{b\to\infty}\dfrac{b^2+1}{(2b+3)^\alpha} = 1/4$ and hence

$\int_1^\infty\left(\dfrac{x}{x^2+1} - \dfrac{2}{2x+3}\right)dx = (1/2)[\ln(1/4)-\ln(2/25)] = (1/2)\ln(25/8).$

33a. $A = \int_1^\infty \dfrac{dx}{x^p}$ exists and is therefore finite for $p > 1$ by Ex. (4).

b. $V = \pi\int_1^\infty\left(\dfrac{1}{x^p}\right)^2 dx = \pi\int_1^\infty\left(\dfrac{dx}{x^{2p}}\right)$ exists and is therefore finite for

$2p > 1$ or $p > 1/2.$

33c. $V = 2\pi \int_1^\infty x\left(\frac{1}{x^p}\right)dx = 2\pi \int_1^\infty \frac{dx}{x^{p-1}}$ exists and is therefore finite

for p-1 > 1 or p > 2.

d. From (a) p > 1 and from (c) the volume will be infinite if p ≤ 2. Thus 1 < p ≤ 2, where the rotation is about the y axis.

e. From (a) we find p ≤ 1 yields an infinite area and from (b) the volume will be finite if p > 1/2 and thus 1/2 < p ≤ 1. In this case the volume is obtained by rotation about the x axis.

35a. $\lim_{R\to\infty} \int_{-R}^R xdx = \lim_{R\to\infty} x^2/2 \Big|_{-R}^R = \lim_{R\to\infty}(R^2/2 - R^2/2) = 0.$

b. $\int_{-\infty}^\infty xdx = \lim_{a\to-\infty}\int_a^0 xdx + \lim_{b\to\infty}\int_0^b xdx$ and thus $\int_{-\infty}^\infty xdx$ diverges since both integrals on the right diverge.

c. The integral $\int_{-R}^R xdx$ is a proper integral for all finite R.

Its value is always zero due to the sum of equal negative area for -R ≤ x ≤ 0 and a positive area for 0 ≤ x ≤ R, while

$\int_{-\infty}^\infty xdx$ is an improper integral which is defined as in part(b)

and thus is divergent.

d. $\lim_{R\to\infty}\int_{-R}^R x^2dx = \lim_{R\to\infty} R^3/3 \Big|_{-R}^R = \lim_{R\to\infty} 2R^3/3 = \infty,$ while

$\int_{-\infty}^\infty x^2dx = \lim_{a\to-\infty}\int_a^0 x^2dx + \lim_{b\to\infty}\int_0^b x^2dx$ diverges since both

integrals on the right diverge. In this case the two results

agree due to the divergence of $\int_{-R}^R x^2dx,$ which represents two

diverging positive areas as R →∞.

35e. $\displaystyle\lim_{R\to\infty}\int_{-R}^{R}\frac{dx}{1+x^2} = \lim_{R\to\infty}\arctan x\,\Big|_{-R}^{R} = \lim_{R\to\infty} 2\arctan R = \pi$, while

$$\int_{-\infty}^{\infty}\frac{dx}{1+x^2} = \lim_{a\to-\infty}\int_{a}^{0}\frac{dx}{1+x^2} + \lim_{b\to\infty}\int_{0}^{b}\frac{dx}{1+x^2} =$$

$\displaystyle\lim_{a\to-\infty}(-\arctan a) + \lim_{b\to\infty}\arctan b = -(-\pi/2)+\pi/2 = \pi$. In this case both integrals on the right converge, thereby agreeing with the proper integral as $R\to\infty$.

Section 11.4, Page 573

1. Since $\dfrac{x}{4+x^3} \le \dfrac{x}{x^3} = \dfrac{1}{x^2}$ and since $\displaystyle\int_{1}^{\infty}\frac{dx}{x^2}$ converges we have that

$\displaystyle\int_{1}^{\infty}\frac{x\,dx}{4+x^3}$ converges by Thm.11.4.1.

3. Since $\dfrac{x^{3/2}}{4+x^2} \cong \dfrac{1}{x^{1/2}}$ for large x we use Thm.11.4.2 with

$g(x) = 1/x^{1/2}$. In this case $\displaystyle\lim_{x\to\infty}\frac{x^{3/2}/(4+x^2)}{1/x^{1/2}} = \lim_{x\to\infty}\frac{x^2}{4+x^2} = 1$

and thus $\displaystyle\int_{1}^{\infty}\frac{x^{3/2}dx}{4+x^2}$ diverges since $\displaystyle\int_{1}^{\infty}x^{-1/2}\,dx$ diverges.

5. Since $\displaystyle\lim_{x\to0}\frac{\sin x}{x} = 1$ we have $\dfrac{\sin x}{x^{3/2}} \cong \dfrac{1}{x^{1/2}}$ for $x\to0$. Thus

$\displaystyle\lim_{x\to0}\frac{\sin x/x^{3/2}}{1/x^{1/2}} = \lim_{x\to0}\frac{\sin x}{x} = 1$ and Thm.11.4.4 tells us that

$\displaystyle\int_{0}^{\pi/2}\frac{\sin x}{x^{3/2}}\,dx$ converges, since we know $\displaystyle\int_{0}^{\pi/2}x^{-1/2}\,dx$ converges.

7. Since $\dfrac{x}{x-2} \ge \dfrac{2}{x-2}$ for $2<x\le5$ and since $\displaystyle\int_{2}^{5}\frac{2dx}{x-2}$ diverges, then

$\displaystyle\int_{2}^{5}\frac{xdx}{x-2}$ diverges by Thm.11.4.3.

9.   Since $\dfrac{\sin^2 x}{\sqrt{4+x^3}} \le \dfrac{1}{\sqrt{4+x^3}} \le \dfrac{1}{x^{3/2}}$ and since $\displaystyle\int_1^\infty x^{-3/2}\,dx$ converges

then $\displaystyle\int_1^\infty \dfrac{\sin^2 x\,dx}{\sqrt{4+x^3}}$ converges.

11. $\displaystyle\int_{-\infty}^\infty \dfrac{e^x dx}{e^{2x}+e^{-2x}} = \int_{-\infty}^0 \dfrac{e^x\,dx}{e^{2x}+e^{-2x}} + \int_0^\infty \dfrac{e^x dx}{e^{2x}+e^{-2x}}$. The first integral

converges since $\dfrac{e^x}{e^{2x}+e^{-2x}} \le \dfrac{e^x}{e^{-2x}} = e^{3x}$ and $\displaystyle\int_{-\infty}^0 e^{3x}dx$ converges.

The second integral converges since $\dfrac{e^x}{e^{2x}+e^{-2x}} \le \dfrac{e^x}{e^{2x}} = e^{-x}$ and

$\displaystyle\int_0^\infty e^{-x}dx$ converges. Thus the given integral converges.

13. $\displaystyle\int_1^\infty \dfrac{dx}{(x^2-1)^{2/3}} = \int_1^2 \dfrac{dx}{(x^2-1)^{2/3}} + \int_2^\infty \dfrac{dx}{(x^2-1)^{2/3}}$. The first integral

converges since $\dfrac{1}{(x^2-1)^{2/3}} = \dfrac{1}{(x+1)^{2/3}} \dfrac{1}{(x-1)^{2/3}} \le \dfrac{1}{2^{2/3}} \dfrac{1}{(x-1)^{2/3}}$ and

$\displaystyle\int_1^2 \dfrac{dx}{(x-1)^{2/3}}$ converges. The second integral converges since

$\displaystyle\lim_{x\to\infty} \dfrac{(x^2-1)^{-2/3}}{x^{-4/3}} = \lim_{x\to\infty} \dfrac{x^{4/3}}{(x^2-1)^{2/3}} = 1$ and $\displaystyle\int_1^\infty x^{-4/3}\,dx$ converges. Thus

the given integral converges.

15. Since $\dfrac{e^{-x}}{\sqrt{1-x^4}} \le \dfrac{1}{\sqrt{1-x^4}} = \dfrac{1}{\sqrt{1+x^2}} \dfrac{1}{\sqrt{1+x}} \dfrac{1}{\sqrt{1-x}} \le \dfrac{1}{\sqrt{1-x}}$ for $0\le x<1$

and since $\displaystyle\int_0^1 (1-x)^{-1/2}\,dx$ converges then $\displaystyle\int_0^1 (1-x^4)^{-1/2}\,e^{-x}\,dx$

converges.

17. Since $\lim\limits_{x\to\infty} x^n e^{-x/2} = 0$, then by the definition of the limit as
     $x\to\infty$ we have $x^n \le e^{x/2}$ for x sufficiently large. Thus
     $\dfrac{x^n}{e^x} \le \dfrac{e^{x/2}}{e^x} = e^{-x/2}$ for x sufficiently large. Since $\displaystyle\int_0^\infty e^{-x/2}\,dx$
     converges we conclude that $\displaystyle\int_0^\infty x^n e^{-x}\,dx$ converges.

19. Since $\lim\limits_{x\to\infty} x^{-1/2}\ln x = 0$ then $\ln x \le \sqrt{x}$ for x sufficiently large
     (as in Prob.17) and thus $\dfrac{\ln x}{x^2} \le \dfrac{\sqrt{x}}{x^2} = x^{-3/2}$ for large x.
     Since $\displaystyle\int_1^\infty x^{-3/2}\,dx$ converges we conclude that $\displaystyle\int_1^\infty x^{-2}\ln x\,dx$
     converges.

21. Since $\lim\limits_{x\to 0+} \dfrac{e^x - 1}{x} = 1$ the integral is not improper and exists by
     Thm.6.2.1.

23. $\displaystyle\int_0^\infty \dfrac{1-e^{-x}}{x^p}\,dx = \int_0^1 \dfrac{1-e^{-x}}{x^p}\,dx + \int_1^\infty \dfrac{1-e^{-x}}{x^p}\,dx$. Since $\dfrac{1-e^{-x}}{x} \to 1$ as
     $x\to 0^+$ we try $g(x) = 1/x^{p-1}$ in Thm.11.4.4. Since
     $\lim\limits_{x\to 0+} \dfrac{(e^x-1)/x^p}{1/x^{p-1}} = \lim\limits_{x\to 0+} \dfrac{e^x-1}{x} = 1$ and since $\displaystyle\int_0^1 x^{-(p-1)}\,dx$ converges
     for p-1<1 or p<2 we conclude that $\displaystyle\int_0^1 \dfrac{1-e^{-x}}{x^p}\,dx$ converges for
     p<2. For the second integral $\dfrac{1-e^{-x}}{x^p} \cong \dfrac{1}{x^p}$ for large x and
     thus $\lim\limits_{x\to\infty} \dfrac{(1-e^{-x})/x^p}{1/x^p} = \lim\limits_{x\to\infty} (1-e^{-x}) = 1$. Thus $\displaystyle\int_1^\infty \dfrac{1-e^{-x}}{x^p}\,dx$
     converges for p>1 since $\displaystyle\int_1^\infty x^{-p}\,dx$ converges for p>1. Thus the
     given integral converges for 1<p<2.

25.   Since p may be negative we must consider $\displaystyle\int_3^\infty \frac{(x^2-9)^p}{1+x^4}\,dx =$

$\displaystyle\int_3^4 \frac{(x^2-9)^p}{1+x^4}\,dx + \int_4^\infty \frac{(x^2-9)^p}{1+x^4}\,dx.$ The first integral converges for

p>-1 as follows: $\displaystyle\frac{(x^2-9)^p}{1+x^4} = \frac{(x+3)^p(x-3)^p}{1+x^4} \cong \frac{6^p}{82}(x-3)^p$ as x→3+ and

thus $\displaystyle\lim_{x\to3+} \frac{(x^2-9)^p/(1+x^4)}{6^p(x-3)^p/82} = 1$ and we know $\displaystyle\int_1^\infty (x-3)^p\,dx$ converges

for p>-1.   The second integral converges for p<3/2 as

follows: $\displaystyle\frac{(x^2-9)^p}{1+x^4} \le \frac{(x^2-9)^p}{x^4} = \left(\frac{x^2}{x^{4/p}} - \frac{9}{x^{4/p}}\right)^p \le \frac{x^{2p}}{x^4} = x^{2p-4}$ and we

know $\displaystyle\int_1^\infty x^{2p-4}\,dx$ converges for 2p-4<-1 or p<3/2.   Thus the

original integral converges for -1<p<3/2.

27.   Since $\displaystyle\lim_{x\to0+} x\ln x = 0$, the given integral is not an improper
integral and thus exists by Thm.6.2.1.

29. $\displaystyle\int_{-\infty}^\infty \frac{dx}{1+x^2} = \int_{-\infty}^{-1} \frac{dx}{1+x^2} + \int_{-1}^1 \frac{dx}{1+x^2} + \int_1^\infty \frac{dx}{1+x^2}.$ The last

integral converges since $\displaystyle\frac{1}{1+x^2} \le \frac{1}{x^2}$ and we know $\displaystyle\int_1^\infty x^{-2}\,dx$

converges.   The first integral on the right is the same as
the last, when the substitution x = -y is made, and hence
converges and the middle integral exists by Thm.6.2.1. Thus
the given integral converges.

31. $\displaystyle\int_{-\infty}^\infty \frac{e^{-x}}{1+x^2}\,dx = \int_{-\infty}^0 \frac{e^{-x}}{1+x^2}\,dx + \int_0^\infty \frac{e^{-x}}{1+x^2}\,dx.$ Since $\displaystyle\frac{e^{-x}}{1+x^2} \ge e^{-x}$ and

since $\displaystyle\int_{-\infty}^0 e^{-x}\,dx$ diverges, we conclude the first integral on

the right diverges and hence the given integral diverges.
Note that the second integral on the right converges, but
that does not affect the conclusion.

33. Since $\dfrac{x^p}{(4+x^3)^q} \cong \dfrac{x^p}{x^{3q}}$ for large x we have $\lim\limits_{x\to\infty} \dfrac{x^p/(4+x^3)^q}{x^p/x^{3q}} =$

$\lim\limits_{x\to\infty} \dfrac{x^{3q}}{(4+x^3)^q} = 1$ and thus the given integral converges or

diverges according to whether $\displaystyle\int_1^\infty (x^p/x^{3q})\,dx$ converges or

diverges. Since this latter integral converges only for
$(3q-p)>1$, the same is true for the given integral.

35. $S = 2\pi \displaystyle\int_1^\infty x^{-p}\sqrt{1+(-p/x^{p+1})^2}\ dx = 2\pi \int_1^\infty \dfrac{\sqrt{x^{2p+2}+p^2}}{x^{2p+1}}\,dx.$ Now, the

integrand is approximated by $\dfrac{x^{p+1}}{x^{2p+1}} = x^{-p}$ for large x and since

$\displaystyle\int_1^\infty x^{-p}dx$ converges for $p > 1$ we conclude, by Thm.11.4.2, that

the surface area is finite only for $p > 1$. The explanation of
the apparent contradiction that is referred to in the note to
this problem involves an understanding of the behavior of
functions at infinity. Note also that one really cannot
compare surface area and volume since the dimensions are not
the same.

37. Let $q = -p$ and thus $\displaystyle\int_0^\infty \dfrac{dx}{x^p\sqrt{1+x^p}} = \int_0^\infty \dfrac{dx}{x^{-q}\sqrt{1+x^{-q}}} = \int_0^\infty \dfrac{x^{3q/2}dx}{\sqrt{x^q+1}}.$

Now $\dfrac{x^{3q/2}}{\sqrt{x^q+1}} \cong \dfrac{x^{3q/2}}{x^{q/2}} = x^q$ and hence, by Thm.11.4.2, the given

integral diverges for $p \le 0$ since $\displaystyle\int_0^\infty x^q dx$ diverges for $q \ge 0$.

39a. The conditions of Theorem 11.4.1a hold and thus for any $b > a$
$\displaystyle\int_a^b f(x)dx$ exists since f is integrable on every interval [a,b]
and thus S is not empty. In addition, since $0 \le f(x) \le g(x)$,
we have $\displaystyle\int_a^b f(x)dx \le \int_a^b g(x)dx \le \int_a^\infty g(x)dx$ and thus S is bounded above.

39b. Since I is the least upper bound of S and since every element

of S is of the form $s = \int_a^b f(x)dx$ for some b > a, we have that

$\int_a^b f(x)dx \leq I$ for each b > a.

c. If $\int_a^B f(x)dx \leq I - \varepsilon$ for all B > a then $I - \varepsilon$ would be the least

upper bound, contradicting the result of part (a). Thus there

exists a B > a such that $\int_a^B f(x)dx \geq I - \varepsilon$.

d. From parts (b) and (c) we have, for each $\varepsilon > 0$, a b > a such

that $I - \varepsilon \leq \int_a^b f(x)dx \leq I < I + \varepsilon$ or $\left| \int_a^b f(x)dx - I \right| \leq \varepsilon$ which

yields $\lim_{b \to \infty} \int_a^b f(x)dx = I$ by the definition of limit.

41a. Suppose $f(x)/g(x) \to 0$ as $x \to \infty$. Thus, for given $\epsilon > 0$ there is

an N such that for x > N, $|f(x)/g(x)| < \epsilon$. Since $f(x) \geq 0$, $g(x) > 0$,

$|f(x)/g(x)| = f(x)/g(x)$, so $0 < f(x) < \epsilon g(x)$ for x > N.

Therefore, by Thm. 11.4.1, if $\int_a^\infty g(x)dx$ converges, so does

$\int_a^\infty \epsilon g(x)dx$ and finally $\int_a^\infty f(x)dx$ converges.

b. For $g(x) = x^{-1/2}$, $f(x) = x^{-1}$, $f(x)/g(x) = x^{-1/2}$, so
$\lim_{x \to \infty} f(x)/g(x) = 0$. Finally, by Ex. (4) Section 11.3, both

$\int_a^\infty dx/\sqrt{x}$ and $\int_a^\infty dx/x$ diverge.

c. In this case $\lim_{x \to \infty} f(x)/g(x) = \lim_{x \to \infty} 1/x^{3/2} = 0$. Again by Ex. (4)

of Section 11.3, $\int_a^\infty g(x)dx$ diverges while $\int_a^\infty f(x)dx$ converges.

41d. Suppose $\lim\limits_{x\to\infty} f(x)/g(x)\to\infty$ as $x\to\infty$. Thus for given $M > 0$ there

is an N such that for $x>N$, $f(x)/g(x)>M$. Therefore for $x>N$,

$Mg(x) < f(x)$. By Thm.11.4.1b if $\int_a^{\infty} g(x)\,dx$ diverges so do

$\int_a^{\infty} Mg(x)\,dx$ and $\int_a^{\infty} f(x)\,dx$.

  e. Let $g(x) = x^{-4}$, $f(x) = x^{-2}$ then $f(x)/g(x) = x^2\to\infty$ and both

$\int_a^{\infty} g(x)\,dx$ and $\int_a^{\infty} f(x)\,dx$ converge, but if $g(x) = x^{-4}$, $f(x) = x^{-1}$,

$f(x)/g(x) = x^3\to\infty$, then $\int_a^{\infty} g(x)\,dx$ converges while $\int_a^{\infty} f(x)\,dx$

diverges.

Chapter 11 Review, Page 575

1.   (0/0) $L = \lim\limits_{x\to 0} \dfrac{-2\sin(x-\pi/2)\cos(x-\pi/2)}{1} = 0$.

3.   (0/0). Since the derivative of the denominator is 1 for $x > \pi$
      and -1 for $x < \pi$, the limit does not exist as the right hand
      limit is 2 while the left hand limit is -2.

5.   (0/0) $L = \lim\limits_{x\to 0} \dfrac{2\ln(x+1)}{2(x+1)^2} = 0$.

7.   (0/0) $L = \lim\limits_{x\to 2} \dfrac{2x}{2/(2x-3)} = 2$.

9. $(0\cdot\infty)$ $L = \lim\limits_{x\to 0} \dfrac{\ln(x^{-2}+1)}{x^{-2}} = \lim\limits_{u\to\infty} \dfrac{\ln(u^2+1)}{u^2} = \lim\limits_{u\to\infty} \dfrac{2u}{2u(u^2+1)} = 0$.

11.   (0/0) $L = \lim\limits_{x\to\pi-} (2\cos x\,\sin x\sqrt{\pi^2 - x^2})/x = 0$.

13. $L^- = \lim\limits_{x\to 0-} \dfrac{\exp(1/x)}{\exp(e^{1/x})} = 0$ directly, while $L^+ = \lim\limits_{x\to 0+} \dfrac{\exp(1/x)}{\exp(e^{1/x})}$ has

      form ∞/∞, thus $L^+ = \lim\limits_{x\to 0+} \dfrac{(-1/x^2)\exp(1/x)}{(-1/x^2)\exp(1/x)\exp(e^{1/x})} = 0$. Since

      $L^+ = L^-$, the limit is 0.

15.   (0/0) $L = \lim\limits_{x \to 0} \dfrac{\arcsin x + \left(x/\sqrt{1-x^2}\right)}{2\,(\pi/2 - \arccos x)/\sqrt{1-x^2}}$

$\qquad = \lim\limits_{x \to 0} \dfrac{\sqrt{1-x^2}\,\arcsin x + x}{2\,(\pi/2 - \arccos x)} = \lim\limits_{x \to 0} \dfrac{2\sqrt{1-x^2} - x\arcsin x}{2} = 1.$

17.   ($\infty - \infty$) $L = \lim\limits_{x \to 0} \dfrac{\sin x - x\cos x}{x\sin x} = \lim\limits_{x \to 0} \dfrac{x\sin x}{\sin x + x\cos x}$

$\qquad = \lim\limits_{x \to 0} \dfrac{x\cos x + \sin x}{2\cos x - x\sin x} = 0.$

19.   (∞/∞) $L = \lim\limits_{x \to 0+} \exp\left[\dfrac{1}{\sqrt{x}} - \dfrac{1}{\arcsin x}\right].$ Now

$\lim\limits_{x \to 0+} \dfrac{\arcsin x - \sqrt{x}}{\sqrt{x}\,\arcsin x} = \lim\limits_{x \to 0+} \dfrac{2\sqrt{x} - \sqrt{1-x^2}}{2x + \sqrt{1-x^2}\,\arcsin x} = -\infty.$ Thus $L = 0.$

21.   (1^∞) $L = \exp(\lim\limits_{x \to \infty}[x\ln|\cos(1/x)|] = \exp[\lim\limits_{u \to 0+} \dfrac{\ln\cos u}{u}]$

$\qquad = \exp[\lim\limits_{u \to 0+} -\dfrac{\sin u}{\cos u}] = \exp(0) = 1.$

23.   (∞^0) $L = \exp(\lim\limits_{x \to 1+}\left[\dfrac{-\ln(x^2 - 1)}{\csc \pi x}\right]) = \exp(\lim\limits_{x \to 1+}\left[\dfrac{2x}{\pi\cos \pi x}\dfrac{\sin^2 \pi x}{x^2 - 1}\right])$

$\qquad = \exp\left[\dfrac{-2}{\pi}\lim\limits_{x \to 1+}\dfrac{2\pi\sin \pi x\cos \pi x}{2x}\right] = \exp(0) = 1.$

25.   (1^∞) $L = \exp(\lim\limits_{x \to 1}\left[\dfrac{(3/2)\ln x}{1 - x^2}\right]) = \exp(\lim\limits_{x \to 1}[-3/4x^2]) = e^{-3/4}.$

27.   If $p \le 0$, let $s = -p$ and then $L = \lim\limits_{x \to \infty} \dfrac{1}{x^s \ln(x^2 + 1)} = 0.$

$\qquad$ For $p > 0$, $L = \lim\limits_{x \to \infty} \dfrac{p\,x^{p-1}\,(x^2 + 1)}{2x} = \lim\limits_{x \to \infty} (p/2)\,(x^p + x^{p-2}) = \infty.$

29.   (1^∞) $L = \exp\{\lim\limits_{x \to 0}[(\ln\cos x)/x^2]\} = \exp\{\lim\limits_{x \to 0}\left[\dfrac{\sin x}{x}\left(-\dfrac{1}{2\cos x}\right)\right]\} = e^{-1/2}.$

31. $I = \lim\limits_{h \to 2-} \int_0^h \dfrac{x}{(x^2 - 4)^{1/3}}\,dx = (3/4)\lim\limits_{h \to 2-}(x^2-4)^{2/3}\Big|_0^h = (-3/4)(16)^{1/3}$,

so the integral converges.

33. Letting $u = x^{-1/2}$, $\displaystyle\int_0^1 x^{-3/2}e^{x^{-1/2}}dx = 2\int_1^\infty e^u du$, which diverges.

35. Since $\sin 2x/\sin^{3/2}x = 2\cos x(\sin x)^{-1/2}$,

$$\int_0^\pi [\sin 2x/\sin^{3/2}x]\,dx = \int_0^{\pi/2} 2\cos x(\sin x)^{-1/2}dx + \int_{\pi/2}^\pi 2\cos x(\sin x)^{-1/2}dx.$$

Both integrals converge to 0, thus the original integral
converges to 0. Early printings of the text had incorrect
limits of integration.

37. $I = \lim\limits_{h \to \infty}\int_1^h \dfrac{x^{-3/2}}{(x^{-1/2} + 1)^2}\,dx = 2\lim\limits_{h \to \infty}(x^{-1/2} + 1)^{-1}\Big|_1^h = 1$, guaranteeing

the convergence of the given integral.

39. $I = \displaystyle\int_0^2 \dfrac{x-2}{x^p(4-x)^p}\,dx + \int_2^4 \dfrac{x-2}{x^p(4-x)^p}\,dx$. By comparing with $\displaystyle\int_0^2 x^{-p}dx$

and $\displaystyle\int_2^4 (4-x)^{-p}dx$ respectively, both integrals converge for

$p < 1$ and diverge for $p \geq 1$. Early printings of the text had
an incorrect integrand.

41. $I = \displaystyle\int_0^\pi x^{-1/2}dx - \int_0^\pi x^{-3/2}\sin x\,dx$. Since $(\sin x)/x^{3/2} \cong x^{-1/2}$ for x

near 0, I converges since $\displaystyle\int_0^\pi x^{-1/2}dx$ converges.

43. $x^{3/2}(x^{5/2}-1)^{-2} \cong x^{-7/2}$ for large x, thus the given integral

converges since $\displaystyle\int_2^\infty x^{-7/2}\,dx$ converges.

45. By L'Hospital's rule, $\lim\limits_{x \to 1+} \dfrac{x + \cos \pi x}{x - 1} = 1$. Thus, since

$\displaystyle\int_1^2 \dfrac{x^{3/2}}{(x-1)^2}\,dx \geq \int_1^2 \dfrac{dx}{(x-1)^2}$, which diverges, the original integral

diverges by comparison.

47. Note that $e^{1/(x^2-1)} = e^{(1/2)[1/(x-1) - 1/(x+1)]} = e^{1/2(x-1)}e^{-1/2(x+1)}$ and
$1/2(x-1) \leq e^{1/2(x-1)}$. Thus I diverges since

$$I = \int_1^2 e^{1/(x^2-1)}\ dx + \int_2^\infty e^{1/(x^2-1)}dx,\ \text{and the first integral}$$

diverges by comparison with $\displaystyle\int_1^2 (x-1)^{-1}dx$.

49. $\displaystyle I = \int_1^\infty \frac{\sin x\ dx}{(x^3-1)^{1/2}} + \int_1^\infty \frac{x\ dx}{(x^3-1)^{1/2}}$.   Since $x(x^3-1)^{-1/2} \cong x^{-1/2}$ as

$x \to \infty$, the second integral diverges by comparison with $\displaystyle\int_1^\infty \frac{dx}{x^{1/2}}$,

and thus I diverges.

CHAPTER 12

1a. $a_1 = 3$, $a_2 = 5/2$, $a_3 = 7/3$, $a_4 = 9/4$, $a_5 = 11/5$.

 b. a_n is bounded since $2 < a_n \leq 3$.

 c. Since for $m > n$, $1/m < 1/n$, we have $a_m < a_n$ so a_n is monotone decreasing.

 d. Since a_n is a bounded monotone sequence it converges by Thm.12.1.6, and $\lim_{n \to \infty} a_n = 2$.

3a. $a_1 = 0$, $a_2 = 1/3$, $a_3 = 1/2$, $a_4 = 3/5$, $a_5 = 2/3$.

 b. $a_n = \dfrac{n-1}{n+1} = \dfrac{(n+1)-2}{n+1} = 1 - 2/(n+1)$ which is bounded below by 0 and above by 1.

 c. For $m > n$, $m + 1 > n+1$, $\dfrac{1}{m+1} < \dfrac{1}{n+1}$, $-\dfrac{1}{m+1} > -\dfrac{1}{n+1}$ so $1 - 2/(m+1) > 1 - 2/(n+1)$. Thus the sequence is monotone increasing.

 d. By Thm.12.1.6, the sequence a_n converges. Finally $\lim_{n \to \infty} a_n = 1$.

5a. $a_1 = 1$, $a_2 = -2^{1/3}$, $a_3 = 3^{1/3}$, $a_4 = -4^{1/3}$, $a_5 = 5^{1/3}$.

 b. The sequence is not bounded since for any k, if $n = (k+1)^3$, $|a_n| = [(k+1)^3]^{1/3} = k + 1$.

 c. The sequence is not monotone, since $a_n < 0$ for n even and $a_n > 0$ for n odd.

 d. The sequence diverges by part(b).

7a. $a_1 = -3/2$, $a_2 = -2/5$, $a_3 = -1/10$, $a_4 = 0$, $a_5 = 1/26$.

 b. Let $f(x) = (x-4)/(x^2+1)$, then $f'(x) = (-x^2+8x+1)/(x^2+1)^2$, from which we can conclude that a_n is increasing for $n \leq 8$ and decreasing for $n \geq 8$. Thus the sequence is bounded by a_8 and a_1.

 c. From part(b), the sequence is montone decreasing for $n \geq 8$.

 d. By Thm.12.1.6, the sequence converges and
$$\lim_{n \to \infty} a_n = \lim_{x \to \infty} \frac{x-4}{x^2+1} = \lim_{x \to \infty} \frac{1}{2x} = 0, \text{ by Thm.12.1.4.}$$

9a. $a_1 = 1/2$, $a_2 = 1/4$, $a_3 = 2/9$, $a_4 = 1/4$, $a_5 = 8/25$.

 b. In Section 11.2 it was shown that exponential growth dominates algebraic growth. Thus 2^n grows much more rapidly than $4n^2$ so that $2^n/4n^2$ is unbounded.

9c.   Since $a_{n+1} - a_n = \dfrac{2^n}{4}\left(\dfrac{n^2 - 2n - 1}{n^2(n+1)^2}\right) > 0$ for $n \geq 3$, the sequence is

monotone increasing.

d.   Since $\lim\limits_{n \to \infty} \dfrac{2^n}{4n^2} = \lim\limits_{x \to \infty} \dfrac{2^x}{4x^2} = \lim\limits_{x \to \infty} \dfrac{(\ln 2)^2\, 2^x}{8} = \infty$ the sequence

diverges.

11. $\lim\limits_{n \to \infty} \dfrac{n^2 + 2n - 3}{2n^2 - 3n + 1} = \lim\limits_{x \to \infty} \dfrac{x^2 + 2x - 3}{2x^2 - 3x + 1} = \lim\limits_{x \to \infty} \dfrac{2}{4} = \dfrac{1}{2}$ by applying

L'Hospital's rule twice.

13. $\lim\limits_{n \to \infty} \dfrac{n^2 + 1}{n^{3/2} + n} = \lim\limits_{n \to \infty} \dfrac{n^{1/2}(1 + 1/n^2)}{(1 + 1/n^{1/2})} = \infty$, so the sequence diverges to ∞.

15. $\lim\limits_{n \to \infty} \dfrac{\ln n}{n} = \lim\limits_{x \to \infty} \dfrac{\ln x}{x} = \lim\limits_{x \to \infty} \dfrac{1}{x} = 0$, by L'Hospital's rule. Thus

the sequence converges to 0.

17. $\lim\limits_{n \to \infty} n^{1/p} = \lim\limits_{x \to \infty} x^{1/p} = \lim\limits_{x \to \infty} \exp[(1/p)\ln x] = \infty$.   Thus the

sequence diverges to ∞.

19. $\lim\limits_{n \to \infty} (1.01)^n = \lim\limits_{x \to \infty} (1.01)^x = \lim\limits_{x \to \infty} \exp[x\ln(1.01)] = \infty$.   The

sequence diverges to ∞.

21. $\lim\limits_{n \to \infty} n^3/3^n = \lim\limits_{x \to \infty} x^3/3^x = \lim\limits_{x \to \infty} 3!/(\ln 3)^3 3^x = 0$ applying

L'Hospital's rule three times.   The sequence converges to 0.

23. $a_n = (2^{n+1} + n^2)/3^n = 2(2/3)^n + n^2/3^n$.   Thus

$\lim\limits_{n \to \infty} a_n = \lim\limits_{n \to \infty} 2(2/3)^n + \lim\limits_{n \to \infty} n^2/3^n = 0 + 0 = 0$ and the sequence

converges to 0.

25. $a_n = \sqrt{n}(\sqrt{n+1} - \sqrt{n}) = \sqrt{n}(\sqrt{n+1} - \sqrt{n})(\sqrt{n+1} + \sqrt{n})/(\sqrt{n+1} + \sqrt{n})$

so that $a_n = \sqrt{n}/(\sqrt{n+1} + \sqrt{n})$.   Thus

$\lim\limits_{n \to \infty} a_n = \lim\limits_{n \to \infty} \dfrac{1}{\sqrt{(n+1)/n} + 1} = \lim\limits_{n \to \infty} \dfrac{1}{\sqrt{1 + 1/n} + 1} = 1/2$.

27. $\lim\limits_{n \to \infty} \ln(n+1) - \ln n = \lim\limits_{n \to \infty} \ln\dfrac{n+1}{n} = \lim\limits_{n \to \infty} \ln(1 + 1/n) = 0$ so the

sequence converges to 0.

29. $\lim\limits_{n \to \infty}$ (ln bn)/(ln cn) = $\lim\limits_{n \to \infty}$ (ln b + ln n)/(ln a + ln n) =

$\lim\limits_{n \to \infty}$ [(ln b/ln n) + 1]/[(ln a/ln n) + 1] = 1.

31.   Note that $a_n = (2n)!/n^n = [1 \cdot 2 \cdot 3 \cdots n \cdot (n+1)(n+2) \cdots 2n]/n^n$.   So

$a_n = [n!] \dfrac{(n+1)}{n} \cdot \dfrac{(n+2)}{n} \cdots \left(\dfrac{2n}{n}\right) = n!(1+1/n)(1+2/n) \cdots (2) > n!$,

thus $\lim\limits_{n \to \infty} a_n = \infty$.

33. $a_n = 1 + (-1)^n/n$ converges to 1, but for n even $a_n > 1$ while
for n odd, $a_n < 1$, thus a_n is not monotone. Similar results
hold for $(-1)^{n+1}/n$.

35. $a_n = 2^n$ is monotone since $2^{n+1} = 2 \cdot 2^n = 2a_n$ so $a_{n+1} > a_n$, but
2^n is not bounded. Similar results hold for $\sqrt{n}$.

37. $a_n = 1 + 1/n$ is monotone decreasing and $\lim\limits_{n \to \infty} a_n = 1$ so the
sequence is convergent. Similar results hold for 1/n.

39.   Since 1/n is small when n is large, it is reasonable to
conjecture $\lim\limits_{n \to \infty} a_n = 2$.   Given $\epsilon > 0$ we must find N such that
$|a_n - 2| < \epsilon$ when n > N.   But $|a_n - 2| = |1/n|$ and $1/n < \epsilon$ when
n > $1/\epsilon$. Thus we choose N to be any integer for which $N \geq 1/\epsilon$,
then for all n > N, $1/n < 1/N \leq \epsilon$ so $|a_n - 2| < \epsilon$, verifying
that $\lim\limits_{n \to \infty} a_n = 2$.

41.   As in Prob.3, we may write $a_n = 1 - 2/(n+1)$ and conjecture
that the sequence converges to 1.   For $\epsilon > 0$ we must find N so
that $|a_n - 1| < \epsilon$ when n > N.   Since $|a_n - 1| = 2/(n+1)$, choose N
to be any integer for which $N \geq 2/\epsilon + 1$, then for n > N ,
$2/(n+1) < \epsilon$ and thus for n > N, $|a_n - 1| < \epsilon$.

43. $a_n = 2^{-n} = 1/2^n$ becomes small for large n, thus L = 0 seems
reasonable.   We must find N so that for $\epsilon > 0$ and for all
n > N, $|a_n| < \epsilon$. $|a_n| = |1/2^n| = 1/2^n < \epsilon$ so choosing
$0 < \epsilon < 1$ assures $1/\epsilon > 1$.   Now, since $2^n > 1/\epsilon$,
$n \ln 2 > \ln(1/\epsilon)$.   Thus by choosing N to be any integer for
which $N \geq \ln(1/\epsilon)/(\ln 2)$ we have that whenever n > N, $2^{-n} < \epsilon$.

45a. Using induction we have
   i. $a_2 = 1 + \sqrt{a_1} = 1 + \sqrt{1} = 2 > 1.$ Therefore $a_2 > a_1.$
   ii. Suppose $a_{k+1} > a_k$ for some k then

$$a_{k+2} - a_{k+1} = (1 + \sqrt{a_{k+1}}) - (1 + \sqrt{a_k}) = \sqrt{a_{k+1}} - \sqrt{a_k} > 0 \text{ by}$$

   the hypothesis. Thus $a_{k+2} > a_{k+1}$, when $a_{k+1} > a_k$ and by
   mathematical induction the sequence is monotone increasing.
  b. Trying 4 as the upper bound we have
   i. $a_1 = 1$ so $a_1 < 4.$
   ii. Suppose $a_k \leq 4$ for some k, then

$$a_{k+1} = 1 + \sqrt{a_k} \leq 1 + \sqrt{4} = 3 < 4, \text{ so } a_{k+1} < 4. \quad \text{Thus by}$$

   mathematical induction, $a_n < 4$ for all $n \geq 1.$
  c. $L = \lim\limits_{n \to \infty} a_n = \lim\limits_{n \to \infty}(1 + \sqrt{a_{n-1}}) = 1 + \sqrt{L}.$ Thus $L = 1 + \sqrt{L}$ or $L - 1 = \sqrt{L}$
   or $L^2 - 2L + 1 = L$ or $L^2 - 3L + 1 = 0$, so $L = (3 \pm \sqrt{5})/2.$ Since
   $2 < L < 4$, we must choose $L = (3 + \sqrt{5})/2 \cong 2.6180.$

47a. If $|a_n - L| < \epsilon$, then $-\epsilon < a_n - L < \epsilon$ so $L - \epsilon < a_n < L + \epsilon.$ In any
   event, $|a_n|$ is less than the larger of $|L - \epsilon|$ and $|L + \epsilon|.$ But
   $|L - \epsilon| \leq |L| + \epsilon$ and $|L + \epsilon| \leq |L| + \epsilon$ so that $|a_n| < |L| + \epsilon$ for $n > N.$
  b. $|a_1|, |a_2| \dots |a_N|$ is a finite collection of bounded real
   numbers and so has a maximum, call it $M_1.$
  c. Let M be the maximum of M_1 and $|L| + \epsilon.$ Then for $n > N,$
   $|a_n| \leq M$ and for $1 \leq n \leq N$, $|a_n| \leq M$ so that for all $n \geq 1$, $|a_n| \leq M.$

49. Let $\epsilon > 0$ be given, and $c_n = a_n b_n.$ Thus $|c_n| = |a_n b_n| =$
   $|a_n||b_n| \leq |a_n|B$, when $|b_n| \leq B$ for all $n.$ As a_n converges to
   0, there is an N such that for $n > N$, $|a_n| < \epsilon/B.$ Thus
   $|a_n b_n| < (\epsilon/B)(B)$, so $a_n b_n$ converges to 0.

51a. Since L is the least upper bound of $\{a_n\}$, for $\epsilon > 0$, $L - \epsilon < L$
   and hence for some integer N, $a_N > L - \epsilon.$ Otherwise $L - \epsilon$
   would be an upper bound for $\{a_n\}$ less than the least upper
   bound.
  b. Since $\{a_n\}$ is monotone nondecreasing then $a_N \leq a_n \leq L$ for all
   $n \geq N$, where N is as in part (a). Thus, from part (a),
   $L - \epsilon < a_N \leq a_n < L < L + \epsilon$ so $|a_n - L| < \epsilon$, or $\lim\limits_{n \to \infty} a_n = L.$

Section 12.2, Page 598

1. $\displaystyle\sum_{k=1}^{\infty} a_k = a_1+a_2+a_3+\ldots$. $\displaystyle\sum_{i=1}^{\infty} a_i = a_1+a_2+a_3+\ldots$. Thus $\displaystyle\sum_{k=1}^{\infty} a_k = \sum_{i=1}^{\infty} a_i$.

3. $\displaystyle\sum_{k=1}^{\infty} a_k = a_1+a_2+a_3+\ldots$. $\displaystyle\sum_{k=2}^{\infty} a_{k-1} = a_1+a_2+a_3+\ldots$. Thus $\displaystyle\sum_{k=1}^{\infty} a_k = \sum_{k=2}^{\infty} a_{k-1}$.

5. $\displaystyle\sum_{k=0}^{\infty}\left(\frac{1}{2}\right)^{k+1} = \left(\frac{1}{2}\right) + \left(\frac{1}{2}\right)^2 + \left(\frac{1}{2}\right)^3 + \ldots$. $\displaystyle\sum_{k=2}^{\infty}\left(\frac{1}{2}\right)^{k-1} = \left(\frac{1}{2}\right) + \left(\frac{1}{2}\right)^2 + \left(\frac{1}{2}\right)^3 + \ldots$.

7. $\displaystyle\sum_{k=1}^{\infty} \frac{1}{2^k} = \sum_{k=1}^{\infty}(1/2)^k = 1/2 + (1/2)^2 + (1/2)^3 + \cdots$

    $= [1+(1/2)+(1/2)^2+(1/2)^3+ \cdots] - 1 = \dfrac{1}{1-1/2} - 1 = 1$ since

    $1 + 1/2 + (1/2)^2 + \cdots$ is a geometric series with $r = 1/2$.

9. $\displaystyle\lim_{n\to\infty} a_n = \lim_{n\to\infty} 2^{n/2} = \infty$. Since $\displaystyle\lim_{n\to\infty} a_n \neq 0$, the series diverges.

11. We have $\displaystyle\sum_{k=1}^{\infty}\frac{(-1)^{k+1}}{2^{k/2}} + 1 = \sum_{k=0}^{\infty}\frac{(-1)^k}{2^{k/2}} = \sum_{k=0}^{\infty}\left(-\frac{1}{\sqrt{2}}\right)^k$. The last series

    converges to $\dfrac{1}{1+1/\sqrt{2}}$ by Eq.(17). Thus $\displaystyle\sum_{k=1}^{\infty}\frac{(-1)^{k+1}}{2^{k/2}}$ has sum

    $\dfrac{1}{1+1/\sqrt{2}} - 1 = 1/(\sqrt{2} + 1)$.

13. $\displaystyle\sum_{k=0}^{\infty} \frac{1}{(k+1)(k+2)} = \sum_{k=1}^{\infty} \frac{1}{k(k+1)} = 1$, as was shown in Ex.2.

15. Since $\dfrac{1}{k(k+1)(k+2)} = \dfrac{1/2}{k} - \dfrac{1}{k+1} + \dfrac{1/2}{k+2} = \dfrac{1}{2}\left(\dfrac{1}{k} - \dfrac{1}{k+1}\right) - \dfrac{1}{2}\left(\dfrac{1}{k+1} - \dfrac{1}{k+2}\right)$

    we have that $\displaystyle\sum_{k=1}^{n} \frac{1}{k(k+1)(k+2)} = \sum_{k=1}^{n}\frac{1}{2}\left(\frac{1}{k} - \frac{1}{k+1}\right) - \frac{1}{2}\left(\frac{1}{k+1} - \frac{1}{k+2}\right) =$

    $(1/2)[1 - 1/2] = 1/4$ by Ex.2 and Prob.13. Thus the series
    converges to the sum of $1/4$.

17. Since $\{a_n\}$ diverges (the terms oscillate between 2^{-k} and 1),

     then $\displaystyle\sum_{k=1}^{\infty} a_k$ diverges by Thm.12.2.1 and its extensions.

19. Since $\displaystyle\sum_{k=1}^{\infty} \frac{1}{k\,(k+1)}$ and $\displaystyle\sum_{k=1}^{\infty} \frac{1}{2^k}$ both converge then

$$\sum_{k=1}^{\infty} \left(\frac{3}{k\,(k+1)} - \frac{1}{2^k} \right) = 3\sum_{k=1}^{\infty} \frac{1}{k\,(k+1)} - \sum_{k=1}^{\infty} \frac{1}{2^k} = 3 - 1 = 2.$$

21. $\displaystyle\sum_{k=0}^{\infty} (\alpha-2)^k$ converges for $|\alpha-2| < 1$ or $1 < \alpha < 3$, since the

     series is geometric, and $s = \dfrac{1}{1-(\alpha-2)} = \dfrac{1}{3-\alpha}$.

23. Since $\displaystyle\sum_{k=0}^{\infty}(2\alpha-3)^k$ is a geometric series, it converges for

     $|2\alpha-3|<1$ or $-1<2\alpha-3<1$ or $1<\alpha<2$ and $s = \dfrac{1}{1-(2\alpha-3)} = \dfrac{1}{4-2\alpha}$.

25. Using Eqs.(14) and (17) we have $|s-s_n| = |r|^{n+1}/(1-r)$ where in
     this case $r = 1/3$ so for $|s-s_n| < 0.0001$, we must have

     $(1/3)^n < 0.0002$ or $-n\ln 3 < \ln(.0002)$. That is,
     $n > |\ln(.0002)/\ln 3|$, or $N = 8$.

27. $\displaystyle\sum_{n=1}^{\infty} \frac{1}{k\,(k+1)} = 1 = s$ and $s_n = 1 - \dfrac{1}{n+1}$ from Ex.2. Therefore,

     for $|s-s_n| <.0001$, we must have $\dfrac{1}{n+1} < \dfrac{1}{10^4}$ or $n+1 > 10^4$ so

     $N = 10^4$.

29. The ball is dropped from a height h and bounces to a height
     of αh. It then falls the height αh and bounces to a height
     $\alpha(\alpha h)$. Then it falls the height $\alpha(\alpha h)$ and bounces to a height
     $\alpha(\alpha^2 h)$. It continues bouncing, so that the total distance
     traveled is

$$d = h + \alpha h + \alpha h + \alpha(\alpha h) + \alpha(\alpha h) + \cdots$$

$$= h + 2\alpha h + 2\alpha^2 h + 2\alpha^3 h + \cdots = h + 2h\alpha \sum_{k=0}^{\infty} \alpha^k$$

$$= h + 2h\alpha[1/(1-\alpha)] = h(1+\alpha)/(1-\alpha).$$

31. $s = .5 + .05 + .005 + .0005 + \cdots$

$$= 5/10 + 5/10^2 + 5/10^3 + 5/10^4 + \cdots$$

$$= (5/10) \sum_{k=0}^{\infty} (1/10)^k = (5/10)[1/(1-1/10)] = 5/9.$$

33. $s = 3/10 + 9/10^2 + 9/10^3 + 9/10^4 + \cdots$

$$= 3/10 + (9/10^2) \sum_{k=0}^{\infty} (1/10)^k$$

$$= 3/10 + (9/100)\left(\frac{1}{1-1/10}\right) = 3/10 + 1/10 = 4/10 = 2/5.$$

35. $s = 81/100 + 81/(100)^2 + 81/(100)^3 + \cdots$

$$= (81/100) \sum_{k=0}^{\infty} (1/100)^k = (81/100)[1/(1 - 1/100)] = 9/11.$$

37. Since $a_n = s_n - s_{n-1}$, then $a_n = \left(1 - \frac{1}{(n+1)^2}\right) - \left(1 - \frac{1}{n^2}\right) =$

$(1/n^2) - [1/(n+1)^2] = (2n+1)/n^2(n+1)^2.$

39. Let $s_n = 1+r+r^2+\cdots+r^n$ then $rs_n = r+r^2+\cdots+r^n+r^{n+1}$ and

$s_n - rs_n = 1-r^{n+1}$ or $s_n(1-r) = 1-r^{n+1}$, so $s_n = (1-r^{n+1})/(1-r).$

Section 12.3, Page 606

1. Since $\dfrac{1}{k^2+4} < 1/k^2$ for all k and $\sum_{k=1}^{\infty} 1/k^2$ converges, then

$\sum_{k=0}^{\infty} \dfrac{1}{k^2+4}$ converges by Thm.12.3.1.

3.   Let $a_k = \sqrt{k}/(k+2)$ and $b_k = 1/\sqrt{k}$. Then $a_k/b_k = k/(k+2)$ and

$\lim_{k\to\infty} k/(k+2) = 1$. Thus, since $\sum\limits_{k=1}^{\infty} 1/\sqrt{k}$ diverges by the p-test,

$\sum\limits_{k=1}^{\infty} \sqrt{k}/(k+2)$ diverges by Thm.12.3.2.

5.   For $a_k = (2k-1)^2/(k^2+4)^{3/2}$, $b_k = 1/k$ we have

$\lim_{k\to\infty} a_k/b_k = \lim_{k\to\infty} \dfrac{(2-1/k)^2}{(1+4/k^2)^{3/2}} = 4$. Thus, since $\sum\limits_{k=1}^{\infty} \dfrac{1}{k}$ diverges,

we have that $\sum\limits_{k=0}^{\infty} \dfrac{(2k-1)^2}{(k^2+4)^{3/2}}$ diverges by Thm.12.3.2.

7.   For $a_k = (k+1)^{4/3}/(2k^2-1)^{1/2}$, $b_k - 1/k^{2/3}$,

$a_k/b_k = (1 + 1/k)^{4/3}/(2 - 1/k^2)^2$ so $\lim_{k\to\infty} a_k/b_k = 1/4$. Thus by

Thm.12.3.2, and since $\sum\limits_{k=1}^{\infty} 1/k^{2/3}$ diverges by the p-test, one

must conclude that $\sum\limits_{k=2}^{\infty} \dfrac{(k+1)^{4/3}}{(2k^4-1)^{1/2}}$ diverges.

9.   Let $a_k = \sin(1/k)$, $b_k = (1/k)$. Thus $a_k/b_k = \sin(1/k)/(1/k)$
     and $\lim_{k\to\infty}(a_k/b_k) = \lim_{x\to 0^+}(\sin x)/x = 1$. Thus, by Thm.12.3.2, since

$\sum\limits_{k=1}^{\infty} 1/k$ diverges, $\sum\limits_{k=1}^{\infty} \sin(1/k)$ diverges.

11. Note that $a_k = e^k/(4+e^k)^2 = 1/e^k(4e^{-k}+1)^2 < 1/e^k = (1/e)^k$.

Since $(1/e) < 1$, the geometric series $\sum\limits_{k=0}^{\infty} 1/e^k$ converges so

that by Thm.12.3.1, $\sum\limits_{k=0}^{\infty} e^k/(4+e^k)^2$ converges.

13. Note that $a_k = \dfrac{1}{3} \cdot \dfrac{2}{5} \cdot \dfrac{3}{7} \cdots \dfrac{k}{(2k+1)} < \left(\dfrac{1}{2}\right)^k$. Since $\sum\limits_{k=0}^{\infty} (1/2)^k$

converges, $\sum\limits_{k=1}^{\infty} \dfrac{k!}{3 \cdot 5 \cdot 7 \cdots (2k+1)}$, also converges by Thm.12.3.1.

15. Note that for large k, $a_k = (k^2+1)^p/(k^q+4) = \dfrac{k^{2p}(1+1/k^2)^p}{k^q(1+4/k^q)}$ is

close to k^{2p-q}. Then by choosing $b_k = k^{2p-q}$,

$\lim\limits_{k\to\infty} a_k/b_k = \lim\limits_{k\to\infty} \dfrac{(1+1/k^2)^p}{(1+4/k^q)} = 1$. Now $\sum\limits_{k=1}^{\infty} k^{2p-q}$ converges for

q-2p > 1, by the p-test, and diverges otherwise, so

$\sum\limits_{k=1}^{\infty} \dfrac{(k^2+1)^p}{k^q+1}$ converges when q-2p > 1 and diverges for q-2p ≤ 1

by the limit comparison test.

17. Since $|\sin k| \le 1$ for all k, $|\sin k|/k^2 \le 1/k^2$ for all k > 1,

so $\sum\limits_{k=1}^{\infty} |\sin k|/k^2$ converges by comparison with $\sum\limits_{k=1}^{\infty} (1/k^2)$.

19. $a_k = \dfrac{1}{k(4/k^2 \ln k + 1)}$, so $b_k = 1/k$ gives $\lim\limits_{k\to\infty} a_k/b_k = 1$. Thus

$\sum\limits_{k=2}^{\infty} \dfrac{k \ln k}{4 + k^2 \ln k}$ diverges by Thm.12.3.2 since $\sum\limits_{\ln=1}^{\infty} (1/k)$ diverges.

21. With $b_k = 1/k$, we have $\lim\limits_{k\to\infty} a_k/b_k = \lim\limits_{k\to\infty} k a_k = A \ne 0$, so

$\sum\limits_{k=1}^{\infty} a_k$ diverges by Thm.12.3.2 since $\sum\limits_{k=1}^{\infty} (1/k)$ diverges.

23a. If $a_k > 0$ and $\sum\limits_{k=1}^{\infty} a_k$ converges then $\lim\limits_{k\to\infty} a_k = 0$ by Thm.12.2.1.

For k > N, for some N, $0 < a_k < 1$ so $a_k^2 < a_k$ and by

Thm.12.3.1, $\sum\limits_{k=1}^{\infty} a_k^2$ converges.

23b. Let $a_k = 1/k$, then $\displaystyle\sum_{k=1}^{\infty} a_k^2 = \sum_{k=1}^{\infty} 1/k^2$ converges by the p-test,

while $\displaystyle\sum_{k=1}^{\infty} 1/k$ diverges. If $a_k = 1/k^2$, then $a_k^2 = 1/k^4$ and both

$\displaystyle\sum_{k=1}^{\infty} a_k$ and $\displaystyle\sum_{k=1}^{\infty} a_k^2$ converge.

25. If $p = 1+2r$, $(r > 0)$, then $(\ln k)/k^p = (\ln k/k^r)\cdot(1/k^{1+r})$. By
Ex.1 of Section 11.2, $\displaystyle\lim_{x\to\infty} \ln x/x^p = 0$ for $r > 0$. Thus, for all

$k > N$, and for some fixed integer N, $(\ln k/k^r) < 1$ so that

$(\ln k)/k^p < 1/k^{1+r}$. Since $\displaystyle\sum_{k=2}^{\infty} 1/k^{1+r}$ converges, by Thm.12.3.1,

$\displaystyle\sum_{k=2}^{\infty} (\ln k)/k^p$ converges.

27. If $c_k \le a_k \le b_k$ then $0 \le b_k - c_k$ and $0 \le a_k - c_k$ and further

$(a_k - c_k) \le (b_k - c_k)$. Since $\displaystyle\sum_{k=1}^{\infty} b_k$ and $\displaystyle\sum_{k=1}^{\infty} c_k$ both converge, then

$\displaystyle\sum_{k=1}^{\infty} (b_k - c_k)$ converges by Thm.12.2.2, and thus $\displaystyle\sum_{k=1}^{\infty} (a_k - c_k)$

converges by Thm.12.3.1. Let $\displaystyle\sum_{k=1}^{\infty} (a_k - c_k) = s$ and $\displaystyle\sum_{k=1}^{\infty} c_k = S$, so

that $\displaystyle\sum_{k=1}^{\infty} a_k = \sum_{k=1}^{\infty} [(a_k - c_k) + c_k] = \sum_{k=1}^{\infty} (a_k - c_k) + \sum_{k=1}^{\infty} c_k = s + S$, again

by Thm.12.2.2. Therefore $\displaystyle\sum_{k=1}^{\infty} a_k$ converges.

Section 12.4, Page 611

1. If $a_k = 2k/k!$, then $a_{k+1} = 2^{k+1}/(k+1)!$ so

$\dfrac{a_{k+1}}{a_k} = \dfrac{2^{k+1}}{(k+1)!} \cdot \dfrac{k!}{2^k} = \dfrac{2}{k+1}$. Thus $\lim\limits_{k\to\infty}(a_{k+1}/a_k) = \lim\limits_{k\to\infty}\dfrac{2}{k+1} = 0$, so

by the ratio test $\sum\limits_{k=1}^{\infty} 2k/k!$ converges.

3. If $a_k = k!/2^{2k}$, then $a_{k+1} = (k+1)!/2^{2(k+1)}$ so

$\dfrac{a_{k+1}}{a_k} = \dfrac{(k+1)!}{2^{2(k+1)}} \cdot \dfrac{2^{2k}}{k!} = (k+1)/4$. Thus $\lim\limits_{k\to\infty} a_{k+1}/a_k = \lim\limits_{k\to\infty}(k+1)/4 = \infty$,

so $\sum\limits_{k=1}^{\infty}(k!/2^{2k})$ diverges by the ratio test.

5. If $a_k = \dfrac{k!}{1\cdot3\cdot5\cdots(2k-1)}$, then $a_{k+1} = \dfrac{(k+1)!}{1\cdot3\cdot5\cdots(2k-1)(2k+1)}$ so

$\dfrac{a_{k+1}}{a_k} = \dfrac{k+1}{2k+1}$ and $\lim\limits_{k\to\infty}(a_{k+1}/a_k) = \lim\limits_{k\to\infty}(k+1)/(2k+1) = 1/2$ so the

series converges.

7. If $a_k = (k!)^2/(2k)!$, then $a_{k+1} = [(k+1)!]^2/(2k+2)!$ so

$a_{k+1}/a_k = (k+1)^2/(2k+1)(2k+2)$ and $\lim\limits_{k\to\infty} a_{k+1}/a_k = 1/4$. Thus, by

the ratio test, $\sum\limits_{k=1}^{\infty}(k!)^2/(2k)!$ converges.

9. If $a_k = 3^k k!/k^k$, $a_{k+1} = 3^{k+1}(k+1)!/(k+1)^{k+1}$

so $\dfrac{a_{k+1}}{a_k} = 3\left(\dfrac{k}{k+1}\right)^k$, as in Ex.5.   Thus

$\lim\limits_{k\to\infty}\dfrac{a_{k+1}}{a_k} = \lim\limits_{k\to\infty} 3\left(\dfrac{k}{k+1}\right)^k = 3/e$, from Ex.5.   Therefore the series

diverges, since $3/e > 1$.

11.   If $a_k = (\ln k)^3/(\ln 3)^k$, then $a_{k+1} = [\ln(k+1)]^3/(\ln 3)^{k+1}$ so

$$\frac{a_{k+1}}{a_k} = \frac{1}{\ln 3}\left[\frac{\ln(k+1)}{\ln k}\right]^3 \text{ and } \lim_{k\to\infty} a_{k+1}/a_k = 1/\ln 3. \text{ Since } 3 > e,$$

$\ln 3 > \ln e = 1$, $1/\ln 3 < 1$, and the series $\displaystyle\sum_{k=2}^{\infty}(\ln k)^3/(\ln 3)^k$

converges.

13.   If $a_k = \dfrac{2\cdot 4\cdot 6\cdots(2k)}{1\cdot 4\cdot 7\cdots(3k-2)}$ then $a_{k+1} = a_k(2k+2)/(3k+1)$,

$a_{k+1}/a_k = (2k+2)/(3k+1)$, and $\displaystyle\lim_{k\to\infty} a_{k+1}/a_k = 2/3$, so the series

converges.

15.   With $a_k = 5^k/(k^2 2^{2k})$, $a_{k+1} = 5^{k+1}/(k+1)^2 2^{2(k+1)}$ so

$$\frac{a_{k+1}}{a_k} = \left(\frac{k}{k+1}\right)^2\cdot\frac{5}{4} \text{ and } \lim_{k\to\infty}\left(\frac{k}{k+1}\right)^2\frac{5}{4} = \frac{5}{4}. \text{ Therefore the series}$$

diverges.

17.   If $a_k = (\ln k)^{10}/(\ln a)^k$, then $a_{k+1} = [\ln(k+1)]^{10}/(\ln a)^{k+1}$ and

$$\frac{a_{k+1}}{a_k} = \left(\frac{1}{\ln a}\right)\left(\frac{\ln(k+1)}{\ln k}\right)^{10} \text{ so } \lim_{k\to\infty}(a_{k+1}/a_k) = 1/\ln a, \text{ since}$$

$\displaystyle\lim_{k\to\infty}\frac{\ln(k+1)}{\ln k} = 1$. If $a > e$, then $\ln a > \ln e = 1$ and $1/\ln a < 1$

so the series $\displaystyle\sum_{k=2}^{\infty}\frac{(\ln k)^{10}}{(\ln a)^k}$ converges. If $1 < a < e$, $1/\ln a > 1$ and

the series diverges. If $a = e$ then $\displaystyle\sum_{k=2}^{\infty}\frac{(\ln k)^{10}}{(\ln a)^k} = \sum_{k=2}^{\infty}(\ln k)^{10}$

diverges, since $\displaystyle\lim_{k\to\infty} a_k = \lim_{k\to\infty}(\ln k)^{10} = \infty$.

19.   If $a_k = (2k)!/(2k)^k$, then $a_{k+1} = (2k+2)!/(2k+2)^{k+1}$,

$a_{k+1}/a_k = (2k+1)[k/(k+1]^k$, and

$\displaystyle\lim_{k\to\infty} a_{k+1}/a_k = \lim_{k\to\infty}(2k+1)[k/(k+1)]^k = \infty$ so the series diverges.

21. $(2^{k+1})! = 1 \cdot 2 \cdot 3 \cdots 2^k \cdot (2^k+1) \cdot (2^k+2) \cdots (2^{k+1})$

$\qquad = (2^k)!(2^k+1)(2^k+2) \cdots 2^{k+1}$ and

$2^{(k+1)!} = 2^{1 \cdot 2 \cdot 3 \cdots k(k+1)} = (2^{k!})^{k+1} = (2^{k!})(2^{k!})^k$. Thus

$$\frac{a_{k+1}}{a_k} = \frac{(2^k)!(2^k+1)(2^k+2) \cdots 2^{k+1} \cdot 2^{k!}}{(2^{k!})(2^{k!})^k(2^k)!}$$

$$= \frac{(2^k+1)(2^k+2) \cdots 2^{k+1}}{2^{k!} \cdot 2^{k!} \cdots 2^{k!}}.$$ The numerator can be written as

$(2^k+1)(2^k+2) \cdots (2^k+2^k)$ so there are 2^k terms, each of which

is less than or equal to 2^k+2^k.   Thus

$(2^k+1)(2^k+2) \cdots (2^k+2^k) \le (2^{k+1})^{2^k} = 2^{(k+1)2^k}$.   Therefore

$a_{k+1}/a_k \le 2^{(k+1)2^k}/2^{k \cdot k!} = 2^{(k+1)2^k - k(k)!}$.   The exponent,

$(k+1)2^k - k(k)!$, is less than zero for large k, since $2^k < k!$

and $k+1 \cong k$. Thus $\lim\limits_{k \to \infty} a_{k+1}/a_k = 0$ and the series converges.

23. If k is odd, say k = 2n-1, then $a_k = a^n$ and $a_{k+1} = b^n$ so

$a_{k+1}/a_k = (b/a)^n$ and, since b > a, $(b/a)>1$ and $\lim\limits_{k \to \infty} a_{k+1}/a_k = \infty$.

On the other hand, if k=2n is even, then $a_k = b^n$ and $a_{k+1} = a^{n+1}$

so $\lim\limits_{k \to \infty} a_{k+1}/a_k = \lim\limits_{k \to \infty} a(a^n/b^n) = 0$, so the ratio test fails.

Note that $\sum\limits_{k=1}^{\infty} a^k$ and $\sum\limits_{k=1}^{\infty} b^k$ are convergent geometric series

with $0<a<b<1$ and thus $\sum\limits_{k=1}^{\infty}(a^k + b^k) = \sum\limits_{k=1}^{\infty} a^k + \sum\limits_{k=1}^{\infty} b^k$ converges.

25. For $\sum\limits_{k=1}^{\infty} \dfrac{2^k k!}{k^k}$, $\dfrac{a_{k+1}}{a_k} = \dfrac{2^{k+1}(k+1)!}{(k+1)^{k+1}} \dfrac{k^k}{2^k k!} = \dfrac{2(k+1)k^k}{(k+1)^{k+1}} = 2\left(\dfrac{k}{k+1}\right)^k$ and

hence the series converges by the ratio test. Thus we know

that $\lim\limits_{k \to \infty} a_k = \lim\limits_{k \to \infty} \dfrac{2^k k!}{k^k} = 0$ by the kth term test.

27. For $\displaystyle\sum_{k=1}^{\infty}\frac{(2k)!}{(k!)^2 2^{2k}}$ we have $\displaystyle\frac{a_{k+1}}{a_k} = \frac{(2k+2)!}{[(k+1)!]^2 2^{k+1}}\frac{(k!)^2 2^k}{(2k)!} = \frac{(2k+2)(2k+1)}{2(k+1)^2}$

and thus $\displaystyle\lim_{k\to\infty} a_{k+1}/a_k = 2$. Thus for k sufficiently large, say

k>N, for some N, a_{k+1}/a_k is close to 2. In particular, for k>N,

$a_{k+1}/a_k > 1.5$ or $a_{k+1} > 1.5a_k$. Thus $a_{N+1} > 1.5a_N$, $a_{N+2} > (1.5)^2 a_N$,

and $a_{N+j} > (1.5)^j a_N$ and hence $\displaystyle\lim_{j\to\infty} a_{N+j} = \lim_{j\to\infty}(1.5)^j a_N = \infty$, so

$\displaystyle\lim_{k\to\infty}\frac{(2k)!}{(k!)^2 2^{2k}} = \infty$.

29a. Since $\displaystyle\lim_{k\to\infty} a_{k+1}/a_k$ does not exist (it is 1/2 for k odd, 1/3 for

k even) the ratio test, Thm.12.4.1, does not apply.

   b. Since for all $k \geq 1$, $a_{k+1}/a_k \leq 1/2 < 1$, then by Prob.28(a), the
series must converge.

Section 12.5, Page 619

1. Let $f(x) = 1/(1+x^2)$ so f is continuous, positive and monotone
decreasing and $f(k) = a_k = 1/(1+k^2)$ and Thm.12.5.1 applies.

Now $\displaystyle\int_1^{\infty} f(x)\,dx = \lim_{b\to\infty}\int_1^b dx/(1+x^2) = \lim_{b\to\infty} \arctan x\Big|_1^b$

$= \lim_{b\to\infty} (\arctan b - \arctan 1) = \pi/2 - \pi/4 = \pi/4$.

Thus, since $\displaystyle\int_1^{\infty} dx/(1+x^2)$ converges so does $\displaystyle\sum_{k=1}^{\infty} 1/(1+k^2)$.

3. For $a_k = k/(1+k^2)$, let $f(x) = x/(1+x^2)$. f(x) satisfies all
hypothesis of Thm.12.5.1 and

$\displaystyle\int_1^{\infty} f(x)\,dx = \lim_{b\to\infty}\int_1^b x\,dx/(1+x^2) = \lim_{b\to\infty}(1/2)[\ln(1+b^2)-\ln(2)] = \infty$.

Thus $\displaystyle\sum_{k=1}^{\infty} k/(1+k^2)$ diverges.

5. Let $f(x) = x^3/(1+x^4)$, which satisfies the hypothesis of Thm.12.5.1. Since

$$\int_1^\infty f(x)\,dx = \int_1^\infty x^3\,dx/(x^4+1) = \lim_{b\to\infty} \int_1^b x^3\,dx/(1+x^4)$$

$$= (1/4)\lim_{b\to\infty}[\ln(1+b^4) - \ln 2] = \infty, \text{ the series diverges.}$$

7. Let $f(x) = 1/[x \ln x \ln(\ln x)]$, which satisfies the hypothesis of Thm.12.5.1 for all $x \geq 4$. Thus

$$\int_4^\infty \frac{dx}{x \ln x \ln(\ln x)} = \lim_{b\to\infty} \int_4^b \frac{dx}{x \ln x \ln(\ln x)}$$

$$= \lim_{b\to\infty} \ln[\ln(\ln b)] - \ln[\ln(\ln 4)] = \infty, \text{ so the series diverges.}$$

9. Let $f(x) = (\arctan x)/(1+x^2)$. Since

$$\int_1^\infty f(x)\,dx = \lim_{b\to\infty} \int_1^b [(\arctan x)/(1+x^2)]\,dx$$

$$= \lim_{b\to\infty}[(\arctan^2 b - \arctan^2 1]/2 = 3\pi^2/32, \text{ the series converges.}$$

11a. For $f(x) = 1/x^4$ and $n = 6$, Eq.(20) yields

$$\int_6^\infty dx/x^4 - 1/6^4 \leq R_6 \leq \int_6^\infty dx/x^4. \quad \text{Now} \int_6^\infty dx/x^4 = \lim_{b\to\infty} \int_6^b dx/x^4 =$$

$$\lim_{b\to\infty}(-1/3)(1/b^3 - 1/6^3) \text{ so } \int_6^\infty dx/x^4 = 1/3 \cdot 6^3. \quad \text{Therefore}$$

$1/3 \cdot 6^3 - 1/6^4 \leq R_6 \leq 1/3 \cdot 6^3$ and $1/6^4 \leq R_6 \leq 2/6^4$ or
$0.0007716 \leq R_6 \leq 0.0015432$.

b. $s_6 = \sum_{k=1}^{6} 1/k^4 = 1 + 1/2^4 + 1/3^4 + 1/4^4 + 1/5^4 + 1/6^4 = 1.0811235$.

c. An approximation to R_6 may be obtained by computing the mean of the upper and lower bounds computed in part(a). Thus

$\bar{R}_6 = (1/2)(0.0007716 + 0.0015432) = 0.0011574$. This gives

$s \cong s_6 + \bar{R}_6 = 1.0822809$.

d. Using Eq.(22), the maximum error associated with s_n is $a_n/2$.

Thus we must determine n so that $a_n/2 = 1/2n^4 \leq 10^{-6}$ or

$n^4 \geq 10^6/2$ or $n \geq 27$.

13a. $\ln n = \int_1^n dx/x = \int_1^2 dx/x + \int_2^3 dx/x + \cdots + \int_{n-1}^n dx/x$

　　　　$< 1 + 1/2 + 1/3 + \cdots + 1/(n+1)$ by using the upper bound on each interval and since $1/x$ is decreasing on each interval. Adding $1/n$ to both sides we obtain

　　　　$1/n + \ln n < \sum_{k=1}^{n} 1/k = s_n$. Using the lower bound on each

interval we have $\ln n > 1/2 + 1/3 + \cdots + 1/n$, so adding 1 to both sides we have $s_n < 1 + \ln n$. Thus, putting the two results together we have $1/n + \ln n < s_n < 1 + \ln n$.

　b. $s_{10^4} < 1 + \ln 10^4 = 1 + 4\ln 10 = 10.22$; $s_{10^8} < 1 + 8\ln 10 = 19.43$;
　　　$s_{10^{12}} < 1 + 12\ln 10 = 28.64$.

15. 　Note that $\sigma_n = 1 + 1/2 + 1/3 + \cdots + 1/n - \ln n = s_n - \ln n$.
From the inequality of Prob.13(a) we have, by subtracting $\ln n$, $1/n < s_n - \ln n < 1$ or $0 < \sigma_n < 1$. To show $\{\sigma_n\}$ is monotone decreasing we must show $\sigma_n - \sigma_{n+1} > 0$. Now

$\sigma_{n+1} = s_{n+1} - \ln(n+1) = s_n + 1/(n+1) - \ln(n+1)$ so

$\sigma_n - \sigma_{n+1} = \ln(n+1) - \ln n - 1/(n+1) = \ln(1 + 1/n) - 1/(n+1)$.

Thus we need $\ln(1 + 1/n) > 1/(n+1)$ or $(n+1)\ln(1 + 1/n) > 1$. By

L'Hospital's rule, $\lim_{n \to \infty}(n+1)\ln(1 + 1/n) = \lim_{x \to \infty}(x+1)\ln(1 + 1/x) =$

$\lim_{x \to \infty}(x+1)^2/(x^2+x) = 1$. Further, since $(x+1)^2/(x^2+x) = (x+1)/x > 1$,

we find $(n+1)\ln(1 + 1/n) > 1$ for all n (refer to the proof of

L'Hospital's rule). Thus $\sigma_n - \sigma_{n+1} > 0$ and $\{\sigma_n\}$ is monotone

decreasing. Finally, $\sigma_1 = s_1 - \ln 1 = 1$,

$\sigma_5 = s_5 - \ln 5 = .6738954$, $\sigma_{10} = s_{10} - \ln 10 = .6263832$.

Section 12.6, Page 625

1. 　The series is alternating, with $a_k = 1/k^2$. $\{a_k\}$ is monotone
decreasing, $a_k > 0$ and $\lim_{k \to \infty} a_k = 0$. Thus by Thm.12.6.1 the
series converges.

3.   The series is alternating, with $a_k = (3/2)^k$. Since $\lim\limits_{k\to\infty} a_k = \infty$,
      Thm.12.6.1 does not apply, but by Thm.12.2.1 the series
      diverges.

5.   The series is alternating, with $a_k = 10^k/k!$ For $k > 10$,
      $a_{k+1} < a_k$, and for all k, $a_k > 0$. Further, $\lim\limits_{k\to\infty} a_k = 0$ (see Ex.5,
      Section 12.1). Thus by Thm.12.6.1, the series converges.

7.   The series is alternating with $a_k = k!/k^k$. Now, $a_k > 0$, and
      $a_{k+1} = (k+1)!/(k+1)^{k+1} = k!/(k+1)^k < k!/k^k = a_k$. Also
      $a_k = \dfrac{k(k-1)\cdots(2)(1)}{k\cdot k\cdots k\cdot k}$, where the numerator and the denominator
      each have k factors. Thus $a_k = 1(1-1/k)(1-2/k)\cdots(2/k)(1/k) < 1/k$
      and hence $\lim\limits_{k\to\infty} a_k = 0$. Thus the series converges.

9.   Since $\sin[(2k-1)\pi/2] = (-1)^{k+1}$, the series can be written as
      $\sum\limits_{k=1}^{\infty} 1/\sqrt{2k-1}$, which diverges by comparison with $\sum\limits_{k=1}^{\infty} 1/k$.

11.   By Eq.(14) we need $1/(n+1)^2 < 10^4$, or $n = 100$ terms if s_n is
      the approximation. Using Eq.(16) we need $1/2(n+1)^2 \le 10^{-4}$ or
      $(n+1)^2 > 10/2$ or $(n+1) > 100/\sqrt{2} = 70.71$. Thus $n = 70$ terms
      are needed if $\bar{s}$ is the approximation.

13.   For s_n, by Eq.(14), $1/(n+1)^{5/4} < 10^{-4}$ or $(n+1)^{5/4} > 10^4$ or
      $n+1 > 1584.89$. Thus 1584 terms are needed. For $\bar{s}$, by Eq.(16),
      $1/2(n+1)^{5/4} < 1/10^4$ or $(n+1)^{5/4} > 10^4/2$ or $n+1 > 910.28$.
      Thus 910 terms are needed.

15.   We want $1/\ln(n+1) < 10^{-4}$ or $\ln(n+1) > 10^4$. Since
      $\ln(n+1) = \log(n+1)/\log e$, we then must have $\log(n+1) > 10^4 \log e$,
      or $n+1 > 10^{10^4 \log e} \cong 10^{4342.9448} = 8.81 \times 10^{4342}$. Thus choose
      $n \cong 8.81 \times 10^{4342}$ for s_n. For $\bar{s}$ we need $1/2\ln(n+1) < 10^{-4}$ or
      $\log(n+1) > 10^4[(\log e)/2] = (2171.472)10^4$. So choose
      $n \cong 2.97 \times 10^{2171}$.

17a. For $|x| < 1$ we have $0 < x^2 < 1$ and thus there is a number r
such that $0 < x^2 \le r < 1$ so $-1 < -r \le -x^2 < 0$,

$0 < 1-r \le 1-x^2 < 1$, and $\dfrac{1}{1-x^2} \le \dfrac{1}{1-r}$. This gives, by Eq.(ii),

$|E_n(x)| \le \dfrac{r^n}{(2n+1)(1-r)}$ $(r<1)$ and $\lim\limits_{n\to\infty} |E_n(x)| = 0$, so $\lim\limits_{n\to\infty} E_n(x) = 0$.

b. To compute $\ln 2$, we solve $\dfrac{1+x}{1-x} = 2$, which gives $x = 1/3$.

c. For $E_n(1/3) \le 10^{-4}$, we must choose n so that

$[(1/3)^{2n+1}/(2n+1)(8/9)] \le 10^{-4}$ or $3^{2n-1}(2n+1)8 \ge 10^4$.
Letting $d_n = 3^{2n-1}(2n+1)8$, we have $d_1 = 72$, $d_2 = 1080$, and
$d_3 = 13608$.   Thus using only the first 3 terms gives $\ln 2$ with
error less than 10^{-4}.

19.   The series can be written as $1 + \displaystyle\sum_{k=1}^{\infty}\left(\dfrac{1}{3k-1} - \dfrac{1}{3k} + \dfrac{1}{3k+1}\right)$.

Letting $b_n = \dfrac{1}{3n-1} - \dfrac{1}{3n} + \dfrac{1}{3n+1} = \dfrac{1}{(3n-1)\,3n} + \dfrac{1}{3n+1}$ we see that

$b_n > \dfrac{1}{3n+1}$.   Thus, since $\displaystyle\sum_{k=1}^{\infty}\dfrac{1}{3k+1}$ diverges, the given series

diverges.

Section 12.7, Page 631

1. $\displaystyle\sum_{k=1}^{\infty}\dfrac{1}{k^2+4}$ converges by comparison with $\displaystyle\sum_{k=1}^{\infty}\dfrac{1}{k^2}$. Thus $\displaystyle\sum_{k=0}^{\infty}\dfrac{(-1)^k}{k^2+4}$
converges absolutely.

3. $\displaystyle\sum_{k=1}^{\infty}\dfrac{\sqrt{k}}{k+2}$ diverges by comparison with $\displaystyle\sum_{k=1}^{\infty}1/\sqrt{k}$ and thus

$\displaystyle\sum_{k=1}^{\infty}(-1)^k\sqrt{k}/(k+2)$ does not converge absolutely. However, it
does converge by the alternating series test and hence it
converges conditionally.

5. Since $\displaystyle\sum_{k=0}^{\infty} k^4/2^k$ converges by the ratio test, $\displaystyle\sum_{k=0}^{\infty}(-1)^k k^4/2^k$ converges absolutely.

7. Since $\displaystyle\sum_{k=1}^{\infty}\frac{(k!)^2}{(2k)!}$ converges (see Prob.7, Sect.12.4), $\displaystyle\sum_{k=1}^{\infty}\frac{(-1)^k(k!)^2}{(2k)!}$ converges absolutely.

9. From Prob.8, Sect.12.4, $\displaystyle\sum_{k=1}^{\infty}\frac{2^k k!}{k^k}$ converges, and since

$$\frac{k!}{k^k} \le \frac{2^k k!}{k^k}, \quad \sum_{k=1}^{\infty}\frac{k!}{k^k}$$ converges by the comparison test.

Therefore the given series converges absolutely.

11. Let $a_n = (-1)^{n+1}(\arctan n)/n^2$. Since $|\arctan n| \le \pi/2$,

$|a_n| \le \pi/2n^2$. Therefore, since $\displaystyle\sum_{k=1}^{\infty} 1/k^2$ converges, we have

that the given series converges absolutely.

13. From Sect.12.4, Prob.5, $\displaystyle\sum_{k=1}^{\infty}\frac{k!}{1\cdot 3\cdot 5\cdots(2k-1)}$ converges so the

given series converges absolutely.

15. $\displaystyle\sum_{k=1}^{\infty}\frac{(-1)^{k+1}(k+1)^2}{k^{1/3}(k^2+4)}$ converges by the alternating series test.

However, $\displaystyle\sum_{k=1}^{\infty}\frac{(k+1)^2}{k^{1/3}(k^2+4)}$ diverges by comparison with $\displaystyle\sum_{k=1}^{\infty} 1/k^{1/3}$.

Therefore the series converges conditionally.

17. Note that $|(\sin k)/k^2| \le 1/k^2$. Thus, since $\displaystyle\sum_{k=1}^{\infty} 1/k^2$ converges,

$\displaystyle\sum_{k=1}^{\infty}(\sin k)/k^2$ converges absolutely.

19. $a_{k+1}/a_k = \alpha k/(k+1)$ so $\lim\limits_{k\to\infty} |a_{k+1}/a_k| = |\alpha|$.   Thus, by the ratio

test, $\displaystyle\sum_{k=1}^{\infty} \frac{(-1)^k \alpha^k}{k}$ converges absolutely for $|\alpha| < 1$ and diverges

for $|\alpha| > 1$. At $\alpha = 1$ we have $\displaystyle\sum_{k=1}^{\infty} \frac{(-1)^k}{k}$, which converges

conditionally.   At $\alpha = -1$, we have $\displaystyle\sum_{k=1}^{\infty} \frac{1}{k}$, which diverges.

21. $a_k = \alpha^k/2^k = (\alpha/2)^k$, so the series converges absolutely for
$|\alpha/2| < 1$, or $|\alpha| < 2$, by the ratio test.   If $\alpha = 2$, we have

$\displaystyle\sum_{k=0}^{\alpha}(-1)^k$ and if $\alpha = -2$ we have $\displaystyle\sum_{k=0}^{\alpha}(1)$, which both diverge, and

thus the given series diverges for $|\alpha| \geq 2$.

23. $a_{k+1}/a_k = 2\alpha/k+1$ and $\lim\limits_{k\to\infty} \dfrac{2\alpha}{k+1} = 0$ imply that the series

converges absolutely for all α.

25a. The series (ii) diverges by comparison with $\displaystyle\sum_{k=1}^{\infty} 1/k$ and thus

has sum ∞, since all terms are positive.   The series (iii)

can be written as $-(1/2) \displaystyle\sum_{k=1}^{\infty} 1/k$ and thus has sum $-\infty$, again

since $\displaystyle\sum_{k=1}^{\infty} 1/k$ diverges to ∞.

b. We form an alternating series $\displaystyle\sum_{k=1}^{\infty} (-1)^k a_k$ as follows:

$a_1 = \displaystyle\sum_{k=1}^{n_1} p_k$ so that $\displaystyle\sum_{k=1}^{n_1} p_k > \sigma$ and $\displaystyle\sum_{k=1}^{n_1-1} p_k < \sigma.$

$$a_2 = \sum_{k=1}^{n_2} q_k \quad \text{so that} \quad a_1 - \sum_{k=1}^{n_2-1} q_k > \sigma \quad \text{and} \quad a_1 - \sum_{k=1}^{n_2} q_k < \sigma.$$

$$a_3 = \sum_{k=n_1+1}^{n_3} p_k \quad \text{so that} \quad a_1 - a_2 + \sum_{k=n_1+1}^{n_3-1} p_k < \sigma \quad \text{and} \quad a_1 - a_2 + \sum_{k=n_1+1}^{n_3} p_k > \sigma.$$

Continuing, $a_{2i+1} = \displaystyle\sum_{k=n_{2i-1}+1}^{n_{2i+1}} p_k$, so that $a_1 - a_2 + a_3 - \cdots + a_{2i+1} > \sigma$

and $a_1 - a_2 + a_3 + \cdots + \displaystyle\sum_{k=n_{2i-1}+1}^{n_{2i+1}-1} p_k < \sigma$, and $a_{2i} = \displaystyle\sum_{k=n_{2i-2}+1}^{n_{2i}} q_k$, so that

$a_1 - a_2 + a_3 + \cdots - a_{2i} < \sigma$ and $a_1 - a_2 + a_3 + \cdots - \displaystyle\sum_{k=n_{2i-2}+1}^{n_{2i}-1} q_k > \sigma$. The

series $\displaystyle\sum_{k=1}^{\infty} (-1)^{k+1} a_k$ is alternating since $a_k > 0$ for all k.

Defining $s_m = \displaystyle\sum_{k=1}^{m} (-1)^{k+1} a_k$, we have, from above, $s_{2j} < \sigma$,

$s_{2j+1} > \sigma$, $|\sigma - s_{2j}| < q_{n_{2j}}$, and $|\sigma - s_{2j+1}| < p_{n_{2j+1}}$. Since

$\lim_{k \to \infty} p_k = \lim_{k \to \infty} q_k = 0$, we have that, for given $\varepsilon > 0$ we can find

an N so that $|\sigma - s_m| < \varepsilon$ whenever $m > N$. This means that

$$\sum_{k=1}^{\infty} (-1)^{k+1} a_k = \sigma.$$

25c. If $\sigma < 0$, start by taking just enough terms from the series
(iii) to form a sum less than σ, then take just enough terms
from (ii) to form a sum greater than σ, and so on, as was
done in part (b).

  d. Taking all the postive terms of (ii) first, the series
diverges to $+\infty$. Taking all the terms of (iii) first, the
series diverges to $-\infty$.

25e. Assuming $\sigma_1 > \sigma_2$ and $\sigma_1 > 0$. Take just enough terms of series (ii) to form a sum greater than σ_1, then take just enough terms of (iii) to form a sum less than σ_2, then take just enough terms of (ii) to form a sum greater then σ_1, and so on. In a manner similar to that of part (b), we can show that $\lim_{m\to\infty} s_{2m+1} = \sigma_1$, while $\lim_{m\to\infty} s_{2m} = \sigma_2$.

Chapter 12 Review, Page 633

1. $\displaystyle\sum_{k=1}^{\infty} \frac{1}{k^2+9}$ converges by comparison with $\displaystyle\sum_{k=1}^{\infty} 1/k^2$.

3. $\displaystyle\sum_{k=1}^{\infty} \frac{\sqrt{k}}{k^2+k+1}$ converges by the limit comparison test, comparing with the series $\displaystyle\sum_{k=1}^{\infty} 1/k^{3/2}$.

5. $a_{k+1}/a_k = k^4/2(k+1)^3$ so $\lim_{k\to\infty} a_{k+1}/a_k = \infty$ and the series diverges by the ratio test.

7. $a_{k+1}/a_k = 4/(2k+1)(2k+2)$. Thus $\lim_{k\to\infty} a_{k+1}/a_k = 0$ and the given series converges by the ratio test.

9. $\lim_{k\to\infty} a_{k+1}/a_k = 3/2$ so the series diverges, by the ratio test.

11. $\lim_{k\to\infty} a_{k+1}/a_k = 2/3$, so the series converges by the ratio test.

13. $a_{k+1}/a_k = (1 + 1/k)^k/2$ so $\lim_{k\to\infty} a_{k+1}/a_k = e/2 > 1$. Therefore the series diverges by the ratio test.

15. $a_{k+1}/a_k = \dfrac{(3k+1)}{2(k+1)}$, thus $\lim_{k\to\infty} a_{k+1}/a_k = 3/2 > 1$ and the series diverges by the ratio test.

17. For $k > 4$, $2^k > k^2$ and $\lim\limits_{k\to\infty} a_k = \lim\limits_{k\to\infty} 2^k/k^2 \neq 0$. Thus the series diverges.

19. By comparison with $\sum\limits_{k=1}^{\infty} 1/k$, the series diverges by the limit comparison test.

21. k^2 is odd when k is odd and even when k is even so that $\sum\limits_{k=1}^{\infty} (-1)^{k^2}/\sqrt{k} = \sum\limits_{k=1}^{\infty} (-1)^k/\sqrt{k}$. The latter series converges conditionally, so $\sum\limits_{k=1}^{\infty} (-1)^{k^2}/\sqrt{k}$ converges conditionally.

23. Since $\dfrac{1}{\sqrt{k+1}-\sqrt{k}} = \dfrac{1}{\sqrt{k+1}-\sqrt{k}} \dfrac{\sqrt{k+1}+\sqrt{k}}{\sqrt{k+1}+\sqrt{k}} = \sqrt{k+1}+\sqrt{k}$,

    $\lim\limits_{k\to\infty} \dfrac{(-1)^{k+1}}{\sqrt{k+1}-\sqrt{k}} \neq 0$ and the series diverges.

25. $\sum\limits_{k=1}^{\infty} (-1)^{k+1} [\sqrt{k+1}-\sqrt{k}] = \sum\limits_{k=1}^{\infty} (-1)^{k+1}/(\sqrt{k+1}+\sqrt{k})$. Since

    $1/(\sqrt{k+1}+\sqrt{k}) < \dfrac{1}{2\sqrt{k}}$ and $\sum\limits_{k=1}^{\infty} (-1)^{k+1}/2\sqrt{k}$ converges conditionally (see Prob.21), the series converges conditionally.

27. $a_{k+1}/a_k = \dfrac{k(k+2)}{(2k+1)(2k+2)}$ so $\lim\limits_{k\to\infty} a_{k+1}/a_k = 1/4 < 1$ and the series converges absolutely by the ratio test.

29. The series can be written as

    $(1 + 1/4 + 1/9 + 1/25 + 1/36 + \cdot) + (1/2 + 1/4 + 1/8 + 1/16 + 1/32 + \cdot)$

    or $\sum\limits_{k=1}^{\infty} (1/k^2 + 1/2^k)$, which converges, since it's the sum of two converging series.

31. $\lim_{k\to\infty} |a_{k+1}/a_k| = \lim_{k\to\infty} 25/4\,(k+1)^2 = 0$.   Thus, by the tatio test, the given series converges absolutely.

33.   Note that $\dfrac{3\cdot 7\cdot 11\cdots(4k-1)}{2^k\,k!} = \left(\dfrac{3}{2}\right)\left(\dfrac{7}{4}\right)\left(\dfrac{11}{6}\right)\cdots\left(\dfrac{4k-1}{2k}\right) > 1$ for all $k > 1$.   Thus $\lim_{k\to\infty} a_k \neq 0$ and the given series diverges.

35. $\displaystyle\sum_{k=1}^{\infty} k^2/e^{k^3}$ converges by the ratio test since

$a_{k+1}/a_k = [(k+1)/k]^2 e^{k^3-(k+1)^3}$ so $\lim_{k\to\infty} a_{k+1}/a_k = 0$.

37.   With $a_k = (\tanh k)/k = (e^k-e^{-k})/k(e^k+e^{-k})$ and for $b_k = 1/k$, $\lim_{k\to\infty} a_k/b_k = \lim_{k\to\infty} \tanh k = 1$.   Thus, by the limit comparison

test, $\displaystyle\sum_{k=1}^{\infty}(\tanh k)/k$ diverges since $\displaystyle\sum_{k=1}^{\infty} 1/k$ diverges.

39.   Since $1/2^{\ln k} > 1/e^{\ln k} = 1/k$ and $\displaystyle\sum_{k=1}^{\infty} 1/k$ diverges, $\displaystyle\sum_{k=1}^{\infty} 1/2^{\ln k}$

diverges also by the comparison test.

41.   The statement is true, for suppose to the contrary that $\{b_n\}$ converged.   Then $\lim_{n\to\infty} a_n b_n = \lim_{n\to\infty} a_n \lim_{n\to\infty} b_n$. As both limits on the right exist, then $\{a_n b_n\}$ converges, contrary to the assumption.

43.   Let $s_n = a_1+a_2+a_3+\cdots+a_n$ converge to s.   That is, for each $\epsilon > 0$, there is an N such that for $n > N$, $|s-s_n| < \epsilon$.   Now suppose n is even, say $n = 2k$, so that $s_n = a_1+a_2+\cdots+a_{2k-1}+a_{2k}$. Let $b_j = a_{2j-1}+a_{2j}$.   Thus $s_n = b_1+b_2+\cdots+b_k = \sigma_k$.   Now, given $\epsilon > 0$ there is N such that for $2k = n > N$, $|s-s_n| = |s-\sigma_k| < \epsilon$. Thus $(a_1+a_2)+(a_3+a_4)+\cdots$ converges.   If $a_k > 0$ for each k then $a_1+a_2+a_3+\cdots$ converges absolutely, and thus any rearrangement will converge, and to the same sum.

45. False, for if $a_k = (-1)^k/k^{1/3}$, then $(-1)^k a_k^3 = 1/k$ and $\sum\limits_{k=1}^{\infty} 1/k$

diverges while $\sum\limits_{k=1}^{\infty} (-1)^k/k^{1/3}$ converges.

47. False, for suppose $a_k = (-1)^k/\sqrt{k}$, then $\sum\limits_{k=1}^{\infty} 1/k$ diverges while

$\sum\limits_{k=1}^{\infty} (-1)^k/\sqrt{k}$ converges. If $\sum\limits_{k=1}^{\infty} a_k$ converges absolutely, then

$\sum\limits_{k=1}^{\infty} |a_k|$ converges, and for all j sufficiently large, $|a_j| < 1$

so $a_j^2 < |a_j|$. Thus $\sum\limits_{k=1}^{\infty} a_k^2$ converges.

49. Let $f(x) = x - (n-1/n)$, $n - 1/n \le x \le n$
and $f(x) = -x + (n+1/n)$, $n \le x \le n+1/n$
and $f(x) = 0$ otherwise.   Thus

$f(n) = 1/n$, so $\sum\limits_{n=1}^{\infty} f(n)$ diverges,

while $\int_1^{\infty} f(x)\,dx = \sum\limits_{n=1}^{\infty} 1/n^2$ converges.

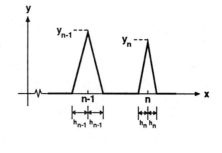

51. From the figure in Prob.49, let $y_n = 1$ and $h_n = 1/n^2$. Then the

area of each triangle is $1/n^2$ so $\int_1^{\infty} f(x)\,dx = \sum\limits_{n=1}^{\infty} 1/n^2$, which

converges and $f(x)$ does not approach zero as $x \to \infty$.

CHAPTER 13

<u>Section 13.1, Page 641</u>

1. $f' = \cos x$, $f'' = -\sin x$, $f''' = -\cos x$ so that Eq.(3) yields
    $a_0 = \sqrt{2}/2$, $a_1 = \sqrt{2}/2$, $a_2 = \dfrac{-\sqrt{2}/2}{2!}$, and $a_3 = \dfrac{-\sqrt{2}/2}{3!}$. Thus,
    $P_3(x) = (\sqrt{2}/2)[1 + (x-\pi/4) - (x-\pi/4)^2/2! - (x-\pi/4)^3/3!]$.

3. $f' = -2\sin 2x$, $f'' = -4\cos 2x$, $f''' = 8\sin 2x$, $f^{(iv)} = 16\cos 2x$ so
    that $a_0 = -1$, $a_1 = 0$, $a_2 = 4/2!$, $a_3 = 0$, $a_4 = -16/4!$ and
    $P_4(x) = -1 + 2(x-\pi/2)^2 - (2/3)(x-\pi/2)^4$.

5. $f' = -2xe^{-x^2}$, $f'' = -2e^{-x^2} + 4x^2e^{-x^2}$, $f''' = 12xe^{-x^2} - 8x^3e^{-x^2}$,
    $f^{(iv)} = 12e^{-x^2} - 48x^2e^{-x^2} + 16x^4e^{-x^2}$ so $a_0 = 1$, $a_1 = 0$,
    $a_2 = -2/2!$, $a_3 = 0$, $a_4 = 12/4!$, and $P_4(x) = 1 - x^2 + x^4/2$.

7. $f' = (1/2)(1+x)^{-1/2}$, $f'' = (-1/4)(1+x)^{-3/2}$,
    $f''' = (3/8)(1+x)^{-5/2}$ so that $a_0 = 1$, $a_1 = 1/2$, $a_2 = \dfrac{-1/4}{2!}$,
    $a_3 = \dfrac{3/8}{3!}$, and $P_3(x) = 1 + x/2 - x^2/8 + x^3/16$.

9. $f' = x^{-1/2}/2$, $f'' = -x^{-3/2}/4$, $f''' = 3x^{-5/2}/8$,
    $f^{(iv)} = -15x^{-7/2}/16$ so that $a_0 = 1$, $a_1 = 1/2$, $a_2 = \dfrac{-1/4}{2!}$,
    $a_3 = \dfrac{3/8}{3!}$, $a_4 = \dfrac{-15/16}{4!}$, and
    $P_4(x) = 1 + (x-1)/2 - (x-1)^2/8 + (x-1)^3/16 - 15(x-1)^4/128$.

11. $f' = (1-x^2)^{-1/2}$, $f'' = x(1-x^2)^{-3/2}$,
    $f''' = (1-x^2)^{-3/2} + 3x^2(1-x^2)^{-5/2}$ so that $a_0 = 0$, $a_1 = 1$,
    $a_2 = 0$, $a_3 = 1/3!$, and $P_3(x) = x + x^3/6$.

13. The first five derivatives of $\cos x$ are $-\sin x$, $-\cos x$, $\sin x$,
    $\cos x$, $-\sin x$ so that $a_0 = 1$, $a_1 = 0$, $a_2 = -1/2!$, $a_3 = 0$,
    $a_4 = 1/4!$, and $a_5 = 0$. From this pattern, we see that the
    odd coefficients are always zero and the even coefficients
    are given by $a_{2k} = (-1)^k/(2k)!$ for $k = 0,1,\ldots,n$. Thus,
    $P_{2n}(x) = 1 - x^2/2! + x^4/4! - \ldots + (-1)^n x^{2n}/(2n)!$.

15. $f' = -1/(x-a)^2$, $f'' = 2/(x-a)^3$, $f''' = -3!/(x-a)^4$,
$f^{(iv)} = 4!/(x-a)^5$, and in general, $f^{(k)} = (-1)^k k!/(x-a)^{k+1}$ so
$a_k = (-1)^k/(x-a)^{k+1}\Big|_{x=0} = -1/a^{k+1}$, $k = 0,1,\ldots,n$. Thus,
$P_n(x) = -(1/a)[1 + (x/a) + (x/a)^2 +\ldots+ (x/a)^n]$.

17. $f' = 1/(x-1)$, $f'' = -1/(x-1)^2$, $f''' = 2/(x-1)^3$, and in general,
$f^{(k)} = (-1)^{k-1}(k-1)!/(x-1)^k$ so $a_0 = 0$ and $a_k = -1/k$ for
$k = 1,2,\ldots,n$. Thus $P_n(x) = -(x + x^2/2 + x^3/3 +\ldots+ x^n/n)$.

19. $f' = (1/2)(1+x)^{-1/2}$, $f'' = (-1/4)(1+x)^{-3/2}$,
$f''' = (1\cdot3/8)(1+x)^{-5/2}$, $f^{(iv)} = (1\cdot3\cdot5/16)(1+x)^{-7/2}$, and in
general, $f^{(k)} = (-1)^{k+1}\dfrac{1\cdot3\cdot5\cdots(2k-3)}{2^k}(1+x)^{-(2k-1)/2}$ so $a_0 = 1$,
$a_1 = 1/2$, and $a_k = (-1)^{k+1}\dfrac{1\cdot3\cdot5\cdots(2k-3)}{k!2^k}$, $k = 2,3,\ldots,n$. Thus
$P_n(x) = 1+x/2 - x^2/8 + x^3/16 -\ldots+ [(-1)^n 1\cdot3\cdot5\cdots(2n-3)]x^n/2^n n!$.

21. $f' = -1/(x+1)^2$, $f'' = 2!/(x+1)^3$, $f''' = -3!/(x+1)^4$, and in
general, $f^{(k)} = (-1)^k k!/(x+1)^{k+1}$ so $a_k = (-1)^k/(x+1)^{k+1}$,
$k = 0,1,\ldots,n$. Thus,
$P_n(x) = 1/2 - (x-1)/4 + (x-1)^2/8 -\ldots+ (-1)^n(x-1)^n/2^{n+1}$.

23. The first several derivatives are $2\cos 2x$, $-4\sin 2x$, $-8\cos 2x$,
$16\sin 2x$, which have the values 0, -4, 0, 16. Hence, the odd
coefficients are zero and $a_{2k} = (-1)^k 2^{2k}/(2k)!$, $k = 0,1,\ldots,n$.
Thus,
$P_{2n}(x) = 1 - 2(x-\pi/4)^2 + 2(x-\pi/4)^4/3 -\ldots+ (-1)^n 2^{2n}(x-\pi/4)^{2n}(2n)!$.

25. The first several derivatives are $\cosh x$, $\sinh x$, $\cosh x$, $\sinh x$,
which have the values 1, 0, 1, 0. Hence, $a_0 = 0$, $a_1 = 1$,
$a_2 = 0$, $a_3 = 1/3!$, $a_4 = 0$, $a_5 = 1/5!$, and thus,
$P_{2n+1}(x) = x + x^3/3! + x^5/5! +\ldots+ x^{2n+1}/(2n+1)!$.

27. $f' = -2x/(1+x^2)^2$, $f'' = -2/(1+x^2)^2 + 8x^2/(1+x^2)^3$,
$f''' = 24/(1+x^2)^3 - 288x^2/(1+x^2)^4 + 384x^4/(1+x^2)^5$, so $a_0 = 1$,
$a_1 = 0$, $a_2 = -2/2!$, $a_3 = 0$, $a_4 = 24/4!$, and thus
$P_4(x) = 1 - x^2 + x^4$.

Section 13.2, Page 649

In Problems 1 through 19, the derivative is found directly from
the corresponding problem of Section 13.1.

1. $f^{(iv)}(x) = \sin x$ so $R_4(x) = (\sin c)(x-\pi/4)^4/4!$ from Eq.(9).

3. $f^{(v)}(x) = -32\sin 2x$, so $R_5(x) = (-32\sin 2c)(x-\pi/2)^5/5!$.

5. $f^{(v)}(x) = -120xe^{-x^2} + 160x^3e^{-x^2} - 32x^5e^{-x^2}$, so
      $R_5(x) = (-32c^5 + 160c^3 - 120c)e^{-c^2}x^5/5!$.

7. $f^{(iv)}(x) = (-15/16)(1+x)^{-7/2}$, so $R_4(x) = -15x^4/(16)4!(1+c)^{7/2}$.

9. $f^{(iv)}(x) = (9/16)(4-x)^{-5/2}$, so $R_4(x) = 9(4-c)^{-5/2}x^4/(16)(4!)$.

11. $f^{(2k)}(x) = (-1)^k\cos x$, so $f^{(2k+1)}(x) = (-1)^{k+1}\sin x$ and thus
      $R_{2n+1}(x) = (-1)^{n+1}(\sin c)x^{2n+1}/(2n+1)!$.

13. $f^{(n+1)}(x) = (n+1)!a^{n+1}(1-ax)^{-(n+2)}$, so
      $R_{n+1}(x) = a^{n+1}(1-ac)^{-(n+2)}x^{n+1}$.

15. $f^{(n+1)}(x) = (-2)^{n+1}e^{-2x}$, $R_{n+1}(x) = (-2)^{n+1}e^{-2c}(x+1)^{n+1}/(n+1)!$.

17. $f^{(2k)}(x) = (-1)^k2^{2k}\sin 2x$, so $f^{(2k+1)}(x) = (-1)^k2^{2k+1}\cos 2x$ and
      thus $R_{2n+1}(x) = (-1)^n2^{2n+1}(\cos 2c)(x-\pi/4)^{2n+1}/(2n+1)!$

19. $f^{(2k+1)}(x) = \cosh x$, so $f^{(2k+2)}(x) = \sinh x$ and thus
      $R_{2n+2}(x) = (\sinh c)x^{2n+2}/(2n+2)!$.

21a. $f'(x) = (7/3)x^{4/3}$, $f''(x) = (28/9)x^{1/3}$, and
      $f'''(x) = (28/27)x^{-2/3}$, so $P_1(x) = 0$ with
      $R_2(x) = (28/9)c^{1/3}x^2/2! = 14c^{1/3}x^2/9$, and $P_2(x) = 0$ with
      $R_3(x) = (28/27)c^{-2/3}x^3/3! = 14c^{-2/3}x^3/81$.
   b. No, since $f'''(0)$ does not exist.
   c. Evaluating the derivatives found in part(a) at $x = 1$, we find
      $P_3(x) = 1 + 7(x-1)/3 + 14(x-1)^2/9 + 14(x-1)^3/81$.

23.   Since $(\pi/4 - .1)$ is near $\pi/4$, we expand cosx in a Taylor polynomial about $\pi/4$, in which case the remainder after n terms is $R_{n+1}(x) = f^{(n+1)}(c)(x-\pi/4)^{n+1}/(n+1)!$.   Thus,

$|R_{n+1}(x)| = |f^{(n+1)}(c)(0.1)^{n+1}/(n+1)!| \leq (0.1)^{n+1}/(n+1)!$

since $f^{(n+1)}(x)$ is either $\pm$cosx or $\pm$sinx so $|f^{(n+1)}(c)| \leq 1$. Hence, $|R_{n+1}(x)| \leq 10^{-4}$ if n = 3   and therefore

$\cos x \cong (\sqrt{2}/2)[1 - (x-\pi/4) - (x-\pi/4)^2/2 + (x-\pi/4)^3/6] = .7742$

when $x = \pi/4 - .1$.

25.   Following the pattern of Prob.23, we expand ln x about $x_o = 1$ to find $|R_{n+1}(1.2)| = \left|\dfrac{(-1)^n n!(.2)^{n+1}}{c^{n+1}(n+1)!}\right| = \left|\dfrac{(-1)^n(.2)^{n+1}}{c^{n+1}(n+1)}\right| \leq \dfrac{(.2)^{n+1}}{n+1}$.

Thus we must choose n = 4   and hence
$\ln x \cong (x-1) - (x-1)^2/2 + (x-1)^3/3 - (x-1)^4/4 = .1823$ when x = 1.2.

27. $f^{(2k)}(x) = (-1)^k \cos x$ and $f^{(2k+1)}(x) = (-1)^{k+1}\sin x$ so that

$$\cos x = \sum_{k=0}^{\infty} \frac{(-1)^k x^{2k}}{(2k)!}, \text{ provided}$$

$R_{2k+1}(x) = (-1)^{k+1}(\sin c)x^{2k+1}/(2k+1)! \to 0$ as $k \to \infty$.   To prove

this latter point, consider $\displaystyle\sum_{k=0}^{\infty} \frac{x^{2k+1}}{(2k+1)!}$, which

converges by the ratio test for all x and thus the kth term $(x^{2k+1}/(2k+1)!) \to 0$ as $k \to \infty$.   Hence, $R_{2k+1}(x) \to 0$ for all x and the above series is a Taylor series for cosx for all x.

29. $f^{(2k)}(x) = \sinh x$, $f^{(2k+1)}(x) = \cosh x$ and thus

$$\sinh x = \sum_{k=0}^{\infty} \frac{x^{2k+1}}{(2k+1)!} \text{ provided } R_{2k+2}(x) = (\sinh c)x^{2k+2}/(2k+2)! \to 0$$

as $k \to \infty$.   As in Prob.27 consider $\displaystyle\sum_{k=0}^{\infty} \frac{x^{2k+2}}{(2k+2)!}$, which converges

for all x by the ratio test, and thus $R_{2k+2}(x) \to 0$ as $R \to \infty$, and the above series is a Taylor series for all x.

31. $f^{(2k)}(x) = (-1)^k 2^{2k} \cos 2x$, $f^{(2k+1)}(x) = (-1)^{k+1} 2^{2k+1} \sin 2x$ and

thus $\cos 2x = \displaystyle\sum_{k=0}^{\infty} \frac{(-1)^k 2^{2k} x^{2k}}{(2k)!}$ provided

$R_{2k+1}(x) = (-1)^{k+1} 2^{2k+1} (\sin 2c) x^{2k+1} / (2k)! \to 0$ as $k \to \infty$. Since

$|R_{2k+1}(x)| \le \dfrac{2^{2k+1} x^{2k+1}}{(2k)!} = \dfrac{(2x)^{2k+1}}{(2k)!}$, we know $R_{2k+1}(x) \to 0$ as in

Prob.27 and thus the above series for $\cos 2x$ is a Taylor
series for all x.

33. $f^{(2k)}(x) = (-1)^k \cos x$, $f^{(2k+1)}(x) = (-1)^{k+1} \sin x$ and thus

$\cos x = \displaystyle\sum_{k=0}^{\infty} (-1)^{k+1} (x - \pi/2)^{2k+1} / (2k+1)!$ provided

$R_{2k+1}(x) = (-1)^{k+1} (\cos c)(x - \pi/2)^{2k+2} / (2k+2)! \to 0$ as $k \to \infty$.

Since $|R_{2k+2}(x)| \le \dfrac{|x - \pi/2|^{2k+2}}{(2k+2)!}$, consider $\displaystyle\sum_{k=0}^{\infty} \frac{(x-\pi/2)^{2k+2}}{(2k+2)!}$, which

converges for all x by the ratio test. Thus, $R_{2k+2}(x) \to 0$ as
$k \to \infty$, and the above Taylor series converges to $\cos x$ for all x.

35. $f^{(k)}(x) = 3^k e^{3k}$ so $e^{3x} = \displaystyle\sum_{k=0}^{\infty} \frac{3^k x^k}{k!} = \sum_{k=0}^{\infty} \frac{(3x)^k}{k!}$ provided

$R_{k+1}(x) = 3^{k+1} e^{3c} x^{k+1} / (k+1)! \to 0$ as $k \to \infty$. Since $\displaystyle\sum_{k=0}^{\infty} \frac{(3x)^{k+1}}{(k+1)!}$

converges for all x by the ratio test, we conclude that
$R_{k+1}(x) \to 0$ as $k \to \infty$ and thus the above Taylor series converges
to e^{3x} for all x.

37a. $f'(0) = \displaystyle\lim_{h \to 0} \frac{f(h) - f(0)}{h} = \lim_{h \to 0} \frac{e^{-1/h^2}}{h} = \lim_{h \to 0} \frac{1/h}{e^{1/h^2}} = \lim_{h \to 0} \frac{-1/h^2}{-2e^{1/h^2}/h^3}$

$= \displaystyle\lim_{h \to 0} \frac{h}{e^{1/h^2}} = 0.$

b. $f'(x) = 2e^{-1/x^2} / x^3$.

37c. $f''(0) = \lim_{h\to 0}\dfrac{f'(h)-f'(0)}{h} = \lim_{h\to 0}\dfrac{2e^{-1/h^2}/h^3-0}{h} = \lim_{h\to 0}\dfrac{2e^{-1/h^2}}{h^4} = \lim_{h\to 0}\dfrac{2/h^4}{e^{1/h^2}}$

$\qquad = \lim_{h\to 0}\dfrac{4/h^2}{e^{1/h^2}} = \lim_{h\to 0}\dfrac{4}{e^{1/h^2}} = 0.$

d. Since $f^{(n)}(0) = 0$ for all integer values of n, we have
$P_n(x) = 0$ and $R_{n+1}(x) = f^{(n+1)}(c)x^{n+1}/(n+1)!$ for all n. Thus
$\lim_{n\to\infty} P_n(x) = 0$ for all x and hence $\lim_{n\to\infty} P_n(x) \neq f(x)$ unless
$x = 0$ since $f(x) > 0$ for $x \neq 0$.

39. Assuming $y = \phi(t)$ has suffcient differentiability, we have
$\phi(t) = \phi(t_n) + \phi'(t_n)(t-t_n) + R_2(t)$. Letting
$t = t_{n+1}$, then $t - t_n = t_{n+1} - t_n = h$ and
$\phi(t_{n+1}) = \phi(t_n) + \phi'(t_n)h + R_2(t_{n+1})$. By the Corollary to

Theorem 13.2.1, $R_2(t_n+1) = \dfrac{\phi''(c)h^2}{2!}$, where $t_n < c < t_n+h$.

Setting $c = \bar{t}_n$, we have $\phi(t_{n+1}) = \phi(t_n) + \phi'(t_n)h + \phi''(\bar{t}_n)h^2/2!$,

where $t_n < \bar{t}_n < t_n + h$. Further, if

$y_{n+1} = y_n + hf(t_n,y_n) = \phi(t_n) + h\phi'(t_n)$, then

$\phi(t_{n+1}) - y_{n+1} = [\phi(t_n) + \phi'(t_n)h + \phi''(\bar{t}_n)h^2/2] - [\phi(t_n) + h\phi'(t_n)]$

or $\phi(t_{n+1}) - y_{n+1} = \phi''(\bar{t}_n)h^2/2.$

Section 13.3, Page 656

1. Applying the ratio test we have
$\lim_{k\to\infty}\left|\dfrac{x^{k+1}/(k+1)^2}{x^k/k^2}\right| = \lim_{k\to\infty}\left(\dfrac{k}{k+1}\right)^2|x| = |x|$ and thus $\rho = 1$. Setting

$x = 1$, we have $\displaystyle\sum_{k=1}^{\infty} 1/k^2$, and setting $x = -1$, we have

$\displaystyle\sum_{k=1}^{\infty}(-1)^k/k^2$ which both converge, and hence, the series
converges absolutely for $-1 \leq x \leq 1$ and diverges otherwise.
ρ could also be found using Eq.(10).

3. Using Eq.(10), we have $\rho = \lim\limits_{k\to\infty}\left|\dfrac{1/2^k}{1/2^{k+1}}\right| = 2$ and thus we know the series converges absolutely for $-2 < x+3 < 2$ or

$-5 < x < -1$. For $x = -1$, we have $\sum\limits_{k=0}^{\infty} 1^k$, and for $x = -5$, we

have $\sum\limits_{k=0}^{\infty} (-1)^k$ which both diverge and thus the given series

diverges for $x \geq -1$ and for $x \leq -5$.

5. $\rho = \lim\limits_{k\to\infty}\left|\dfrac{1/\sqrt{k}\,2^{2k}}{1/\sqrt{k+1}\,2^{2k+2}}\right| = \lim\limits_{k\to\infty}\sqrt{\dfrac{k+1}{k}}\,2^2 = 4$. For $x-2 = 4 = 2^2$,

we have $\sum\limits_{k=1}^{\infty} (-1)^{k+1}/\sqrt{k}$, which converges, and for

$x-2 = -4 = -2^2$, we have $\sum\limits_{k=1}^{\infty} -1/\sqrt{k}$, which diverges. Thus the

series converges absolutely for $-2 < x < 6$, conditionally for $x = 6$, and diverges otherwise.

7. We have $\lim\limits_{k\to\infty}\left|\dfrac{x^{2k+3}/(2k+3)!}{x^{2k+1}/(2k+1)!}\right| = \lim\limits_{k\to\infty}\dfrac{x^2}{(2k+3)(2k+2)} = 0$, so $\rho = \infty$ and the series converges absolutely for all x.

9. The ratio test yields $\lim\limits_{k\to\infty}\left|\dfrac{x^{2k+2}/(k+2)^2}{x^{2k}/(k+1)^2}\right| = \lim\limits_{k\to\infty}\left(\dfrac{k+1}{k+2}\right)^2 x^2 = x^2$ and

thus $\rho = 1$. For $x = \pm 1$, we have $\sum\limits_{k=0}^{\infty} 1/(k+1)^2$, and hence the

series converges absolutely for $-1 \leq x \leq 1$ and diverges otherwise.

11. The ratio test yields $\lim\limits_{k\to\infty}\left|\dfrac{x^{2k+2}/(k+1)!}{x^{2k}/k!}\right| = \lim\limits_{k\to\infty} x^2/(k+1) = 0$.

Hence $\rho = \infty$ and the series converges absolutely for all x.

13. Since $\lim\limits_{k \to \infty}\left|\dfrac{(2x-1)^{k+1}/(k+1)}{(2x-1)^k/k}\right| = \lim\limits_{k \to \infty}\left(\dfrac{k}{k+1}\right)|2x-1| = |2x-1|$ the series

converges absolutely for $-1 < 2x-1 < 1$ or $0 < x < 1$, with

$\rho = 1/2$. For $x = 0$ we have $\sum\limits_{k=1}^{\infty}(-1)^k/k$, which converges

conditionally, and for $x = 1$ we have $\sum\limits_{k=1}^{\infty}1/k$, which diverges

and thus the given series diverges for $x \geq 1$ and for $x < 0$.

15. Since $\lim\limits_{k \to \infty}\left|\dfrac{2^{k+1}(k+1)!\,x^{2k+3}/(2k+3)!}{2^k k!\,x^{2k+1}(2k+1)!}\right| = \lim\limits_{k \to \infty}\dfrac{2(k+1)x^2}{(2k+3)(2k+1)} = 0,\ \rho = \infty$

and the series converges absolutely for all x.

17. Since $\lim\limits_{k \to \infty}\left|\dfrac{x^{2k+3}/(2k+3)(k+1)!}{x^{2k+1}/(2k+1)k!}\right| = \lim\limits_{k \to \infty}\dfrac{(2k+1)x^2}{(2k+3)(k+1)} = 0$, we have

$\rho = \infty$ and the series converge absolutely for all x.

19. Since $\lim\limits_{k \to \infty}\left|\dfrac{(k+1)^2(3x+2)^{k+1}/2^{k+1}}{k^2(3x+2)^k/2^k}\right| = \lim\limits_{k \to \infty}\dfrac{1}{2}\left(\dfrac{k+1}{k}\right)^2|3x+2| = \dfrac{1}{2}|3x+2|$,

the series converges absolutely for $-1 < (3x+2)/2 < 1$ or

$-4/3 < x < 0$, with $\rho = 2/3$. For $x = 0$, we have $\sum\limits_{k=1}^{\infty}k^2$ and for

$x = -4/3$, we have $\sum\limits_{k=1}^{\infty}(-1)^k k^2$ which both diverge and thus the

given series diverges for $x \geq 0$ and for $x \leq -4/3$.

21a. From Eq.(10), $\rho = \lim\limits_{k \to \infty}\dfrac{k!/k^k}{(k+1)!/(k+1)^{k+1}} = \lim\limits_{k \to \infty}\dfrac{(k+1)^{k+1}}{(k+1)k^k} = \lim\limits_{k \to \infty}\left(\dfrac{k+1}{k}\right)^k = e.$

b. For $x = e$ we have $\sum\limits_{k=1}^{\infty}k!\,(e/k)^k$. From Stirling's formula for

large k, we have $k! \cong \sqrt{2\pi k}\,(k/e)^k$ or $k!\,(e/k)^k \cong \sqrt{2\pi k}$. Thus

the kth term does not go to zero and the series will diverge.
(The limit comparison test could also be used with the series

$\sum\limits_{k=1}^{\infty}\sqrt{2\pi k}$). Likewise at $x = -e$, $\sum\limits_{k=1}^{\infty}(-1)^k k!\,(e/k)^k$ also diverges.

Section 13.4, Page 664

1.   From Prob.33 of Section 13.2, we have $\cos t = \sum\limits_{k=0}^{\infty} \dfrac{(-1)^k t^{2k}}{(2k)!}$,

$-\infty < t < \infty$.   Since $\cos x = -\cos(x-\pi)$, we set $t = x-\pi$ in the
above series and multiply by -1 to obtain

$\cos x = \sum\limits_{k=0}^{\infty} \dfrac{(-1)^{k+1}(x-\pi)^{2k}}{(2k)!}$, $-\infty < x-\pi < \infty$ or $-\infty < x < \infty$.

3.   Since $\cos x = -\sin(x-\pi/2)$, we set $t = x-\pi/2$ in Eq.(3) and
multiply that series by -1 to obtain

$\cos x = \sum\limits_{k=0}^{\infty} \dfrac{(-1)^{k+1}(x-\pi/2)^{2k+1}}{(2k+1)!}$, $-\infty < x-\pi/2 < \infty$ or $-\infty < x < \infty$.

5. $f(x) = \dfrac{1}{1-(-2x)}$ and thus we set $t = -2x$ in Eq.(10) to obtain

$\dfrac{1}{1+2x} = \sum\limits_{k=0}^{\infty} (-2x)^k = \sum\limits_{k=0}^{\infty} (-1)^k 2^k x^k$, $|-2x| < 1$ or $-1/2 < x < 1/2$.

7.   By partial fractions we have

$\dfrac{3x+4}{(1+2x)(2-x)} = \dfrac{1}{1+2x} + \dfrac{2}{2-x} = \dfrac{1}{1-(-2x)} + \dfrac{1}{1-x/2} = \sum\limits_{k=0}^{\infty} (-2x)^k + \sum\limits_{k=0}^{\infty} (x/2)^k$

$= \sum\limits_{k=0}^{\infty} [(-1)^k 2^k + 2^{-k}] x^k$.   The first series has

$\rho_1 = 1/2$ and the second series has $\rho_2 = 2$, and thus the sum
has $\rho = 1/2$ (at least), so the series converges at least for
$-1/2 < x < 1/2$.

9.   By partial fractions we have

$\dfrac{x-4}{(x-1)(x-2)} = \dfrac{3}{x-1} - \dfrac{2}{x-2} = \dfrac{1}{1-x/2} - 3\dfrac{1}{1-x} = \sum\limits_{k=0}^{\infty} (x/2)^k - 3\sum\limits_{k=0}^{\infty} x^k$

$= \sum\limits_{k=0}^{\infty} (2^{-k} - 3) x^k$.   Again, $\rho_1 = 2$ and $\rho_2 = 1$, so the
sum converges at least for $-1 < x < 1$.

11. Setting $t = x/2$ in the cost series of Prob.1 we obtain

$$\cos(x/2) = \sum_{k=0}^{\infty} \frac{(-1)^k (x/2)^{2k}}{(2k)!} = \sum_{k=0}^{\infty} \frac{(-1)^k x^{2k}}{2^{2k}(2k)!}, \quad -\infty < x/2 < \infty, \text{ or all } x.$$

13. $\cos^2 x - \sin^2 x = \cos 2x = \sum_{k=0}^{\infty} \frac{(-1)^k (2x)^{2k}}{(2k)!} = \sum_{k=0}^{\infty} \frac{(-1)^k 2^{2k} x^{2k}}{(2k)!}$, for all x.

15. Setting $t = -x^2$ in Eq.(18), we obtain

$$e^{-x^2/2} = \sum_{k=0}^{\infty} \frac{(-x^2)^k}{k!} = \sum_{k=0}^{\infty} \frac{(-1)^k x^{2k}}{k!}, \quad -\infty < -x^2 \le 0 \text{ or } -\infty < x < \infty.$$

17. Since $\cosh t = (e^t + e^{-t})/2$, we have

$$\cosh t = (1/2)[\sum_{k=0}^{\infty} t^k/k! + \sum_{k=0}^{\infty} (-t)^k/k!]$$

$$= (1/2)[\sum_{k=0}^{\infty} t^k/k! + \sum_{k=0}^{\infty} (-1)^k t^k/k!] = \sum_{k=0}^{\infty} t^{2k}/(2k)! \text{ for all } t.$$

Thus, we set $t = 3x$ to obtain

$$\cosh 3x = \sum_{k=0}^{\infty} \frac{(3x)^{2k}}{(2k)!} = \sum_{k=0}^{\infty} \frac{3^{2k} x^{2k}}{(2k)!}, \quad \text{for } -\infty < 3x < \infty \text{ or } -\infty < x < \infty.$$

The series for $\cosh t$ was also found in Prob.34, Section 13.2.

19. $\dfrac{1}{2-x} = \dfrac{1}{-1-(x-3)} = \dfrac{-1}{1-[-(x-3)]} = -\sum_{k=0}^{\infty} [-(x-3)]^k = \sum_{k=0}^{\infty} (-1)^{k+1}(x-3)^k$,

for $-1 < -(x-3) < 1$ or $2 < x < 4$.

21. Eq.(3), $\Rightarrow \sin^2 x = (x - x^3/3! + x^5/5! - x^7/7! + ...)^2 =$
$x^2 - (2/3!)x^4 + [2/5!+1/(3!)^2]x^6 - [2/7!+2/(3!)(5!)]x^8 + ...$,
using polynomial multiplication.

23. We have $e^{-2x} = \sum_{k=0}^{\infty} \dfrac{(-1)^k 2^k x^k}{k!} = 1 - 2x + \dfrac{4x^2}{2!} - \dfrac{8x^3}{3!} + ...$ and

$$\sin 3x = \sum_{k=0}^{\infty} \frac{(-1)^k 3^{2k+1} x^{2k+1}}{(2k+1)!} = 3x - \frac{27x^3}{3!} + ...\ .$$

Thus, $f(x) = (1 - 2x + 2x^2 - 4x^3/3 + ...)(3x - 9x^3/2 + ...)$
$= 3x - 6x^2 + (6-9/2)x^3 + (9-4)x^4 + ...$
$= 3x - 6x^2 + 3x^3/2 + 5x^4 + ...\ .$

25. $[f(x)]^2 = (a_0+a_1x+a_2x^2+a_3x^3+a_4x^4+\ldots)^2 =$
       $a_0^2+2a_0a_1x+(2a_0a_2+a_1^2)x^2+(2a_0a_3+2a_1a_2)x^3+(2a_0a_4+2a_1a_3+a_2^2)x^4+.$

27a. $f(x) = \dfrac{1}{1-x} = \displaystyle\sum_{k=0}^{\infty} x^k = 1+x+x^2+x^3+\ldots, |x| < 1,$ so

       $g(x) = x^2f(x) = \displaystyle\sum_{k=0}^{\infty} x^{k+2} = x^2+x^3+x^4+\ldots, |x| < 1.$

   b. From Part (a), $f(x) - g(x) = 1+x$, which converges for all x
       since it is a polynomial of degree 1.

Section 13.5, Page 670

1.   Since $\dfrac{d}{dx}\ln(1-x) = -\dfrac{1}{1-x}$, we have $\ln(1-x) = -\displaystyle\int_0^x \dfrac{dt}{1-t} = -\displaystyle\int_0^x \sum_{k=0}^{\infty} t^k dt$

      $= -\displaystyle\sum_{k=0}^{\infty} \int_0^x t^k dt = -\sum_{k=0}^{\infty} \dfrac{x^{k+1}}{k+1}$, at least for

      $-1 < x < 1$. For $x = -1$, the series converges by the
      alternating series test, while for $x = 1$, the series diverges.
      Thus the series converges to $\ln(1-x)$ for $-1 \le x < 1$.

3. $\ln\left(\dfrac{1+x}{1-x}\right) = \ln(1+x) - \ln(1-x)$ and thus

      $\left[\ln\left(\dfrac{1+x}{1-x}\right)\right]' = \dfrac{1}{1+x} + \dfrac{1}{1-x} = \dfrac{2}{1-x^2} = 2\displaystyle\sum_{k=0}^{\infty} x^{2k}$. Hence,

      $\ln\left(\dfrac{1+x}{1-x}\right) = 2\displaystyle\sum_{k=0}^{\infty} \dfrac{x^{2k+1}}{2k+1}; -1 < x < 1.$ The series diverges for $x = \pm1.$

5. $f(x) = \displaystyle\int_0^x e^{-t^2} dt = \int_0^x \sum_{k=0}^{\infty} \dfrac{(-1)^k t^{2k}}{k!} dt = \sum_{k=0}^{\infty} \dfrac{(-1)^k x^{2k+1}}{k!(2k+1)}$, $-\infty < x < \infty.$

7.   Note that f(x) is proportional to the second derivative of

$$1/(1+2x). \quad \text{Thus, if } g(x) = \frac{1}{1+2x} = \sum_{k=0}^{\infty} (-1)^k 2^k x^k, \text{ then}$$

$$g'(x) = \frac{-2}{(1+2x)^2} = \sum_{k=1}^{\infty} (-1)^k 2^k k x^{k-1} \text{ and}$$

$$g''(x) = \frac{8}{(1+2x)^3} = \sum_{k=2}^{\infty} (-1)^k 2^k k(k-1) x^{k-2}. \quad \text{Hence,}$$

$$f(x) = (1/8) \sum_{n=2}^{\infty} (-1)^n 2^n n(n-1) x^{n-2} = \sum_{k=0}^{\infty} (-1)^k 2^{k-1} (k+2)(k+1) x^k,$$

where we have set n-2 = k. Since the series for g(x)
converges for -1/2 < x < 1/2, so will the series for f(x).

9.   From Eq. (13), we have $\arctan(x/2) = \sum_{k=0}^{\infty} \frac{(-1)^k x^{2k+1}}{2^{2k+1}(2k+1)}$ at least for

-2 < x < 2. Setting x = ±2, we obtain a convergent
alternating series and thus the series converges for |x| ≤ 2.

11. Setting x = t^2 in Eq. (9) yields

$$f(x) = \int_0^x \ln(1+t^2)\, dt = \int_0^x \sum_{k=0}^{\infty} \frac{(-1)^k t^{2k+2}}{k+1}\, dt = \sum_{k=0}^{\infty} \frac{(-1)^k x^{2k+3}}{(k+1)(2k+3)} \text{ at least}$$

for -1 < x < 1. Setting x = ±1 yields a convergent
alternating series and thus the series converges for |x| ≤ 1.

13. Following the procedure of Ex.4, we have

$$\cos x = \sum_{k=0}^{\infty} \frac{(-1)^k x^{2k}}{(2k)!} = 1 - x^2/2! + x^4/4! - \ldots \text{ and thus for}$$

$$x \neq 0, (1-\cos x)/x^2 = 1/2! - x^2/4! + x^4/6! - \ldots = \sum_{k=0}^{\infty} \frac{(-1)^k x^{2k}}{(2k+2)!}.$$

Note that the series does converge for x = 0 and has value of
1/2 when x = 0 and thus it represents f(x) for all x.

15. From Ex.2, we have $\ln(1+x) = \sum_{k=0}^{\infty} \frac{(-1)^k t^{k+1}}{k+1}$ so $\frac{\ln(1+t)}{t} = \sum_{k=0}^{\infty} \frac{(-1)^k t^k}{k+1}$

and $f(x) = \sum_{k=0}^{\infty} \frac{(-1)^k t^{k+1}}{(k+1)^2} \Big|_0^{x/2} = \sum_{k=0}^{\infty} \frac{(-1)^k x^{k+1}}{2^{k+1}(k+1)^2}$ for -2 ≤ x ≤ 2, since

the series does converge at both end points.

17a. $\pi \cong 4(1 - 1/3 + 1/5 - 1/7 + 1/9 - 1/11 + 1/13 - 1/15) = 3.017$.

b. The error in an alternating series is no more than the next term and thus we want $4/(2k+1) \le .01$ or $k \ge 200$. If the average of two succeeding partial sums is used, then we want $2/(2k+1) \le .01$ or $k \ge 100$.

19. From Prob.13, we have

$$I = \int_0^1 \sum_{k=0}^\infty \frac{(-1)^k x^{2k}}{(2k+2)!} = \sum_{k=0}^\infty \frac{(-1)^k x^{2k+1}}{(2k+2)!(2k+1)}\Big|_0^1 = \sum_{k=0}^\infty \frac{(-1)^k}{(2k+2)!(2k+1)}$$

$$= \frac{1}{2!} - \frac{1/3}{4!} + \frac{1/5}{6!} - \frac{1/7}{8!} + \frac{1/9}{10!} - \cdots .$$

Note that $\frac{1/9}{10!} > 10^{-8}$ while $\frac{1/11}{12!} < 10^{-8}$, so $I \cong .48638538$ using five terms.

21. Setting $x = 1$ in the solution to Prob.5, we get

$$I = \sum_{k=0}^\infty \frac{(-1)^k}{k!(2k+1)} \cong 1 - \frac{1}{3} + \frac{1/5}{2!} - \frac{1/7}{3!} + \frac{1/9}{4!} - \frac{1/11}{5!} + \frac{1/13}{6!} - \frac{1/15}{7!} = .74682.$$

Since $\frac{1/15}{7!} > 10^{-5}$ while $\frac{1/17}{8!} < 10^{-5}$, eight terms are needed to ensure $|E| \le 10^{-5}$.

23. Setting $x = 1$ in the solution to Prob.15, we get

$$I = \sum_{k=0}^\infty \frac{(-1)^k}{2^{k+1}(k+1)^2}$$

$$\cong 1/2 - 1/2^2 2^2 + 1/2^3 3^2 - 1/2^4 4^2 + 1/2^5 5^2 - 1/2^6 6^2 + 1/2^7 7^2$$

$$\cong .4485.$$

Seven terms are needed since $1/2^7 7^2 > 10^{-4}$ while $1/2^8 8^2 < 10^{-4}$.

25. By the ratio test, the radius of convergence is ∞. Now

$$xJ_0 = x + \sum_{k=1}^\infty \frac{(-1)^k x^{2k+1}}{2^{2k}(k!)^2},$$

$$J_0' = \sum_{n=1}^\infty \frac{(-1)^n 2n x^{2n-1}}{2^{2n}(n!)^2} = -\frac{x}{2} + \sum_{k=1}^\infty \frac{(-1)^{k+1} x^{2k+1}}{2^{2k+1}(k+1)!(k!)}, \quad \text{and}$$

$$xJ_0'' = \sum_{n=1}^{\infty} \frac{(-1)^n(2n)(2n-1)x^{2n-1}}{2^{2n}(n!)^2} = -\frac{x}{2} + \sum_{k=1}^{\infty} \frac{(-1)^{k+1}(2k+1)x^{2k+1}}{2^{2k+1}(k!)(k+1)!} .$$

Thus, $xJ_0'' + J_0' + xJ_0 = \sum_{k=1}^{\infty} \frac{(-1)^k x^{2k+1}}{2^{2k}k!}\left[\frac{-(2k+1)}{2(k+1)!} - \frac{1}{2(k+1)!} + \frac{1}{k!}\right] = 0.$

27a. Set $x = 0$ and thus $f(0) = a_0 = 1$. Also,

$$f'(x) = \sum_{k=1}^{\infty} ka_k x^{k-1} = \sum_{k=0}^{\infty} (k+1)a_{k+1}x^k = a_1 + 2a_2 x + \sum_{k=2}^{\infty} (k+1)a_{k+1}x^k,$$

and $xf(x) = \sum_{k=0}^{\infty} a_k x^{k+1} = x + \sum_{k=2}^{\infty} a_{k-1}x^k.$

b. From part (a) we have

$$f'(x) - xf(x) - 1 = a_1 + 2a_2 x - x + \sum_{k=2}^{\infty} [(k+1)a_{k+1} - a_{k-1}]x^k - 1 =$$

$(a_1 - 1) + (2a_2 - 1)x + (3a_3 - a_1)x^2 + \ldots + [(k+1)a_{k+1} - a_{k-1}]x^k + \ldots = 0.$

c. From Eq.(vi), we have $a_1 = 1$, $2a_2 = 1$ or $a_2 = 1/2$, $3a_3 = a_1$ or $a_3 = 1/3$, $4a_4 = a_2$ or $a_4 = 1/(2)(4)$, and $5a_5 = a_3$ or $a_5 = 1/(3)(5)$. Thus, we see that $a_{2m+1} = 1/[(1)(3)(5)\ldots(2m+1)]$ and $a_{2m} = 1/[(2)(4)(6)\ldots(2m)].$

d. Using the ratio test, with a_{2m+1} and a_{2m} as in part (c), gives $\rho = \infty$ for both series and thus the series for $f(x)$ has $\rho = \infty$.

Section 13.6, Page 676

1. Set $\alpha = 1/2$ in Eq.10 to obtain
   $(1+x)^{1/2} = 1 + x/2 + (1/2)(-1/2)x^2/2! + (1/2)(-1/2)(-3/2)x^3/3!$
   $+ (1/2)(-1/2)(-3/2)(-5/2)x^4/4! + \ldots$

   $$= 1 + x/2 + \sum_{k=2}^{\infty} \frac{(-1)^k 1\cdot3\cdot5\cdots(2k-3)}{2^k k!}x^k \qquad \text{for } |x| < 1 \text{ or}$$

   $\rho = 1$, by Eq.12.

3. $f(x) = 1/(4+x)^{1/2} = (1/2)(1+x/4)^{-1/2}$ and thus we replace x by x/4 in Eq.15 to obtain

$$f(x) = (1/2)[1 + \sum_{k=1}^{\infty} \frac{(-1)^k 1 \cdot 3 \cdot 5 \cdots (2k-1)}{2^k k!} (x/4)^k]$$

$$= (1/2)[1 + \sum_{k=1}^{\infty} \frac{(-1)^k 1 \cdot 3 \cdot 5 \cdots (2k-1) x^k}{2^{3k} k!}], \quad |x/4| < 1 \text{ or } \rho = 4$$

by Eq. 12.

5.   Note that $f'(x) = \dfrac{(1/2)(1+x^2)^{-1/2} 2x + 1}{(1+x^2)^{1/2} + x} = (1+x^2)^{-1/2}$ and thus

$$f(x) = \int_0^x (1+t^2)^{-1/2} dt = \int_0^x [1 + \sum_{k=1}^{\infty} \frac{(-1)^k 1 \cdot 3 \cdot 5 \cdots (2k-1) t^{2k}}{2^k k!}] dt$$

$$= x + \sum_{k=1}^{\infty} \frac{(-1)^k 1 \cdot 3 \cdot 5 \cdots (2k-1) x^{2k+1}}{2^k k!(2k+1)} \quad \text{for } |x^2| < 1 \text{ or } \rho = 1.$$

Note that Eq.15 has been used for $(1+t^2)^{-1/2}$.

7.   From Ex.3, we have

$$\arcsin(x/a) = x/a + \sum_{k=1}^{\infty} \frac{1 \cdot 3 \cdot 5 \cdots (2k-1)(x/a)^{2k+1}}{2^k k!(2k+1)}$$

$$= x/a + \sum_{k=1}^{\infty} \frac{1 \cdot 3 \cdot 5 \cdots (2k-1) x^{2k+1}}{2^k k!(2k+1) a^{2k+1}} \quad \text{for } |x/a| < 1 \text{ or } \rho = a.$$

9. $f(x) = \dfrac{\sqrt{1+x} - \sqrt{1-x}}{2x} = (1/2x)\{1 + x/2 + \displaystyle\sum_{k=2}^{\infty} \frac{(-1)^{k-1} 1 \cdot 3 \cdots (2k-3) x^k}{2^k k!}$

$$- [1 - x/2 + \sum_{k=2}^{\infty} \frac{(-1)^{k-1} 1 \cdot 3 \cdots (2k-3)(-x)^k}{2^k k!}]\}$$

$$= (1/2x)[x + 2(3x^3/2^3 3!) + 2(3 \cdot 5 \cdot 7 \cdot x^5/2^5 5!) + \cdots$$
$$= 1/2 + 3x^2/2^3 3! + 3 \cdot 5 \cdot 7 x^4/2^5 5! + \cdots$$

$$= 1/2 + \sum_{k=1}^{\infty} \frac{1 \cdot 3 \cdot 5 \cdots (4k-1) x^{2k}}{2^{2k+1}(2k+1)!}, \quad \text{for } |x| < 1 \text{ or } \rho = 1.$$

11.   Note that $\sqrt{3+x} = 2\sqrt{1+(x-1)/4}$. Using the results of Prob.1, with x replaced by (x-1)/4, we have

$$\sqrt{3+x} = 2\{1 + \frac{(x-1)/4}{2} + \sum_{k=2}^{\infty}\frac{(-1)^{k-1}1\cdot3\cdot5\cdots(2k-3)[(x-1)/4]^{k}}{2^{k}k!}\}$$

$$= 2 + (x-1)/4 + \sum_{k=2}^{\infty}\frac{(-1)^{k-1}1\cdot3\cdot5\cdots(2k-3)(x-1)^{k}}{2^{3k-1}k!}$$

for $|(x-1)/4| < 1$ or $\rho = 4$.

13.   From Eq.10, we have
$$(1+x)^{2/3} = 1 + 2x/3 + (2/3)(-1/3)x^2/2! + (2/3)(-1/3)(-4/3)x^3/3!$$
$$+ (2/3)(-1/3)(-4/3)(-7/3)x^4/4! + \ldots$$

$$= 1 + 2x/3 + 2\sum_{k=2}^{\infty}\frac{(-1)^{k+1}1\cdot4\cdot7\cdots(3k-5)x^{k}}{3^{k}k!}, \quad |x| < 1 \text{ or } \rho = 1.$$

15a. If $f(t) = (1+t)^{\alpha}$, then $f'(t) = \alpha(1+t)^{\alpha-1}$. Assume

$f^{(n)}(t) = \alpha(\alpha-1)\ldots[\alpha-(n-1)](1+t)^{\alpha-n}$. Then

$f^{(n+1)}(t) = \alpha(\alpha-1)\ldots(\alpha-n)(1+t)^{\alpha-(n+1)}$ and thus the form of

the nth derivative is valid by mathematical induction.
Substituting the expression for $f^{(n+1)}(t)$ into $R_{n+1}(x)$ gives
the desired result.

  b. For $0 \leq t \leq x < 1$, we have $\frac{|x-t|}{1+t} = \frac{x-t}{1+t} \leq \frac{x}{1+t} \leq x$. For

$-1 < x \leq t \leq 0$, we have $t \leq (-x)t \leq 0$ and hence

$-x+t \leq -x-xt = -x(1+t)$ and thus $\frac{t-x}{1+t} \leq -x$. In this case,

$|x-t| = t-x$ and therefore we conclude that $\frac{|x-t|}{1+t} \leq |x|$ for

$0 \leq t \leq x < 1$ or for $-1 < x \leq t \leq 0$.

  c. From Corollary 2 of Theorem 6.3.4, we have
$$|R_{n+1}(x)| \leq \frac{|\alpha(\alpha-1)\cdots(\alpha-n)|}{n!}\int_{0}^{x}|\frac{x-t}{1+t}|^{n}(1+t)^{\alpha-1}dt$$
$$\leq \frac{|\alpha(\alpha-1)\cdots(\alpha-n)|}{n!}x^{n}\int_{0}^{x}(1+t)^{\alpha-1}dt,$$

where the results of part(b) have been used and $|x| = x$ for
$0 \leq x < 1$.

15d. Now $\int_0^x (1+t)^{\alpha-1} dt = \dfrac{(1+x)^\alpha - 1}{\alpha} = \dfrac{|(1+x)^\alpha - 1|}{|\alpha|}$ since α can be either

   positive or negative. Substituting this in Eq.(i), yields
   Eq.(ii).

e. Since $|(1+x)^\alpha - 1|$ is independent of n, we consider

   $$\sum_{k=1}^\infty \frac{|(\alpha-1)\cdots(\alpha-k)|}{k!} x^k,$$ which converges for $|x| < 1$ by the ratio

   test and thus the kth term goes to zero as $k \to \infty$, from which
   we conclude that $R_{n+1}(x) \to 0$ as $n \to \infty$.

f. For $-1 < x < 0$, we have

   $$R_{n+1}(x) = \frac{-\alpha(\alpha-1)\cdots(\alpha-n)}{n!} \int_x^0 \left(\frac{x-t}{1+t}\right)^n (1+t)^{\alpha-1} dt \quad \text{from part(a)}.$$

   Therefore $|R_{n+1}(x)| \le \dfrac{|\alpha(\alpha-1)\cdots(\alpha-n)| \, |x|^n}{n!} \displaystyle\int_x^0 (1+t)^{\alpha-1} dt$, as in

   part(c). As in part(d), we have

   $$\int_x^0 (1+t)^{\alpha-1} dt = \frac{(1+t)^\alpha}{\alpha} \Big|_x^0 = \frac{1-(1+x)^\alpha}{\alpha} = \frac{|1-(1+x)^\alpha|}{|\alpha|}, \quad \text{yielding}$$

   $$|R_{n+1}| \le \frac{|(\alpha-1)\cdots(\alpha-n)| \, |x|^n}{n!} |1-(1+x)^\alpha|,$$ which goes to zero as

   $n \to \infty$ just as in part(e), with x^n replaced by $|x|^n$ or $(-x)^n$.

17a. $s = 4\displaystyle\int_0^{\pi/2} \sqrt{[x'(t)]^2 + [y'(t)]^2}\ dt = 4\int_0^{\pi/2} \sqrt{a^2\cos^2 t + b^2 \sin^2 t}\ dt$

   $= 4a\displaystyle\int_0^{\pi/2} (1-k^2\sin^2 t)^{1/2}\, dt$, where $k^2 = 1 - b^2 a^2$.

b. Setting $x = -k^2 \sin^2 t$ in the solution to Prob.1, we get

   $$(1 - k^2\sin^2 t)^{1/2} = 1 - k^2\sin^2 t/2 - \sum_{n=2}^\infty \frac{1\cdot 3\cdot 5 \cdots (2k-3)k^{2n}\sin^{2n} t}{2^n n!} \quad \text{so,}$$

   $$E(k) = \pi/2 - (k^2/2)\int_0^{\pi/2} \sin^2 t\, dt - \sum_{n=2}^\infty \frac{1\cdot 3\cdot 5\cdots(2n-3)k^{2n}}{2^n n!} \int_0^{\pi/2} \sin^{2n} t\, dt.$$

17c. Since $\sin^2 t = 1/2 - (\cos 2t)/2$ and $\cos^2 t = 1/2 + (\cos 2t)/2$,
      we have $\sin^4 t = (1/4)(1 - 2\cos t + \cos^2 t)$
                      $= (1/4)[3/2 - 2\cos 2t + (\cos 4t)/2]$.

Therefore, $\displaystyle\int_0^{\pi/2} \sin^2 t \, dt = \pi/4$, $\displaystyle\int_0^{\pi/2} \sin^4 t \, dt = 3\pi/16$, and

$E(k) \cong \pi/2 - \pi k^2/8 - 3\pi k^4/128 = (\pi/2)[1 - k^2/4 - 3k^4/64]$.

d. $E(k) = \pi/2 - \pi k^2/8 - \displaystyle\sum_{n=2}^{\infty} \frac{1\cdot 3\cdots(2n-3)k^{2n}}{2^n n!} \cdot \frac{1\cdot 3\cdots(2n-1)}{2\cdot 4\cdots(2n)} \cdot \frac{\pi}{2}$

$= (\pi/2)[1 - k^2/4 - \displaystyle\sum_{n=2}^{\infty} \frac{1^2\cdot 3^2\cdots(2n-1)^2 k^{2n}}{2^{2n}(n!)^2(2n-1)}]$

$= (\pi/2)\{1 - k^2/4 - \displaystyle\sum_{n=2}^{\infty} \frac{[(2n)!]^2 k^{2n}}{[2^n n!]^2 2^2\cdot 4^2\cdots(2n)^2(2n-1)}\}$

$= (\pi/2)\{1 - k^2/4 - \displaystyle\sum_{n=2}^{\infty} \left[\frac{(2n)!}{(2^n n!)^2}\right]^2 \frac{k^{2n}}{2n-1}\}$.

Chapter 13 Review, Page 678

1. $\rho = \displaystyle\lim_{k\to\infty} \frac{(2k+1)^2}{(2k+3)^2} = 1$, so the series converges for $-1 < x < 1$ by

the ratio test. At $x = -1$, $\displaystyle\sum_{k=0}^{\infty} \frac{(-1)^k}{(2k+1)^2}$ converges absolutely.

Similarly the series converges absolutely for $x = 1$.

3.   Since $\displaystyle\lim_{k\to\infty} \frac{2^k}{2^{k+1}} = \frac{1}{2}$, the series converges for $|t+1| < \sqrt{2}$, or

$-1 - \sqrt{2} < t < -1 + \sqrt{2}$, by the ratio test. At both end points,

$-1 \pm\sqrt{2}$, the series is $\displaystyle\sum_{k=0}^{\infty} 1$, which diverges.

5.   As $\displaystyle\lim_{k\to\infty} \frac{k(k^2+1)}{(k-1)[(k+1)^2+1]} = 1$, the series converges for $|x| < 1$.

At $x = -1$, $-\displaystyle\sum_{k=0}^{\infty} \frac{k-1}{k^2+1}$ diverges by comparison with $\displaystyle\sum_{1}^{\infty} 1/k$. At

$x = 1$ the series is alternating and converges conditionally.

In Problems 7-11 we use Eq. (10), Section 13.3.

7. $\rho = \lim\limits_{k \to \infty} \left| \dfrac{a_k}{a_{k+1}} \right| = \lim\limits_{k \to \infty} \dfrac{k!}{(k+1)!} \dfrac{2^{k+1}}{2^k} = \lim\limits_{k \to \infty} \dfrac{2}{k+1} = 0.$

9. $\rho = \lim\limits_{k \to \infty} \dfrac{2^k}{\sqrt{k+1}} \dfrac{\sqrt{k+2}}{2^{k+1}} = 1/2.$

11. $\rho = \lim\limits_{k \to \infty} \dfrac{2^k}{k\,(\ln k)^2} \dfrac{(k+1)\,[\ln(k+1)]^2}{2^{k+1}} = \lim\limits_{k \to \infty} \left(\dfrac{k+1}{k} \right)\left(\dfrac{\ln(k+1)}{\ln(k)} \right)^2 \dfrac{1}{2} = \dfrac{1}{2}.$

13. $f(x) = \dfrac{1}{(1+x^3/2)} = \sum\limits_{k=0}^{\infty} (-1)^k x^{3k}/2^k$, which converges for

$-2^{1/3} < x < 2^{1/3}$, from Ex.5, Section 13.2.

15. $f(x) = -1 + 2/(1-x^2) = 1 + 2\sum\limits_{k=1}^{\infty} x^{2k}$, which converges for $-1 < x < 1$.

17. $f(x) = \ln a + \ln(1 + bx/a)$ with $a, b > 0$, and replacing x with (bx/a) in Ex.2 of Section 13.5 gives

$f(x) = \ln a + \sum\limits_{k=1}^{\infty} (-1)^{k+1} (b/a)^k x^k/(k)$ for $-(a/b) < x < (a/b)$.

19. From Eq. (10), Section 13.6, with $\alpha = 1/3$,

$f(x) = 1 + (1/3)x + \sum\limits_{k=2}^{\infty} \dfrac{(1/3)(-2/3)(-5/3)\cdots(1/3 - k + 1)}{k!} x^k$

$= 1 + (1/3)x + \sum\limits_{k=2}^{\infty} (-1)^{k-1} \dfrac{2 \cdot 5 \cdots (3k-4)}{3^k\, k!} x^k$ for $-1 < x < 1$.

21a. For the given function, $f'(x) = [-1/(1+x)^2]\cos[1/(1+x)]$ and
$f''(x) = [2/(1+x)^3]\cos[1/(1+x)] - [1/(1+x)^4]\sin[1/(1+x)]$,   so
$f(0) = \sin(1)$, $f'(0) = -\cos(1)$ and $f''(0) = 2\cos(1) - \sin(1)$.
Thus $p_2(x) = \sin(1) - \cos(1)x + (1/2)[2\cos(1) - \sin(1)]x^2$.

   b. Substituting the series for $1/(1+x)$ into the series for $\sin x$
   gives
$f(x) = (1-x+x^2- \cdots) - (1-x+x^2- \cdots)^3/3! + (1-x+x^2- \cdots)^5/5! + \cdots$
$= \sin(1) - \cos(1)x + \cdots.$

23a. $f'(x) = \cos x e^{\sin x}$ and $f''(x) = (\cos^2 x - \sin x)e^{\sin x}$, so
    $f(0) = f'(0) = f''(0) = 1$ and $p_2(x) = 1 + x + x^2/2$.

  b. From the series for e^x, $f(x) = 1 + \sin x + (\sin x)^2/2! + \cdots$, so
    $f(x) = 1 + (x - x^3/3! + x^5/5! - \cdots) + (x - x^3/3! + x^5/5! - \cdots)^2/2!$
       $= 1 + x + x^2/2 + \cdots$.

25. Since $f^{(iv)}(x) = (-40/81)(x-1)^{-11/3}$, Eq. (9) of Section 13.2
    gives $R_4(x) = -(10/243)(c-1)^{-11/3}(x-2)^4$.

27. Since $f^{iv}(x) = 3!(x+1)^{-4}$, we have $R_4(x) = (1/4)(c+1)^{-4}x^4$.

29. For the given function $f(0) = f''(0) = f^{iv}(0) = f^{vi}(0) = 0$
    while $f'(0) = 1$, $f'''(0) = -2$, $f^v(0) = 16$ and $f^{vii}(0) = -272$.
    Thus $f(x) = x - x^3/3! + (2/15)x^5 - (17/315)x^7 + \cdots$.

31. $f(x) = [1 + x^2 + x^4/2! + x^6/3! + \cdots][1 + x + x^2 + \cdots]$
       $= 1 + x + 2x^2 + \cdots$ for $-1 < x < 1$.

33. $\arctan x = \displaystyle\sum_{n=0}^{\infty} (-1)^n x^{2n+1}/(2n+1)$, for $|x| < 1$. If $x = 1/\sqrt{3}$, then

    $\pi/6 = \displaystyle\sum_{n=0}^{\infty} (-1)^n/3^n\sqrt{3}(2n+1)$, since $(\sqrt{3})^{2n+1} = 3^n\sqrt{3}$.

35. Assume for some $\rho \neq 0$, the solution can be written as
    $y = f(x) = \displaystyle\sum_{n=0}^{\infty} a_n x^n = a_0 + a_1 x + a_2 x^2 + a_3 x^3 + \cdots + a_n x^n$. Then
    $y' = a_1 + 2a_2 x + 3a_3 x^2 + \cdots + na_n x^{n-1} + \cdots$. This gives
    $0 = y' - 2xy = a_1 + (2a_2 - 2a_0)x + (3a_3 - 2a_1)x^2 + \cdots + (na_n - 2a_{n-2})x^{n-1} + \cdots$.
    Thus $a_1 = 0$, $a_2 = a_0$, $3a_3 = 2a_1 = 0$, $\cdots$ $a_n = (2/n)a_{n-2}$. So
    $a_{2k-1} = 0$, while $a_4 = a_0/2$, $a_6 = a_0/6$, $\cdots$. From $y(0) = y_0$ we
    then have $a_0 = y_0$, so $y = y_0(1 + x^2 + x^4/2 + x^6/6 + \cdots) = y_0 e^{x^2}$.

37. Using the series for e^x and proceeding as in Prob.35, we have
     $(a_1-a_0)+(2a_2-a_1)x+(3a_3-a_2)x^2+\cdots = 1+x+x^2/2!+x^3/3!+\cdots$. Thus
     $a_1-a_0 = 1$, $2a_2-a_1 = 1$, $3a_3-a_2 = 1/2!,\cdots$. With $y(0) = 1 = a_0$,
     $a_1 = 2$, $a_2 = 3/2$, $a_3 = 2/3,\cdots$, so that
     $y = 1+2x+(3/2)x^2+(2/3)x^3+\cdots= (1+x)e^x$, as in Chapter 10.

39. If $x = a_0+a_1\epsilon+a_2\epsilon^2+\cdots$ then $x^3 = a_0^3+3a_0^2a_1\epsilon+3(a_0^2a_2+a_0a_1^2)\epsilon^2+\cdots$
     and hence substitution into the equation gives
     $(a_0^3-8) + (3a_0^2a_1+a_0)\epsilon + [a_1+3(a_0^2a_2+a_0a_1^2)]\epsilon^2 + \cdots = 0$. Thus
     $a_0 = 2$, $a_1 = -1/6$, and $a_2 = 0$, so that $x = 2 - (1/6)\epsilon + \cdots$.

41. $x^2 = a_0^2+2a_0a_1\epsilon+(2a_0a_2+a_1^2)\epsilon^2$ and x^3 as in Prob.39 yields
     $(a_0^3-3a_0+2) + (3a_0^2a_1-a_0^2-3a_1+2)\epsilon + [3(a_0^2a_2+a_0a_1^2) - 2a_0a_1-3a_2]\epsilon^2+\cdots = 0$.
     From $a_0^3-3a_0+2 = 0$ we choose the root $a_0 = -2$, then from the
     coefficient of ϵ we find $a_1 = 2/9$ and from the coefficient of
     ϵ^2 we find $a_2 = -16/243$. Thus $x = -2 + (2/9)\epsilon - (16/243)\epsilon^2 + \cdots$.

43. Since $\lim\limits_{x \to 0+} \dfrac{\tan x}{x} = 1$, the integral exists. Using
     $\dfrac{\tan x}{x} = 1 + x^2/3 + 2x^4/15 + \cdots$, we have that
     $I = x + x^3/9 + 2x^5/75 + \cdots \Big|_0^{0.1} \cong 0.100111$.

45. Since $\lim\limits_{x \to 0+} \dfrac{\ln(\cos x)}{x} = 0$, the integral exists. Integrating the
     series for $-\tan x$ gives $\ln(\cos x) = -[x^2/2+(1/12)x^4+(1/45)x^6+\cdots]$
     and so $I = -(x^2/4 + x^4/48 + x^6/270+\cdots)\Big|_0^{0.1} \cong -0.002502$.

CHAPTER 14

1.

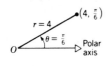

3.

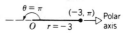

5.

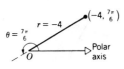

7.

9.

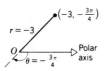

11. $(2,\pi/3)$ is the same point as $(-2,\pi + \pi/3) = (-2,4\pi/3)$. Also, $(2,\pi/3)$ is the same point as $(2,\pi/3 - 2\pi) = (2,-5\pi/3)$.

13. $(-2,\pi/4)$ is the same point as $(2,\pi + \pi/4) = (2,5\pi/4)$ and $(-2,\pi/4 - 2\pi) = (-2,-7\pi/4)$.

15. $(4,-\pi/4)$ is the same point as $(-4,-\pi/4 + \pi) = (-4,3\pi/4)$ and $(4,-\pi/4 + 2\pi) = (4,7\pi/4)$.

17. $(-1,-\pi/4)$ is the same point as $(1,-\pi/4 + \pi) = (1,3\pi/4)$ and as $(-1,-\pi/4 + 2\pi) = (-1,7\pi/4)$.

19. $x = 4\cos(\pi/6) = 4\sqrt{3}/2 = 2\sqrt{3}$; $y = 4\sin(\pi/6) = 4(1/2) = 2$.

21. $x = 2\cos(-\pi/4) = \sqrt{2}$; $y = 2\sin(-\pi/4) = -\sqrt{2}$.

23. $x = 5\cos\pi = -5$; $y = 5\sin\pi = 0$.

25. $x = -4\cos(\pi/3) = -2$; $y = -4\sin(\pi/3) = -2\sqrt{3}$.

27. $x = -4\cos(4\pi/3) = 2$; $y = -4\sin(4\pi/3) = 2\sqrt{3}$.

29. $r = \sqrt{x^2+y^2} = \sqrt{12+4} = 4$; $\tan\theta = y/x = 1/\sqrt{3}$, so $\theta = \pi/6$.

31. $r = \sqrt{x^2+y^2} = \sqrt{4+12} = 4$; $\tan\theta = y/x = \sqrt{3}$, so $\theta = 4\pi/3$ since the point is in the third quadrant.

33. $r = \sqrt{4+12} = 4$; $\tan\theta = -\sqrt{3}$, so $\theta = 2\pi/3$ since the point is in the second quadrant.

35. $r = \sqrt{1+4} = \sqrt{5}$; $\tan\theta = y/x = 2$, $\theta = \arctan 2$.

37. 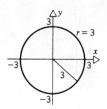   39.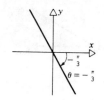

41. Convert $r = 3\cos\theta$ to rectangular

coordinates: $\sqrt{x^2+y^2} = 3\dfrac{x}{\sqrt{x^2+y^2}}$, or

$x^2+y^2 = 3x$, or $(x-3/2)^2+y^2 = (3/2)^2$,
which is a circle of radius 3/2
and center at $(3/2,0)$.

43. Converting to rectangular
coordinates gives $x^2+y^2 = -6y$ or
$x^2+(y+3)^2 = 9$, which is a circle
with center at $(0,-3)$ and radius 3.

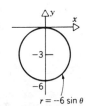

45. Sketch the graph by simply plotting
the points. Note that as θ increases
so does r, yielding a spiral.

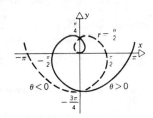

47. $r = e^{|\theta|/2\pi}$ is a spiral with
      $r \geq 1$ for all θ.

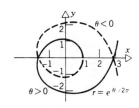

49. $r = 4\cos(\theta - \pi/6)$
      $= 4[\cos\theta\cos(\pi/6) + \sin\theta\sin(\pi/6)]$
      $= 4\left(\dfrac{x}{r}\right)\dfrac{\sqrt{3}}{2} + 4\left(\dfrac{y}{r}\right)\dfrac{1}{2}$, so
      $x^2 + y^2 = 2\sqrt{3}x + 2y$ or
      $(x^2 - 2\sqrt{3}x + 3) + (y^2 - 2y + 1) = 4$ or
      $(x - \sqrt{3})^2 + (y - 1)^2 = 2$. In polar coordinates the center is at
      $(2, \pi/6)$,

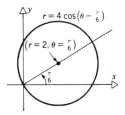

51. $r = 4\sin(\theta + 5\pi/6)$
      $= 4\sin\theta\cos(5\pi/6) + 4\cos\theta\sin(5\pi/6)$ so
      $x^2 + y^2 = -2\sqrt{3}y + 2x$ as, in Prob.49.
      Thus $(x - 1)^2 + (y + \sqrt{3})^2 = 2^2$. In
      polar coordinates, the center is
      at $(2, -\pi/3)$.

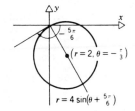

53. $r = 2\sec\theta$ becomes $2 = r\cos\theta$ or
      $2 = r\left(\dfrac{x}{r}\right) = x$.   Thus, the graph has
      equation $x = 2$.

55. $r = 2\csc\theta$ becomes $2 = r\sin\theta$ or
      $2 = (r)y/r = y$.   The rectangular
      form of the equation is $y = 2$.

57. $4 = r\cos(\theta - \pi/4)$

   $= r\cos\theta\cos(\pi/4) + r\sin\theta\sin(\pi/4)$

   $= \dfrac{\sqrt{2}}{2}[r\left(\dfrac{x}{r}\right)+r\left(\dfrac{y}{r}\right)] = \dfrac{\sqrt{2}}{2}x + \dfrac{\sqrt{2}}{2}y$, which

   is a straight line. When $\theta = 0$,
   $r = 4\sqrt{2}$ and when $\theta = \pi/2$, $r = 4\sqrt{2}$.

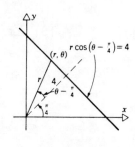

59. $r = 3\sec(\theta - \pi/2)$ can be written as
   $3 = r\cos(\theta - \pi/2)$, and as in Prob.57,
   the graph is a straight line.
   However, we may rewrite $\cos(\theta - \pi/2)$ as
   $\sin\theta$ so $3 = r\sin\theta = y$, a clearer
   form of the equation.

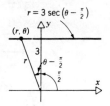

61. $r = 2|\cos\theta|$ differs from $r = 2\cos\theta$
   in that in the former, $r \geq 0$ for all
   θ. Thus, while $r = 2\cos\theta$ traverses
   the circle $(x-1)^2+y^2 = 1$ twice as θ
   goes from 0 to 2π, the equation $r = 2|\cos\theta|$ defines 2
   circles: $(x-1)^2+y^2 = 1$ $(-\pi/2 \leq \theta \leq \pi/2)$ and
   $(x+1)^2+y^2 = 1$ $(\pi/2 \leq \theta \leq 3\pi/2)$.

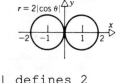

63. $r|\cos\theta| = 2$ becomes

   $r\left|\dfrac{x}{r}\right| = r\dfrac{|x|}{r} = |x| = 2$. Thus, $x = 2$,
   and $x = -2$, which gives two
   vertical straight lines.

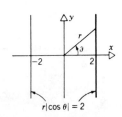

65. We have $x^2 + (y-2)^2 = 4$ or $x^2 + y^2 = 4y$. But $x^2 + y^2 = r^2$
   and $y = r\sin\theta$. Thus, $r^2 = 4r\sin\theta$, or $r = 4\sin\theta$.

67. We have $(x-1)^2 + (y-1)^2 = 2$, or $x^2 + y^2 = 2x + 2y$. Thus,
   $r^2 = 2r\cos\theta + 2r\sin\theta$ or $r = 2\cos\theta + 2\sin\theta = 2\sqrt{2}\cos(\theta - \pi/4)$,
   by Prob.27, Section 1.6

69. $2x - y = 5$ may be written as $2r\cos\theta - r\sin\theta = 5$ or
   $5 = r(2\cos\theta - \sin\theta)$.

71. $x^2-xy+y^2 = (x^2+y^2)-(xy) = r^2 - r^2\sin\theta\cos\theta = 3.$ Thus
     $6 = r^2(2 - 2\sin\theta\cos\theta)$ or $6 = r^2(2 - \sin2\theta).$

73. $r = 2(1-\sin\theta)$ becomes $\sqrt{x^2+y^2} = 2(1-y/\sqrt{x^2+y^2})$ or, multiplying
     by $\sqrt{x^2+y^2}$, $x^2+y^2 = 2[\sqrt{x^2+y^2} - y]$. Thus, $x^2+y^2+2y = 2\sqrt{x^2+y^2}$.
     Squaring both sides gives $(x^2+y^2+2y)^2 = 4(x^2+y^2).$

75. $r = 3\sin2\theta = 6\sin\theta\cos\theta$, or $\sqrt{x^2+y^2} = 6xy/(x^2+y^2).$ Thus,
     $(x^2+y^2)^3 = 36x^2y^2.$

77. $d = \sqrt{(x_1 - x_2)^2 + (y_1 - y_2)^2}$

     $= \sqrt{(r_1\cos\theta_1 - r_2\cos\theta_2)^2 + (r_1\sin\theta_1 - r_2\sin\theta_2)^2}$

     $= \sqrt{r_1^2 + r_2^2 - 2r_1r_2(\cos\theta_1\cos\theta_2 + \sin\theta_1\sin\theta_2)}$

     $= \sqrt{r_1^2 + r_2^2 - 2r_1r_2\cos(\theta_1 - \theta_2)}.$

Section 14.2, Page 699

1. $r = 2(1 - \cos\theta)$ is a cardioid,
    as in Eq.(7) with $a = 2$.

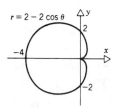

3. $r = 1 - 2\cos\theta$ is a limaçon, $a = 1$,
    $b = -2$ and so is similar to Eq.(12).

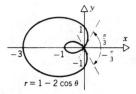

5. $r = 3(-1+\sin\theta)$ is a cardioid with
    vertical axis of symmetry. Plot
    by computing r for $-\pi/2 \le \theta \le \pi/2$
    and reflecting across the vertical
    axis. Note that $r \le 0$ for all θ.

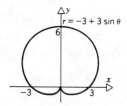

7.   Since $r^2 = 2\sin2\theta \geq 0$, we need only
      consider values of θ for which
      $\sin2\theta \geq 0$. That is $0 \leq \theta \leq \pi/2$.
      The graph will be a lemniscate, as
      in Eq.(16).

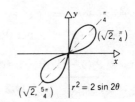

9. $r = 3(1 + \sin\theta)$ is a cardioid with
      vertical axis of symmetry and has
      $r \geq 0$. It has a graph the same as
      that of Prob.5, with points
      labeled differently.

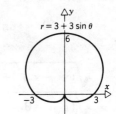

11. $r^2 = 4\cos2\theta$ is a lemniscate by
       Eq.(16), with $a = \sqrt{2}$ and graph
       similar to Fig.14.2.12.

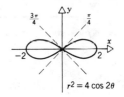

13. $r = 4\sin4\theta$ is a leaf curve. For
       $\theta \leq \theta \leq \pi/4$, 4θ increases from 0 to π
       so r goes from 0 to 4 and back to 0
       in the first quadrant. For
       $\pi/4 \leq \theta \leq \pi/2$, 4θ increases to 2π
       with r negative, so the graph is in
       the third quadrant. Continuing for
       $\pi/2 \leq \theta \leq 2\pi$, we get the eight leaves.

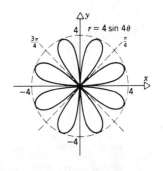

15. $r = 2 - 4\sin\theta$ is of the form of Eq.(12),
       and hence is a limaçon, with
       symmetry about the vertical axis,
       $\theta = \pi/2$. Plot the values for
       $-\pi/2 \leq \theta \leq \pi/2$ and then reflect
       across $\theta = \pi/2$.

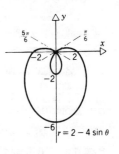

17. $r = 4(1 + \cos\theta)$ intersects $r = 6$ when $6 = 4(1 + \cos\theta)$, or $\cos\theta = 1/2$, so that $\theta = \pm\pi/3$. Thus, the points of intersection are $(6,\pi/3)$ and $(6,-\pi/3)$.

19. The circles $r = 3\sin\theta$ and $r = 3\cos\theta$ intersect when $\cos\theta = \sin\theta$, or $\tan\theta = 1$, or $\theta = \pi/4, 5\pi/4$. The points of intersection are then $(3\sqrt{2}/2,\pi/4)$ and $(-3\sqrt{2}/2,5\pi/4)$, which are the same. Thus only $(3\sqrt{2},\pi/4)$ satisfies both equations. However, $r = 0$ is a point on both curves, so the pole is also a point of intersection.

21. $r = 2 - \cos\theta$ and $r = 1 + \cos\theta$ intersect when $2 - \cos\theta = 1 + \cos\theta$ or $\cos\theta = 1/2$, so $\theta = \pm\pi/3$. Thus, the points $(3/2,\pi/3)$ and $(3/2,-\pi/3)$ are on both graphs. They are the only points of intersection.

23. When $\theta = 2\pi/3$, $r = 2 + 3\cos\theta = 2 - 3/2 = 1/2$. Thus $(1/2,2\pi/3)$ is on both curves. But since the ray $2\pi/3$ is the same as the ray $-\pi/3$, and $r = 2 + 3\cos(-\pi/3) = 7/2$, then $(7/2,-\pi/3)$ is also on the graph of both equations. Finally, since both curves contain the pole, that is third intersection.

25. The curves intersect for $\sin 3\theta = \sqrt{2}/2$, or when $3\theta = \pi/4, 3\pi/4$. These give the points of intersection as $(\sqrt{2},\pi/12)$ and $(\sqrt{2},\pi/4)$. Other intersections lie outside the first quadrant.

27. $r = (3/2 - \sin\theta)^{-1}$ and $r = 3/2 - \sin\theta$ intersect when $3/2 - \sin\theta = 1$ or when $3/2 - \sin\theta = -1$. Since $3/2 - \sin\theta \geq 1/2$ for all θ, the latter case need not be considered. So, $3/2 - \sin\theta = 1$ for $\theta = \pi/6, 5\pi/6$. Thus, $(1,\pi/6)$ and $(1,5\pi/6)$ are the intersection points.

29. We have $4\cos\theta = 1 + 2\cos\theta$, or $\theta = \pi/3, 5\pi/3$, which yield $(2,\pi/3)$ and $(2,5\pi/3)$ as the intersections. Both graphs also pass through the pole.

31a. $|OP| = e|DP|$, where $|OP| = r$ and $|DP| = r\sin\theta + p$, so that $r = e(r\sin\theta + p)$. Solving for r, we find $r - er\sin\theta = ep$, or $r = ep/(1 - e\sin\theta)$.

31b. In this case, as $\theta < 0$,
      $|PC| = -r\sin\theta > 0$ so $r = e(-r\sin\theta + p)$.
      Solving for r gives $r = ep/(1 + e\sin\theta)$.

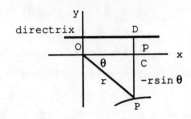

33. $r = \dfrac{4}{2-\cos\theta} = \dfrac{2}{1-(1/2)\cos\theta}$ so $e = 1/2$
     and $p = 4$. Thus the conic section
     is an ellipse, with directrix 4 units
     to the left of the pole and
     perpendicular to the polar axis. To
     determine the second focus, we find
     the center of the ellipse, as in
     Prob.32, by $h = pe^2/(1-e^2) = 4/3$. Thus,
     the second focus is 4/3 to the right, or at (8/3,0), with the
     second directrix 4 units to its right, or at $x = 20/3$.

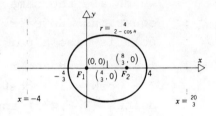

35. The equation may be written
     $r = \dfrac{4(2/3)(1/2)}{1+(2/3)\cos\theta}$, so $e = 2/3 < 1$ and
     the directrix is $x = 2 = p$. The
     center of the ellipse will be at
     $h = -pe^2/(1-e^2) = -8/5$. Thus the
     second focus will be at (-16/5,0)
     and the second directrix is 2 units
     to its left at $x = -26/5$.

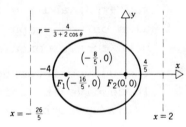

37. $r = \dfrac{15}{2-3\cos\theta} = \dfrac{(3/2)5}{1-(3/2)\cos\theta}$. Thus
     $p = 5$ and $e = 3/2 > 1$, so the graph is
     a hyperbola, $\dfrac{(x-h)^2}{a^2} - \dfrac{y^2}{b^2} = 1$, where

     $h = pe^2/(1-e^2) = -9$, $a^2 = \dfrac{p^2e^2}{(1-e^2)^2} = 36$,

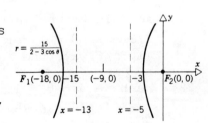

     and $b^2 = \dfrac{p^2e^2}{e^2-1} = 45$. Thus one focus
     is at (0,0) and one directrix is 5 units to the left at
     $x = -5$. The center is at (-9,0), with second focus 9 units
     to it's left at (-18,0). The second directrix is at $x = -13$,
     symmetric with the first with respect to the center.

39. $r = \dfrac{2}{1+\sin\theta}$ has $e = 1$, and so is a
     parabola. Further, the directrix
     is parallel to the polar axis and 2
     units above it because the
     denominator is $1+\sin\theta$. Finally, the
     rectangular form of the equation is
     $y-h = -\dfrac{1}{2p}x^2$, where $h = p/2$. Thus, $y-1 = -(1/4)x^2$, with
     vertex $(0,1)$ and directrix $y = 2$.

41. From the discussion in Probs.30-32, $e = 3/2$, the equation has
     the form $r = ep/(1+e\cos\theta)$, and $p = 2$. Thus,
     $r = (3/2)(2)/[1+(3/2)\cos\theta] = 6/(2+3\cos\theta)$. The center of the
     hyperbola is at $x = h = \dfrac{-pe^2}{1-e^2} = \dfrac{18}{5}$. The second focus and
     directrix are symmetric about the center and are thus
     $(36/5,0)$ and $x = 26/5$ respectively.

Section 14.3, Page 708

1.   The lemniscate will complete one loop for $-\pi/4 \leq \theta \leq \pi/4$.
     Thus, by Eq. 7, $A = (1/2)\displaystyle\int_{-\pi/4}^{\pi/4} 8\cos 2\theta\, d\theta = 2\sin 2\theta\Big|_{-\pi/4}^{\pi/4} = 4$.

3.   The lemniscate is defined for $0 \leq \theta \leq 2\pi$, so Eq.(7) yields
     $$A = (1/2)\int_0^{2\pi} (1 + 0.5\sin\theta)^2 d\theta = (1/2)\int_0^{2\pi} (1 + \sin\theta + 0.25\sin^2\theta)\, d\theta$$
     $$= (1/2)\int_0^{2\pi} [1+\sin\theta+(1/8)(1-\cos 2\theta)]\, d\theta = (1/2)[2\pi + \pi/4] = 9\pi/8.$$

5. $A = (1/2)\displaystyle\int_0^{\pi} (3\theta)^2\, d\theta = 3\pi^3/2$.

7a. Since the limaçon has symmetry across the polar axis and
     passes through the pole for $\theta = 2\pi/3$ and $4\pi/3$, we have
     $$A = 2[(1/2)\int_0^{2\pi/3} (1+2\cos\theta)^2 d\theta] = \int_0^{2\pi/3} (1+4\cos\theta+4\cos^2\theta)\, d\theta$$
     $$= \int_0^{2\pi/3} (3+4\cos\theta+2\cos 2\theta)\, d\theta = 2\pi + 3\sqrt{3}/2.$$

7b.   Likewise, the area of the smaller loop is given by

$$A = (1/2)\int_{2\pi/3}^{4\pi/3} (3+4\cos\theta+2\cos2\theta)\,d\theta = \pi - 3\sqrt{3}/2.$$

c. $A = (2\pi + 3\sqrt{3}/2) - (\pi - 3\sqrt{3}/2) = \pi + 3\sqrt{3}.$

9.   The curves intersect at $\theta = \pi/6, 5\pi/6$, so by symmetry,

$$A = 2\{(1/2)\int_{\pi/6}^{\pi/2} [a^2(1+\sin\theta)^2 - (3a/2)^2]\,d\theta\}$$

$$= a^2\int_{\pi/6}^{\pi/2} [1+2\sin\theta+\sin^2\theta - 9/4]\,d\theta = a^2[9\sqrt{3}/8 - \pi/4].$$

11.   The curves intersect at $\theta = \pm\pi/3$, so by symmetry we have

$$A = 2\{(1/2)\int_{0}^{\pi/3} [(1+\cos\theta)^2 - (2-\cos\theta)^2]\,d\theta\} = \int_{0}^{\pi/3} (6\cos\theta-3)\,d\theta$$

$$= 3\sqrt{3} - \pi.$$

13.   The area common to both is the area inside the limaçon for
       $-\pi/3 \le \theta \le \pi/3$ plus the area inside the cardioid for
       $\pi/3 \le \theta \le 5\pi/3$.   Using symmetry, the integrals then become

$$A = 2[(1/2)\int_{0}^{\pi/3} (2-\cos\theta)^2 d\theta + (1/2)\int_{\pi/3}^{\pi} (1+\cos\theta)^2 d\theta]$$

$$= \int_{0}^{\pi/3} [9/2 - 4\cos\theta + (1/2)\cos2\theta]\,d\theta + \int_{\pi/3}^{\pi} [3/2 + 2\cos\theta + (1/2)\cos2\theta]\,d\theta$$

$$= 5\pi/2 - 3\sqrt{3}.$$

15.   From Eq.(20), with $dr/d\theta = -\sin\theta$,

$$s = \int_{0}^{\pi/2} [1+2\cos\theta+\cos^2\theta + \sin^2\theta]^{1/2}d\theta = 2\int_{0}^{\pi/2} |\cos(\theta/2)|\,d\theta \text{ by}$$

the half-angle formula. For $0 \le \theta \le \pi/2$, $\cos(\theta/2) > 0$, thus

$|\cos(\theta/2)| = \cos(\theta/2)$. Therefore, $s = 2\int_{0}^{\pi/2} \cos(\theta/2)\,d\theta = 2\sqrt{2}.$

17.   By Eq.(20), with $r = a\cos\theta$, and $dr/d\theta = -a\sin\theta$,

$$s = \int_{0}^{\pi} \sqrt{(a\cos\theta)^2+(-a\sin\theta)^2}\,d\theta = a\int_{0}^{\pi} d\theta = a\pi, \text{ which is the}$$

circumference of a circle of radius $a/2$.

19. The graph of r = asecθ, 0 ≤ θ ≤ π/4, is a part of a vertical
     line, so its length is that of a straight line segment, a. Also

$$s = \int_0^{\pi/4} \sqrt{a^2\sec^2\theta + a^2\sec^2\theta\tan^2\theta}\,d\theta = \int_0^{\pi/4} a\sec^2\theta\,d\theta = a\tan\theta\Big|_0^{\pi/4} = a.$$

21. Since dr/dθ = ke^{kθ}, we have

$$s = \int_0^{\pi} \sqrt{e^{2k\theta} + k^2e^{2k\theta}}\,d\theta = \sqrt{1+k^2}\int_0^{\pi} e^{k\theta}\,d\theta = \sqrt{1+k^2}\,(e^{k\pi} - 1)/k.$$

23. With dr/dθ = -acosθ and -π/2 ≤ θ ≤ π/2,

$$s = \int_{-\pi/2}^{\pi/2} [a^2(1-\sin\theta)^2 + a^2\cos^2\theta]\,d\theta = a\sqrt{2}\int_{-\pi/2}^{\pi/2} (1+\sin\theta)^{1/2}\,d\theta$$

$$= a\sqrt{2}\int_{-\pi/2}^{\pi/2} [1+\cos(\theta-\pi/2)]^{1/2}\,d\theta, \text{ since } \sin\theta = \cos(\theta-\pi/2).$$

     Using the half-angle formula then gives

$$s = a(\sqrt{2})^2\int_{-\pi/2}^{\pi/2} \cos[(\theta-\pi/2)/2]\,d\theta = 4a\sin[(\theta-\pi/2)/2]\Big|_{-\pi/2}^{\pi/2} = 4a.$$

25. $r^2 + (dr/d\theta)^2 = 2a^2\cos2\theta + \dfrac{4a^4\sin^2 2\theta}{2a^2\cos2\theta} = 2a^2\sec2\theta$ so that

$$s = \int_{-\pi/4}^{\pi/4} \sqrt{2a^2\sec2\theta}\,d\theta = \sqrt{2}\,a\int_{-\pi/4}^{\pi/4} (\sec2\theta)^{1/2}\,d\theta. \quad \text{By the symmetry}$$

     of sec2θ this becomes s = $2\sqrt{2}\,a\displaystyle\int_0^{\pi/4} (\sec2\theta)^{1/2}\,d\theta.$

27. Substituting for r in Eq. (i) and integrating gives

$$S = 2\pi\int_0^{\pi/2} a\sin\theta\cos\theta\sqrt{a^2\cos^2\theta + a^2\sin^2\theta}\,d\theta$$

$$= 2\pi a^2\int_0^{\pi/2} [(\sin2\theta)/2]\,d\theta = \pi a^2(-\cos2\theta)/2\Big|_0^{\pi/2} = \pi a^2.$$

29. $S = 2\pi\displaystyle\int_0^{\pi} a(1+\cos\theta)\sin\theta[a^2(1+\cos\theta)^2 + a^2\sin^2\theta]^{1/2}\,d\theta$

$$= 2\pi a^2\int_0^{\pi} (1+\cos\theta)\sin\theta[2(1+\cos\theta)]^{1/2}\,d\theta = 8\pi a^2\int_0^{\pi} \cos^3(\theta/2)\sin\theta\,d\theta$$

$$= 16\pi a^2\int_0^{\pi} \cos^4(\theta/2)\sin(\theta/2)\,d\theta = 32\pi a^2/5.$$

31. As in Prob. 25, $[r^2+(dr/d\theta)^2]^{1/2} = \sqrt{2}a(\sec2\theta)^{1/2}$, so that

$$S = \int_{-\pi/4}^{\pi/4} 4\pi a^2\cos\theta\,d\theta = 4\sqrt{2}\pi a^2, \text{ by Eq.(ii)}.$$

33. With $r = 1+\cos\theta$, $dr/d\theta = -\sin\theta$, so $\tan\psi = -(\sqrt{2}+1)$ at $\pi/4$ and

thus the slope $= \dfrac{1-(\sqrt{2}+1)}{1+(\sqrt{2}+1)} = \dfrac{-\sqrt{2}}{2+\sqrt{2}} = 1 - \sqrt{2}$, from Prob.32c.

35. $dr/d\theta = 2\sin\theta$ so $\tan\psi = 4/\sqrt{3}$ at $\pi/3$, so the

slope $= \dfrac{\sqrt{3}+4/\sqrt{3}}{1-(\sqrt{3})(4/\sqrt{3})} = -7/3\sqrt{3}$, from Prob.32c.

37. $dr/d\theta = -3\sin\theta$, so $\tan\psi = -\cot(\pi/3) = -1/\sqrt{3}$,   so the

slope $= \dfrac{\sqrt{3}-1/\sqrt{3}}{1+(\sqrt{3})(1/\sqrt{3})} = \sqrt{3}/3$.

39. Let $r_1 = 3\cos\theta$, then $\tan\psi_1 = r_1/(dr_1/d\theta) = -\cot\theta$ and

$m_1 = \dfrac{\tan\theta-\cot\theta}{2} = 0$ at $\theta = \pi/4$. Thus the tangent to r_1 is

horizontal. Similarly, for $r_2 = 3\sin\theta$, $\tan\psi_2 = \tan\theta$ so

$m_2 = \dfrac{2\tan\theta}{1-\tan^2\theta}$, which for $\theta = \pi/4$ is undefined. Thus the

tangent to r_2 is vertical and the two tangents are
perpendicular so that the angle between them is $\pi/2$ radians.

41. $\dfrac{dy}{dx} = \dfrac{dy/d\theta}{dx/d\theta} = \dfrac{f'(\theta)\sin\theta+f(\theta)\cos\theta}{f'(\theta)\cos\theta-f(\theta)\sin\theta}$. Factoring $f'(\theta)\cos\theta$ from

numerator and denominator, we have $\dfrac{dy}{dx} = \dfrac{\tan\theta+\tan\psi}{1-\tan\theta\tan\psi}$, where

$\tan\psi = r/(dr/d\theta) = f(\theta)/f'(\theta)$, provided $\cos\theta \neq 0$ and $f'(\theta) \neq 0$.

43. P has coordinates $(a,0)$, so the line PQ has equation

$y = m(x-a)$, where $m = \dfrac{\tan\theta+\tan\psi}{1-\tan\theta\tan\psi} = \tan\psi$, when $\theta = 0$, and

$\tan\psi = r/(dr/d\theta) = ae^{-b\theta}/-abe^{-b\theta} = -1/b$. Thus $m = -1/b$, Q has
coordinates $(0,a/b)$, and PQ has length
$|PQ| = (a^2 + a^2/b^2)^{1/2} = (a/b)(b^2+1)^{1/2}$. Now from Eq.(20),
the required arclength is

$$s = \lim_{k\to\infty}\int_0^k [a^2e^{-2b\theta} + a^2b^2e^{-2b\theta}]d\theta = \lim_{k\to\infty} a(b^2+1)^{1/2}\int_0^k e^{-b\theta}d\theta$$

$$= a(b^2+1)^{1/2}\lim_{k\to\infty}(-e^{-b\theta}/b)\Big|_0^k = (a/b)(b^2+1)^{1/2}.$$

Thus the length of the spiral equals the length of the
tangent segment PQ.

Chapter 14 Review, Page 710

1. $(-2,5\pi/2) \to (2,5\pi/2 - \pi) = (2,3\pi/2)$
     $(-2,5\pi/2) \to (-2,5\pi/2 - 4\pi) = (-2,-3\pi/2)$.

3. $(-3,\pi/3)$, $(3,4\pi/3)$.

5. $x = 5\cos(-\pi/3) = 5/2$, $y = 5\sin(-\pi/3) = -5\sqrt{3}/2$.

7. $x = 3\cos(-2\pi/3) = -3/2$, $y = 3\sin(-2\pi/3) = -3\sqrt{3}/2$.

9. $r = (12+4)^{1/2} = 4$, $\theta = \arctan(2/\sqrt{12}) = \arctan(1/\sqrt{3}) = \pi/6$.

11. $r = (16)^{1/2} = 4$, $\theta = \arctan(\sqrt{3}/-1) = 2\pi/3$.

13. In rectangular form the equation
     becomes $(x+3/4)^2+(y-3\sqrt{3}/4)^2 = 9/4$,
     which is a circle, center at
     $(-3/4,3\sqrt{3}/4)$, radius $3/2$.

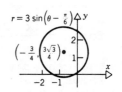

15. The graph is part of a spiral,
     starting at the origin and ending
     at $(r,\theta) = (2\pi,2\pi)$.

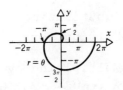

17. The graph will be a cardioid with
     horizontal axis of symmetry and $r\geq0$.

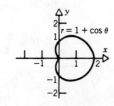

19. The graph will be a lemniscate. We
     only need consider θ for which
     $\sin 2\theta \le 0$, or $-\pi/2 \le \theta \le 0$.

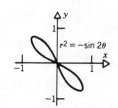

21. Following the discussion of
     Prob.30-32, Section 14.2, we see
     that e = 3 (a hyperbola), one
     focus is at the pole, and one
     directrix is y = 3.   The
     rectangular form of the equation
     is $\dfrac{(y - 27/8)^2}{81/64} - \dfrac{x^2}{81/8} = 1$. Thus
     the center is at (0,27/8), the
     second focus is at (0,27/4) and the second directrix is
     y = 27/8 + 3/8 = 15/4.

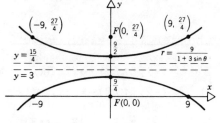

23. Following the results of Prob.32(a),
     Section 14.2, and expressing the
     equation as $r = \dfrac{(1/4)(6)}{[1 - (1/4)\cos\theta]}$, we
     have e = 1/4 < 1, giving an ellipse
     with one focus at the pole and one
     directrix at x = -6.   The rectangular
     form is $\dfrac{(x - 2/5)^2}{(24/15)^2} + \dfrac{y^2}{36/15} = 1$.   Thus
     the center is at (2/5,0), the second
     focus is at x = 2/5 + 2/5 = 4/5, y = 0, while the second
     directrix is x = 4/5 + 6 = 34/5.

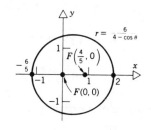

25. Substituting gives $3 = r^2\cos\theta/(r\cos\theta + r)$, simplifying and
     solving for r then gives $r = 3(1+\cos\theta)/\cos\theta = 3\sec\theta + 3$.

27. Multiplying by 1-x and substituting yields
     $r^2\sin^2\theta - r^3\sin^2\theta \cos\theta = r^3\cos^3\theta$, or
     $r = \sin^2\theta/\cos\theta = (1-\cos^2\theta)/\cos\theta = \sec\theta - \cos\theta$.

29. $y/x = \tan\theta = \tan(11\pi/6) = -1/\sqrt{3}$, or $y = (-1/\sqrt{3})x$.

31. $\sqrt{x^2 + y^2} = 2x/y$ gives $x^2y^2 + y^4 - 4x^2 = 0$.

33. $3\sec\theta + 1 = 4$ when $\sec\theta = 1$.   Thus $\theta = 0$ and $r = 4$.

35.   When $\theta = 3\pi/4$, $r^2 = 1$ so $r = \pm1$, so the points of intersection
      are $(-1, 3\pi/4)$, $(1, 3\pi/4)$ and the pole.

37.   The lemniscate is described for $0 \le \theta \le 2\pi$.   Thus, by Eq. (7)
      of Section 14.3, $A = (1/2)\displaystyle\int_0^{2\pi} (2+\sin\theta)^2 d\theta = 9\pi/2$.

39. $A = (1/2)\displaystyle\int_0^{2\pi} 4\sin2\theta\, d\theta = 2$.

41. $dr/d\theta = -3\csc\theta\cot\theta$ and, from Eq. (20) of Section 14.3,
      $s = \displaystyle\int_{\pi/4}^{3\pi/4} (9\csc^2\theta + 9\csc^2\theta\cot^2\theta)^{1/2} d\theta = 3\int_{\pi/4}^{3\pi/4} \csc^2\theta\, d\theta$.

43.   Differentiating implicitly yields $2r\,dr/d\theta = -8\sin2\theta$ so that
      $dr/d\theta = -4\sin2\theta/r$ and $(dr/d\theta)^2 = 16\sin^2 2\theta/r^2 = 4\sin^2 2\theta/\cos2\theta$.
      Thus $s = \displaystyle\int_0^{\pi/6} (4\cos2\theta + 4\sin^2 2\theta/\cos2\theta)^{1/2} d\theta = 2\int_0^{\pi/6} (\sec2\theta)^{1/2} d\theta$.

45.   From Prob. 32, Section 14.3, slope $= (\tan\theta + \tan\psi)/(1 - \tan\theta\tan\psi)$
      and $\tan\psi = r/(dr/d\theta) = (3-\cos\theta)/\sin\theta = 5/\sqrt{3}$ for $\theta = \pi/3$ and
      thus the slope $= -2\sqrt{3}/3$.

47.   Again, $\tan\psi = (\sec\theta-\cos\theta)/(\sec\theta\tan\theta+\sin\theta) = 3/5\sqrt{3}$, for
      $\theta = \pi/3$, and the slope $= [\sqrt{3} + (3/5\sqrt{3})]/[1 - (3/5)] = 3\sqrt{3}$.

CHAPTER 15

1. $\mathbf{a}+\mathbf{b}$ = (-2,1)+(3,-2) = (-2+3,1-2) = (1,-1),
    $\mathbf{a}-\mathbf{b}$ = (-2-3,1+2) = (-5,3), $2\mathbf{a}-3\mathbf{b}$ = (-4,2)+(-9,6) = (-13,8).

3. $\mathbf{a}+\mathbf{b}$ = (1,1)+(1,-1) = (1+1,1-1) = (2,0),
    $\mathbf{a}-\mathbf{b}$ = (1-1,1+1) = (0,2), and $2\mathbf{a}-3\mathbf{b}$ = (2,2)+(-3,3) = (-1,5).

5. From Eq.(17), $\overrightarrow{PQ}$ = (4,-2)-(2,1) = (4-2,-2-1) = (2,-3).

7. $\overrightarrow{PQ}$ = (-1,0)-(2,4) = (-1-2,0-4) = (-3,-4).

9. Note that $\mathbf{b}$ = $\mathbf{a}$ + $\overrightarrow{PQ}$ so $\overrightarrow{PQ}$ = (3,-2)-(2,1) = (1,-3) and
    $\overrightarrow{QP}$ = $-\overrightarrow{PQ}$ = (-1,3).

11. If $\mathbf{a}$ = $\mathbf{b}$, then x+y = 2 and x-y = 3, by Eq.(3). Solving these
    two equations yields x = 5/2 and y = -1/2.

13. $\|\mathbf{a}\|$ = $\sqrt{2^2+1^2}$ = $\sqrt{5}$, by Eq.(2).

15. $\|\mathbf{a}\|$ = $\sqrt{(-3)^2+(-4)^2}$ = $\sqrt{9+16}$ = 5.

17. $\|\mathbf{a}\|$ = $\sqrt{(x^2+2xy+y^2)+(x^2-2xy+y^2)}$ = $\sqrt{2}\sqrt{x^2+y^2}$.

19. The vector $\mathbf{a}$ lies along the line that has slope 2 and thus $\mathbf{b}$
    must lie along the line with slope -1/2. Hence, $\mathbf{b}$ = c(-2,1),
    where c is chosen so that $\|\mathbf{b}\|$ = $|c|\sqrt{4+1}$ = $|c|\sqrt{5}$ = 5.
    Therefore, $|c|$ = $\sqrt{5}$ and $\mathbf{b}$ = $\pm\sqrt{5}$(-2,1).

21. By Eq.(2), we must have $\sqrt{2^2+(-3)^2}$ = $\sqrt{\lambda^2+1^2}$ or $4+9 = \lambda^2+1$, by
    squaring both sides. Thus λ^2 = 12 or λ = $\pm2\sqrt{3}$.

23. Let $\mathbf{a}$ = (a_1,a_2), then $(\lambda\mu)\mathbf{a}$ = $(\lambda\mu a_1,\lambda\mu a_2)$ and $\mu\mathbf{a}$ = $(\mu a_1,\mu a_2)$,
    by Eq.(8). Thus $\lambda(\mu\mathbf{a})$ = $(\lambda[\mu a_1],\lambda[\mu a_2])$ = $(\lambda\mu a_1,\lambda\mu a_2)$ = $(\lambda\mu)\mathbf{a}$.

25. Let $\mathbf{T}_1$ = $(-T\cos\theta,T\sin\theta)$, which is the vector to the left in
    Fig.15.1.14, $\mathbf{T}_2$ = $(T\cos\theta,T\sin\theta)$, which is the vector to the
    right, and $\mathbf{W}$ = (0,-W), which is the downward vector. Thus,
    we have $\mathbf{T}_1+\mathbf{T}_2+\mathbf{W}$ = $\mathbf{0}$ or $(-T\cos\theta+T\cos\theta,T\sin\theta+T\sin\theta-W)$ = (0,0),
    so $-T\cos\theta+T\cos\theta$ = 0 and $T\sin\theta+T\sin\theta-W$ = 0 or T = $W/2\sin\theta$.

Section 15.2, Page 725

1. By Eq.(1), $\mathbf{a\cdot b}$ = (1)(2)+(1)(-3) = -1 and thus
    $\cos\theta = -\dfrac{1}{\sqrt{1+1}\sqrt{4+9}}$ = $-1/\sqrt{26}$ $\cong$ -.1961 by Eq.(13).

3. $\mathbf{a\cdot b}$ = (1)(2)+(-2)(1) = 0 and thus $\cos\theta$ = 0 also.

5. $\mathbf{a\cdot b}$ = (1)(4)+(-1)(1) = 3 so $\cos\theta = \dfrac{3}{\sqrt{2}\sqrt{17}}$ = $3/\sqrt{34}$ $\cong$ .5145.

7. Since $2\mathbf{i}+2\mathbf{j}$ and $\mathbf{i}+3\mathbf{j}$ are also the vectors (2,2) and (1,3)
    respectively, we have from Eq.(1), $\mathbf{a\cdot b}$ = (2)(1)+(2)(3) = 8
    and thus $\cos\theta = \dfrac{8}{\sqrt{8}\sqrt{10}}$ = $2/\sqrt{5}$ $\cong$ .8944.

9. Since $\|\mathbf{a}\|$ = $\sqrt{9+16}$ = 5, we have $\mathbf{u}$ = (-3/5,-4/5), by Eq.(8).

11. $\|\mathbf{a}\|$ = $\sqrt{9x^2+16x^2}$ = 5|x|, so $\mathbf{u}$ = (3x/5|x|,-4x/5|x|).

13. $\|\mathbf{a}\|$ = $\sqrt{25+144}$ = 13, so $\mathbf{u}$ = $(-5\mathbf{i}+12\mathbf{j})/13$ = -(5/13)$\mathbf{i}$+(12/13)$\mathbf{j}$.

15. $\|\mathbf{a}\|$ = $\sqrt{1+9}$ = $\sqrt{10}$,so $\mathbf{u}$ = $(-1/\sqrt{10},-3/\sqrt{10})$.

17. The vectors $\mathbf{b}$ = $\pm(2\mathbf{i}+\mathbf{j})$ are perpendicular to $\mathbf{a}$ since
    $\mathbf{a\cdot b}$ = $\pm[(1)(2)-(2)(1)]$ = 0. Thus, $\mathbf{u}$ = $\pm[(2/\sqrt{5})\mathbf{i}+(1/\sqrt{5})\mathbf{j}]$.

19. If $\mathbf{u}$ = $\mathbf{a}/\|\mathbf{a}\|$, then from Eq.(16), $\mathbf{b\cdot u}$ is the component of $\mathbf{b}$ in
    the direction of $\mathbf{a}$. Thus, $\mathbf{u}$ = $(1/\sqrt{2},1/\sqrt{2})$ and
    $\mathbf{b\cdot u}$ = = $2/\sqrt{2}$ + $1/\sqrt{2}$ = $3/\sqrt{2}$.

21. As in Prob.19, $\mathbf{u}$ = $(2/\sqrt{5},1/\sqrt{5})$ so $\mathbf{b\cdot u}$ = $2/\sqrt{5}$ - $3/\sqrt{5}$ = $-1/\sqrt{5}$.

23. As in Prob.19, $\mathbf{u}$ = $(1/\sqrt{5})\mathbf{i}$ - $(2/\sqrt{5})\mathbf{j}$ so
    $\mathbf{b\cdot u}$ = $x/\sqrt{5}$ - $2x/\sqrt{5}$ = $-x/\sqrt{5}$.

25. From Eq.(16), $\text{proj}_\mathbf{a}\mathbf{b}$ = $(\mathbf{b\cdot u})\mathbf{u}$. In this case,
    $\mathbf{u}$ = $\mathbf{a}/\|\mathbf{a}\|$ = $(2/\sqrt{5},-1/\sqrt{5})$, $\mathbf{b\cdot u}$ = $2/\sqrt{5}$ + $3/\sqrt{5}$ = $5/\sqrt{5}$, and thus
    $(\mathbf{b\cdot u})\mathbf{u}$ = (2,-1).

27. As in Prob.25, we have $\mathbf{u}$ = $\mathbf{a}/\|\mathbf{a}\|$ = $(1/\sqrt{2})\mathbf{i}+(1/\sqrt{2})\mathbf{j}$,
    $\mathbf{b\cdot u}$ = $2/\sqrt{2}$ + $3/\sqrt{2}$ = $5/\sqrt{2}$ and thus $(\mathbf{b\cdot u})\mathbf{u}$ = (5/2)$\mathbf{i}$+(5/2)$\mathbf{j}$.

29. From Eq.(1), $\mathbf{a\cdot b}$ = $a_1b_1+a_2b_2$ = $b_1a_1+b_2a_2$ = $\mathbf{b\cdot a}$, also from
    Eq.(1). Thus, $\mathbf{a\cdot b}$ = $\mathbf{b\cdot a}$.

31. $\mathbf{a}+\mathbf{b} = (4,1)$ so $\|\mathbf{a}+\mathbf{b}\| = \sqrt{17} \cong 4.123$, and
      $\|\mathbf{a}\| + \|\mathbf{b}\| = \sqrt{5} + \sqrt{10} \cong 5.398$ and thus $\|\mathbf{a}+\mathbf{b}\| < \|\mathbf{a}\| + \|\mathbf{b}\|$.

33. $\mathbf{a}+\mathbf{b} = 5\mathbf{i}-3\mathbf{j}$ so $\|\mathbf{a}+\mathbf{b}\| = \sqrt{34} \cong 5.831$ and
      $\|\mathbf{a}\| + \|\mathbf{b}\| = \sqrt{5} + \sqrt{25} \cong 7.236$ and thus $\|\mathbf{a}+\mathbf{b}\| < \|\mathbf{a}\| + \|\mathbf{b}\|$.

35.   As in Eq.(22), we have:
$$\|c_1\mathbf{a}+c_2\mathbf{b}\|^2 = (c_1\mathbf{a}+c_2\mathbf{b}) \cdot (c_1\mathbf{a}+c_2\mathbf{b})$$
$$= c_1^2\|\mathbf{a}\|^2 + 2c_1c_2\mathbf{a}\cdot\mathbf{b} + c_2^2\|\mathbf{b}\|^2$$
$$\leq c_1^2\|\mathbf{a}\|^2 + 2c_1c_2\|\mathbf{a}\|\,\|\mathbf{b}\| + c_2^2\|\mathbf{b}\|^2$$
$$= [|c_1|\|\mathbf{a}\| +|c_2|\|\mathbf{b}\|]^2.$$ Taking the positive square
root of both sides yields the desired inequality.

37a. From the figure shown, we see that
      $\mathbf{r} = \mathbf{a} + \overrightarrow{PR}$. But $\overrightarrow{PR} = t\overrightarrow{PQ} = t(\mathbf{b}-\mathbf{a})$,
      where $0 \leq t \leq 1$. Thus, $\mathbf{r} = \mathbf{a} + t(\mathbf{b}-\mathbf{a})$.
   b. If $t < 0$ or $t > 1$, then the points
      given by Eq.(i) lie outside the
      interval PQ on the straight line
      through PQ.

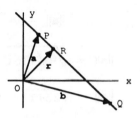

Section 15.3, Page 733

1.   Let $\mathbf{a}$ and $\mathbf{b}$ be the position vectors of P and Q respectively.
     Then $\mathbf{a} = (2,1,-1)$, $\mathbf{b} = (3,-2,4)$ and hence
     $\overrightarrow{PQ} = \mathbf{b}-\mathbf{a} = (3,-2,4) - (2,1,-1) = (1,-3,5)$.

3.   If Q has position vector $\mathbf{b}$, then
     $\mathbf{b} = (2\mathbf{i}-\mathbf{j}+2\mathbf{k}) + (-\mathbf{i}+4\mathbf{j}+2\mathbf{k}) = \mathbf{i}+3\mathbf{j}+4\mathbf{k}$.

5. $3\mathbf{a}-2\mathbf{b} = 3(2,-1,3)-2(-3,2,1) = (6,-3,9)+(6,-4,-2) = (12,-7,7)$.

7. $3\mathbf{a}-2\mathbf{b} = (6\mathbf{i}+3\mathbf{j}-3\mathbf{k})-(-2\mathbf{i}+6\mathbf{k}) = 8\mathbf{i}+3\mathbf{j}-9\mathbf{k}$.

9. $\|\mathbf{a}\| = \sqrt{3^2+(-2)^2+1^2} = \sqrt{9+4+1} = \sqrt{14}$.

11. $\|\mathbf{a}\| = \sqrt{2^2+1^2+(-4)^2} = \sqrt{4+1+16} = \sqrt{21}$.

13. $\overrightarrow{PQ} = (3,-2,4)-(2,1,-1) = (1,-3,5)$ so $\|\overrightarrow{PQ}\| = \sqrt{1+9+25} = \sqrt{35}$.

15. $\mathbf{a}\cdot\mathbf{b} = (3)(2)+(1)(-4)+(2)(1) = 4$, so
     $$\cos\theta = \frac{4}{\sqrt{9+1+4}\sqrt{4+16+1}} = 4/7\sqrt{6} \cong .2333.$$

17. $\mathbf{a}\cdot\mathbf{b} = (1)(1)+(-2)(-1)+(-3)(1) = 0$, so $\cos\theta = 0$.

19. $\|\mathbf{a}\| = \sqrt{4+1+9} = \sqrt{14}$, $\mathbf{u} = (2/\sqrt{14}, 1/\sqrt{14}, -3/\sqrt{14})$ and thus
     $\mathbf{b}\cdot\mathbf{u} = (2+1-6)/\sqrt{14} = -3/\sqrt{14} \cong -.8018$.

21. $\|\mathbf{a}\| = \sqrt{4+9+16} = \sqrt{29}$, $\mathbf{u} = (2\mathbf{i}+3\mathbf{j}+4\mathbf{k})/\sqrt{29}$ and thus
     $\mathbf{b}\cdot\mathbf{u} = (-2+3+4)/\sqrt{29} = 5/\sqrt{29} \cong .9285$.

23. The given vectors will be perpendicular if $\mathbf{a}\cdot\mathbf{b} = 0$. Thus,
     $\mathbf{a}\cdot\mathbf{b} = 6\lambda+\lambda-3 = 7\lambda-3$ and hence we need $\lambda = 3/7$.

25. $\|\mathbf{a}\| = \sqrt{9+1+4} = \sqrt{14}$ so $\mathbf{u} = (3\mathbf{i}+\mathbf{j}-2\mathbf{k})/\sqrt{14}$.

27. The vector from P to Q is given by
     $(1,-1,1)-(-2,0,4) = (3,-1,-3)$, so $(3/\sqrt{19}, -1/\sqrt{19}, -3/\sqrt{19})$ is
     the desired unit vector. The vector may also be written as
     $(3\mathbf{i}-\mathbf{j}-3\mathbf{k})/\sqrt{19}$.

29. Let $\mathbf{a}$ and $\mathbf{b}$ represent two sides of
     the parallelogram, then the
     diagonals are given by $\mathbf{a}+\mathbf{b}$ and $\mathbf{a}-\mathbf{b}$
     as shown. For the diagonals to be
     perpendicular, we need
     $(\mathbf{a}+\mathbf{b})\cdot(\mathbf{a}-\mathbf{b}) = 0$, but
     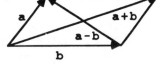
     $(\mathbf{a}+\mathbf{b})\cdot(\mathbf{a}-\mathbf{b}) = \mathbf{a}\cdot\mathbf{a} - \mathbf{a}\cdot\mathbf{b} + \mathbf{b}\cdot\mathbf{a} - \mathbf{b}\cdot\mathbf{b} = \|\mathbf{a}\|^2 - \|\mathbf{b}\|^2$. Thus, the
     diagonals are perpendicular if and only if $\|\mathbf{a}\| = \|\mathbf{b}\|$.

31. Let a be the length of one side of the cube, which is
     situated so that one corner is at the origin. Then one side
     is represented by the vector $\mathbf{a} = a\mathbf{i}$ and the diagonal is
     represented by the vector $\mathbf{b} = a\mathbf{i}+a\mathbf{j}+a\mathbf{k}$. Thus
     $$\cos\theta = \frac{\mathbf{a}\cdot\mathbf{b}}{\|\mathbf{a}\|\|\mathbf{b}\|} = \frac{a^2}{a(\sqrt{3}a)} = 1/\sqrt{3}, \text{ so that}$$
     $\theta = \arccos(1/\sqrt{3}) \cong .9533$ radians.

33. From Prob.31, we have that $\mathbf{c} = a\mathbf{i}+a\mathbf{j}$ represents the diagonal
     of one side and thus $\cos\theta = \dfrac{\mathbf{b}\cdot\mathbf{c}}{\|\mathbf{b}\|\|\mathbf{c}\|} = \dfrac{2a^2}{(\sqrt{3}a)(\sqrt{2}a)} = \sqrt{2/3}$ so that
     $\theta = \arccos(\sqrt{2/3}) \cong .6155$.

Section 15.4, Page 741

1. $\mathbf{a} \times \mathbf{b} = \begin{vmatrix} \mathbf{i} & \mathbf{j} & \mathbf{k} \\ 2 & -1 & -1 \\ 1 & 2 & 4 \end{vmatrix} = (-4+2)\mathbf{i} - (8+1)\mathbf{j} + (4+1)\mathbf{k} = -2\mathbf{i} - 9\mathbf{j} + 5\mathbf{k}.$

3. $\mathbf{a} \times \mathbf{b} = \begin{vmatrix} \mathbf{i} & \mathbf{j} & \mathbf{k} \\ 1 & 1 & 1 \\ -2 & -3 & 1 \end{vmatrix} = (1+3)\mathbf{i} - (1+2)\mathbf{j} + (-3+2)\mathbf{k} = 4\mathbf{i} - 3\mathbf{j} - \mathbf{k}.$

5. $\mathbf{a} \times \mathbf{b} = \begin{vmatrix} \mathbf{i} & \mathbf{j} & \mathbf{k} \\ a_1 & a_2 & 0 \\ b_1 & b_2 & 0 \end{vmatrix} = 0\mathbf{i} - 0\mathbf{j} + (a_1 b_2 - a_2 b_1)\mathbf{k}.$

7. $\mathbf{b} + \mathbf{c} = -\mathbf{i} + 5\mathbf{j} + 4\mathbf{k}$ and thus $\mathbf{a} \times (\mathbf{b} + \mathbf{c}) = \begin{vmatrix} \mathbf{i} & \mathbf{j} & \mathbf{k} \\ 2 & -1 & -1 \\ -1 & 5 & 4 \end{vmatrix} = \mathbf{i} - 7\mathbf{j} + 9\mathbf{k}.$

9. $\mathbf{a} \cdot (\mathbf{b} \times \mathbf{c}) = \begin{vmatrix} 2 & -1 & -1 \\ 1 & 2 & 3 \\ -2 & 3 & 1 \end{vmatrix} = 2(2-9) + 1(1+6) - 1(3+4) = -14.$

11. $\mathbf{b} \times \mathbf{c} = \begin{vmatrix} \mathbf{i} & \mathbf{j} & \mathbf{k} \\ 1 & 2 & 3 \\ -2 & 3 & 1 \end{vmatrix} = -7\mathbf{i} - 7\mathbf{j} + 7\mathbf{k}$ and thus

$\mathbf{a} \times (\mathbf{b} \times \mathbf{c}) = \begin{vmatrix} \mathbf{i} & \mathbf{j} & \mathbf{k} \\ 2 & -1 & -1 \\ -7 & -7 & 7 \end{vmatrix} = -14\mathbf{i} - 7\mathbf{j} - 21\mathbf{k}.$

13. $\mathbf{a} \times \mathbf{b} = \begin{vmatrix} \mathbf{i} & \mathbf{j} & \mathbf{k} \\ 1 & -2 & 1 \\ 3 & 1 & -4 \end{vmatrix} = 7\mathbf{i} + 7\mathbf{j} + 7\mathbf{k}.$ Thus, $\|\mathbf{a} \times \mathbf{b}\| = 7\sqrt{3}$ and
$\mathbf{u} = \pm(\mathbf{i} + \mathbf{j} + \mathbf{k})/\sqrt{3}$ since both $\mathbf{a} \times \mathbf{b}$ and $\mathbf{b} \times \mathbf{a} = -(\mathbf{a} \times \mathbf{b})$ are perpendicular to $\mathbf{a}$ and $\mathbf{b}$.

15. Let $\mathbf{a} = \overrightarrow{PQ} = 3\mathbf{j}$ and $\mathbf{b} = \overrightarrow{PR} = 2\mathbf{i}+5\mathbf{j}+8\mathbf{k}$. Then

$$\mathbf{a} \times \mathbf{b} = \begin{vmatrix} \mathbf{i} & \mathbf{j} & \mathbf{k} \\ 0 & 3 & 0 \\ 2 & 5 & 8 \end{vmatrix} = 24\mathbf{i}-6\mathbf{j} \text{ and thus, by Eq.(8),}$$

$A = \|\mathbf{a} \times \mathbf{b}\|/2 = \sqrt{24^2+6^2}/2 = \sqrt{153} \cong 12.369$.

17. Let $\mathbf{a} = \overrightarrow{PQ} = \mathbf{i}+\mathbf{j}$, $\mathbf{b} = \overrightarrow{PR} = -\mathbf{i}+3\mathbf{j}$, $\mathbf{c} = \overrightarrow{PS} = \mathbf{i}+4\mathbf{k}$ and then

$$\mathbf{a} \cdot (\mathbf{b} \times \mathbf{c}) = \begin{vmatrix} 1 & 1 & 0 \\ -1 & 3 & 0 \\ 1 & 0 & 4 \end{vmatrix} = 16. \quad \text{Since } \mathbf{a} \cdot (\mathbf{b} \times \mathbf{c}) \neq 0, \text{ the points}$$

are not coplanar and the volume $= 16$ by Eq.(23).

19. Let $\mathbf{a} = \overrightarrow{PQ} = (-2, \lambda-4, 2)$, $\mathbf{b} = \overrightarrow{PR} = (-4,-3,3)$,

$\mathbf{c} = \overrightarrow{PS} = (-1,-3,1)$ and then

$$\mathbf{a} \cdot (\mathbf{b} \times \mathbf{c}) = \begin{vmatrix} -2 & \lambda-4 & 2 \\ -4 & -3 & 3 \\ -1 & -3 & 1 \end{vmatrix} = -2(-3+9)-(\lambda-4)(-4+3)+2(12-3) = \lambda+2.$$

The points are coplanar if $\mathbf{a} \cdot (\mathbf{b} \times \mathbf{c}) = 0$ and thus $\lambda = -2$.

21. $\mathbf{a} = \overrightarrow{P_1P_2} = (x_2-x_1)\mathbf{i}+(y_2-y_1)\mathbf{j}$, $\mathbf{b} = \overrightarrow{P_1P_3} = (x_3-x_1)\mathbf{i}+(y_3-y_1)\mathbf{j}$, so

$\mathbf{a} \times \mathbf{b} = [x_2y_3-x_2y_1-x_1y_3+x_1y_1-(x_3y_2-x_3y_1-x_1y_2+x_1y_1)]\mathbf{k}$

$\qquad = [(x_2y_3-x_3y_2)-(x_1y_3+x_3y_1)+(x_1y_2-x_2y_1)]\mathbf{k}$

$$\qquad = \begin{vmatrix} 1 & 1 & 1 \\ x_1 & x_2 & x_3 \\ y_1 & y_2 & y_3 \end{vmatrix} \mathbf{k}$$

and thus, by Eq.(8), we have the desired result. Note that abs must be used since the last determinant can be positive or negative.

23a. $(\alpha\mathbf{a}) \times \mathbf{b} = \begin{vmatrix} \mathbf{i} & \mathbf{j} & \mathbf{k} \\ \alpha a_1 & \alpha a_2 & \alpha a_3 \\ b_1 & b_2 & b_3 \end{vmatrix} = \alpha \begin{vmatrix} \mathbf{i} & \mathbf{j} & \mathbf{k} \\ a_1 & a_2 & a_3 \\ b_1 & b_2 & b_3 \end{vmatrix} = \alpha(\mathbf{a} \times \mathbf{b})$.

23b. $\mathbf{a} \times (\alpha \mathbf{b}) = \begin{vmatrix} \mathbf{i} & \mathbf{j} & \mathbf{k} \\ a_1 & a_2 & a_3 \\ \alpha b_1 & \alpha b_2 & \alpha b_3 \end{vmatrix} = \alpha \begin{vmatrix} \mathbf{i} & \mathbf{j} & \mathbf{k} \\ a_1 & a_2 & a_3 \\ b_1 & b_2 & b_3 \end{vmatrix} = \alpha (\mathbf{a} \times \mathbf{b}).$

25a. $\mathbf{b} \times \mathbf{c} = (b_2 c_3 - b_3 c_2) \mathbf{i} - (b_1 c_3 - b_3 c_1) \mathbf{j} + (b_1 c_2 - b_2 c_1) \mathbf{k}$ from Eq.(1).
   Using the distributive law, Eq.(15a), and the appropriate
   terms from Eqs.(12) to (14), we then have
   $\mathbf{i} \times (\mathbf{b} \times \mathbf{c}) = \mathbf{0} - (b_1 c_3 - b_3 c_1) \mathbf{k} - (b_1 c_2 - b_2 c_1) \mathbf{j}.$ Now
   $c_1 \mathbf{b} - b_1 \mathbf{c} = (c_1 b_1 - b_1 c_1) \mathbf{i} + (c_1 b_2 - b_1 c_2) \mathbf{j} + (c_1 b_3 - b_1 c_3) \mathbf{k}$ and thus
   $\mathbf{i} \times (\mathbf{b} \times \mathbf{c}) = c_1 \mathbf{b} - b_1 \mathbf{c}.$

   b. Likewise, $\mathbf{j} \times (\mathbf{b} \times \mathbf{c}) = -(b_2 c_3 - b_3 c_2) \mathbf{k} + (b_1 c_2 - b_2 c_1) \mathbf{i} = c_2 \mathbf{b} - b_2 \mathbf{c}$
      and $\mathbf{k} \times (\mathbf{b} \times \mathbf{c}) = (b_2 c_3 - b_3 c_2) \mathbf{j} + (b_1 c_3 - b_3 c_1) \mathbf{i} = c_3 \mathbf{b} - b_3 \mathbf{c}.$

   c. Using Eq.(15b) with $\mathbf{a} = a_1 \mathbf{i} + a_2 \mathbf{j} + a_3 \mathbf{k}$, we obtain
      $\mathbf{a} \times (\mathbf{b} \times \mathbf{c}) = a_1 (c_1 \mathbf{b} - b_1 \mathbf{c}) + a_2 (c_2 \mathbf{b} - b_2 \mathbf{c}) + a_3 (c_3 \mathbf{b} - b_3 \mathbf{c})$
      $$= (a_1 c_1 + a_2 c_2 + a_3 c_3) \mathbf{b} - (a_1 b_1 + a_2 b_2 + a_3 b_3) \mathbf{c}$$
      $$= (\mathbf{c} \cdot \mathbf{a}) \mathbf{b} - (\mathbf{b} \cdot \mathbf{a}) \mathbf{c}.$$

27a. $\mathbf{d} \cdot \mathbf{a} = d_1 a_1 + d_2 a_2 + d_3 a_3 = 0$ and $\mathbf{d} \cdot \mathbf{b} = d_1 b_1 + d_2 b_2 + d_3 b_3 = 0.$
   Treating d_2 and d_3 as unknowns, these two equations may be
   written as: $a_2 d_2 + a_3 d_3 = -a_1 d_1$
   $$b_2 d_2 + b_3 d_3 = -b_1 d_1$$
   which have the solutions shown in Eqs.(i) and (ii).

   b. $\|\mathbf{d}\|^2 = \|\mathbf{a}\|^2 \|\mathbf{b}\|^2 \sin^2 \theta = \|\mathbf{a}\|^2 \|\mathbf{b}\|^2 - \|\mathbf{a}\|^2 \|\mathbf{b}\|^2 \cos^2 \theta$
      $$= \|\mathbf{a}\|^2 \|\mathbf{b}\|^2 - (\mathbf{a} \cdot \mathbf{b})^2$$
      $$= (a_1^2 + a_2^2 + a_3^2)(b_1^2 + b_2^2 + b_3^2) - (a_1 b_1 + a_2 b_2 + a_3 b_3)^2$$
      $$= (a_1^2 b_2^2 - 2 a_1 a_2 b_1 b_2 + a_2^2 b_1^2) + (a_1^2 b_3^2 - 2 a_1 a_3 b_1 b_3 + a_3^2 b_1^2)$$
      $$+ (a_2^2 b_3^2 - 2 a_2 a_3 b_2 b_3 + a_3^2 b_2^2)$$
      $$= (a_1 b_2 - a_2 b_1)^2 + (a_3 b_1 - a_1 b_3)^2 + (a_2 b_3 - a_3 b_2)^2.$$

   c. If $\|\mathbf{d}\| \neq 0$, then at least one of the squared terms in
      Eq.(iii) is non-zero. Hence, assume $a_2 b_3 - a_3 b_2 \neq 0$ and thus
      Eqs.(i) and (ii) can be solved for d_2 and d_3 so that

$$\|\mathbf{d}\|^2 = d_1{}^2 + d_2{}^2 + d_3{}^2 = d_1{}^2 + \frac{(a_3b_1 - a_1b_3)^2 d_1^2}{(a_2b_3 - a_3b_2)^2} + \frac{(a_1b_2 - a_2b_1)^2 d_1^2}{(a_2b_3 - a_3b_2)^2}$$

$$= \frac{d_1^2}{(a_2b_3 - a_3b_2)^2} [(a_2b_3 - a_3b_2)^2 + (a_3b_1 - a_1b_3)^2 + (a_1b_2 - a_2b_1)^2].$$

Comparing this expression for $\|\mathbf{d}\|^2$ with that of part(b), we conclude that $d_1{}^2 = (a_2b_3 - a_3b_2)^2$ or $d_1 = \pm(a_2b_3 - a_3b_2)$. For $d_1 \neq 0$, substituting the +value into Eqs.(i) and (ii) yields the +values in Eqs.(v) and (vi), and similarly for the -values.

27d. If $\mathbf{d} = \mathbf{i} \times \mathbf{j}$, then the right-hand rule requires $\mathbf{d} = \mathbf{k}$ and thus $d_3 = +1$ and hence the plus sign must be used in Eqs.(iv),(v), and (vi).

e. Rewrite the equations in part(a) as: $a_1d_1 + a_2d_2 = -a_3d_3$
$$b_1d_1 + b_2d_2 = -b_3d_3$$
which have the solutions $(a_1b_2 - a_2b_1)d_1 = (a_2b_3 - a_3b_2)d_3$
$$(a_1b_2 - a_2b_1)d_2 = (a_3b_1 - a_1b_3)d_3$$
In this case assume $a_1b_2 - a_2b_1 \neq 0$ and thus

$$\|\mathbf{d}\|^2 = d_2{}^2 + d_3{}^2 = \frac{d_3^2}{(a_1b_2 - a_2b_1)^2}[(a_1b_2 - a_2b_1)^2 + (a_3b_1 - a_1b_3)^2] \text{ since}$$

$d_1 = a_2b_3 - a_3b_2 = 0$. Comparing to $\|\mathbf{d}\|$, as given in part(b), we conclude that $d_3 = \pm(a_1b_2 - a_2b_1)$ and $d_2 = \pm(a_3b_1 - a_1b_3)$. Finally, if $d_1 = d_3 = 0$, we conclude that $d_2 = \pm(a_3b_1 - a_1b_3)$ since $\|\mathbf{d}\|^2 = d_2{}^2 = (a_3b_1 - a_1b_3)^2$ from part(b).

Section 15.5, Page 750

1. We have $\mathbf{r} = x\mathbf{i} + y\mathbf{j} + z\mathbf{k}$ and $\mathbf{r}_0 = 2\mathbf{i} - \mathbf{j} + 3\mathbf{k}$ and hence $\mathbf{r} - \mathbf{r}_0 = (x-2)\mathbf{i} + (y+1)\mathbf{j} + (z-3)\mathbf{k}$. Thus Eq.(1) yields $(\mathbf{r} - \mathbf{r}_0) \cdot \mathbf{n} = (x-2)(-1) + (y+1)(4) + (z-3)(5) = 0$ or $-x + 4y + 5z = 9$.

3. $\overrightarrow{PQ} = -3\mathbf{i} + 2\mathbf{j} - 2\mathbf{k}$ and $\overrightarrow{PR} = -\mathbf{i} + 5\mathbf{j} - \mathbf{k}$ so $\mathbf{n} = \overrightarrow{PQ} \times \overrightarrow{PR} = 8\mathbf{i} - \mathbf{j} - 13\mathbf{k}$. Using P to find $\mathbf{r}_0$ we have $\mathbf{r} - \mathbf{r}_0 = (x-1)\mathbf{i} + (y+2)\mathbf{j} + (z-3)\mathbf{k}$ and thus $\mathbf{n} \cdot (\mathbf{r} - \mathbf{r}_0) = 8(x-1) - (y+2) - 13(z-3) = 0$ or $8x - y - 13z = -29$.

5. The normal to the plane is $\mathbf{n} = 2\mathbf{i} - \mathbf{j} + 5\mathbf{k}$ and $\mathbf{r} - \mathbf{r}_0 = (x-3)\mathbf{i} + (y-1)\mathbf{j} + (z+2)\mathbf{k}$, so we need $\mathbf{n} \cdot (\mathbf{r} - \mathbf{r}_0) = 2(x-3) - (y-1) + 5(z+2) = 0$ or $2x - y + 5z = -5$.

7.   The point $(2,-1,0)$ lies on both lines and hence they
     intersect there. The vectors $3\mathbf{i}-2\mathbf{j}-\mathbf{k}$ and $-\mathbf{i}+4\mathbf{j}+2\mathbf{k}$ are
     parallel to each of the lines respectively and hence their
     cross product, $-5\mathbf{j}+10\mathbf{k}$, is perpendicular to the desired
     plane. Thus $\mathbf{n}\cdot(\mathbf{r}-\mathbf{r}_0) = 0-5(y+1)+10(z-0) = 0$ or $y-2z = -1$.

9.   To find a point on the given line, choose $z = -1$ and solve
     for x and y to find $(-2,3,-1)$ as a point on the line. The
     vector between this point and the given point is $4\mathbf{i}-\mathbf{j}-2\mathbf{k}$,
     which lies in the desired plane, and the vector $-\mathbf{i}+2\mathbf{j}+3\mathbf{k}$ is
     parallel to the desired plane and hence their cross product,
     $\mathbf{i}-10\mathbf{j}+7\mathbf{k}$, is perpendicular to the desired plane. Thus,
     $\mathbf{n}\cdot(\mathbf{r}-\mathbf{r}_0) = (x+2)-10(y-3)+7(z+1) = 0$ or $x-10y+7z = -39$.

11. Setting each term equal to t, we obtain $\dfrac{x-2}{3} = t$, $\dfrac{y+1}{2} = t$,

     $\dfrac{z-4}{-1} = t$, or $x = 2+3t$, $y = -1+2t$, and $z = 4-t$.

13. We have $\mathbf{r}_0 = 2\mathbf{i}-\mathbf{j}+3\mathbf{k}$ and thus, from Eqs.(12), we have
     $x = 2+t$, $y = -1+2t$, and $z = 3-t$.

15. $\mathbf{a} = (2-4)\mathbf{i}+(-1-2)\mathbf{j}+(6-3)\mathbf{k} = -2\mathbf{i}-3\mathbf{j}+3\mathbf{k}$, $\mathbf{r}_0 = 4\mathbf{i}+2\mathbf{j}+3\mathbf{k}$ and
     thus Eqs.(13) become $\dfrac{x-4}{-2} = \dfrac{y-2}{-3} = \dfrac{z-3}{3}$.

17. The vector $\mathbf{j}+2\mathbf{k}$ is perpendicular to the plane. Thus,
     $\mathbf{a} = \mathbf{j}+2\mathbf{k}$, $\mathbf{r}_0 = -2\mathbf{i}+6\mathbf{j}+\mathbf{k}$ and Eqs.(16) become $x = -2, \dfrac{y-6}{1} = \dfrac{z-1}{2}$.

19. The vectors $\mathbf{n}_1 = \mathbf{i}-2\mathbf{k}$ and $\mathbf{n}_2 = 3\mathbf{j}+\mathbf{k}$ are perpendicular to the
     two planes respectively and hence $\mathbf{a} = \mathbf{n}_1\times\mathbf{n}_2 = 6\mathbf{i}-\mathbf{j}+3\mathbf{k}$. If
     $z = 0$, then $x = 4$ and $y = -3$ and thus $\mathbf{r}_0 = 4\mathbf{i}-3\mathbf{j}$ and Eqs.(12)
     become $x = 4+6t$, $y = -3-t$, and $z = 3t$.

21. Setting $x = 0$, $y = 0$, we have $z = 7$ and thus $(0,0,7)$ is a
     point on the plane. The vector from this point to the given
     point is $2\mathbf{i}+\mathbf{j}-12\mathbf{k}$ and hence Eq.(26) yields
     $d = |(2\mathbf{i}+\mathbf{j}-12\mathbf{k})\cdot(\mathbf{i}+\mathbf{j}+\mathbf{k})/\sqrt{3}| = 3\sqrt{3} \cong 5.196$.

23. $\mathbf{a} = -3\mathbf{i}+2\mathbf{j}+4\mathbf{k}$ is parallel to the line and $\mathbf{n} = 2\mathbf{i}-\mathbf{j}+2\mathbf{k}$ is
     perpendicular to the plane. Since $\mathbf{a}\cdot\mathbf{n} = 0$, the line is
     parallel to the plane.

25. The parametric form of the line is x = 2-2t, y = -1+3t, and
    z = 2-t. Substituting these into the equation of the plane
    yields 2(2-2t)+3(-1+3t)-(2-t) = 11, or t = 2. Thus, (-2,5,0)
    is the point of intersection.

27. Since $\mathbf{a} \cdot \mathbf{n}$ = -3(2)+2(-1)+4(2) = 0, the line is parallel to the
    plane. P(2,0,-1) is a point on the line and Q(10,0,0) is a
    point on the plane. Thus, $\overrightarrow{QP}$ = 8$\mathbf{i}$+$\mathbf{k}$ and d = $|\overrightarrow{QP} \cdot \mathbf{n}/3|$ = 6.

29. We have $\mathbf{a}$ = $\mathbf{i}$+3$\mathbf{j}$-2$\mathbf{k}$ and $\mathbf{b}$ = 4$\mathbf{i}$-$\mathbf{j}$+2$\mathbf{k}$ and thus
    $\mathbf{n}$ = $\mathbf{a} \times \mathbf{b}/\|\mathbf{a} \times \mathbf{b}\|$ = (4$\mathbf{i}$-10$\mathbf{j}$-13$\mathbf{k}$)/$\sqrt{285}$ is a unit vector
    perpendicular to the two lines. The points (2,-1,1) and
    (-1,2,-3) lie on the two lines respectively and thus
    $\mathbf{c}$ = 3$\mathbf{i}$-3$\mathbf{j}$+4$\mathbf{k}$ is a vector joining the two lines. The distance
    between the lines is then the projection of $\mathbf{c}$ on $\mathbf{n}$ and hence
    d = $|\mathbf{c} \cdot \mathbf{n}|$ = 10/$\sqrt{285}$ ≅ .5923.

31. The point A(1,-1,0) lies on the line and hence the vector
    from A to the given point P is $\overrightarrow{AP}$ = $\mathbf{i}$+2$\mathbf{j}$+$\mathbf{k}$. $\mathbf{a}$ = 2$\mathbf{i}$+3$\mathbf{j}$-$\mathbf{k}$ is a
    vector parallel to the line and thus $\mathbf{b}$ = $\overrightarrow{AP} \times \mathbf{a}$ = -5$\mathbf{i}$+3$\mathbf{j}$-$\mathbf{k}$ is
    a vector perpendicular to the plane of the point and line.
    Therefore, $\mathbf{n}$ = $\mathbf{a} \times \mathbf{b}/\|\mathbf{a} \times \mathbf{b}\|$ = ($\mathbf{j}$+3$\mathbf{k}$)/$\sqrt{10}$ is a unit vector
    perpendicular to the line and d = $|\overrightarrow{AP} \cdot \mathbf{n}|$ = $\sqrt{10}/2$ ≅ 1.581.

33a. Letting $\mathbf{r}$ = x$\mathbf{i}$+y$\mathbf{j}$+z$\mathbf{k}$, we have $\mathbf{a} \cdot \mathbf{r}$ = $a_1 x + a_2 y + a_3 z$ = 0, which
     is a plane through the origin perpendicular to the vector $\mathbf{a}$.

  b. $|\mathbf{a} \cdot \mathbf{r}|$ = c$\|\mathbf{a}\|$ ⟹ $|a_1 x + a_2 y + a_3 z|$ = c$\sqrt{a_1^2 + a_2^2 + a_3^2}$. If $\mathbf{a} \cdot \mathbf{r}$ > 0,
     then $a_1 x + a_2 y + a_3 z$ = c$\sqrt{a_1^2 + a_2^2 + a_3^2}$, which is a plane
     perpendicular to $\mathbf{a}$. Using Eq.(26), it can be shown that the
     distance from the origin to the plane is c. If $\mathbf{a} \cdot \mathbf{r}$ < 0,
     then $a_1 x + a_2 y + a_3 z$ = -c$\sqrt{a_1^2 + a_2^2 + a_3^2}$, which is also a plane
     perpendicular to $\mathbf{a}$ at a distance c from the origin, in the
     opposite direction.

33c. $(\mathbf{r}-\mathbf{a})\cdot\mathbf{r} = (x-a_1)x + (y-a_2)y + (z-a_3)z = 0$ or
      $(x - a_1/2)^2 + (y - a_2/2)^2 + (z - a_3/2)^2 = (a_1^2 + a_2^2 + a_3^2)/4$,
      which is a sphere of radius $\|\mathbf{a}\|/2$ and center at $\mathbf{a}/2$.

35.   If the lines intersect then $\mathbf{r}_1 + \lambda\mathbf{a} = \mathbf{r}_2 + \mu\mathbf{b}$ or
      $\mathbf{r}_1-\mathbf{r}_2 = \mu\mathbf{b} - \lambda\mathbf{a}$ and hence $(\mathbf{r}_1-\mathbf{r}_2)\cdot(\mathbf{a}\times\mathbf{b}) = (\mu\mathbf{b} - \lambda\mathbf{a})\cdot(\mathbf{a}\times\mathbf{b}) = 0$.
      Conversely, if $(\mathbf{r}_1-\mathbf{r}_2)\cdot(\mathbf{a}\times\mathbf{b}) = 0$, then $\mathbf{r}_1-\mathbf{r}_2$ must lie in the
      plane of $\mathbf{a}$ and $\mathbf{b}$ and hence $\mathbf{r}_1-\mathbf{r}_2 = \alpha\mathbf{a} + \beta\mathbf{b}$ for some choice
      of α and β. Letting $\alpha = -\lambda$ and $\beta = \mu$ then gives
      $\mathbf{r}_1 = \lambda\mathbf{a} = \mathbf{r}_2 = \mu\mathbf{b}$, and hence the lines intersect.

Section 15.6, Page 762

1.   Using Eq.(13), we have
      $\lim\limits_{t\to 0}[e^{-t}\mathbf{i}+2\cos t\,\mathbf{j}+(t^2-1)\mathbf{k}]$

$$= (\lim\limits_{t\to 0}e^{-t})\mathbf{i}+(\lim\limits_{t\to 0}2\cos t)\mathbf{j}+[\lim\limits_{t\to 0}(t^2-1)]\mathbf{k} = \mathbf{i}+2\mathbf{j}-\mathbf{k}.$$

3. $\lim\limits_{t\to 1}(\dfrac{1}{t}\mathbf{i}+\dfrac{t-2}{t+1}\mathbf{j}+\dfrac{2t}{t-1}\mathbf{k}) = (\lim\limits_{t\to 1}\dfrac{1}{t})\mathbf{i}+(\lim\limits_{t\to 1}\dfrac{t-2}{t+1})\mathbf{j}+(\lim\limits_{t\to 1}\dfrac{2t}{t-1})\mathbf{k}$, which

      does not exist due to the third term not existing.

5. $\lim\limits_{t\to 0}\dfrac{1}{t}(\sin 2t\,\mathbf{i}+3t\mathbf{j}+\tan t\,\mathbf{k}) = (\lim\limits_{t\to 0}\dfrac{\sin 2t}{t})\mathbf{i}+(\lim\limits_{t\to 0}3)\mathbf{j}+\lim\limits_{t\to 0}\dfrac{\sin t}{t\cos t}\mathbf{k}$

$$= (\lim\limits_{t\to 0}\dfrac{2\cos 2t}{1})\mathbf{i}+3\mathbf{j}+(\lim\limits_{t\to 0}\dfrac{\cos t}{\cos t-t\sin t})\mathbf{k}$$

$$= 2\mathbf{i}+3\mathbf{j}+\mathbf{k}, \text{ using L'Hospital's rule.}$$

7.   Let $\mathbf{u}(t) = u_1(t)\mathbf{i}+u_2(t)\mathbf{j}+u_3(t)\mathbf{k}$, $\mathbf{v}(t) = v_1(t)\mathbf{i}+v_2(t)\mathbf{j}+v_3(t)\mathbf{k}$,
      then $\lim\limits_{t\to t_0}[c_1\mathbf{u}(t)+c_2\mathbf{v}(t)]$

$$= \lim\limits_{t\to t_0}[c_1u_1+c_2v_1)\mathbf{i}+(c_1u_2+c_2v_2)\mathbf{j}+(c_1u_3+c_2v_3)\mathbf{k}]$$

$$= \lim\limits_{t\to t_0}(c_1u_1+c_2v_1)\mathbf{i}+\lim\limits_{t\to t_0}(c_1u_2+c_2v_2)\mathbf{j}+\lim\limits_{t\to t_0}(c_1u_3+c_2v_3)\mathbf{k}$$

$$= c_1\lim\limits_{t\to t_0}(u_1\mathbf{i}+u_2\mathbf{j}+u_3\mathbf{k})+c_2\lim\limits_{t\to t_0}(v_1\mathbf{i}+v_2\mathbf{j}+v_3\mathbf{k})$$

$$= c_1\lim\limits_{t\to t_0}\mathbf{u}(t)+c_2\lim\limits_{t\to t_0}\mathbf{v}(t).$$

9. $\lim_{t \to t_0} [\mathbf{u}(t) \times \mathbf{v}(t)]$

$$= \lim_{t \to t_0} (u_2 v_3 - u_3 v_2)\mathbf{i} + \lim_{t \to t_0} (u_3 v_1 - u_1 v_3)\mathbf{j} + \lim_{t \to t_0} (u_1 v_2 - u_2 v_1)\mathbf{k}$$

$$= (\lim_{t \to t_0} u_2 \lim_{t \to t_0} v_3 - \lim_{t \to t_0} u_3 \lim_{t \to t_0} v_2)\mathbf{i} + (\lim_{t \to t_0} u_3 \lim_{t \to t_0} v_1 - \lim_{t \to t_0} u_1 \lim_{t \to t_0} v_3)\mathbf{j}$$
$$+ (\lim_{t \to t_0} u_1 \lim_{t \to t_0} v_2 - \lim_{t \to t_0} u_2 \lim_{t \to t_0} v_1)\mathbf{k}$$

$$= (\lim_{t \to t_0} u_1 \mathbf{i} + \lim_{t \to t_0} u_2 \mathbf{j} + \lim_{t \to t_0} u_3 \mathbf{k}) \times (\lim_{t \to t_0} v_1 \mathbf{i} + \lim_{t \to t_0} v_2 \mathbf{j} + \lim_{t \to t_0} v_3 \mathbf{k})$$

$$= [\lim_{t \to t_0} \mathbf{u}(t)] \times [\lim_{t \to t_0} \mathbf{v}(t)].$$

11. $\lim_{t \to 0} [\mathbf{u}(t) \cdot \mathbf{v}(t)]$

$$= [\lim_{t \to 0} \{e^{-t}\mathbf{i} + 2\cos t\mathbf{j} + (t^2-1)\mathbf{k}\}] \cdot [\lim_{t \to 0} (e^{2t}\mathbf{i} - \cos t\mathbf{j} + t^3\mathbf{k})]$$
$$= (\mathbf{i} + 2\mathbf{j} - \mathbf{k}) \cdot (\mathbf{i} - \mathbf{j}) = -1.$$

13. By Eq. (19), $\mathbf{F}'(t) = \dfrac{d}{dt}(t^2)\mathbf{i} + \dfrac{d}{dt}(\cos t)\mathbf{j} + \dfrac{d}{dt}(2\sin t)\mathbf{k}$
$$= 2t\mathbf{i} - \sin t\mathbf{j} + 2\cos t\mathbf{k}.$$

15. $\mathbf{F}'(t) = (2\sin 2t + 4t\cos 2t)\mathbf{i} + 3\sin t\mathbf{j} + 6(2t-1)(2)\mathbf{k}$
$$= (2\sin 2t + 4t\cos 2t)\mathbf{i} + 3\sin t\mathbf{j} + 12(2t-1)\mathbf{k}.$$

17. By Eq. (26),
$$\frac{d}{dt}[\mathbf{u}(t) + \mathbf{v}(t)] = \frac{d}{dt}[e^{-t}\mathbf{i} + 2\cos t\mathbf{j} + (t^2-1)\mathbf{k}] + \frac{d}{dt}(e^{2t}\mathbf{i} - \sin t\mathbf{j} + t^3\mathbf{k})$$

$$= (-e^{-t}\mathbf{i} - 2\sin t\mathbf{j} + 2t\mathbf{k}) + (2e^{2t}\mathbf{i} - \cos t\mathbf{j} + 3t^2\mathbf{k})$$
$$= (2e^{2t} - e^{-t})\mathbf{i} - (2\sin t + \cos t)\mathbf{j} + (3t^2 + 2t)\mathbf{k}.$$

19. $\dfrac{d}{dt}[3\mathbf{u}(t) - 2\mathbf{v}(t)] = 3\dfrac{d\mathbf{u}}{dt} - 2\dfrac{d\mathbf{v}}{dt}$

$$= 3(-e^{-t}\mathbf{i} - 2\sin t\mathbf{j} + 2t\mathbf{k}) - 2(2e^{2t}\mathbf{i} - \cos t\mathbf{j} + 3t^2\mathbf{k})$$
$$= -(3e^{-t} + 4e^{2t})\mathbf{i} + 2(\cos t - 3\sin t)\mathbf{j} + 6t(1-t)\mathbf{k}.$$

21. $\dfrac{d}{dt}[\mathbf{u}(t) \cdot \mathbf{v}(t)] = \dfrac{d\mathbf{u}}{dt} \cdot \mathbf{v} + \mathbf{u} \cdot \dfrac{d\mathbf{v}}{dt}$

$$= (2\mathbf{i} - 2t\mathbf{j} + 4t^3\mathbf{k}) \cdot (t^2\mathbf{i} + 6t\mathbf{j} + 5t\mathbf{k}) + (2t\mathbf{i} - t^2\mathbf{j} + t^4\mathbf{k}) \cdot (2t\mathbf{i} + 6\mathbf{j} + 5\mathbf{k})$$
$$= (2t^2 - 12t^2 + 2t^4) + (4t^2 - 6t^2 + 5t^4) = 25t^4 - 12t^2.$$

23. $\dfrac{d}{dt}[\mathbf{v}(t) \cdot \mathbf{u}(t)] = \dfrac{d\mathbf{v}}{dt} \cdot \mathbf{u} + \mathbf{v} \cdot \dfrac{d\mathbf{u}}{dt}$

$$= (2t\mathbf{i} + 6\mathbf{j} + 5\mathbf{k}) \cdot (2t\mathbf{i} - t^2\mathbf{j} + t^4\mathbf{k}) + (t^2\mathbf{i} + 6t\mathbf{j} + 5t\mathbf{k}) \cdot (2\mathbf{i} - 2t\mathbf{j} + 4t^3\mathbf{k})$$
$$= 25t^4 - 12t^2.$$

25a. We have x = 2t-3 and y = -4t^2 so
      t = (x+3)/2 and thus y = -(x+3)^2,
      which is the parabola shown.

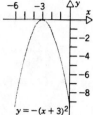

$y = -(x+3)^2$

  b. **r'** = 2**i**-8t**j** and thus a unit vector
      tangent to the curve, when t = -1,
      is (**i**+4**j**)/$\sqrt{17}$.

  c. If t = -1, then **r** = -5**i**-4**j** and thus
      the position vector is at the point
      (-5,-4) when t = -1. From part(b),
      the direction of the tangent is given by **i**+4**j** and hence
      x = -5+τ, y = -4+4τ, z = 0.

  d. The desired tangent plane has a normal vector **i**+4**j** and goes
      through the point (-5,-4). Thus 1(x+5)+4(y+4) = 0 or
      x+4y = -21.

27a. **r'** = -2sint**i**+3cost**j**+4**k** and thus a unit vector tangent to the
      curve, when t = 1, is (3**j**+4**k**)/5.
  b. The direction of the tangent line is 3**j**+4**k** and **r** = 2**i**+0**j**+0**k**
      when t = 1, and thus x = 2, y = 3τ, and z = 4τ.
  c. A normal vector to the plane is 3**j**+4**k** and thus the plane is
      given by 0(x-2)+3(y-0)+4(z-0) = 0 or 3y+4z = 0.

29.   Note that t = 1 and τ = 2 at the point of intersection. Thus
      2**i**-3**j**+4**k** is tangent to the first curve and 2**i**+**j**+2**k** is
      tangent to the second curve, at the point of intersection.
      Hence cosθ = $\dfrac{(2\mathbf{i}-3\mathbf{j}+4\mathbf{k})}{\sqrt{29}} \cdot \dfrac{(2\mathbf{i}+\mathbf{j}+2\mathbf{k})}{\sqrt{9}}$ = 3/$\sqrt{29}$ and

      θ = arccos(3/$\sqrt{29}$) ≅ .9799 radians.

31a. Note that x = acosωt and y = bsinωt
      and thus (x/a)^2+(y/b)^2 = 1, which
      is the ellipse shown.

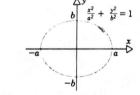

  b. The velocity **v** is given by
      v = **r'**=-aωsinωt**i**+bωcosωt**j**.
  c. Since -**r** is directed toward the
      origin, the desired component is

$$-(\mathbf{v}\cdot\mathbf{r})/\|\mathbf{r}\| = -(-a^2\omega\sin\omega t\cos\omega t+b^2\omega\cos\omega t\sin\omega t)/\sqrt{a^2\cos^2\omega t + b^2\sin^2\omega t}$$

$$= \omega(a^2-b^2)\sin\omega t\cos\omega t/\sqrt{a^2\cos^2\omega t + b^2\sin^2\omega t}.$$

33. If **u** and **v** are as in Prob.7, then $\mathbf{u}\cdot\mathbf{v} = u_1v_1+u_2v_2+u_3v_3$ and

$$\frac{d}{dt}(\mathbf{u}\cdot\mathbf{v}) = \frac{du_1}{dt}v_1 + u_1\frac{dv_1}{dt} + \frac{du_2}{dt}v_2 + u_2\frac{dv_2}{dt} + \frac{du_3}{dt}v_3 + u_3\frac{dv_3}{dt}$$

$$= (\frac{du_1}{dt}v_1+\frac{du_2}{dt}v_2+\frac{du_3}{dt}v_3) + (u_1\frac{dv_1}{dt}+u_2\frac{dv_2}{dt}+u_3\frac{dv_3}{dt})$$

$$= \frac{d\mathbf{u}}{dt}\cdot\mathbf{v} + \mathbf{u}\cdot\frac{d\mathbf{v}}{dt}.$$

Section 15.7, Page 771

1. We have $\mathbf{r}' = -\sin t\,\mathbf{i}+\cos t\,\mathbf{j}+3\mathbf{k}$ and hence Eq.(6) yields
$$1 = \int_1^4 \sqrt{\sin^2 t+\cos^2 t+9}\,dt = \int_1^4 \sqrt{10}\,dt = 3\sqrt{10} \cong 9.487.$$

3. We have $\mathbf{r}' = -2\sin t\,\mathbf{i}-2\cos t\,\mathbf{j}+t^{1/2}\mathbf{k}$ and hence
$$1 = \int_0^5 \sqrt{4\sin^2 t+4\cos^2 t+t}\,dt = \int_0^5 \sqrt{4+t}\,dt = (2/3)(4+t)^{3/2}\Big|_0^5 = 38/3.$$

5. We have $\mathbf{r}' = e^t\mathbf{i} - e^{-t}\mathbf{j} + \sqrt{2}\mathbf{k}$ and hence
$$1 = \int_0^2 \sqrt{e^{2t}+e^{-2t}+2}\,dt = \int_0^2 \sqrt{(e^t+e^{-t})^2}\,dt = 2\int_0^2 \cosh t\,dt = 2\sinh t\Big|_0^2$$
$$= 2\sinh 2 \cong 7.254.$$

7. Since $t = 0$ for the point $(0,0,0)$ and $t = 1$ for the
    point $(2\sqrt{2}/3,1,1/2)$, we conclude that $t > 0$, $|t| = t$, and
$$\mathbf{r}' = \sqrt{2}t^{1/2}\mathbf{i}+\mathbf{j}+t\mathbf{k}. \text{ Thus } 1 = \int_0^1 \sqrt{2t+1+t^2}\,dt = \int_0^1 (t+1)\,dt = 3/2.$$

9. $\mathbf{v} = \mathbf{r}' = -\sin t\,\mathbf{i} + \cos t\,\mathbf{j} + 3\mathbf{k}$, $\|\mathbf{v}\| = \sqrt{\sin^2 t+\cos^2 t+9} = \sqrt{10}$,
    and $\mathbf{a} = \mathbf{v}' = -\cos t\,\mathbf{i} - \sin t\,\mathbf{j}$.

11. $\mathbf{v} = \mathbf{r}' = t\mathbf{i} - 4\mathbf{j} + 9t^2\mathbf{k}$, $\|\mathbf{v}\| = (t^2+16+81t^4)^{1/2}$, and
     $\mathbf{a} = \mathbf{v}' = \mathbf{i} + 18t\mathbf{k}$.

13. $\mathbf{T} = \mathbf{r}'/\|\mathbf{r}'\| = (-\sin t\,\mathbf{i} + \cos t\,\mathbf{j} + 3\mathbf{k})/\sqrt{10}$, from Prob.9.

15. $\mathbf{T} = \mathbf{r}'/\|\mathbf{r}'\| = (t\mathbf{i}-4\mathbf{j}+9t^2\mathbf{k})/(16+t^2+81t^4)^{1/2}$, from Prob.11.

17. From Prob.13, we have $\mathbf{r}'' = -\cos t\,\mathbf{i} - \sin t\,\mathbf{j}$ and hence
     $\mathbf{r}'\times\mathbf{r}'' = 3\sin t\,\mathbf{i} - 3\cos t\,\mathbf{j} + \mathbf{k}$.   Thus
$$\kappa = \frac{\|\mathbf{r}'\times\mathbf{r}''\|}{\|\mathbf{r}'\|^3} = (9\sin^2 t + 9\cos^2 t + 1)^{1/2}/(\sqrt{10})^3 = 1/10.$$

19.   From Prob.15, we have $\mathbf{r}'' = \mathbf{i} + 18t\mathbf{k}$ and hence
        $\mathbf{r}' \times \mathbf{r}'' = -72t\mathbf{i} - 9t^2\mathbf{j} + 4\mathbf{k}$. Thus
        $\kappa = (5184t^2+81t^4+16)^{1/2}/(16+t^2+81t^4)^{3/2}$.

21. $\mathbf{r}' = -a\sin t\mathbf{i} + b\cos t\mathbf{j}$, $\|\mathbf{r}'\| = \sqrt{a^2\sin^2 t+b^2\cos^2 t}$, and
        $\mathbf{r}'' = -a\cos t\mathbf{i} - b\sin t\mathbf{j}$. Hence, $\mathbf{r}' \times \mathbf{r}'' = ab\mathbf{k}$ and
        $\kappa = |ab|/(a^2\sin^2 t+b^2\cos^2 t)^{3/2}$.

23. $\mathbf{r}' = a(1-\cos t)\mathbf{i} + a\sin t\mathbf{j}$,
        $\|\mathbf{r}'\| = a\sqrt{1-2\cos t+\cos^2 t+\sin^2 t} = \sqrt{2}a\sqrt{1-\cos t}$, and
        $\mathbf{r}'' = a\sin t\mathbf{i} + a\cos t\mathbf{j}$. Hence $\mathbf{r}' \times \mathbf{r}'' = a^2(\cos t-1)\mathbf{k}$ and
        $\kappa = a^2(1-\cos t)/2^{3/2}a^3(1-\cos t)^{3/2} = 1/2^{3/2}a(1-\cos t)^{1/2}$.

25. $y' = -e^{-x}$ and $y'' = e^{-x}$ and hence Eq.(35) yields
        $\kappa = e^{-x}/[1+(-e^{-x})^2]^{3/2} = e^{-x}/(1+e^{-2x})^{3/2}$.

27. $y' = 1/x^2$ and $y'' = -2/x^3$ and hence
        $\kappa = (2/x^3)/[1+(1/x^2)^2]^{3/2} = 2x^3/(x^4+1)^{3/2}$.

29.   From Ex.5, we know $\mathbf{r}(t) = b\cos\omega t\mathbf{i} + b\sin\omega t\mathbf{j}$ is a circle of
        radius b, center at the origin. Thus
        $\mathbf{v} = \mathbf{r}' = -b\omega\sin\omega t\mathbf{i} + b\omega\cos\omega t\mathbf{j}$ and
        $\mathbf{a} = -b\omega^2\cos\omega t\mathbf{i} - b\omega^2\sin\omega t\mathbf{j} = -\omega^2\mathbf{r}$ and therefore the
        acceleration vector $\mathbf{a}$ is in the opposite direction to the
        position vector; in other words, towards the center of the
        circle.

31a.   Taking the derivative of the right side of Eq.(i), we have
        $\dfrac{d\mathbf{T}}{dt} = \dfrac{\mathbf{r}''}{s'} + \mathbf{r}'\dfrac{d}{dt}(1/s') = \mathbf{r}''/s' - \mathbf{r}'s''/(s')^2$, so division by s'
        yields Eq.(ii).

   b.   Take the derivative of both sides of the given equation to get
        $2\|\mathbf{r}'\|\dfrac{d}{dt}\|\mathbf{r}'\| = \mathbf{r}''\cdot\mathbf{r}' + \mathbf{r}'\cdot\mathbf{r}'' = 2\mathbf{r}'\cdot\mathbf{r}''$, or $\dfrac{d}{dt}\|\mathbf{r}'\| = (\mathbf{r}'\cdot\mathbf{r}'')/\|\mathbf{r}'\|$.

   c.   From Eq.(20), $s' = \|\mathbf{r}'\|$ so Eq.(iii) becomes $s'' = (\mathbf{r}'\cdot\mathbf{r}'')\|\mathbf{r}'\|$
        and Eq.(ii) becomes
        $$\frac{d\mathbf{T}}{ds} = \frac{1}{\|\mathbf{r}'\|}\left[\frac{\mathbf{r}''}{\|\mathbf{r}'\|} - \frac{\mathbf{r}'(\mathbf{r}'\cdot\mathbf{r}'')/\|\mathbf{r}'\|}{\|\mathbf{r}'\|^2}\right] = \frac{1}{\|\mathbf{r}'\|^2}\left[\mathbf{r}'' - \frac{\mathbf{r}'\cdot\mathbf{r}''}{\|\mathbf{r}'\|^2}\mathbf{r}'\right].$$

31d. Thus, $\dfrac{d\mathbf{T}}{ds}\cdot\dfrac{d\mathbf{T}}{ds} = \dfrac{1}{\|\mathbf{r}'\|^4}[\mathbf{r}''\cdot\mathbf{r}'' - \dfrac{2(\mathbf{r}'\cdot\mathbf{r}'')^2}{\|\mathbf{r}'\|^2} + \dfrac{(\mathbf{r}'\cdot\mathbf{r}'')^2(\mathbf{r}'\cdot\mathbf{r}')}{\|\mathbf{r}'\|^4}]$.

Therefore, $\|\dfrac{d\mathbf{T}}{ds}\|^2 = \dfrac{1}{\|\mathbf{r}'\|^4}[\|\mathbf{r}''\|^2 - \dfrac{2(\mathbf{r}'\cdot\mathbf{r}'')^2}{\|\mathbf{r}'\|^2} + \dfrac{(\mathbf{r}'\cdot\mathbf{r}'')^2}{\|\mathbf{r}'\|^2}]$

$= \dfrac{1}{\|\mathbf{r}'\|^6}[\|\mathbf{r}'\|^2\|\mathbf{r}''\|^2 - (\mathbf{r}'\cdot\mathbf{r}'')^2]$, which yields the desired result.

e. If θ is the angle between $\mathbf{r}'$ and $\mathbf{r}''$, then
$\mathbf{r}'\cdot\mathbf{r}'' = \|\mathbf{r}'\|\,\|\mathbf{r}''\|\cos\theta$ and thus

$\kappa = \|\dfrac{d\mathbf{T}}{ds}\| = \dfrac{\|\mathbf{r}'\|\,\|\mathbf{r}''\|(1-\cos^2\theta)^{1/2}}{\|\mathbf{r}'\|^3} = \dfrac{\|\mathbf{r}'\|\,\|\mathbf{r}''\|\sin\theta}{\|\mathbf{r}'\|^3} = \dfrac{\|\mathbf{r}'\times\mathbf{r}''\|}{\|\mathbf{r}'\|^3}$.

Section 15.8, Page 779

1. From Eq. (13), we have $a_n = \kappa(ds/dt)^2$, where $ds/dt = \|\mathbf{v}\|$ from Eq. (3), and $\kappa = \|\mathbf{v}\times\mathbf{a}\|/\|\mathbf{v}\|^3$ from Eq. (25) of Section 15.7. Thus,

$a_n = \dfrac{\|\mathbf{v}\times\mathbf{a}\|\,\|\mathbf{v}\|^2}{\|\mathbf{v}\|^3} = \|\mathbf{v}\times\mathbf{a}\|/\|\mathbf{v}\|$.

3. $\mathbf{v} = \mathbf{r}' = -a\omega\sin\omega t\,\mathbf{i} + b\omega\cos\omega t\,\mathbf{j}$, $\mathbf{a} = \mathbf{v}' = -a\omega^2\cos\omega t\,\mathbf{i} - b\omega^2\sin\omega t\,\mathbf{j}$,
$\dfrac{ds}{dt} = \|\mathbf{v}\| = (a^2\omega^2\sin^2\omega t + b^2\omega^2\cos^2\omega t)^{1/2}$, and
$\mathbf{v}\times\mathbf{a} = (ab\omega^3\sin^2\omega t + ab\omega^3\cos^2\omega t)\mathbf{k} = ab\omega^3\mathbf{k}$. Thus, Eq. (13)

yields $a_t = \dfrac{d^2s}{dt^2} = \dfrac{2a^2\omega^3\sin\omega t\cos\omega t - 2b^2\omega^3\cos\omega t\sin\omega t}{2(a^2\omega^2\sin^2\omega t + b^2\omega^2\cos^2\omega t)^{1/2}}$

$= (a^2 - b^2)\omega^2\sin\omega t\cos\omega t/(a^2\sin^2\omega t + b^2\cos^2\omega t)^{1/2}$ and
$a_n = \kappa(ds/dt)^2 = \|\mathbf{v}\times\mathbf{a}\|/\|\mathbf{v}\| = ab\omega^2/(a^2\sin^2\omega t + b^2\cos^2\omega t)^{1/2}$.

5. $\mathbf{v} = \mathbf{r}' = -[\sin t + (1/2)\sin 2t]\mathbf{i} + [\cos t + (1/2)\cos 2t]\mathbf{j}$,
$\mathbf{a} = \mathbf{v}' = -[\cos t + \cos 2t]\mathbf{i} - [\sin t + \sin 2t]\mathbf{j}$,
$\dfrac{ds}{dt} = \|\mathbf{v}\| = ([\sin t + (1/2)\sin 2t]^2 + [\cos t + (1/2)\cos 2t]^2)^{1/2}$

$= (5/4 + \sin t\sin 2t + \cos t\cos 2t)^{1/2} = (5/4 + \cos t)^{1/2}$, and

$\mathbf{v} \times \mathbf{a} = \{[\sin t + (1/2)\sin 2t]^2 + [\cos t + (1/2)\cos 2t]^2\}\mathbf{k}$

$\quad\quad = [3/2 + (3/2)\cos t]\mathbf{k}$. Thus, Eq.(13) yields

$a_t = \dfrac{d^2 s}{dt^2} = (1/2)(5/4 + \cos t)^{-1/2}(-\sin t) = -\sin t/2(5/4 + \cos t)^{1/2}$

and $a_n = \kappa (ds/dt)^2 = \|\mathbf{v} \times \mathbf{a}\|/\|\mathbf{v}\| = (3/2)(1 + \cos t)/(5/4 + \cos t)^{1/2}$.

7a. We have $\mathbf{r} = a(\omega t - \sin\omega t)\mathbf{i} + a(1 - \cos\omega t)\mathbf{j}$ and hence

$\mathbf{v} = a(\omega - \omega\cos\omega t)\mathbf{i} + a\omega\sin\omega t\mathbf{j}$ and therefore

$\|\mathbf{v}\| = a\omega[(1 - \cos\omega t)^2 + \sin^2\omega t]^{1/2} = a\omega(2 - 2\cos\omega t)^{1/2}$

$\quad\quad = 2a\omega\sin(\omega t/2)$.

Thus, the particle is moving most rapidly when $\theta = \omega t = \pi$, which, from Fig.5.5.4, is at the top of the arch.

b. $\mathbf{a} = \mathbf{v}' = a\omega^2\sin\omega t\,\mathbf{i} + a\omega^2\cos\omega t\,\mathbf{j} = a\omega^2(\sin\omega t\,\mathbf{i} + \cos\omega t\,\mathbf{j})$ and $\|\mathbf{a}\| = a\omega^2$. From Fig.5.5.4, we see that the vector from the center of the circle to the point P is given by $-a\sin\theta\mathbf{i} - a\cos\theta\mathbf{j} = -a(\sin wt\,\mathbf{i} + \cos wt\,\mathbf{j})$. Thus, $\mathbf{a}$ is directed towards the center of the circle.

c. From Fig.5.5.4, the point at the top of the circle has coordinates $(a\theta, 2a)$ and the point P has coordinates $(a\theta - a\sin\theta, a - a\cos\theta)$, and thus the vector from P to the top of the circle is $\mathbf{b} = a\sin\theta\mathbf{i} + a(1 + \cos\theta)\mathbf{j} = a\sin\omega t\,\mathbf{i} + a(1 + \cos\omega t)\mathbf{j}$. Since $\mathbf{b} \times \mathbf{v} = \mathbf{0}$ and $0 < \omega t < \pi$, both $\mathbf{b}$ and $\mathbf{v}$ have positive $\mathbf{i}$ and $\mathbf{j}$ components, so we conclude that $\mathbf{v}$ points towards the top of the circle.

d. From parts (a) and (b) we have $\dfrac{ds}{dt} = \|\mathbf{v}\| = 2a\omega\sin(\omega t/2)$ and

$\mathbf{v} \times \mathbf{a} = a^2\omega^3(\cos\omega t - 1)\mathbf{k}$, and thus $a_t = \dfrac{d^2 s}{dt^2} = a\omega^2\cos(\omega t/2)$ and

$a_n = \dfrac{a^2\omega^3(1 - \cos\omega t)}{2a\omega\sin(\omega t/2)} = \dfrac{2a\omega^2\sin^2(\omega t/2)}{2\sin(\omega t/2)} = a\omega^2\sin(\omega t/2)$.

9a. Setting $g = 0$, $R = R+h$, and $\alpha = 30$ degrees in Eq.(30) and using values from Table 15.2, we obtain $GMT^2 = 3\pi^2(R+h)^3$, or $h = (GMT^2/3\pi^2)^{1/3} - R = (2979 \times 10^{21}/29.61)^{1/3} - 6.37 \times 10^6$

$\quad\quad = 40.1 \times 10^6\,\text{m} \cong 24,900$ miles.

b. For $\alpha = 45$ degrees we get $GMT^2 = 2\pi^2(R+h)^3$, which yields $h \cong 29,100$ miles.

Section 15.9, Page 787

1. We have $\dfrac{dr}{dt}$ = $b\cos\theta\,(d\theta/dt)$ = $b\omega\cos\theta$ and d^2r/dt^2 = $-b\omega^2\sin\theta$ and
    hence Eq.(9) gives
    a_r = $r'' - r(\theta')^2$ = $-b\omega^2\sin\theta - (b\sin\theta)\omega^2$ = $-2b\omega^2\sin\theta$ and
    a_θ = $r\theta'' + 2r'\theta'$ = $0 + (2b\omega\cos\theta)\omega$ = $2b\omega^2\cos\theta$. Since $\mathbf{r}$ = $b\sin\theta\mathbf{u}_r$,
    $\mathbf{v}$ = $b\cos\theta\,(d\theta/dt)\mathbf{u}_r + b\sin\theta\,(d\theta/dt)\mathbf{u}_\theta$ = $b\omega\cos\theta\mathbf{u}_r + b\omega\sin\theta\mathbf{u}_\theta$ and
    thus $\mathbf{h}$ = $m(\mathbf{r}\times\mathbf{v})$ = $m(b^2\sin^2\theta)(\omega)(\mathbf{u}_r\times\mathbf{u}_\theta)$ = $m\omega b^2\sin^2\theta\mathbf{k}$.

3. r' = $f'(\theta)(d\theta/dt)$ = $\omega f'(\theta)$ and r'' = $\omega f''(\theta)(d\theta/dt)$ = $\omega^2 f''(\theta)$,
    and thus a_r = $\omega^2 f''(\theta) - \omega^2 f(\theta)$ = $\omega^2[f''(\theta) - f(\theta)]$,
    a_θ = $0 + 2\omega f'(\theta)\omega$ = $2\omega^2 f'(\theta)$, and
    $\mathbf{h}$ = $m(\mathbf{r}\times\mathbf{v})$ = $mr^2\omega\mathbf{k}$ = $m\omega[f(\theta)]^2\mathbf{k}$.

5a. From Eq.(23), dA/dt = $(1/2)r^2(d\theta/dt)$, so $dA/dt > 0$ if $d\theta/dt > 0$,
    as in the text. Thus Eq.(26) yields dA/dt = $\|\mathbf{h}\|/2m$, which when
    substituted above gives us $d\theta/dt$ = $(\|\mathbf{h}\|/m)(1/r^2)$ = H/r^2.
  b. If $u = 1/r$, then $r = 1/u$ and $dr/du = -1/u^2 = -r^2$ and hence
    dr/dt = $(-r^2)(du/d\theta)(H/r^2)$ = $-H\,du/d\theta$. Thus,
    $$\dfrac{d^2r}{dt^2} = \dfrac{d(dr/dt)}{d\theta}\dfrac{d\theta}{dt} = -H\dfrac{d^2u}{d\theta^2}\dfrac{d\theta}{dt} = \dfrac{-H^2}{r^2}\dfrac{d^2u}{d\theta^2} = -H^2u^2\dfrac{d^2u}{d\theta^2}.$$
  c. Substituting Eqs.(ii) and (v) into Eq.(i), we get
    $-H^2u^2(d^2u/d\theta^2) - H^2/r^3$ = $-GM/r^2$. Multiplying by $-r^2$, dividing
    by H^2, and setting $u = 1/r$, then yields $\dfrac{d^2u}{d\theta^2} + u = GM/H^2$.
  d. We have $du/d\theta$ = $-C\sin(\theta-\alpha)$ and $\dfrac{d^2u}{d\theta^2}$ = $-C\cos(\theta-\alpha)$ and thus

    $\dfrac{d^2u}{d\theta^2} + u$ = $-C\cos(\theta-\alpha) + C\cos(\theta-\alpha) + GM/H^2$ = GM/H^2.
  e. Setting $\alpha = 0$ and $u = 1/r$ in Eq.(vii) yields
    $1/r$ = $C\cos\theta + GM/H^2$, and thus p/r = $\cos\theta + 1/\varepsilon$ or
    r = $\varepsilon p/(1 + \varepsilon\cos\theta)$, where p and ε are defined as in Eq.(ix).

7a.   Kepler's second law says that $\dfrac{dA}{dt} = \dfrac{1}{2}r^2\dfrac{d\theta}{dt}$ = constant and thus

differentiation yields $r\dfrac{dr}{dt}\dfrac{d\theta}{dt} + \dfrac{1}{2}r^2\dfrac{d^2\theta}{dt^2} = 0$.   Multiply by 2

and divide by r to get $2\dfrac{dr}{dt}\dfrac{d\theta}{dt} + r\dfrac{d^2\theta}{dt^2} = 0$, and thus $a_\theta = 0$.

b.   Eq.(i) yields $r = \dfrac{\varepsilon p}{1+\varepsilon\cos\theta}$ so

$\dfrac{dr}{dt} = \dfrac{\varepsilon^2 p\sin\theta(d\theta/dt)}{(1+\varepsilon\cos\theta)^2} = \dfrac{\varepsilon^2 p\sin\theta}{(1+\varepsilon\cos\theta)^2}\dfrac{H}{r^2} = (\dfrac{H}{p})\sin\theta$ and

$\dfrac{d^2r}{dt^2} = (\dfrac{H}{p})\cos\theta\dfrac{d\theta}{dt} = (\dfrac{H}{p})(\dfrac{p}{r} - \dfrac{1}{\varepsilon})(\dfrac{H}{r^2}) = \dfrac{H^2}{r^3} - \dfrac{H^2}{r^2\varepsilon p}$, where $\dfrac{d\theta}{dt}$ is

given in Prob.5(a) and $\cos\theta$ is found from Eq.(i).

c. $a_r = \dfrac{d^2r}{dt^2} - r(\dfrac{d\theta}{dt})2 = (\dfrac{H^2}{r^3} - \dfrac{H^2}{r^2\varepsilon p}) - r\dfrac{H^2}{r^4} = \dfrac{-(H^2/\varepsilon p)}{r^2}$.

d.   Eq.(30) follows from Kepler's second law and thus
      $\pi ab = \|\mathbf{h}\|T/2m$. From Prob.(5a), we have $H = \|\mathbf{h}\|/m$ and hence

$H = 2\pi ab/T$, or $H^2 = \dfrac{4\pi^2 a^2 b^2}{T^2} = \dfrac{4\pi^2 a^2 b^2}{4\pi^2 a^3}MG = \dfrac{a^2(1-\varepsilon^2)}{a}MG = \varepsilon pMG$ and

therefore $H^2/\varepsilon p = MG$.

Section 15.10, Page 799

1.   Since no x appears in the equation,
      the graph is a cylinder parallel to
      the x-axis. The cross section is a
      circle of radius 2, center at y = 0,
      z = 0.

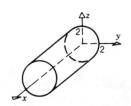

3.   Since no z appears in the equation,
      the graph is a cylinder parallel to
      the z-axis. The cross section is a
      hyperbola with asymptotes y = ±x.

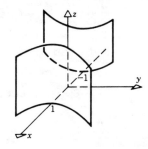

5. Since no y appears in the equation, the graph is a cylinder parallel to the y-axis. The cross section is the sine curve $x = \sin z$.

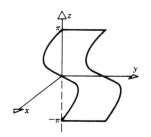

7. Dividing by 36, we obtain $x^2/4 - y^2/9 + z^2 = 0$, which is of the form of Eq. (22) and thus the graph is a cone opening about the y-axis (due to the $-y^2/9$ term). Setting $y = y_0$, we find the elliptic cross sections $x^2/4 + z^2 = y_0^2/4$.

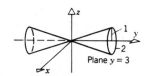

9. Solving for z we find $z = x^2/(1/9) - y^2/(1/2)$, which is in the form of Eq. (24) and thus the graph is a hyperbolic paraboloid. If $y = y_0$, then $z = 9x^2 - 2y_0^2$, which are parabolas opening in the positive z direction, and if $x = x_0$, then $z = 9x_0^2 - 2y^2$, which are parabolas opening in the negative z direction.

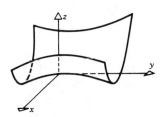

11. Solving for y we find $y = -x^2/(1/4) - z^2/(1/9)$, which is in the form of Eq. (23) and thus the graph is an elliptic paraboloid opening about the negative y-axis. For $y = -y_0^2$, the cross section is the ellipse $x^2/(1/4) + z^2/(1/9) = y_0^2$.

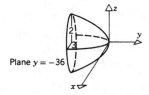

13. Divide by 16 to obtain $x^2/16 + y^2/4 + z^2/4 = 1$, which is the form of Eq. (18) and thus the graph is an ellipsoid. Since $b^2 = c^2 = 4 < a^2 = 16$, the surface is an ellipsoid of revolution about the x-axis and is also known as a prolate spheroid.

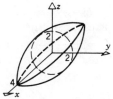

15. Solving for $z^2/3$, we find
$z^2/3 = x^2+y^2$, which by Eq.(22),
is a cone and also a surface of
revolution about the z-axis.

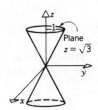

17. Since there is no z term, the graph
is a cylinder parallel to the
z-axis. Completing the square in
both x and y, we obtain
$9(x^2-2x+1)+4(y^2+4y+4) = 11+9+16$ or
$(x-1)^2/4 + (y+2)^2/9 = 1$ and thus the
cross sections are ellipses
centered at $(1,-2,0)$.

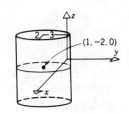

19. Completing the square in both x
and z, we obtain
$6(x^2+4x+4)-3y^2+2(z^2-2z+1) = 6$, or
$(x+2)^2 - 3y^2 + (z-1)^2/3 = 1$, which
by Eq.(19) is a hyperboloid of one
sheet about an axis through $(x,z) = (-2,1)$ parallel to the
y-axis (due to the $-y^2$ term). The cross sections are the
ellipses $(x+2)^2 + (z-1)^2/3 = 1 + 3y_0^2$.

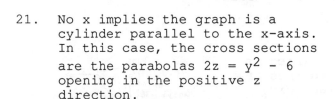

21. No x implies the graph is a
cylinder parallel to the x-axis.
In this case, the cross sections
are the parabolas $2z = y^2 - 6$
opening in the positive z
direction.

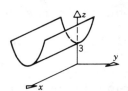

23. Completing the square in x,y, and z,
we obtain
$(x^2-6x+9)+(y^2+2y+1)+(z^2-8z+16) = 16$
or $(x-3)^2 + (y+1)^2 + (z-4)^2 = 16$,
which is the form of Eq.(18) with
$a^2 = b^2 = c^2$ and thus the graph is a
sphere of radius 4 and center at $(3,-1,4)$.

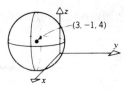

25. If $y = 1-x^2$ is rotated about the y-axis, we need to replace
x^2 by $(x^2 + z^2)$ and thus the surface of revolution is given by
$y = 1 - (x^2 + z^2)$ or $y = 1 - x^2 - z^2$.

27. We must replace x^2 by $x^2 + z^2$, when rotating about the y-axis, and hence we square the given equation twice to obtain $y^4 = x^2$. Thus the surface of revolution is given by $y^4 = x^2+z^2$.

29. We must replace y^2 by $x^2 + y^2$, when rotating about the z-axis, and thus we square the given equation to obtain $y^2 = [z^2/(1+z^2)]^2$. Hence, the surface of revolution is given by $x^2 + y^2 = z^4/(1+z^2)^2$.

31. Cubing the given equation yields $z^3 = x^2$, and replacing x^2 by $x^2 + y^2$ then gives $z^3 = x^2 + y^2$ as the equation for the surface of revolution.

33a. If $u = y-3$, then $y = u+3$ and hence $u+3 = \sqrt{x}$.
  b. From part (a), we have $u^2 = (\sqrt{x}-3)^2$ and thus $u^2 + z^2 = (\sqrt{x}-3)^2$ is the equation of the surface of revolution about $u = 0$.
  c. Setting $u = y-3$ in the solution of part (b), we obtain $(y-3)^2 + z^2 = (\sqrt{x}-3)^2$.

35. Let $u = y-1$, then $u+1 = -x^2$ or $u^2 = (1+x^2)^2$ and hence $u^2 + z^2 = (1+x^2)^2$ is the surface of revolution about $u = 0$. Thus $(y-1)^2 + z^2 = (1+x^2)^2$ is the surface of revolution about $y = 1$.

37. Let $u = z-\pi$, then $u+\pi = \arctan y$ or $u^2 = (\arctan y - \pi)^2$ and hence $x^2 + u^2 = (\arctan y - \pi)^2$ is the surface of revolution about $u = 0$ and $x^2 + (z-\pi)^2 = (\arctan y - \pi)^2$ is the surface of revolution about $z = \pi$.

39. We have $x = 2\sqrt{t}\cosh t$, $y = 3\sqrt{t}\sinh t$, and $z = 4\sqrt{1-t}$ and hence $(x/2)^2 - (y/3)^2 + (z/4)^2 = t\cosh^2 t - t\sinh^2 t + (1-t)$ or $x^2/4 - y^2/9 + z^2/16 = 1$, since $\cosh^2 t - \sinh^2 t = 1$. Thus the surface is a hyperboloid of one sheet.

41a. Since $\|\mathbf{a}\times\mathbf{r}\| = \|\mathbf{a}\|\,\|\mathbf{r}\|\sin\theta$ and $\mathbf{a}\cdot\mathbf{r} = \|\mathbf{a}\|\,\|\mathbf{r}\|\cos\theta$, the given condition requires that $\|\mathbf{a}\|\,\|\mathbf{r}\|\sin\theta = \lambda\|\mathbf{a}\|\,\|\mathbf{r}\|\,|\cos\theta|$ and hence $\sin\theta = \lambda|\cos\theta|$ for all θ. Thus $\theta = \arctan\lambda$ and the surface on which P lies is a cone making a constant angle θ with the vector $\mathbf{a}$ (the axis of the cone). Note that $0 < \theta < \pi/2$.

41b. Let $\mathbf{a} = ai+bj+ck$, then $\mathbf{a}x\mathbf{r} = (bz-cy)\mathbf{i}+(cx-az)\mathbf{j}+(ay-bx)\mathbf{k}$ and
hence $\|\mathbf{a}x\mathbf{r}\| = \tan\alpha|\mathbf{a}\cdot\mathbf{r}|$ gives

$$\sqrt{(bz-cy)^2+(cx-az)^2+(ay-bx)^2} = \lambda(ax+by+cz) \text{ where } \lambda = \tan\alpha.$$

Squaring both sides of the last equation and gathering like
terms yields
$$(-\lambda^2a^2+b^2+c^2)x^2 + (a^2-\lambda^2b^2+c^2)y^2 + (a^2+b^2-\lambda^2c^2)z^2$$
$$- 2(1+\lambda^2)(abxy + acxz + bcyz) = 0.$$

43.   Using the hint we have $\sqrt{(x-c)^2+y^2+z^2} - \sqrt{(x+c)^2+y^2+z^2} = 2a$ or

$\sqrt{(x-c)^2+y^2+z^2} = \sqrt{(x+c)^2+y^2+z^2} + 2a$. Squaring both sides and

subtracting like terms yields $-2cx = 2cx + 4a\sqrt{(x+c)^2+y^2+z^2} + 4a^2$

or $-a\sqrt{(x+c)^2+y^2+z^2} = a^2 + cx$. Squaring again and gathering

like terms yields $(c^2-a^2)x^2 - a^2y^2 - a^2z^2 = a^2(c^2-a^2)$. Setting
$b^2 = c^2-a^2$ and dividing by a^2b^2 gives
$x^2/a^2 - y^2/b^2 - z^2/b^2 = 1$, a two-sheeted hyperboloid of
revolution about the x-axis, by Eq.(20).

Chapter 15 Review, Page 800

1. $\mathbf{a}+\mathbf{b}-2\mathbf{c} = (3,-2,4) + (2,0,-5) - 2(1,3,-2) = (3,-8,3)$.

3. $\mathbf{b}x\mathbf{c} = (0+15)\mathbf{i} - (-4+5)\mathbf{j} + (6-0)\mathbf{k} = 15\mathbf{i} - \mathbf{j} + 6\mathbf{k}$.

5. $\mathbf{a}x\mathbf{b} = (10,23,4)$ and $\mathbf{b}\cdot\mathbf{c} = 12$, so $(\mathbf{a}x\mathbf{b})(\mathbf{b}\cdot\mathbf{c}) = 12(10,23,4)$.

7. $\|\mathbf{a}\| = \|\mathbf{b}\|$ when $4\lambda^2+4+9 = 49+1+\lambda^2$ or $\lambda = \pm\sqrt{37/3}$.

9. $\mathbf{a}$ and $\mathbf{b}$ form an angle of $\pi/4$ radians when
$1/\sqrt{2} = \cos\theta = (\mathbf{a}\cdot\mathbf{b})/\|\mathbf{a}\|\|\mathbf{b}\| = 5/\sqrt{25+\lambda^2}$, or $\lambda = \pm5$.

11.   For $t \neq 1$, $3\mathbf{u}+4\mathbf{v} = (3t+3+4\sin\pi t, 3t+4\cos2\pi t, 3\ln(1+t) + 4e^{t^2})$, so
$\lim_{t\to1}(3\mathbf{u}+4\mathbf{v}) = (6,7,3\ln2 + 4e)$.

13.   Since $\mathbf{u}\cdot\mathbf{v} = (t+1)\sin\pi t + t\cos2\pi t + \ln(t+1)e^{t^2}$, $\lim_{t\to0}\mathbf{u}\cdot\mathbf{v} = 0$.

15. $\mathbf{u}+\mathbf{v} = (\cos t, 3t^4, e^{3t^2} + \ln t)$, so that
$\dfrac{d}{dt}(\mathbf{u}+\mathbf{v}) = (-\sin t, 12t^3, 6te^{3t^2} + 1/t)$.

17.   Since $\|\mathbf{u}(t)\| = (\cos^2 t + e^{6t^2})^{1/2}$,

$\dfrac{d}{dt}\|\mathbf{u}(t)\| = (-\sin t \cos t + 6te^{6t^2})(\cos^2 t + e^{6t^2})^{-1/2}$.

19a. Since $\mathbf{u} = \mathbf{r}'(t)/\|\mathbf{r}'(t)\|$, and $\mathbf{r}'(t) = 2\cos t\,\mathbf{i} - 2\sin t\cos t\,\mathbf{j}$,

$\mathbf{u} = \dfrac{2\cos t(\mathbf{i} - \sin t\,\mathbf{j})}{\sqrt{4\cos^2 t + 4\cos^2 t\,\sin^2 t}} = (\mathbf{i} - \sin t\,\mathbf{j})/(1 + \sin^2 t)^{1/2}$.

   b. $1 = \displaystyle\int_0^{\pi/2} \sqrt{4\cos^2 t + 4\cos^2 t\sin^2 t}\; dt = 2\int_0^{\pi/2} \cos t\sqrt{1+\sin^2 t}\; dt$.

21a. Since $d\mathbf{r}/dt = (\cot t)\mathbf{i} + \mathbf{j}$, $\mathbf{u} = \cos t\,\mathbf{i} + \sin t\,\mathbf{j}$.

   b. $1 = \displaystyle\int_{\pi/4}^{3\pi/4} \csc t\; dt$.

23.   Completing the square yields

$\dfrac{(x-4)^2}{21} - \dfrac{(z-1)^2}{21} + \dfrac{y}{7} = 1$, which is

a hyperbolic paraboloid.

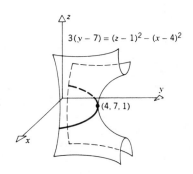

25.   Completing the square gives

$\dfrac{(x+1)^2}{4} - \dfrac{(y-2)^2}{25} - \dfrac{z^2}{100} = 1$, which is

a hyperboloid of two sheets.

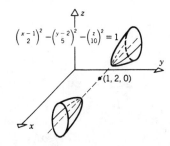

27.   By Eq.(12), Section 15.10, we have $2y^2 + 5(x^2 + z^2) = 10$.

29.   As in Prob.33, Section 15.10, let $u = z-1$, or $z = u+1 = x^3$,
      so $u = x^3 - 1$. Squaring both sides gives $u^2 = (x^3 - 1)^2$, from
      which the rotation gives $u^2 + y^2 = (x^3 - 1)^2$, or in terms of z,
      $(z-1)^2 + y^2 = (x^3 - 1)^2$.

31. The line is parallel to the vector $\mathbf{v} = 6\mathbf{i}+3\mathbf{j}+\mathbf{k}$, the plane has
     normal $\mathbf{n} = \mathbf{i}-2\mathbf{j}+3\mathbf{k}$, and $\mathbf{v}\cdot\mathbf{n} = 3$, so the line is not parallel
     to the plane and must intersect it. From the parametric
     equations of the line, $x = 6t+1$, $y = 3t$, $z = t$, so
     substitution into the plane gives $t = 0$. Thus the point of
     intersection is $(1,0,0)$.

33. With $\mathbf{n} = \overrightarrow{PQ}\times\overrightarrow{PR} = (-2\mathbf{j}-2\mathbf{k})\times(-\mathbf{i}+\mathbf{j}+\mathbf{k}) = 2\mathbf{j}-2\mathbf{k}$, the plane has
     equation $0(x-1) + 2(y-2) - 2(z-3) = 0$, or $y-z = -1$.

35. Choosing $A(2,1,0)$ on the line, $\mathbf{a} = 3\mathbf{i}+2\mathbf{j}+4\mathbf{k}$ as determining

     the direction of the line, and $\overrightarrow{AP} = 3\mathbf{i}+\mathbf{j}-\mathbf{k}$ the vector from A

     to $P(-1,0,1)$, then $\mathbf{u} = \overrightarrow{AP}\times\mathbf{a} = 3(2\mathbf{i}-5\mathbf{j}+\mathbf{k})$ will be normal to
     the required plane, which then has equation
     $2(x-2) - 5(y-1) + z = 0$, or $2x-5y+z = -1$.

37. The two lines given in early printings of the text did not
     intersect. For the lines $(x-2)/3 = (y-6)/2 = (z-2)/4$ and
     $(x-1)/1 = (y-3)/3 = z/2$ (which do intersect) we have
     $\mathbf{u} = 3\mathbf{i}+2\mathbf{j}+4\mathbf{k}$ and $\mathbf{v} = \mathbf{i}+3\mathbf{j}+2\mathbf{k}$ as the directions of the lines
     and thus $\mathbf{u}\times\mathbf{v} = -8\mathbf{i}-2\mathbf{j}+7\mathbf{k}$ will be a normal to the plane.
     Choosing $(2,6,2)$ as a point on the plane, the equation will
     be $-8(x-2) - 2(y-6) + 7(z-2) = 0$, or $8x+2y-7z = 14$.

39. Letting the vertices of the triangle be $A(1,0,0)$, $B(2,3,4)$,
     and $C(5,1,0)$, then the area is given by

     $(1/2)\|\overrightarrow{AB}\times\overrightarrow{AC}\| = (1/2)\|-4\mathbf{i}+16\mathbf{j}-11\mathbf{k}\| = \sqrt{393}/2$, Eq.(8), Sect.15.4.

41a. $dr/dt = -a\sin\theta\dfrac{d\theta}{dt} = -a\omega\sin\theta$, $d^2r/dt^2 = -a\omega\cos\theta\dfrac{d\theta}{dt} = -a\omega^2\cos\theta$,

     so $a_r = -2a\omega^2\cos\theta$ and $a_\theta = -2a\omega^2\sin\theta$, by Eq.(9) of Sect.15.9.

   b. With $\mathbf{r} = r\mathbf{u}_r$ and $\mathbf{v} = r'\mathbf{u}_r + r\omega\mathbf{u}_\theta$, we have $\mathbf{h} = \mathbf{r}\times m\mathbf{v} = m\omega r^2\mathbf{u}_z$,
      by Eq.(13), Section 15.9.

   c. $\mathbf{r} = a\cos^2\theta\mathbf{i} + a\sin\theta\cos\theta\mathbf{j}$, $\mathbf{r}' = -a\omega\sin2\theta\mathbf{i} + a\omega\cos2\theta\mathbf{j}$, and
      $\mathbf{r}'' = -2a\omega^2\cos2\theta\mathbf{i} - 2a\omega^2\sin2\theta\mathbf{j}$, so
      $\kappa = \|\mathbf{r}'\times\mathbf{r}''\|/\|\mathbf{r}'\|^3 = (2a^2\omega^3)/(a^3\omega^3) = 2/a$.

43. We must have $110\cos45° = 800\cos\theta$
     from which $\theta \cong N84.4°W$. The speed
     in the westerly direction is then
     $800\sin84.4° - 110\sin45° \cong 718.4$ km/hr,
     so the flight takes 8.35 hours
     or 8 hours 21 minutes.

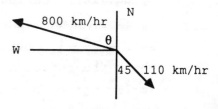

CHAPTER 16

Section 16.1, Page 811

1a. $f(-1,0) = |-1+1|\sqrt{1-0} = 0.$
 b. $f(0,1) = |0+1|\sqrt{1-1} = 0.$
 c. $f(-3,-3) = |-3+1|\sqrt{1-(-3)} = 4.$

3a. $f(-2,2) = 2/2^2 + (-2)\sqrt{2^3} = 1/2 - 4\sqrt{2}.$
 b. $f(a,a^2) = 2a^{2a} + a \cdot a^3 = 2a^{2a} + a^4.$
 c. $f(1/3,8) = 2 \cdot 8^{1/3} + (1/3)\sqrt{8^3} = 4 + (16\sqrt{2})/3.$

5a. $f(2,-\pi/2) = 2\sin(-\pi/2) = -2.$ b. $f(x,x) = x\sin x.$
 c. $f(y,x) = y\sin x.$ d. $f(r,s) = r\sin s.$

7a. $f(a,a,a) = a^3/\sqrt{3a^2} = a^2/\sqrt{3}.$
 b. $f(a^{-1},a^{-1},a^{-1}) = 1/a^3\sqrt{3/a^2} = 1/(a^2\sqrt{3}).$

9. $\phi(x,y)$ is defined when $9 - 2(x^2 + y^2) \geq 0$. Thus the domain of
     $\phi(x,y)$ is $(x^2 + y^2) \leq 9/2.$

11. $\phi(x,y)$ is defined for all x,y; so the domain of ϕ is the x,y
     plane or $-\infty < x < \infty$, $-\infty < y < \infty$.

13. $\phi(x,y,z)$ is defined for all x,y, and z $\geq$ 0.

15. $\phi(x,y)$ is defined for xy > 0. Thus ϕ is defined in the first
     and third quadrants, without the axis.

17. The level curves are $x^2+y^2 = c$, or
     circles, centered at the origin.

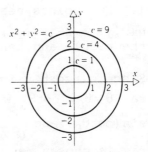

19. The level curves are the parabolas
     $y - x^2 = c$.

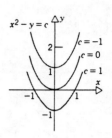

21. The level curves are $e^{x^2-y^2} = k^2$.
     Thus they are the hyperbolas
     $x^2 - y^2 = \ln(k^2) = c$.

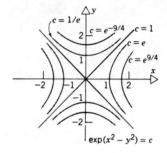

23. Since sine is continuous everywhere and $3x+7y+2z$ is
     continuous everywhere, ϕ is continuous everywhere.

25. f is discontinuous where $x^2 - y^2 = 0$, or $y = \pm x$.

27. $g(x,y)$ is continuous for $x^2+y^2 < 1$ and $x^2+y^2 > 1$. We need to
     consider only points $P_0(x_0,y_0)$ on the circle $x^2+y^2 = 1$. If
     (x_0,y_0) is on this circle, then $g(x_0,y_0) = x_0^2+y_0^2 = 1$. Any
     neighborhood of P_0 contains points in the interior of the
     circle as well as points in the exterior of the circle. For
     a point in the interior and near (x_0,y_0), $g(x,y)$ is close to
     1. For points near (x_0,y_0) and outside the circle,
     $g(x,y) = 2-(x^2+y^2)$ is also near 1. Thus g is continuous on
     the circle and hence is nowhere discontinuous.

29. Since $g(r\cos\theta, r\sin\theta) = \sin(\cot\theta)$, as $(x,y) \to (0,0)$ along any
     ray $\theta = \theta_0$, $\cot\theta_0$ will have a fixed value, so also will
     $\sin(\cot\theta_0)$ and $\lim\limits_{(x,y) \to (0,0)} g(x,y) = \sin(\cot\theta_0)$. If a different
     ray is chosen, a different value is obtained and thus
     $\lim\limits_{(x,y) \to (0,0)} g(x,y)$ does not exist.

31. Since $f(r\cos\theta, r\sin\theta) = r\sin^3\theta$, as $(x,y) \to (0,0)$, and $r \to 0$, so $r\sin^3\theta \to 0$. Thus, for f to be continuous at $(0,0)$, we must define $f(0,0)$ to be 0.

33. $f(r\cos\theta, r\sin\theta) = r^\alpha(\cos\theta+\sin\theta)^\alpha/r^2 = r^{\alpha-2}(\cos\theta+\sin\theta)^\alpha$. For $\alpha < 2$, $\lim_{r\to 0} f(r\cos\theta, r\sin\theta)$ does not exist, so f has no limit

at $(0,0)$. For $\alpha = 2$, $\lim_{r\to 0}(\cos\theta+\sin\theta)^2$ depends on θ, so again f

has no limit at $(0,0)$. For $\alpha > 2$, $\lim_{(x,y)\to(0,0)} f(x,y) = 0$, so the

function has limit 0 at $(0,0)$. Thus define $f(0,0)$ to be 0.

Section 16.2, Page 820

1. $z_x = \partial(2x^2-3xy+y^2)/\partial x = \partial(2x^2)/\partial x - \partial(3xy)/\partial x + \partial(y^2)/\partial x = 4x-3y$
and similarly $z_y = \partial z/\partial y = \partial(2x^2-3xy+y^2)/\partial y = -3x+2y$.

3. $f_x(x,y) = [(x^2+y^2+1)\partial(2xy)/\partial x - 2xy\partial(x^2+y^2+1)/\partial x]/(x^2+y^2+1)^2$
$= 2y(-x^2+y^2+1)/(x^2+y^2+1)^2$.

5. $D_x f(x,y) = D_x(x^3) + D_x(4x^2y) - D_x(6xy^2) + D_x(y^3)$ so that
$\qquad\qquad = 3x^2 + 8xy - 6y^2$. Similarly,
$D_y f(x,y) = 4x^2 - 12xy + 3y^2$.

7. $D_r u = s\cos(r-s)D_r r + rs\,D_r\cos(r-s) = s\cos(r-s) - rs\sin(r-s)$.
Similarly, $D_s u = r\cos(r-s) + rs\sin(r-s)$.

9. $f_x(x,y) = \sin(x^2y^2)D_x(x^2y^2) + x^2y^2 D_x\sin(x^2y^2)$
$\qquad\qquad = 2xy^2\sin(x^2y^2) + x^2y^2\cos(x^2y^2)D_x(x^2y^2)$
$\qquad\qquad = 2xy^2\sin(x^2y^2) + 2x^3y^4\cos(x^2y^2)$.

11. $f_y(x,y,z) = \dfrac{(x^2+y^2+z^2)D_y(x-2y+z)^2-(x-2y+z)^2 D_y(x^2+y^2+z^2)}{(x^2+y^2+z^2)^2}$.

$\qquad = \dfrac{2(x^2+y^2+z^2)(x-2y+z)(-2) - 2y(x-2y+z)^2}{(x^2+y^2+z^2)^2}$

$\qquad = -2(x-2y+z)(2x^2+2z^2+xy+yz)/(x^2+y^2+z^2)^2$.

13. $f_x(x,y,z) = (-1/2)(x^2+y^2+z^2)^{-3/2}D_x(x^2+y^2+z^2) = -x(x^2+y^2+z^2)^{-3/2}$.

15. $z_x = \dfrac{\sqrt{x^2+y^2}\,D_x\ln(x^2+y^2) - \ln(x^2+y^2)D_x\sqrt{x^2+y^2}}{x^2+y^2}$

      $= x[2 - \ln(x^2+y^2)]/(x^2+y^2)^{3/2}$. Since z is symmetric in x
      and y, z_y may be written directly from z_x by interchanging x
      and y. Thus $z_y = y[2 - \ln(x^2+y^2)]/(x^2+y^2)^{3/2}$.

17. $z_x = \dfrac{1}{1+(y/x)^2}D_x(y/x) = \dfrac{-y}{x^2+y^2}$ and $z_y = \dfrac{1}{1+(y/x)^2}D_y(y/x) = \dfrac{x}{x^2+y^2}$.

19. $z_x = 4x-3y$, $z_y = -3x+2y$, $z_{xy} = -3$, $z_{yx} = -3$. Thus $z_{xy} = z_{yx}$.

21.   From Prob.9, $z_x = 2xy^2\sin(x^2y^2) + 2x^3y^4\cos(x^2y^2)$, and by
      symmetry, $z_y = 2x^2y\sin(x^2y^2) + 2x^4y^3\cos(x^2y^2)$. Therefore,
      $z_{xy} = 4xy\sin(x^2y^2) + 4x^3y^3\cos(x^2y^2) + 8x^3y^3\cos(x^2y^2)$
      $\qquad - 4x^5y^5\sin(x^2y^2) = 4xy(1-x^4y^4)\sin(x^2y^2) + 12x^3y^3\cos(x^2y^2)$.
      A similar calculation then shows that $z_{yx} = z_{xy}$.

23. $z_x = \dfrac{x}{x^2+y^2}$, $z_y = \dfrac{y}{x^2+y^2}$, so $xz_x + yz_y = (x^2+y^2)/(x^2+y^2) = 1$.

25.   Since $u_x = -2\sin2x\,\sin4t$, $u_{xx} = -4\cos2x\,\sin4t$,
      $u_t = 4\cos2x\,\cos4t$, and $u_{tt} = -16\cos2x\,\sin4t$ we then have
      $4u_{xx} = -16\cos2x\,\sin4t = u_{tt}$.

27. $u_x = -x(x^2+y^2+z^2)^{-3/2}$ and
      $u_{xx} = -(x^2+y^2+z^2)^{-3/2} + 3x^2(x^2+y^2+z^2)^{-5/2}$. Using variable
      symmetry we obtain
      $u_{yy} = -(x^2+y^2+z^2)^{-3/2} + 3y^2(x^2+y^2+z^2)^{-5/2}$ and
      $u_{zz} = -(x^2+y^2+z^2)^{-3/2} + 3z^2(x^2+y^2+z^2)^{-5/2}$. Adding then yields
      $u_{xx}+u_{yy}+u_{zz} = -3(x^2+y^2+z^2)^{-3/2} + 3(x^2+y^2+z^2)(x^2+y^2+z^2)^{-5/2} = 0$.

29a. Any 2nd partial derivative of $w = f(x,y,z)$ will be of the
      form w_{uv}, where u and v can be any of three choices, x,y,z.
      Thus there are 3·3, or 9, possibilities for uv and hence 9
      second partial derivatives. Given that the order of
      differention is immaterial, then $w_{xy} = w_{yx}$, $w_{xz} = w_{zx}$, and
      $w_{yz} = w_{zy}$, so there are only 9-3 = 6 distinct second partial
      derivatives.

29b. As in part(a), the nth partial derivative of $w = f(x,y,z)$ will be like $w_{u_1 u_2 \ldots u_n}$, where each u_i is one of the three variables x, y, or z. Thus there are $3 \cdot 3 \cdot \ldots \cdot 3 = 3^n$ n^{th} derivatives. If the order of differentiation is immaterial, then we proceed by considering the distinct cases of $0, 1, 2, \ldots n$ partial derivatives with respect to x. If there are no partial derivatives with respect to x, then there can be from 0 to n y partial derivatives. This provides n+1 choices. If there is one x partial, then there may be from 0 to n-1 y partials, the rest being z partial derivatives. This gives n new choices. Continuing this way, there is a total of $(n+1) + (n) + (n-1) + \ldots + 3 + 2 + 1 = (n+2)(n+1)/2$ distinct n^{th} partial derivatives of $w = f(x,y,z)$.

31. Since $u_x = 2x$, $u_y = -2y$, $v_y = 2x$, and $v_x = 2y$, then $u_x = v_y$ and $u_y = -v_x$.

33. $u_x = (y^2-x^2)/(x^2+y^2)^2$, $u_y = -2xy/(x^2+y^2)^2$, $v_y = (y^2-x^2)/(x^2+y^2)^2$, and $v_x = 2xy/(x^2+y^2)^2$, so $u_x = v_y$ and $u_y = -v_x$.

35. In finding z_x we hold y constant. Hence to find $z(x,y)$ from $z_x = 1/y$, we hold y constant and integrate with respect to x. Thus $z = x/y + F(y)$, where $F(y)$ is the "constant" of integration since $\partial F(y)/\partial x = 0$. Thus, any function of the form $z = x/y + F(y)$ will satisfy $z_x = 1/y$.

37. In this case we hold x constant and integrate with respect to y to obtain $z(x,y) = (1/x)e^{xy} + y \ln x + y^2 x^2/2 + F(x)$, where $F(x)$ is the "constant" of integration since $\partial F(x)/\partial y = 0$.

39. Hold y constant and integrate with respect to x to obtain $z(x,y) = x^3 y^2/3 + x^{y+1}/(y+1) + F(y)$.

Section 16.3, Page 829

1. $f(x+h, y+k) = (x+h)^2 + 3(x+h)(y+k)$
$$= x^2 + 3xy + (2x+3y)h + 3xk + (h^2+3hk)$$
$$= (x^2+3xy) + (2x+3y)h + (3x)k + r\sqrt{h^2+k^2}, \text{ where}$$
$r = (h^2+3hk)/\sqrt{h^2+k^2}$. With $h = \rho\cos\theta$, $k = \rho\sin\theta$,
$r = \rho[\cos^2\theta + 3\cos\theta\sin\theta]$, thus $r \to 0$ as $(h,k) \to (0,0)$.

3. $f(x+h,y+k) = (x+h)(y+k)^2 = xy^2+y^2h+2xyk+(x+h)k^2+2yhk$

$= xy^2+y^2h+2xyk+r\sqrt{h^2+k^2}$, where

$r = [(x+h)k^2+2yhk]/\sqrt{h^2+k^2}$. With $h = \rho\cos\theta$, $k = \rho\sin\theta$,

$r = \rho[(x+\rho\cos\theta)\sin^2\theta + 2y\sin\theta\cos\theta]$ and $r \to 0$ as $(h,k) \to (0,0)$.

5. $f(x+h,y+k) = (x+h)\sin(y+k)$

$= x\sin y\cos k + h\sin y\cos k + x\cos y\sin k + h\cos y\sin k$

$= x\sin y + x\sin y(\cos k-1) + h\sin y + h\sin y(\cos k-1)$

$+ (x\cos y)k + (x\cos y)[\sin k-k] + h\cos y\sin k$

$= x\sin y + (\sin y)h + (x\cos y)k + r\sqrt{h^2+k^2}$, where

$r = [(x+h)\sin y(\cos k-1) + x\cos y(\sin k-k) + h\cos y\sin k]/\sqrt{h^2+k^2}$. Now
$\cos k - 1 = -k^2/2! + k^4/4! +\ldots$ and $\sin k - k = -k^3/3! + h^5/5! +\ldots$,
and hence $r = [(x+h)\sin y(-k^2/2! + k^4/4! +\ldots)$

$+ x\cos y(-k^3/3! + k^5/5! +\ldots) + h\cos y\sin k]/\sqrt{h^2+k^2}$.
Now letting $h = \rho\cos\theta$, $k = \rho\sin\theta$, we have
$r = \rho[(x+\rho\cos\theta)\sin y(-\sin^2\theta/2! +\ldots) + x\cos y(-\rho\cos^3\theta/3 +\ldots)$

$+ \cos\theta\cos y(\sin\theta - \rho^2\sin^3\theta/3! +\ldots)$ and thus $r \to 0$ as
$(h,k) \to (0,0)$.

7.   From Eq.(13), with $f_x = 6x/(3x^2-y+2)$ and $f_y = -1/(3x^2-y+2)$,
$dz = (6xdx - dy)/(3x^2-y+2)$.

9.   With $f_x = yx^{y-1}$, $f_y = (\ln x)x^y$, Eq.(13) becomes
$dz = yx^{y-1}dx + (\ln x)x^ydy$.

11.   Using the results of Prob.17, Section 16.2,
$dz = (-ydx + xdy)/(x^2+y^2)$.

13.   With $w_x = \cos 2y$, $w_y = -2x\sin 2y + 3z^{1/2}$, and $w_z = (3/2)yz^{-1/2}$,
the extension of Eq.(13) for three variables gives
$dw = \cos 2ydx + (-2x\sin 2y+3z^{1/2})dy + (3/2)yz^{-1/2}dz$.

15. $dV = V_rdr + V_hdh = (2\pi rh/3)dr + (\pi r^2/3)dh$.

17. With $r = (x^2+y^2+z^2)^{1/2}$, $d(x/r) = dx/r - (x^2dx+xydy+xzdz)/r^3$ or
$d(x/r) = (1/r^3)[(y^2+z^2)dx - xydy - xzdz]$. Similarly,
$d(y/r) = (1/r^3)[-yxdx + (x^2+z^2)dy - yzdz]$ and
$d(z/r) = (1/r^3)[-zxdx - zydy + (x^2+y^2)dz]$. Performing the
indicated multiplications and simplifying gives the desired
result. (Note, for example, that $xd(x/r)$ contains the term
$(xy^2/r^3)dx$, while $yd(y/r)$ contains the term $(-xy^2/r^3)dx$. All
other terms behave similarly.)

19. By the chain rule, $dz/dt = (\partial z/\partial x)(dx/dt) + (\partial z/\partial y)(dy/dt)$.
But, $z_x = 6x+2y^3$, $z_y = 6xy^2$, $dx/dt = 1/2\sqrt{t}$, and $dy/dt = 2t$.
Thus, $dz/dt = (6x+2y^3)/2\sqrt{t} + (6xy^2)(2t)$
$$= (6\sqrt{t}+2t^6)/2\sqrt{t} + 6\sqrt{t}\,t^4(2t) = 3 + 13t^{11/2}.$$

21. $z_x = ye^{1+xy}$, $z_y = xe^{1+xy}$, $dx/dt = 2t$, and $dy/dt = \cos t$, so
$dz/dt = z_x(dx/dt) + z_y(dy/dt) = (2t\sin t + t^2\cos t)\exp(1+t^2\sin t)$.

23. We have $w_x = 4x-6yz = 4t^{-1} - 6t^{5/6}$, $dx/dt = -t^{-2}$,
$w_y = 9y^2 - 6xz = 9t - 6t^{-2/3}$, $dy/dt = (1/2)t^{-1/2}$,
$w_z = -6xy = -6t^{-1/2}$, and $dz/dt = (1/3)t^{-2/3}$. Thus
$dw/dt = (\partial w/\partial x)(dx/dt) + (\partial w/\partial y)(dy/dt) + (\partial w/\partial z)(dz/dt)$
$$= -4t^{-3} + 6t^{-7/6} + (9/2)t^{1/2} - 3t^{-7/6} - 2t^{-7/6}$$
$$= -4t^{-3} + (9/2)\sqrt{t} + t^{-7/6}.$$

25. Since $z_x = 6y^2$, $z_y = 12xy$, $x_r = 2r$, $x_s = -2s$, $y_r = 2s$, and
$y_s = 2r$, $z_r = z_xx_r + z_yy_r = 6y^2(2r) + (12xy)(2s)$
$$= 12r(4r^2s^2) + 12(r^2-s^2)(2rs)2s = 48rs^2[2r^2-s^2].$$
Similarly, $z_s = (6y^2)(-2s) + (12xy)(2r)$
$$= -48r^2s^3 + 48r^4s - 48r^2s^3 = 48r^2s(r^2-2s^2).$$

27. $z_x = \dfrac{2x}{1+(x^2+y^2)^2}$, $z_y = \dfrac{2y}{1+(x^2+y^2)^2}$, $x_s = e^s\cos t$, $y_s = e^s\sin t$,
$x_t = -e^s\sin t$, $y_t = e^s\cos t$, and $x^2+y^2 = e^{2s}$. Thus,
$$\frac{\partial z}{\partial s} = \frac{2e^{2s}\cos^2 t}{1+e^{4s}} + \frac{2e^{2s}\sin^2 t}{1+e^{4s}} = 2e^{2s}/(1+e^{4s}) \text{ and}$$
$$\frac{\partial z}{\partial t} = \frac{-2e^{2s}\sin t\cos t}{1+e^{4s}} + \frac{2e^{2s}\sin t\cos t}{1+e^{4s}} = 0.$$

29. Let $s = x-y$, $t = y-x$ so $w = f(x-y,y-x) = f(s,t)$ and
$w_x = w_s s_x + w_t t_x = w_s - w_t$ and $w_y = w_s s_y + w_t t_y = -w_s + w_t$. Thus,
$w_x + w_y = (w_s - w_t) + (-w_s + w_t) = 0$.

31. Let $s = x/z$, $t = y/z$, so $w = f(x/z,y/z) = f(s,t)$ and
$s_x = 1/z$, $s_y = 0$, $s_z = -x/z^2$, $t_x = 0$, $t_y = 1/z$, $t_z = -y/z^2$.
Thus $w_x = w_s s_x + w_t t_x = w_s/z$, $w_y = w_s s_y + w_t t_y = w_t/z$,
$w_z = w_s s_z + w_t t_z = -x w_s/z^2 - y w_t/z^2$, and hence
$x w_x + y w_y + z w_z = x w_s/z + y w_t/z - x w_s/z - y w_t/z = 0$.

33a. We have $r = \sqrt{x^2 + y^2}$ and $\theta = \arctan(y/x)$ so
$r_x = x/r = \cos\theta$, $r_y = y/r = \sin\theta$, $\theta_x = -y/r^2 = -\sin\theta/r$, and
$\theta_y = x/r^2 = \cos\theta/r$. Thus $u_x = u_r r_x + u_\theta \theta_x = u_r\cos\theta - u_\theta(\sin\theta/r)$.

  b. Likewise, $u_y = u_r r_y + u_\theta \theta_y = u_r\sin\theta + u_\theta(\cos\theta/r)$.

  c. Squaring and adding yields $(u_x)^2 + (u_y)^2 = (u_r)^2 + (u_\theta/r)^2$.

35. If $z = f(x,y)$, $x = g(t)$, and $y = h(t)$, then
$$\frac{dz}{dt} = \frac{\partial z}{\partial x}\frac{dx}{dt} + \frac{\partial z}{\partial y}\frac{dy}{dt} \quad \text{and}$$

$$\frac{d^2z}{dt^2} = \frac{d}{dt}\left[\frac{\partial z}{\partial x}\frac{dx}{dt} + \frac{\partial z}{\partial y}\frac{dy}{dt}\right] = \frac{\partial z}{\partial x}\frac{d}{dt}\left(\frac{dx}{dt}\right) + \frac{d}{dt}\left(\frac{\partial z}{\partial x}\right)\frac{dx}{dt} + \frac{\partial z}{\partial y}\frac{d}{dt}\left(\frac{dy}{dt}\right) + \frac{d}{dt}\left(\frac{\partial z}{\partial y}\right)\frac{dy}{dt}$$

$$= \frac{\partial z}{\partial x}\frac{d^2x}{dt^2} + \left[\frac{\partial^2 z}{\partial x^2}\frac{dx}{dt} + \frac{\partial^2 z}{\partial x \partial y}\frac{dy}{dt}\right]\frac{dx}{dt} + \frac{\partial z}{\partial y}\frac{d^2y}{dt^2} + \left[\frac{\partial^2 z}{\partial y \partial x}\frac{dx}{dt} + \frac{\partial^2 z}{\partial y^2}\frac{dy}{dt}\right]\frac{dy}{dt}$$

$$= \frac{\partial z}{\partial x}\frac{d^2x}{dt^2} + 2\frac{\partial^2 z}{\partial x \partial y}\frac{dx}{dt}\frac{dy}{dt} + \frac{\partial^2 z}{\partial x^2}\left(\frac{dx}{dt}\right)^2 + \frac{\partial^2 z}{\partial y^2}\left(\frac{dy}{dt}\right)^2 + \frac{\partial z}{\partial y}\frac{d^2y}{dt^2}.$$

37. In writing $w = f(x,y,g(x,y))$, the role of x,y is easily
confused. To clarify, let $s(x,y) = x$, $t(x,y) = y$, and
$r(x,y) = g(x,y) = z$. Thus,
$F(x,y) = w = f(x,y,g(x,y)) = f(s,t,r)$. Now, using the chain
rule, $\dfrac{\partial F}{\partial x} = \dfrac{\partial f}{\partial s}\dfrac{\partial s}{\partial x} + \dfrac{\partial f}{\partial t}\dfrac{\partial t}{\partial x} + \dfrac{\partial f}{\partial r}\dfrac{\partial r}{\partial x}$. Since $\dfrac{\partial s}{\partial x} = 1$, $\dfrac{\partial t}{\partial x} = 0$, $\dfrac{\partial r}{\partial x} = \dfrac{\partial g}{\partial x}$,
and $s = x$, $\dfrac{\partial F}{\partial x} = \dfrac{\partial f}{\partial x} + \dfrac{\partial f}{\partial z}\dfrac{\partial g}{\partial x}$. Similarly, $\dfrac{\partial F}{\partial y} = \dfrac{\partial f}{\partial y} + \dfrac{\partial f}{\partial z}\dfrac{\partial g}{\partial y}$.

39a. $f_x(0,0) = \lim\limits_{h\to 0}\dfrac{f(h,0)-f(0,0)}{h} = \lim\limits_{h\to 0}\dfrac{0}{h} = 0$ since $f(h,0) = f(0,0) = 0.$

Similarly, $f_y(0,0) = \lim\limits_{k\to 0}\dfrac{f(0,k)-f(0,0)}{k} = \lim\limits_{k\to 0}\dfrac{0}{k} = 0.$

b. $\Delta f = f(0+h,0+k) - f(0,0) = \dfrac{hk}{h^2+k^2} - 0 = \dfrac{hk}{h^2+k^2}.$

c. From Eq.(11), $f(0+h,0+k) = \dfrac{hk}{h^2+k^2} = r(0,0,h,k)\sqrt{h^2+k^2}.$ Thus,

$r(0,0,h,k) = hk/(h^2+k^2)^{3/2}.$ For f to be differentiable at (0,0), $\lim\limits_{(h,k)\to(0,0)} r(0,0,h,k)$ must be zero. But, letting

$h = R\cos\theta,\ k = R\sin\theta,\ r = \dfrac{R^2\cos\theta\sin\theta}{R^3} = (\cos\theta\sin\theta)/R$ and

$\lim\limits_{(h,k)\to(0,0)} r = \lim\limits_{R\to 0}(\cos\theta\sin\theta)/R,$ which does not exist. Thus,

although $f_x(0,0)$ and $f_y(0,0)$ both exist, f is not differentiable at (0,0).

Section 16.4, Page 840

1. From Eq.(5), $\nabla f = f_x(\mathbf{r})\mathbf{i} + f_y(\mathbf{r})\mathbf{j} = (x+2y)\mathbf{i} + 2(x+y)\mathbf{j}.$

3. Since $f_x = y^2-6zx,\ f_y = 2(xy+z^2),$ and $f_z = 4yz-3x^2,$ then $\nabla f = (y^2-6zx)\mathbf{i} + 2(xy+z^2)\mathbf{j} + (4yz-3x^2)\mathbf{k}$ by Eq.(5).

5. $f_x = \cosh x\tan y$ and $f_y = \sinh x\sec^2 y$ so $\nabla f = (\cosh x\tan y)\mathbf{i} + (\sinh x\sec^2 y)\mathbf{j}.$

7. By Eq.(10), $D_\lambda f(\mathbf{r}) = \nabla f\cdot\lambda.$ From Prob.1, $\nabla f = (x+2y)\mathbf{i} + 2(x+y)\mathbf{j}$ so $\nabla f(1,1) = 3\mathbf{i}+4\mathbf{j}.$ Also, $\lambda = \mathbf{u}/\|\mathbf{u}\| = (\mathbf{i}+\mathbf{j})/\sqrt{2}.$ Thus $D_\lambda f(1,1) = (3+4)/\sqrt{2} = 7/\sqrt{2}.$

9. Since $\phi_x(x,y) = x/(x^2+y^2),\ \phi_x(a,b) = a/(a^2+b^2)$ and $\phi_y(x,y) = y/(x^2+y^2),$ so $\phi_y(a,b) = b/(a^2+b^2).$ The vector $\mathbf{v}$, from (a,b) to (0,0), is $\mathbf{v} = -a\mathbf{i}-b\mathbf{j}$ so the unit vector λ is $\lambda = -(a\mathbf{i}+b\mathbf{j})/\sqrt{a^2+b^2}.$ Finally,

$D_\lambda\phi(a,b) = \nabla\phi\cdot\lambda = \left(\dfrac{a}{a^2+b^2}\mathbf{i} + \dfrac{b}{a^2+b^2}\mathbf{j}\right)\cdot\dfrac{-a\mathbf{i}-b\mathbf{j}}{\sqrt{a^2+b^2}} = -(a^2+b^2)^{-1/2}.$

11. Since $f_x(x,y,z) = 2xyze^{x^2y}$, $f_y(x,y,z) = x^2ze^{x^2y}$, and
     $f_z(x,y,z) = e^{x^2y}$, then $f_x(2,1/2,2) = 4e^2$, $f_y(2,1/2,2) = 8e^2$,
     and $f_z(2,1/2,2) = e^2$. The vector from $(2,1/2,2)$ toward
     $(1,-1/2,3)$ is $\mathbf{v} = -\mathbf{i}-\mathbf{j}+\mathbf{k}$ so $\boldsymbol{\lambda} = (-\mathbf{i}-\mathbf{j}+\mathbf{k})/\sqrt{3}$. Finally,
     $D_\lambda f(2,1/2,2) = \nabla f(2,1/2,2)\cdot\boldsymbol{\lambda} = (-4e^2-8e^2+e^2)/\sqrt{3} = -11e^2/\sqrt{3}$.

13. By the discussion following Eq.(12), the gradient is the
     direction of the maximum increase of f. Thus $\boldsymbol{\lambda} = \nabla f/\|\nabla f\|$.
     From Prob.1, $\nabla f(x,y) = (x+2y)\mathbf{i} + 2(x+y)\mathbf{j}$ so $\nabla f(1,1) = 3\mathbf{i}+4\mathbf{j}$
     and $\boldsymbol{\lambda} = (3\mathbf{i}+4\mathbf{j})/5$. The greatest change is $\|\nabla f\| = 5$.

15. $\nabla f = 3(x\mathbf{i}+y\mathbf{j}+z\mathbf{k})(x^2+y^2+z^2)^{1/2}$ so $\nabla f(3,-4,12) = 39(3\mathbf{i}-4\mathbf{j}+12\mathbf{k})$
     and thus $\boldsymbol{\lambda} = (3\mathbf{i}-4\mathbf{j}+12\mathbf{k})/13$ is the direction of greatest
     change and $\|\nabla f\| = 507$ is the greatest change.

17. Using the results of Prob.9, grad $\phi(x,y) = (x\mathbf{i}+y\mathbf{j})/(x^2+y^2)$ or,
     letting $\mathbf{r} = x\mathbf{i}+y\mathbf{j}$ and $r = \|\mathbf{r}\| = \sqrt{x^2+y^2}$, grad $\phi(x,y) = \mathbf{r}/r^2$.

19. The question asks that we find the change of $f(x,y)$ in the
     direction of the gradient of $g(x,y)$ at the point $(2,-1)$,
     that is $D_\lambda f(2,-1) = \nabla f(2,-1)\cdot[\nabla g(2,-1)/\|\nabla g(2,-1)\|]$. Now
     $\nabla f(x,y) = (1/y)\mathbf{i} - (x/y^2)\mathbf{j} = -\mathbf{i}-2\mathbf{j}$ and $\nabla g(x,y) = x\mathbf{i}-4\mathbf{j} = 2\mathbf{i}-4\mathbf{j}$
     at $(2,-1)$. Therefore $D_\lambda f(2,-1) = (-\mathbf{i}-2\mathbf{j})\cdot(\mathbf{i}-2\mathbf{j})/\sqrt{5} = 3/\sqrt{5}$.

21. Consider $f(x,y) = b^2x^2 + a^2y^2$. Then $\pm\nabla f = \pm(2b^2x\mathbf{i} + 2a^2y\mathbf{j})$ is
     perpendicular to the curve $f(x,y) = a^2b^2$ at each point. Thus
     $\boldsymbol{\lambda} = \pm(\sqrt{2}ab^2\mathbf{i} + \sqrt{2}a^2b\mathbf{j})/\sqrt{2a^2b^4 + 2a^4b^2} = \pm(b\mathbf{i}+a\mathbf{j})/\sqrt{a^2+b^2}$ are
     the unit normals to the ellipse at $(a/\sqrt{2},b/\sqrt{2})$.

23. The gradient of $\nabla f(x,y) = xy$ will be normal to $xy = 1$ at
     $(a,1/a)$. Since $\nabla f(x,y) = y\mathbf{i}+x\mathbf{j}$, then $\nabla f(a,1/a) = (1/a)\mathbf{i} + a\mathbf{j}$
     and thus $\boldsymbol{\lambda} = \pm\nabla f(a,1/a)/\|\nabla f(a,1/a)\| = \pm(\mathbf{i} + a^2\mathbf{j})/\sqrt{1+a^4}$.

25. We need to find $D_\lambda w$, where λ is a unit tangent vector to the
curve $\mathbf{r}(t) = 2\cos t\,\mathbf{i} + 2\sin t\,\mathbf{j} - \sin 2t\,\mathbf{k}$ at $t = 13\pi/6$. Now
$x = 2\cos t = -\sqrt{3}$, $y = 2\sin t = -1$,
$\nabla w(x,y) = (2x\mathbf{i} + 2y\mathbf{j})e^{x^2+y^2} = (-2\sqrt{3}\mathbf{i} - 2\mathbf{j})e^4$, and
$d\mathbf{r}/dt = -2\sin t\,\mathbf{i} + 2\cos t\,\mathbf{j} - 2\cos 2t\,\mathbf{k} = \mathbf{i} - \sqrt{3}\mathbf{j} + \mathbf{k}$ at $t = 13\pi/6$.
Thus $D_\lambda w = \nabla w \cdot \lambda = e^4(-2\sqrt{3}\mathbf{i} - 2\mathbf{j}) \cdot (\mathbf{i} - \sqrt{3}\mathbf{j} + \mathbf{k})/\sqrt{5} = 0$.

27. $\mathbf{r}(-1) = -\mathbf{i} - 2\mathbf{j}$, so $\nabla w = (y/x + 2y^2)\mathbf{i} + (\ln|x| + 4yx)\mathbf{j} = 10\mathbf{i} + 8\mathbf{j}$,
and $d\mathbf{r}/dt = 3\mathbf{i} - 4t\mathbf{j} = 3\mathbf{i} + 4\mathbf{j}$ when $t = -1$. Thus
$D_\lambda w(-1,-2) = (10\mathbf{i} + 8\mathbf{j}) \cdot (3\mathbf{i} + 4\mathbf{j})/5 = 62/5$.

29. At $t = 0$, $x = 1$, $y = z = 0$, so
$$\nabla w = 2x(y+3)^2 e^{x^2}\tan(z+\pi/4)\mathbf{i} + 2(y+3)e^{x^2}\tan(z+\pi/4)\mathbf{j}$$
$$+ (y+3)^2 e^{x^2}\sec^2(z+\pi/4)\mathbf{k}$$
$= 6(3e\mathbf{i} + \mathbf{j} + 3e\mathbf{k})$ and $\mathbf{r}'(t) = -e^{-t}\mathbf{i} + (te^t + e^t)\mathbf{j} + 4\mathbf{k} = -\mathbf{i} + \mathbf{j} + 4\mathbf{k}$.
Thus $D_\lambda w(1,0,0) = 6e(3\mathbf{i} + \mathbf{j} + 3\mathbf{k}) \cdot (-\mathbf{i} + \mathbf{j} + 4\mathbf{k})/3\sqrt{2} = 20e/\sqrt{2}$.

31. The tangent vector will be perpendicular to the gradient, ∇T.
Now, $\nabla T(x,y) = (6xy - 3x^2)\mathbf{i} + 3x^2\mathbf{j}$ so $\nabla T(1,-1) = 3(-3\mathbf{i} + \mathbf{j})$. If we
let $\mathbf{u} = a\mathbf{i} + b\mathbf{j}$ be a tangent to the isotherm at $(1,-1)$, then
$0 = 3(-3\mathbf{i} + \mathbf{j}) \cdot (a\mathbf{i} + b\mathbf{j}) = -9a + 3b$. Thus the ratio of b to a must
be 3 to 1 and hence any scalar multiple of $\mathbf{u} = \mathbf{i} + 3\mathbf{j}$ would do.

33. $f_x(x,y) = x + 2y$, $f_{xx}(x,y) = 1$, $f_y(x,y) = 2x + 2y$, $f_{yy}(x,y) = 2$,
$f_{xy}(x,y) = 2$, and $\lambda = (\mathbf{i} + \mathbf{j})/\sqrt{2}$ (so $\lambda_1 = 1/\sqrt{2} = \lambda_2$). Thus
from Prob.32,
$D_\lambda^2 f(1,1) = (1)(1/\sqrt{2})^2 + 2(2)(1/\sqrt{2})(1/\sqrt{2}) + (2)(1/\sqrt{2})^2 = 7/2$.

35. Since $f_{xx}(x,y) = \sinh x\,\tan y$, $f_{xy}(x,y) = \cosh x\,\sec^2 y$, and
$f_{yy}(x,y) = 2\sinh x\,\sec^2 y\,\tan y$, then $f_{xx}(0,0) = 0$, $f_{xy}(0,0) = 1$,
and $f_{yy}(0,0) = 1$. Thus, with $\lambda = (-\mathbf{i} + 2\mathbf{j})/\sqrt{5}$,
$D_\lambda^2 f(0,0) = 2f_{xy}(0,0)\lambda_1\lambda_2 = (2)(-2/\sqrt{5}) = -4/\sqrt{5}$.

37. If $f(\mathbf{r}) = GMm/r = GMm/\sqrt{x^2 + y^2 + z^2}$, then
$f_x = -GMmx/(x^2 + y^2 + z^2)^{3/2} = -GMmx/r^3$, $f_y = -GMmy/r^3$, and
$f_z = -GMmz/r^3$. Thus $\nabla f = (-GMm/r^3)(x\mathbf{i} + y\mathbf{j} + z\mathbf{k}) = -GMm\mathbf{r}/r^3$.

39. Following the argument of Ex.6, we know the projection in the xy-plane of the paths of steepest descent on the surface $z = 4x^2+12y^2$ will be tangent to grad $z = 8x\mathbf{i}+24y\mathbf{j}$. So if $y = g(x)$ is such a projection, then $dy/dx = 24y/8x = 3y/x$. Solving, as in Section 10.2, we find $y = kx^3$. Since the projection passes through $(4,12)$ in the xy-plane, $12 = 64k$, or $k = 3/16$. Thus $y = 3x^3/16$.

41. Since grad $z = (-x/8)\mathbf{i} + (y/2)\mathbf{j}$, the tangent to the projections of the curves of steepest descent and ascent will have slope $dy/dx = -4y/x$. Solving as a separable equation gives $y = Cx^{-4}$, where C is a constant.

43a. The vector from $\mathbf{a}$ to $\mathbf{b}$ is $\mathbf{b}-\mathbf{a}$ and thus $\mathbf{a} + t(\mathbf{b}-\mathbf{a})$, $0 \le t \le 1$ represents all points on the desired line segment.
  b. Since f is a differentiable function of two variables then F(t) is a differentiable function of one variable on $0 \le t \le 1$. Hence, the mean value theorem of Section 4.1 applies guaranteeing the existence of α in $(0,1)$ such that
$$F(1) - F(0) = F'(\alpha)(1-0).$$
  c. Noting that $F(1) = f(\mathbf{b})$, $F(0) = f(\mathbf{a})$, and
$$F'(\alpha) = F_x(\alpha)(dx/dt)+F_y(\alpha)(dy/dt) = F_x(\alpha)(b_1-a_1)+F_y(\alpha)(b_2-a_2)$$
$$= f_x(\mathbf{c})(b_1-a_1)+f_y(\mathbf{c})(b_2-a_2) = \nabla f(\mathbf{c}) \cdot (\mathbf{b}-\mathbf{a}),\text{ where}$$
$\mathbf{c} = \mathbf{a} + \alpha(\mathbf{b}-\mathbf{a})$ is some point on the line segment joining $\mathbf{a}$ and $\mathbf{b}$. Thus Eq.(iii) becomes $f(\mathbf{b}) - f(\mathbf{a}) = \nabla f(\mathbf{c}) \cdot (\mathbf{b}-\mathbf{a})$.

Section 16.5, Page 849

1. As in Ex.1, $z_x = 6x+2y = 8$ and $z_y = 2x+2y = 4$ at $(1,1,6)$.
    Thus the tangent plane has equation $8(x-1) + 4(y-1) - (z-6) = 0$ or $8x+4y-z = 6$, and the normal line is given parametrically by $x-1 = 8t$, $y-1 = 4t$, $z-6 = -t$.

3. Letting $F(x,y,z) = 3x^2+2y^2+z^2 = 6$, and proceeding as in Ex.2, $F_x = 6x = 0$, $F_y = 4y = 4$, and $F_z = 2z = 4$ at $(0,1,2)$. Thus the tangent plane is given by $4(y-1)+4(z-2) = 0$ or $y+z = 3$, and the normal line is $x = 0$, $y-1 = t$, $z-2 = t$.

5. With $F(x,y,z) = x^{1/2} + y^{1/2} + (-z)^{1/2} = 6$, $F_x = (1/2)x^{-1/2} = 1/2$, $F_y = (1/2)y^{-1/2} = 1/6$, and $F_z = (-1/2)(-z)^{-1/2} = -1/4$ at $(1,9,-4)$. The equation of the tangent plane is $(1/2)(x-1)+(1/6)(y-9)-(1/4)(z+4) = 0$ or $6x+2y-3z = 36$ and the normal line is $x-1 = 6t$, $y-9 = 2t$, $z+4 = -3t$.

7. Since $F_x = 4x-z = 0$, $F_y = 2y-z = 2$, and $F_z = -x-y = -4$ at
    $(1,3,4)$, the equation of the plane is $2(y-3)-4(z-4) = 0$ or
    $y-2z = -5$, and the normal is $x = 1$, $y-3 = t$, $z-4 = -2t$.

9. $z_x = 2x\ln(2y^2-1)+ye^{xy} = e$ and $z_y = 4x^2y/(2y^2-1) + xe^{xy} = 4+e$
    at $(1,1,e)$. Thus the equation of the tangent plane is
    $e(x-1)+(4+e)(y-1)-(z-e) = 0$ or $ex + (4+e)y - z = 4+e$ and the
    normal line is $x-1 = et$, $y-1 = (4+e)t$, $z-e = -t$.

11. Let $F(x,y,z) = x^2/a^2 + y^2/b^2 + z^2/c^2$, then $F_x = 2x/a^2 = 2x_0/a^2$,
    $F_y = 2y_0/b^2$, $F_z = 2z_0/c^2$ at the given point. Thus the tangent
    plane is given by $\dfrac{2x_0}{a^2}(x-x_0) + \dfrac{2y_0}{b^2}(y-y_0) + \dfrac{2z^2}{c^2}(z-z_0) = 0$ or

    $\dfrac{x_0}{a^2}x + \dfrac{y_0}{b^2}y + \dfrac{z_0}{c^2}z = \dfrac{x_0^2}{a^2} + \dfrac{y_0^2}{b^2} + \dfrac{z_0^2}{c^2} = 1$, as (x_0,y_0,z_0) is on the

    ellipsoid.

13. In a manner similar to Probs.11 and 12, the equation of the
    tangent plane at any point is given by $\dfrac{x_0}{a^2}x + \dfrac{y_0}{b^2}y + \dfrac{z_0}{c^2}z = 0$,
    which passes through the origin, $(0,0,0)$.

15a. Let $F = x^{2/3} + y^{2/3} + z^{2/3}$, then at (x_0,y_0,z_0) we find
    $F_x = 2/3x_0^{1/3}$, $F_y = 2/3y_0^{1/3}$, $F_z = 2/3z_0^{1/3}$. Thus the tangent
    plane is $\dfrac{x}{x_0^{1/3}} + \dfrac{y}{y_0^{1/3}} + \dfrac{z}{z_0^{1/3}} = (x_0^{2/3} + y_0^{2/3} + z_0^{2/3}) = a^{2/3}$.

  b. The intercepts of the tangent plane with the coordinate axes
    are $(a^{2/3}x_0^{1/3},0,0)$, $(0,a^{2/3}y_0^{1/3},0)$, and $(0,0,a^{2/3}z_0^{1/3})$, so
    the sum of the squares of the intercepts is
    $a^{4/3}(x_0^{2/3} + y_0^{2/3} + z_0^{2/3}) = a^{4/3}a^{2/3} = a^2$.

17a. Let $F(x,y,z) = x^{1/2} + y^{1/2} + z^{1/2}$, so $F_x = 1/2x_0^{1/2}$,
    $F_y = 1/2y_0^{1/2}$, $F_z = 1/2z_0^{1/2}$ at (x_0,y_0,z_0), and the tangent
    plane is $x/x_0^{1/2} + y/y_0^{1/2} + z/z_0^{1/2} = (x_0^{1/2} + y_0^{1/2} + z_0^{1/2})$ or
    $x/\sqrt{x_0} + y/\sqrt{y_0} + z/\sqrt{z_0} = \sqrt{a}$.

  b. The intercepts are $(\sqrt{ax_0},0,0)$, $(0,\sqrt{ay_0},0)$, $(0,0,\sqrt{az_0})$, so
    their sum is $\sqrt{a}(\sqrt{x_0} + \sqrt{y_0} + \sqrt{z_0}) = \sqrt{a}\sqrt{a} = a$.

19. With $F(x,y,z) = x+y+z$ and $G(x,y,z) = 2x-3y+4z$ we find
     $F(1,-6/7,-1/7) = 0$ and $G(1,-6/7,-1/7) = 4$, thus $(1,-6/7,-1/7)$
     is on the curve of intersection. $\nabla F = \mathbf{i}+\mathbf{j}+\mathbf{k}$, $\nabla G = 2\mathbf{i}-3\mathbf{j}+4\mathbf{k}$,
     so from Prob.18, the tangent vector is $\mathbf{u} = 7\mathbf{i}-2\mathbf{j}-5\mathbf{k}$ for all
     (x_0,y_0,z_0). Thus, at $(1,-6/7,-1/7)$, the tangent line has
     equation $x-1 = 7t$, $y+6/7 = -2t$, $z+1/7 = -5t$.

21. By direct substitution $(2,-1,-8)$ satisfies both equations and
     is thus on the curve of intersection of the surfaces.
     Letting $F(x,y,z) = f(x,y)-z$ and $G(x,y,z) = g(x,y)-z$,
     $\nabla F = 4y\mathbf{j}+4x\mathbf{j}-\mathbf{k} = -4\mathbf{i}+8\mathbf{j}-\mathbf{k}$, $\nabla G = (2y+2)\mathbf{i}+(2x-5)\mathbf{j}-\mathbf{k} = -\mathbf{j}-\mathbf{k}$ at
     $(2,-1,-8)$, so the tangent vector is $\mathbf{u} = \nabla F \times \nabla G = -9\mathbf{i}-4\mathbf{j}+4\mathbf{k}$,
     giving the tangent line $x-2 = -9t$, $y+1 = -4t$, $z+8 = 4t$.

23. By direct substitution, the point $(0,0,2)$ satisfies both
     equations and so is a point of intersection. With
     $F(x,y,z) = x^2 + y^2 + z^2$, $G(x,y,z) = x^2 + y^2 + (z-4)^2$,
     $\nabla F = 2x\mathbf{i}+2y\mathbf{j}+2z\mathbf{k} = 4\mathbf{k}$, $\nabla G = 2x\mathbf{i}+2y\mathbf{j}+2(z-4)\mathbf{k} = -4\mathbf{k}$ at $(0,0,2)$.
     Since the normals are parallel, the method of Prob.18 fails.
     The surfaces are actually spheres of radius 2, with centers 4
     units apart. Therefore, there is a single point of
     intersection, not a curve.

25a. The tangent plane is horizontal when the normal is parallel
     to the z-axis. Since $\mathbf{N} = (x/a^2)\mathbf{i}+(y/b^2)\mathbf{j}+(z/c^2)\mathbf{k}$, $\mathbf{N}$ is
     parallel to the z-axis when $x = y = 0$. That is, when $z^2/c^2 = 1$ or $z = \pm c$. Thus the points with horizontal tangent planes
     are $(0,0,c)$ and $(0,0,-c)$.

   b. In this case $\mathbf{N}$ is parallel to the x-axis, so $x = \pm a$ and
      $y = z = 0$, giving the points $(a,0,0)$, $(-a,0,0)$.

   c. In this instance the normal vector, $\mathbf{N}$, is required to be
      parallel to the vector $\mathbf{i}+\mathbf{j}+\mathbf{k}$, that is $\mathbf{N} = \alpha(\mathbf{i}+\mathbf{j}+\mathbf{k})$ for some
      scalar α. Thus $(x/a^2)\mathbf{i}+(y/b^2)\mathbf{j}+(z/c^2)\mathbf{k} = \alpha\mathbf{i}+\alpha\mathbf{j}+\alpha\mathbf{k}$, from
      part(a). Therefore $x = \alpha a^2$, $y = \alpha b^2$, and $z = \alpha c^2$ and since
      (x,y,z) is on the ellipsoid, it must satisfy the equation
      $\dfrac{\alpha^2 a^4}{a^2}+\dfrac{\alpha^2 b^4}{b^2}+\dfrac{\alpha^2 c^4}{c^2} = 1$ or $\alpha = \pm 1/r$, where $r = (a^2+b^2+c^2)^{1/2}$. Thus
      the points on the ellipsoid having normal parallel to $\mathbf{i}+\mathbf{j}+\mathbf{k}$
      are $\pm(a^2/r,b^2/r,c^2/r)$.

27. A normal vector to the surface is given by
     $\mathbf{N} = z_x(x_0,y_0)\mathbf{i} + z_y(x_0,y_0)\mathbf{j} - \mathbf{k}$ and thus the tangent plane
     becomes $z_x(x_0,y_0)(x-x_0) + z_y(x_0,y_0)(y-y_0) - (z-z_0) = 0$. Since
     $z_x(x_0,y_0) = f(y_0/x_0) - (y_0/x_0)f'(y_0/x_0)$,
     $z_y(x_0,y_0) = f'(y_0/x_0)$, and $z_0 = z(x_0,y_0) = x_0f(y_0/x_0)$, then
     substituting into and simplifying the equation gives
     $[f(y_0/x_0) - (y_0/x_0)f'(y_0/x_0)]x + [f'(y_0/x_0)]y - z = 0$, which
     passes through the origin.

29. To determine the point of intersection of the curve
     $\mathbf{r} = 2\cos\pi t\mathbf{i} + 2\sin\pi t\mathbf{j} + 6t\mathbf{k}$ with the paraboloid, the
     coordinates of the point on the curve $x = 2\cos\pi t$, $y = 2\sin\pi t$,
     $z = 6t$ must satisfy the equation of the paraboloid. That is,
     $6t = 4\cos^2\pi t + 4\sin^2\pi t = 4$, or $t = 2/3$. Thus the point of
     intersection is $x(2/3) = -1$, $y(2/3) = \sqrt{3}$, and $z(2/3) = 4$.
     Now, the tangent vector to the curve is
     $\mathbf{r}'(2/3) = -\pi\sqrt{3}\mathbf{i} - \pi\mathbf{j} + 4\mathbf{k}$ and a normal to the surface is
     $\mathbf{N}(2/3) = 2\mathbf{i} - 2\sqrt{3}\mathbf{j} + \mathbf{k}$. The angle θ between these vectors is
     found from $\cos\theta = \mathbf{r}'\cdot\mathbf{N}/\|\mathbf{r}'\|\,\|\mathbf{N}\| = 4/\sqrt{68(\pi^2+4)} = 0.130248$, so
     $\theta = 1.4402$ radians, and its complement, the angle of
     intersection, is $\beta = \pi/2 - \theta = 0.1306$ radians.

Section 16.6, Page 859

1.   The stationary points are found by solving
     $f_x = -2+4x = 0$ and $f_y = 6+6y = 0$. The only solution is the
     point $(1/2,-1)$. For classification, we find the second
     partial derivatives, $f_{xx} = 4 = A$, $f_{xy} = 0 = B$, and
     $f_{yy} = 6 = C$. Thus $B^2-AC = -24 < 0$ and $A > 0$, $B > 0$ so
     $(1/2,-1)$ is a relative minimum by Theorem 16.6.2.

3.   Since $f_x = 2x-3y = 0$ and $f_y = -3x-4y = 0$, the only stationary
     point is $(0,0)$. As $f_{xx} = 2 = A > 0$, $f_{xy} = -3 = B$,
     $f_{yy} = -4 = C < 0$, $B^2-AC = 16 > 0$ so then $(0,0)$ is a saddle
     point by Theorem 16.6.2.

5. $f_x = 1+x+y = 0$, $f_y = 6+x+6y = 0$, which has solution $(0,-1)$.
     Further, $f_{xx} = 1$, $f_{xy} = 1$, $f_{yy} = 6$, so $B^2-AC = -5 < 0$ and
     $A > 0$, $C > 0$, and thus $(0,-1)$ is a relative minimum.

7. $f_x = 1-2x-2y = 0$, $f_y = -2x-4y = 0$ which has solution
    $(1,-1/2)$. $f_{xx} = -2$, $f_{xy} = -2$, $f_{yy} = -4$, so $B^2-AC = -6 < 0$
    and thus $f(x,y)$ has a relative maximum at $(1,-1/2)$.

9. $f_x = ye^x$ and $f_y = e^x-1$. Thus $f_x = f_y = 0$ only at $(0,0)$.
    $f_{xx} = ye^x = 0$ and $f_{xy} = e^x = 1$ at $(0,0)$, and $f_{yy}(x,y) = 0$ for
    all (x,y), so $B^2 - 4AC > 0$ and $(0,0)$ is a saddle point.

11. $f_x = y/x + 6x = 0$, $f_y = \ln x - 2 = 0$ at the stationary point
    $(e^2,-6e^4)$. $f_{xx} = -y/x^2 + 6$, $f_{xy} = 1/x$, $f_{yy} = 0$, and thus at
    $(e^2,-6e^4)$, $A = 6(e^2+1)$, $B = 1/e^2$, $C = 0$ so
    $B^2-AC = 1/e^4 > 0$ and $(e^2,-6e^4)$ is a saddle point.

13. $f_x = y+2x = 0$, $f_y = x+3y^2 = 0$ may be solved by substituting
    $y = -2x$ from the first into the second equation, giving
    $x+12x^2 = 0$ or $x = 0$, $x = -1/12$. Thus the stationary points
    are $(0,0)$ and $(-1/12,1/6)$. Now $f_{xx} = 2$, $f_{xy} = 1$, and
    $f_{yy} = 6y$, so $(0,0)$ is a saddle point, since $B^2-AC = 1$, and
    $(-1/12,1/6)$ is a relative minimum, since $B^2-AC = -1$.

15. $f_x = 2xe^{x^2}\sin y = 0$, $f_y = e^{x^2}\cos y = 0$ for all points
    $(0,\pm(2n-1)\pi/2)$, with n an integer, so these points are
    stationary points. Since $f_{xx} = (2+4x^2)e^{x^2}\sin y$,
    $f_{xy} = 2xe^{x^2}\cos y$, and $f_{yy} = -e^{x^2}\sin y$, we have
    $B^2 - AC = 4x^2e^{2x^2}\cos^2 y + (2+4x^2)e^{2x^2}\sin^2 y > 0$ for all (x,y)
    and thus all stationary points are saddle points.

17. $f_x = -2xe^{-(x^2+y^2)} = 0$, $f_y = -2ye^{-(x^2+y^2)} = 0$ only at $(0,0)$.
    Now, $f_{xx} = (4x^2-2)e^{-(x^2+y^2)} = -2$, $f_{xy} = 4xye^{-(x^2+y^2)} = 0$,
    and $f_{yy} = (4y^2-2)e^{-(x^2+y^2)} = -2$ at $(0,0)$ and therefore
    $B^2-AC = -4 < 0$, and thus the stationary point $(0,0)$ is a
    relative maximum.

19. $f_x = 2x+2y = 0$, $f_y = 1/y + y + 2x = 0$ may be solved by
    substituting $y = -x$ from the first into the second equation
    to obtain $-1/x - x + 2x = 0$ or $x^2-1 = 0$, giving $x = \pm1$. Thus
    the stationary points are $(1,-1)$ and $(-1,1)$. Now $f_{xx} = 2$,
    $f_{xy} = 2$, and $f_{yy} = 1-1/y^2$ so both stationary points are
    saddle points since $B^2-AC = 4 > 0$.

21. $f_x = 3x^2 + y^2 + 2x = 0$, $f_y = 2xy+2y = 0$. Examining f_y, we see
     $2y(x+1) = 0$ for $y = 0$ and $x = -1$. At $x = -1$, $f_x = y^2+1$ which
     is never 0. Thus $x = -1$ does not yield a stationary point.
     When $y = 0$, $f_x = x(3x+2) = 0$ for $x = 0$ and $x = -2/3$. Thus,
     the stationary points are $(0,0)$ and $(-2/3,0)$. Now, $f_{xx} = 6x+2$,
     $f_{xy} = 2y$, and $f_{yy} = 2x+2$, so $(0,0)$ is a relative minimum,
     since $B^2-AC = -4$, and $(-2/3,0)$ is a saddle point, since
     $B^2-AC = 4/3$.

23. Let x,y,z be the numbers. Then we are asked to minimize
     $x^2 + y^2 + z^2$ subject to the constraint that $x+y+z = a$. From
     the constraint we have $z = a-(x+y)$ and hence we wish to
     minimize $f(x,y) = x^2 + y^2 + [a-(x+y)]^2$. Thus
     $f_x = 2x - 2[a-(x+y)] = 4x+2y-2a = 0$ and $f_y = 2x+4y-2a = 0$,
     which yield $x = y = a/3$. Since $f_{xx} = 4$, $f_{xy} = 2$, and
     $f_{yy} = 4$, $B^2-AC = -12 < 0$, independent of x and y. Thus
     $x = y = z = a/3$ minimizes $x^2 + y^2 + z^2$ subject to $x+y+z = a$.

25. Letting (x,y,z) be the closest point on the plane to the
     origin, we are asked to minimize $\sqrt{x^2+y^2+z^2}$, or more simply
     $x^2+y^2+z^2$, subject to the constraint $ax+by+cz = d$. As in
     Prob.23, we solve the constraint equation, $z = [d-(ax+by)]/c$,
     and minimize $f(x,y) = x^2 + y^2 + [d-(ax+by)]^2/c^2$. Now
     $f_x(x,y) = 2x(c^2+a^2)/c^2 + 2aby/c^2 - 2ad/c^2 = 0$ and
     $f_y(x,y) = 2abx/c^2 + 2y(c^2+b^2)/c^2 - 2bd/c^2 = 0$, which reduce to
     $(c^2+a^2)x+aby = ad$ and $abx+(c^2+b^2)y = bd$, which may be solved
     to give $x = ad/r^2$, $y = bd/r^2$, where $r^2 = a^2+b^2+c^2$, and from
     the constraint equation, $z = cd/r^2$. Thus the point on the
     plane $ax+by+cz = d$ closest to the origin is
     $(ad/r^2,bd/r^2,cd/r^2)$. The position vector of this point is
     $\mathbf{u} = (a\mathbf{i}+b\mathbf{j}+c\mathbf{k})d/r^2$, which is parallel to the vector
     $a\mathbf{i}+b\mathbf{j}+c\mathbf{k}$, a normal to the plane.

27. Let w,h,d be the dimensions of the box. Then $V = whd = 6$
     (so $h = 6/wd$) and the cost of the materials is
     $C = (1.50)wd + 2(1.00)wh + 2(1.00)hd = (3/2)wd + 12/d + 12/w$.
     Thus $C_d = (3/2)w - 12/d^2 = 0$, $C_w = (3/2)d - 12/w^2 = 0$, which
     give $d = w = 2$ and $h = 3/2$. By calculating the second
     derivatives, it can be shown that these values do give a
     minimum cost.

29a. With $S = \sum\limits_{i=1}^{N} (Y_i - y_i)^2 = \sum\limits_{i=1}^{N} [Y_i - (mX_i + b)]^2$, our minimizing

techniques yield $S_m = 2\sum\limits_{i=1}^{N} [mX_i^2 + bX_i - X_iY_i] = 0$ and

$S_b = 2\sum\limits_{i=1}^{N} [Y_i - (mX_i + b)] = 0$, giving $m\sum\limits_{i=1}^{N} X_i^2 + b\sum\limits_{i=1}^{N} X_i = \sum\limits_{i=1}^{N} X_iY_i$

and $m\sum\limits_{i=1}^{N} X_i + Nb = \sum\limits_{i=1}^{N} Y_i$.

b. With $X_1 = 0$, $X_2 = 1$, $X_3 = 2$, $Y_1 = 0$,
$Y_2 = 1$, and $Y_3 = 3/2$, the equations
in part (a) become $5m + 3b = 4$ and
$3m + 3b = 5/2$. Solving these gives
$m = 3/4$, $b = 1/12$ and thus the least
squares line is $y = (3/4)x + 1/12$ or $y = (9x+1)/12$.

c. $\sum\limits_{i=1}^{N} (Y_i - y_i) = \sum\limits_{i=1}^{N} Y_i - \sum\limits_{i=1}^{N} y_i = \left(m\sum\limits_{i=1}^{N} X_i + Nb\right) - \sum\limits_{i=1}^{N} (mX_i + b)$

$= m\sum\limits_{i=1}^{N} X_i + Nb - m\sum\limits_{i=1}^{N} X_i - Nb = 0$. From

$y = (3/4)x + 1/12$, $y_1 = 1/12$, $y_2 = 10/12$ and $y_3 = 19/12$, so

$\sum\limits_{i=1}^{N} (Y_i - y_i) = (0 - 1/12) + (1 - 10/12) + (3/2 - 19/12) = 0$.

31a. With $n = 1$, $E_1 = \int_{-1}^{1} [f(x) - (a_0 + a_1x)]^2 dx$. Now, we can either
expand the integrand so that the a_i's are outside the
integral, or else we can differentiate under the integral as
it stands, which can be shown to be valid. Using this latter
approach we have

$\dfrac{\partial E_1}{\partial a_0} = -2\int_{-1}^{1} [f(x) - a_0 - a_1x] dx = -2\int_{-1}^{1} f(x) dx + 4a_0 = 0$, so

$a_0 = (1/2)\int_{-1}^{1} f(x) dx$, and $\dfrac{\partial E_1}{\partial a_1} = -2\int_{-1}^{1} [xf(x) - a_0x - a_1x^2] dx = 0$ or

$\int_{-1}^{1} xf(x)\,dx - 2a_1/3 = 0$, so $a_1 = (3/2)\int_{-1}^{1} xf(x)\,dx$. Letting

$I_n = \int_{-1}^{1} x^n f(x)\,dx$, we can write $a_0 = I_0/2$, $a_1 = 3I_1/2$.

31b. With $n = 2$, $E_2 = \int_{-1}^{1} [f(x) - (a_0 + a_1 x + a_2 x^2)]^2 dx$. Referring to the discussion following Ex.5, E_2 will be minimized when $\partial E_2/\partial a_i = 0$, $i = 0,1,2$. Thus

$\partial E_2/\partial a_0 = -2\int_{-1}^{1} [f(x) - (a_0 + a_1 x + a_2 x^2)]\,dx = 0$ when $a_0 + a_2/3 = I_0/2$,

$\partial E_2/\partial a_1 = -2\int_{-1}^{1} [f(x) - (a_0 + a_1 x + a_2 x^2)]x\,dx = 0$ when $2a_1/3 = I_1$, and

$\partial E_2/\partial a_2 = -2\int_{-1}^{1} [f(x) - (a_0 + a_1 x + a_2 x^2)]x^2 dx = 0$ when $a_0/3 + a_2/5 = I_2/2$.

From these three equations we find $a_0 = (9I_0 - 15I_2)/8$, $a_1 = 3I_1/2$, and $a_2 = (45I_2 - 15I_0)/8$.

33a. In this case q_2 is constant in p_1 and q_1 is constant in p_2 so we have $\partial p_1/\partial q_1 = 40 - 2q_1 = 0$ for $q_1 = 20$ and $\partial p_2/\partial q_2 = 60 - 4q_2 = 0$ for $q_2 = 15$. Thus $p_1 = \$287.50$, $p_2 = \$250.00$, and $p_1 + p_2 = \$537.50$.

  b. If they work together to maximize total profit, $p_1 + p_2$, then consider $p = p_1 + p_2 = 40q_1 + 60q_2 - 3q_1^2/2 - 5q_2^2/2$. Thus $\partial p/\partial q_1 = 40 - 3q_1 = 0$ for $q_1 = 40/3$ and $\partial p/\partial q_2 = 60 - 5q_2 = 0$ for $q_2 = 12$. At these production levels we find $p_1 = \$283.56$, $p_2 = \$343.11$, and the combined profit is $p_1 + p_2 = \$626.67$.

35a. For $y = mx$ we have $f(x,y) = f(x,mx) = g(x) = (mx - x^2)(mx - 3x^2) = m^2 x^2 - 4mx^3 + 3x^4$, and we may apply the first derivative test for functions of one variable. Thus $g'(x) = 2m^2 x - 12mx^2 + 12x^3 = 0$ for $x = 0$. Now, if $m > 0$, then $x < 0$ gives $g' < 0$ and $x > 0$ gives $g' > 0$, for x sufficiently small $(12mx < 12x^2 + 2m^2)$. A similar argument holds for $m < 0$. Thus $x = 0$ is a relative minimum for $f(x,y)$ along $y = mx$.

  b. For any $\beta \neq 0$, $f(0, \beta) = (\beta - 0)(\beta - 0) = \beta^2 > 0 = f(0,0)$ while $f(\alpha, 2\alpha^2) = (2\alpha^2 - \alpha^2)(2\alpha^2 - 3\alpha^2) = -2\alpha^2 < 0$ for any real $\alpha \neq 0$.

Section 16.7, Page 868

1.   Following Ex.1, we look for values of x, y, and λ such that
      $\nabla f(x,y) = \lambda \nabla g(x,y)$, where $g(x,y) = x^2/a^2 + y^2/a^2 - 1$. Since
      $\nabla f = y\mathbf{i}+x\mathbf{j}$ and $\lambda \nabla g = (2x\lambda/a^2)\mathbf{i}+(2y\lambda/b^2)\mathbf{j}$, we have that
      $y = 2x\lambda/a^2$, $x = 2y\lambda/b^2$, or $\lambda = ya^2/2x = xb^2/2y$. Since $(0,0)$ is
      not on the ellipse, the last equation yields $y^2 = x^2b^2/a^2$, so
      substituting into the constraint gives $x^2/a^2 + x^2/a^2 - 1 = 0$ or
      $x = \pm a/\sqrt{2}$ and $y = \pm b/\sqrt{2}$. Thus we need to check $(a/\sqrt{2}, \pm b/\sqrt{2})$
      and $(-a/\sqrt{2}, \pm b/\sqrt{2})$. Since $f(a/\sqrt{2}, b/\sqrt{2}) = f(-a/\sqrt{2}, -b/\sqrt{2}) = ab/2$
      and $f(-a/\sqrt{2}, b/\sqrt{2}) = f(a/\sqrt{2}, -b/\sqrt{2}) = -ab/2$, the maximum value
      of f is $ab/2$, and the minimum is $-ab/2$.

3.   If $g(x,y) = xy-1$, then $\nabla f = (2x/a^2)\mathbf{i}+(2y/b^2)\mathbf{j} = \lambda y\mathbf{i}+\lambda x\mathbf{j} = \lambda \nabla g$.
      Thus $\lambda = 2x/ya^2 = 2y/xb^2$ so that $y^2 = x^2b^2/a^2$, or $y = \pm bx/a$.
      Substituting into $g(x,y) = 0$ gives $x^2(b/a)-1 = 0$ or $x = \pm\sqrt{a/b}$
      and $y = \pm\sqrt{b/a}$. (Note that x and y are of the same sign so
      that $xy-1 = 0$.) The minimum value for $f(x,y)$ in either case
      is $2/ab$.

5.   Let $g(x,y,z) = x^2 + y^2 + z^2 - 16$, then $2\mathbf{i}+2\mathbf{j}-\mathbf{k} = 2\lambda x\mathbf{i}+2\lambda y\mathbf{j}+2\lambda z\mathbf{k}$.
      This gives $\lambda = 1/x = 1/y = -1/2z$, which yield $y = x$ and
      $z = -x/2$. Substituting into the equation for the sphere
      gives $x^2 + x^2 + x^2/4 - 16 = 0$ or $9x^2 = 64$. Thus $x = \pm 8/3$,
      $y = \pm 8/3$, and $z = -(\pm 4/3)$. At $(8/3, 8/3, -4/3)$, $f(x,y,z) = 12$,
      a maximum, and at $(-8/3, -8/3, 4/3)$, $f(x,y,z) = -12$, a minimum.

7.   Let $g(x,y,z) = x^2/4 + y^2/4 + z^2 - 1$, then
      $\mathbf{i}+\mathbf{j}-\mathbf{k} = (\lambda x/2)\mathbf{i}+(\lambda y/2)\mathbf{j}+(2\lambda z)\mathbf{k}$. From this $x = y = -4z$.
      Substituting into $g(x,y,z) = 0$ we find $z = \pm 1/3$ so
      $x = y = -(\pm 4/3)$. Thus the maximum is 3 at $(4/3, 4/3, -1/3)$ and
      the minimum is -3 at $(-4/3, -4/3, 1/3)$.

9.   Let $g(x,y,z) = x^2 + y^2 + z^2 - a^2$, then
      $yz\mathbf{i}+xz\mathbf{j}+xy\mathbf{k} = 2\lambda x\mathbf{i}+2\lambda y\mathbf{j}+2\lambda z\mathbf{k}$. Thus $\lambda = yz/2x = xz/2y = xy/2z$,
      or $x^2 = y^2 = z^2$. Substituting for y^2 and z^2 in the equation
      of the sphere gives $3x^2 = a^2$ or $x = \pm a/\sqrt{3}$ and thus $y = \pm a/\sqrt{3}$,
      $z = \pm a/\sqrt{3}$. The extrema occur at the eight choices for
      $(\pm a/\sqrt{3}, \pm a/\sqrt{3}, \pm a/\sqrt{3})$. To minimize $f(x,y,z) = xyz$, choose any
      of the three points with one negative and two positive
      coordinates or the one point with all three negatives. For
      these four points $f(x,y,z) = -a^3/3\sqrt{3}$.

11. With $g(x,y,z) = x+y+z-1$, we have $3x^2\mathbf{i}+3y^2\mathbf{j}+3z^2\mathbf{k} = \lambda\mathbf{i}+\lambda\mathbf{j}+\lambda\mathbf{k}$,
     so that $\lambda = 3x^2 = 3y^2 = 3z^2$. Thus $|x| = |y| = |z|$. If two
     are the same sign and the third the opposite, for example
     $x = y = -z$, the equation $x+y+z = 1$ yields $x = 1$, $y = 1$, and
     $z = -1$ so $f(x,y,z) = 1$. On the other hand, if $x = y = z$, the
     equation $x+y+z = 1$ gives $x = 1/3 = y = z$ and
     $f(x,y,z) = 3/27 = 1/9$. Thus the point $(1/3,1/3,1/3)$ gives the
     minimum of $1/9$.

13. From Prob.34, Section 16.6, we want to minimize
     $f(x,y) = x^2 + y^2$ for points on the line $g(x,y) = ax+by+c = 0$.
     Thus $2x\mathbf{i}+2y\mathbf{j} = \lambda a\mathbf{i}+\lambda b\mathbf{j}$ so that $\lambda = 2x/a = 2y/b$, giving
     $y = bx/a$. Substituting into the equation of the line
     produces $ax + b^2x/a + c = 0$, so $x = -ac/(a^2+b^2)$ and
     $y = -bc/(a^2+b^2)$. The minimum distance is thus
     $\sqrt{a^2c^2/(a^2+b^2)^2+b^2c^2/(a^2+b^2)^2}$ or $|c|/(a^2+b^2)^{1/2}$.

15. Let $f(x,y,z) = x+y+z$ and $g(x,y,z) = \mathbf{r}\cdot\mathbf{r} - a^2 = x^2 + y^2 + z^2 - a^2$,
     then $\mathbf{i}+\mathbf{j}+\mathbf{k} = 2x\lambda\mathbf{i}+2y\lambda\mathbf{j}+2z\lambda\mathbf{k}$. This gives
     $x = y = z$ on $x^2+y^2+z^2 = a^2$. Substituting for y and z we find
     $3x^2 = a^2$ or $x = \pm a/\sqrt{3}$. To maximize $x+y+z$, we choose
     $x = y = z = a/\sqrt{3}$.

17. Letting the rectangular parallelepiped have dimensions $x > 0$,
     $y > 0$, and $z > 0$, then we must maximize $V(x,y,z) = xyz$
     subject to $g(x,y,z) = 2xy+2yz+2xz-S_0 = 0$, where S_0 is the
     fixed surface area. $\nabla V = \lambda\nabla g$ gives
     $yz\mathbf{i}+xz\mathbf{j}+xy\mathbf{k} = 2\lambda(y+z)\mathbf{i}+2\lambda(x+z)\mathbf{j}+2\lambda(x+y)\mathbf{k}$. Equating components
     and solving for λ gives $yz/(y+z) = xz/(x+z) = xy/(x+y)$. From
     the first two, $yz(x+z) = xz(y+z)$ or $x = y$. Similarly, the
     first and third give $x = z$. Thus the volume will be
     maximized when $x = y = z$, that is, when the solid is a cube.

19. Since we know that $f(x,y)$, subject to the constraint
     $g(x,y) = 0$, has an extremum at (x_0,y_0), then
     $\nabla f(x_0,y_0) = \lambda\nabla g(x_0,y_0)$ for some scalar λ. Equating the
     $\mathbf{i}$ and $\mathbf{j}$ componenets yields $f_x = \lambda g_x$ and $f_y = \lambda g_y$ so that
     $f_x/g_x = f_y/g_y$ or $f_x g_y - f_y g_x = 0$, where all functions are
     evaluated at (x_0,y_0).

21a. Let $S(x,y) = x+y-C$ and $f(x,y) = (xy)^{1/2}$, so we must maximize
      f subject to $S(x,y) = 0$. Thus
      $[y/2(xy)^{1/2}]\mathbf{i} + [x/2(xy)^{1/2}]\mathbf{j} = \lambda\mathbf{i}+\lambda\mathbf{j}$ and hence
      $\lambda = y/2(xy)^{1/2} = x/2(xy)^{1/2}$ or $x = y$. Substituting this into
      the constraint equation gives $x+x = C$ or $x = C/2$ and then
      $y = C/2$, so the maximum of f is $f(C/2,C/2) = (C^2/4)^{1/2} = C/2$.

   b. From part (a), $(ab)^{1/2} \le$ max of $(xy)^{1/2}$ such that $x+y = a+b$.
      This maximum is exactly $(a+b)/2$ and thus $(ab)^{1/2} \le (a+b)/2$.

23.   In a manner similar to Probs. 21a and 22a, it can be shown
      that the maximum value of $(x_1x_2...x_n)^{1/n}$ subject to
      $x_1+x_2+...x_n = C$ occurs when $x_1 = x_2 = ... = x_n = C/n$, and
      again the maximum value of $(x_1x_2...x_n)^{1/n}$ is C/n. In this
      case, $C = a_1+a_2+a_3+...+a_n$ so that $(a_1a_2...a_n)^{1/n}$ never
      exceeds its maximum, which is $C/n = (a_1+a_2+...+a_n)/n$. Thus
      $(a_1a_2...a_n)^{1/n} \le (a_1+a_2+...+a_n)/n$.

25.   From Section 16.6, $f(x,y) = x^2/a^2 + y^2/b^2$ has a relative
      minimum at $(0,0)$, which is in the domain $x^2 + y^2 \le r^2$. Next,
      we determine the extrema of $f(x,y)$ on the circle
      $g(x,y) = x^2 + y^2 - r^2 = 0$. From $\nabla f = \lambda\nabla g$, we find
      $(2x/a^2)\mathbf{i} + (2y/b^2)\mathbf{j} = 2\lambda x\mathbf{i} + 2\lambda y\mathbf{j}$ so that $x(1/a^2 - \lambda) = 0$ and
      $y(1/b^2 - \lambda) = 0$. Thus either $x = 0$ and $\lambda = 1/b^2$ or $y = 0$ and
      $\lambda = 1/a^2$. In the first case, the constraint $x^2+y^2-r^2 = 0$
      gives $y^2 = r^2$ so $f(x,y) = r^2/b^2$. In the second case, the
      constraint yields $x^2 = r^2$ so $f(x,y) = r^2/a^2$. Since $a^2 > b^2$,
      $r^2/b^2 > r^2/a^2$ and thus $f(x,y)$ has minimum value of 0 and
      maximum value of r^2/b^2.

27.   For $x^2 + y^2/4 < 1$, $f_x(x,y) = 8x-y = 0$ and $f_y(x,y) = -x+2y+1 = 0$,
      so $x = -1/15$, $y = -8/15$, which satisfies the constraint. For
      the boundary, we let $g(x,y) = x^2 + y^2/4 - 1$ and hence $\nabla f = \lambda\nabla g$
      yields $8x-y = \lambda(2x)$ and $-x+2y+1 = \lambda(y/2)$ or $\lambda = (8x-y)/2x =$
      $(4y+2-2x)/y$, $(x,y) \ne (0,0)$. Solving gives $y^2 = 4x^2-4x$, which
      substituted into the constraint gives $2x^2-x-1 = 0$, or $x = -1/2$
      and $x = 1$. At $x = 1$, $y^2 = 4x^2-4x$ gives $y = 0$. Since we required
      $y \ne 0$, the point $(1,0)$ is not considered. When $x = -1/2$, we have
      $y = \pm\sqrt{3}$. Now $f(-1/15,-8/15) = -60/225$, $f(-1/2,\sqrt{3}) = (8+3\sqrt{3})$
      and $f(-1/2,-\sqrt{3}) = (8-3\sqrt{3})$. Thus f has minimum $-4/15$ and
      maximum $(8+3\sqrt{3})/2$ on the domain $x^2 + y^2/4 \le 1$.

29. From Ex.4, Section 16.6, $f(x,y)$ has a relative maximum at
     $(0,0)$ and a relative minimum at $(0,4)$. Both points are in
     the domain $x^2+y^2 \leq 36$ and thus must be considered. Next, we
     use Lagrange multipliers to find the extrema of f on the
     boundary $g(x,y) = x^2+y^2-36 = 0$. From $\nabla f = \lambda \nabla g$, we have
     $2x(y-2)\mathbf{i}+(y^2+x^2-4y)\mathbf{j} = 2\lambda x\mathbf{i}+2\lambda y\mathbf{j}$. Provided $x \neq 0$, $y \neq 0$,
     this gives $\lambda = (y^2+x^2-4y)/2y = y-2$ or $y^2 = x^2$. Substituting
     into $x^2+y^2 = 36$ yields $x = \pm3\sqrt{2}$ and then $y = \pm3\sqrt{2}$. Now
     $f(3\sqrt{2},3\sqrt{2}) = f(-3\sqrt{2},3\sqrt{2}) = 72\sqrt{2} - 66$, $f(0,4) = -14/3$,
     $f(3\sqrt{2},-3\sqrt{2}) = f(-3\sqrt{2},-3\sqrt{2}) = -72\sqrt{2} - 66$, and $f(0,0) = 6$.
     Thus the maximum is $72\sqrt{2} - 66$ at $(\pm3\sqrt{2},3\sqrt{2})$ while the minimum
     is $-72\sqrt{2} - 66$ at $(\pm3\sqrt{2},-3\sqrt{2})$.

Section 16.8, Page 876

1. Assume $y = f(x)$ and differentiate the equation with respect
    to x, getting $e^x\sin y + e^x\cos y(dy/dx) + 2dy/dx\cos x - 2y\sin x = 0$.
    Solving, we get $dy/dx = (2y\sin x-e^x\sin y)/(e^x\cos y+2\cos x)$.

3. Differentiating, $[2xy-x^2(dy/dx)]/y^2 + \dfrac{1}{1+(y/x)^2} \dfrac{x(dy/dx)-y}{x^2} = 0$.

    Thus $dy/dx = -[2xy(x^2+y^2) - y^3]/[xy^2 - x^2(x^2+y^2)]$.

5. Assume $z(x,y)$ and hold y constant to obtain
    $3y - z_x + e^{x+z}(1+z_x) = 0$. Thus, $z_x = -(3y+e^{x+z})/(e^{x+z} - 1)$.
    Next hold x constant to get $3x - z_y + e^{x+z}z_y - 4y = 0$, so
    $z_y = -(3x-4y)/(e^{x+z} - 1)$.

7. Holding x constant and differentiating with respect to y
    gives $\cos(x+y) - \sin(y+z)(1+z_y) + \cos(x+z)(z_y) = 0$ or
    $z_y = -[\cos(x+y)-\sin(y+z)]/[\cos(x+z)-\sin(y+z)]$.

9. Differentiating with respect to y yields
    $2(x^2+2y^2)4y + 4(x^2+z^2)(2zz_y) + 2(2y^2+z^2)(4y+2zz_y) = 0$, so
    $z_y = -2y(x^2+4y^2+z^2)/z(2x^2+2y^2+3z^2)$.

11.   Let $F(x,y,u,v) = x^2-y^2+u^2-v^2 = 0$ and $G(x,y,u,v) = xyuv = 4$.
      Then differentiating F and G with respect to y, holding x
      fixed, gives $F_y = -2y+2uu_y-2vv_y = 0$ and $G_y = xuv+xyvu_y+xyuv_y = 0$.
      This gives two equations in u_y and v_y, $uu_y - vv_y = y$ and

      $yvu_y + yuv_y = -uv$. Solving gives $u_y = u(y^2-v^2)/y(u^2+v^2)$ and
      $v_y = -v(u^2+y^2)/y(u^2+v^2)$.

13a. To verify $f(2) = 1$, substitute $x = 2$, $y = 1$ into the equation,
      getting $4-2+1-4+1 = 0$. Differentiating the equation yields
      $2x - y - x(dy/dx) + 3y^2(dy/dx) - 2 = 0$ so that
      $dy/dx = f'(x) = -(2x-y-2)/(3y^2-x)$. Now substitute $x = 2$,
      $y = 1$ to get $f'(2) = -1$.

  b. Now $d^2y/dx^2 = f''(x) = \dfrac{-(2-dy/dx)(3y^2-x)+(2x-y-2)(6ydy/dx-1)}{(3y^2-x)^2}$, so

      if $x = 2$, $y = 1$ then $dy/dx = -1$, from part(a), and
      $f''(2) = \dfrac{-(2+1)(3-2)+(4-1-2)(-6-1)}{(3-2)^2} = -10$.

15.   Substituting $x = 1$, $y = -1$, $z = 2$ gives
      $(1)^2+(1)(-1)+4(-1)^2-3(-1)(2)-(2)^3 = 1-1+4+6-8 = 2$.   Thus
      $(1,-1,2)$ satisfies the equation. Assuming x and y
      independent and $z = f(x,y)$, we find $2x+y-3yz_x-3z^2z_x = 0$, or
      $z_x = (2x+y)/(3y+3z^2) = f_x(x,y)$, and $x+8y-3z-3yz_y-3z^2z_y = 0$,
      or $z_y = (x+8y-3z)/(3y+3z^2) = f_y(x,y)$. Now compute
      f_{xx} and f_{xy} from the expression for f_x to obtain
      $f_{xx} = [2(3y+3z^2)-6(2x+y)zz_x]/(3y+3z^2)^2$ and
      $f_{xy} = [(3y+3z^2)-(2x+y)(3+6zz_y)]/(3y+3z^2)^2$. By direct
      substitution, $f_x(1,-1) = 1/9$, $f_{xx}(1,-1) = 50/243$,
      $f_y(1,-1) = -13/9$, and $f_{xy}(1,-1) = 70/243$.

17.   If $y = y(x)$, then $dz/dx = f_x(dx/dx) + f_y(dy/dx) = f_x + f_y(dy/dx)$.
      Since $g(x,y) = 0$, then $g_x + g_y(dy/dx) = 0$ or
      $dy/dx = -g_x/g_y$. Thus $dz/dx = f_x + f_y(-g_x/g_y) = (f_xg_y - f_yg_x)/g_y$.

19. Consider $z = f(x,y)$ so that $F(x,y,z) = F(x,y,f(x,y)) = 0$.
    With x and y now thought of as independent, we differentiate
    with respect to x and have
    $$\frac{\partial F(x,y,f(x,y))}{\partial x} = \frac{\partial F}{\partial x}\frac{\partial x}{\partial x} + \frac{\partial F}{\partial y}\frac{\partial y}{\partial x} + \frac{\partial F}{\partial z}\frac{\partial f}{\partial x} = 0.$$ Note that $\partial x/\partial x = 1$
    and $\partial y/\partial x = 0$ and thus $\partial f/\partial x = -(\partial F/\partial x)/(\partial F/\partial z)$. Next,
    considering $y = g(x,z)$ and $F(x,g(x,z),z) = 0$, differentiation
    with respect to z gives $\dfrac{\partial F(x,g(x,z),z)}{\partial z} = \dfrac{\partial F}{\partial y}\dfrac{\partial g}{\partial z} + \dfrac{\partial F}{\partial z} = 0$ or
    $\partial g/\partial z = -(\partial F/\partial z)/(\partial F/\partial y)$. Similarly, by choosing $x = h(x,z)$,
    we have $\partial h/\partial y = -\dfrac{(\partial F/\partial y)}{(\partial F/\partial x)}$. Finally,
    $$\frac{\partial h}{\partial y}\cdot\frac{\partial g}{\partial z}\cdot\frac{\partial f}{\partial y} = [-\frac{(\partial F/\partial y)}{(\partial F/\partial x)}][-\frac{(\partial F/\partial z)}{(\partial F/\partial y)}][-\frac{(\partial F/\partial x)}{(\partial F/\partial z)}] = -1.$$

21a. Differentiating implicitly we have
     $1 = r_x\cos\theta - r(\sin\theta)\theta_x$ and $1 = r_y\sin\theta + r(\cos\theta)\theta_y$,
     $0 = r_x\sin\theta + r(\cos\theta)\theta_x$ and $0 = r_y\cos\theta - r(\sin\theta)\theta_y$. Solving
     yields $r_x = \cos\theta$, $r_y = \sin\theta$, $\theta_x = -\sin\theta/r$, and $\theta_y = \cos\theta/r$.

  b. If $w = f(r,\theta)$, then $w_x = w_r r_x + w_\theta\theta_x = w_r\cos\theta - (w_\theta\sin\theta)/r$ and
     $w_y = w_r r_y + w_\theta\theta_y = w_r\sin\theta + (w_\theta\cos\theta)/r$.

  c. $w_{xx} = [w_r\cos\theta - w_\theta\sin\theta/r]_x = (w_r\cos\theta)_x - (w_\theta\sin\theta/r)_x$
     $\quad = (w_r)_x\cos\theta + w_r(\cos\theta)_x - [r(w_\theta\sin\theta)_x - (w_\theta\sin\theta)r_x]/r^2$
     $\quad = (w_{rr}r_x + w_{r\theta}\theta_x)\cos\theta - w_r\sin\theta(\theta_x)$
     $\qquad - [r(w_{\theta r}r_x + w_{\theta\theta}\theta_x)\sin\theta + rw_\theta\cos\theta(\theta_x) - (w_\theta\sin\theta)r_x]/r^2$.
     After substituting for r_x, θ_x and simplifying, we get
     $w_{xx} = w_{rr}\cos^2\theta - (w_{r\theta}\sin2\theta)/r + (w_r\sin^2\theta)/r + (w_{\theta\theta}\sin^2\theta)/r^2$
     $\qquad\qquad + w_\theta\sin2\theta/r^2$.
     Similarly,
     $w_{yy} = w_{rr}\sin^2\theta + (w_{r\theta}\sin2\theta)/r + (w_r\cos^2\theta)/r + (w_{\theta\theta}\sin^2\theta)/r^2$
     $\qquad\qquad - (w_\theta\sin2\theta)/r^2$.
     Using trigometric identities, we add to get
     $w_{xx} + w_{yy} = w_{rr} + (1/r)w_r + (1/r^2)w_{\theta\theta}$.

23.   Compute the x and y derivatives of $x = f(u,v)$ and
      $y = g(u,v)$ to get the two systems of equations
      $1 = f_u u_x + f_v v_x$ and $0 = f_u u_y + f_v v_y$
      $0 = g_u u_x + g_v v_x,$ $1 = g_u u_y + f_v v_y.$
      Solving these equations gives, for $\Delta = f_u g_v - f_v g_u,$
      $u_x = g_v/\Delta,$ $v_x = -g_u/\Delta,$ $u_y = -f_v/\Delta,$ and $v_y = f_u/\Delta.$

Chapter 16 Review, Page 877

1a.   The domain is all (x,y) for which $|x| > |y|$. $\phi(1,0) = \ln 1 = 0.$

 b. $\phi(5,4) = \ln(25-16)^{3/2} = \ln 27.$

3a.   The domain is all $x,y,$ and z except for $z = x^2/2.$
      $\phi(1,1,1) = \arctan(1/-1) = -\pi/4.$

 b. $\phi(a^{-1},b^{-1},c^{-1}) = \arctan\left[\left(\dfrac{3}{a} - \dfrac{2}{b}\right)\Big/\left(\dfrac{-2}{c} + \dfrac{1}{a^2}\right)\right]$

                    $= \arctan[(ac/b)(3b-2a)/(c-2a^2)].$

5. f is discontinuous when $2x-y = 0,$ or along the line $y = 2x.$

7. f is discontinuous on the lines $y = x$ and $y = x\pm1$. Since
      $\lim\limits_{(x,y)\to(x,x)} f(x,y) = 0,$ $f(x,x) = 0$ makes f continuous on $y = x.$

9. $f(x,y) = (1/2)\ln(x^2+y^2),$ so $f_x = x/(x^2+y^2)$ and $f_y = y/(x^2+y^2).$

11. $g_r = s^2 t^2 \cos(rs-t),$ $g_t = -st^2\cos(rs-t) + 2st\sin(rs-t)$ and
      $g_s = t^2\sin(rs-t) + rst^2\cos(rs-t).$

13. $z_x = -2x\sin(x^2-y^2)$ and $z_y = 2y\sin(x^2-y^2),$ so
      $z_{xy} = 4xy\cos(x^2-y^2) = z_{yx}.$

15. $z_x = ye^{xy}(\sin xy + \cos xy)$ and $z_y = xe^{xy}(\sin xy + \cos xy),$ so
      $z_{xy} = e^{xy}(2xy\cos xy + \sin xy + \cos xy) = z_{yx}.$

17.   Since $u_x = -3x^2+3y^2 = v_y$ and $u_y = 6xy = -v_x,$ the function
      satisfies the Cauchy-Riemann equations for all x and $y.$

19.   Since $u_x = \omega\cos(\omega x-y)$ and $v_y = \pi x^2 e^{\pi y},$ the Cauchy-Riemann
      equations are not satisfied.

21.   Integrating with x constant yields $z(x,y) = x^3y^2 - \sin 2y + g(x)$.

23.   Integrating with x constant yields $z(x,y) = \arctan(y/x) + g(x)$.

25. $f(x+h,y+k) = (x+h)^3 - 3(y+k)^2 = x^3 - 3y^2 + 3x^2h - 6yk + r\sqrt{h^2+k^2}$, where

$r(x,y,h,k) = (3xh^2 + h^3 - 3k^2)/\sqrt{h^2+k^2}$. For $h = \rho\cos\theta$, $k = \rho\sin\theta$,

$r(x,y,h,k) = \rho[3x\cos^2\theta + \rho\cos^3\theta - 3\sin^2\theta]$, so $r \to 0$ as $(h,k) \to (0,0)$.

27. $f(x+h,y+k) = f(x,y) + f_x(x,y)h + f_y(x,y)k + r\sqrt{x^2+k^2}$

$= e^{xy} + ye^{xy}h + xe^{xy}k - x^2y - 2xyh - x^2k + r\sqrt{h^2+k^2}$

$= e^{xy}(1+yh+xk) - (x^2y+2xyh+x^2k) + r\sqrt{h^2+k^2}$.   On the
other hand, direct calculation yields
$f(x+h,y+k) = e^{xy}(e^{xk+yh+hk}) - (x^2y+2xyh+x^2k) - (yh^2+2xhk+h^2k)$.
Comparing the two expressions gives

$f(x+h,y+k) = e^{xy} - x^2y + (ye^{xy} - 2xy)h + (xe^{xy} - x^2)k + r\sqrt{h^2+k^2}$, where

$r(x,y,h,k) = \{e^{xy}[e^{xk+yh+hk} - (1+xk+yh)] - (y+k)h^2 - 2xhk\}/\sqrt{h^2+k^2}$.
Letting $h = \rho\cos\theta$, $k = \rho\sin\theta$, we see that
$e^{xk+yh+hk} - (1+xk+yh) = hk + \alpha^2/2! + \alpha^3/3! + \cdots$, where α has a
factor of ρ. Every other term in the numerator of r also has
a factor of ρ^2, so that $r(x,y,h,k) = \rho R(x,y,h,k)$, where R is
continuous. Thus $r \to 0$ as $(h,k) \to (0,0)$.

29. $dz = z_x dx + z_y dy = -2y\cos x \sin x\, e^{\cos^2 x} dx + e^{\cos^2 x} dy$.

31. $dw = z^2[1-(x-y)^2]^{-1/2} dx - z^2[1-(x-y)^2]^{-1/2} dy + 2z\arcsin(x-y) dz$.

33. $\dfrac{dz}{dt} = y\cosh(xy-y^2)(1) + (x-2y)\cosh(xy-y^2)(1/2)t^{-1/2}$

$= [(3/2)t^{1/2} - 1]\cosh(t^{3/2} - t)$.

35. $\dfrac{dw}{dt} = (2xyz - y^2z + yz^3)(1) + (x^2z - 2xyz + xz^3)(2t)$

$+ (x^2y - xy^2 + 3xyz^2)(3t^2) = 7t^6 - 8t^7 + 12t^{11}$.

37. $z_r = 3x^2ye^r\sin s - x^3e^s\sin r = e^{3r+s}\sin^3 s(3\cos r - \sin r)$,

$z_s = 3x^2ye^r\cos s + x^3e^s\cos r = e^{3r+s}\sin^2 s\cos r(3\cos s + \sin s)$.

39. $w_r = (2x/a^2)(s) - (2y/b^2)(2r) - (1/c)(1)$
      $= 2rs^2/a^2 - (4r^3+4rs)/b^2 - 1/c,$
      $w_s = (2x/a^2)(r) - (2y/b^2)(1) - (1/c)(-1)$
      $= 2r^2s/a^2 - (2r^2+2s)/b^2 + 1/c.$

41.   From Section 16.4, $\nabla f = y^2\mathbf{i} + 2xy\mathbf{j} = \mathbf{i}+2\mathbf{j}$ at $(1,1)$ and
      $\lambda = (2\mathbf{i}-3\mathbf{j})/\sqrt{13}$, so $D_\lambda f(1,1) = -4/\sqrt{13}$.

43. $\nabla g = 2x\arctan(zy)\mathbf{i} + [zx^2/(1+z^2y^2)]\mathbf{j} + [yx^2/(1+z^2y^2)]\mathbf{k} = 2\mathbf{j}$
      at $(1,0,2)$ and $\lambda = (4\mathbf{i}+\mathbf{j}-2\mathbf{k})/\sqrt{21}$, so that $D_\lambda g(1,0,2) = 2/\sqrt{21}$.

45. $\nabla f = y^2\mathbf{i} + 2xy\mathbf{j} = 4\mathbf{i}-12\mathbf{j}$ at $(3,-2)$, so $\lambda = (4\mathbf{i}-12\mathbf{j})/\sqrt{160}$.
      The rate of change is $\|\nabla f\| = \sqrt{160} = 4\sqrt{10}$.

47. $\nabla g = 2x\arcsin(2y)\mathbf{i} + (x^2z/\sqrt{1-z^2y^2})\mathbf{j} + (x^2y/\sqrt{1-z^2y^2})\mathbf{k} = -4\mathbf{j}$
      at $(1,0,-4)$ so $\lambda = -\mathbf{j}$ and the rate of change is 4.

49.   Since $2(1)-(2) = 0$ and $(1)^2+(2)^2+(4) = 9$, $(1,2,4)$ is on the
      curve of intersection of the two surfaces. From Section
      16.5, with $F(x,y,z) = 2x-y$ and $G(x,y,z) = x^2+y^2+z-9$,
      $\mathbf{u} = \nabla F \times \nabla G = (2\mathbf{i}-\mathbf{j}) \times (2x\mathbf{i}+2y\mathbf{j}+\mathbf{k}) = -\mathbf{i}-2\mathbf{j}+10\mathbf{k}$ is a tangent
      vector to the intersection at $(1,2,4)$. Thus the tangent line
      is given by $\dfrac{x-1}{1} = \dfrac{y-2}{2} = \dfrac{z-4}{-10}$.

51.   By direct substitution, $(1,3,1)$ is on both surfaces. If
      $F = x^2/4 + y^2/9 - z^2/4 - 1$ and $G = x+z-2$, then at $(1,3,1)$ the
      tangent vector is given by $\nabla F \times \nabla G = (1/3)(2\mathbf{i}-3\mathbf{j}-2\mathbf{k})$, so the
      tangent line is $\dfrac{x-1}{2} = \dfrac{y-3}{-3} = \dfrac{z-1}{-2}$.

53. $f_x = 2xy-3y^2+4y = 0$ and $f_y = x^2-6xy+4x = 0$ yield $(0,0)$,
      $(0,4/3)$, $(-4,0)$, and $(-4/3,4/9)$ as the stationary points.
      With $f_{xx} = 2, f_{xy} = 2x-6y+4$, and $f_{yy} = -6x$, we find
      $(f_{xy})^2 - f_{xx}f_{yy} > 0$ for the first three and so they are all
      saddle points. At $(-4/3,4/9)$ the conditions of Thm.16.6.2(b)
      are satisfied, hence this point is a relative minimum.

55. Solving $f_x = -2x \exp(-x^2+4y^2) = 0$ and $f_y = 8y \exp(-x^2+4y^2) = 0$
yields $(0,0)$ as the only stationary point. Since
$f_{xx} = (4x^2-2)\exp(-x^2+4y^2)$, $f_{xy} = -16xy \exp(-x^2+4y^2)$ and
$f_{yy} = (64y^2+8)\exp(-x^2+4y^2)$, we find $(f_{xy})^2 - f_{xx}f_{yy} > 0$ at
$(0,0)$, which is thus a saddle point.

57. Solving $2x-3y^2-6xyy'+6y^2y' = 0$ gives $y' = (3y^2-2x)/(6y^2-6xy)$.

59. $z_x-(z_x\cosh x+z\sinh x)\exp(z\cosh x)+(2x-2zz_x)/(x^2-z^2) = 0$, so $z_x = $
$[(x^2-z^2)z\sinh x\exp(z\cosh x)-2x]/\{(x^2-z^2)[1-\cosh x\exp(z\cosh x)]-2z\}$.
Similarly $x_z = 1/z_x$.

61. As in Section 16.5, let $F(x,y,z) = (x^2/4)+(y^2/9)-(z^2/4)-2 = 0$,
then $\nabla F = (x/2)\mathbf{i} + (2y/9)\mathbf{j} - (z/2)\mathbf{k} = (1/2)(3\mathbf{i}-\mathbf{k})$ represents a
normal vector to the surface at $(3,0,1)$. Thus the tangent
plane has equation $3(x-3)-(z-1) = 0$, or $3x-z = 8$, and the
normal line is $(x-3)/3 = (z-1)/(-1)$, $y = 0$.

63. As g increases most rapidly in the direction of ∇g, we must
find $D_\lambda f$ when $\lambda = \nabla g/\|\nabla g\| = (3/5)\mathbf{i}+(4/5)\mathbf{j}$. Thus
$D_\lambda f = \nabla f\cdot\lambda = (3y/5)+(4x/5) = 13/5$ at $(1,3)$.

65a. Let $A(x,y) = 4xy$ and $g(x,y) = x^2/25 + y^2/144 - 1$, so $\nabla A = \lambda\nabla g$
gives $x = (5/12)y$. Substituting into $g(x,y)$ gives $y = \pm6\sqrt{2}$.
Thus the maximum area occurs with the vertex in the first
quadrant at $(5/\sqrt{2},6\sqrt{2})$, so the rectangle has dimensions
$w = 5\sqrt{2}$, $h = 12\sqrt{2}$ and area $A = 120$.
  b. With the vertex of the rectangle in the first quadrant at
   (x,y) on the ellipse, we have $y = (12/5)\sqrt{25-x^2}$ and thus
   $A(x) = (48/5)x\sqrt{25-x^2}$. Then $A'(x) = (48/5)(25-2x^2)/\sqrt{25-x^2}$
   gives $x = 5/\sqrt{2}$ and so $y = 6\sqrt{2}$, as in part(a).

67. If c is the distance between the cars at time T then, by the
law of cosines, $c^2 = a^2 + b^2 - 2ab\cos C$, where $a = 100\,\text{km}$,
$a' = 60\,\text{km/hr}$, $b = 120\,\text{km}$, $b' = 70\,\text{km/hr}$, and $C = 105°$. Thus
$c \cong 175\,\text{km}$ and $cc' = aa'+bb'- (ab'+a'b)\cos C$, which yields
$c' \cong 103.3\,\text{km/hr}$.

Chapter 17

Section 17.1, Page 889

1a. The four centers are located at $(1/2,1/2)$, $(1/2,3/2)$, $(3/2,1/2)$, $(3/2,3/2)$, and $\Delta A_{ij} = 1$ for each rectangle. Thus

$$\sum_{i=1,2;j=1,2} f(P^*_{ij})\Delta A_{ij} = [2(1/2)(1/2) + 2(1/2)(3/2) + 2(3/2)(1/2)$$
$$+ 2(3/2)(3/2)](1) = 8.$$

b. The four P^*_{ij} are $(1,0)$, $(1,1)$, $(2,0)$, $(2,1)$ and thus

$$\sum_{i=1,2;j=1,2} f(P^*_{ij})\Delta A_{ij} = (0+2+0+4)(1) = 6, \text{ since } \Delta A_{ij} = 1.$$

c. The four P^*_{ij} are $(1,1)$, $(1,2)$, $(2,1)$, $(2,2)$ and thus

$$\sum_{i=1,2;j=1,2} f(P^*_{ij})\Delta A_{ij} = (2+4+4+8)(1) = 18, \text{ since } \Delta A_{ij} = 1.$$

3a. $\Delta A_{ij} = 1$ for each rectangle and the centers are $(1/2,1/2)$ and $(3/2,1/2)$. Thus the Riemann sum is
$[(1/2)^2(1/2) + (3/2)^2(1/2)]1 = 5/4 = 1.25$.

b. $\Delta A_{ij} = 1/4$ for each rectangle and the centers are located at $x_i = (2i-1)/4$, $i = 1,2,3,4$ and $y_j = (2j-1)/4$, $j = 1,2$. Thus the Riemann sum is given by
$[(1/64 + 3/64) + 9(1/64 + 3/64) + 25(1/64 + 3/64)$
$$+ 49(1/64 + 3/64)](1/4) = 21/16 = 1.3125.$$

c. $\Delta A_{ij} = 1/16$ for each rectangle and $x_i = (2i-1)/8$, $i = 1,2,\ldots 8$ and $y_j = (2j-1)/8$, $j = 1,2,3,4$. If we add up the four Riemann sums for $x_1 = 1/8$ we get
$(1/8^3+3/8^3+5/8^3+7/8^3)(1/16) = \alpha$. Thus the total Riemann sum is

$$\alpha+9\alpha+25\alpha+\ldots+225\alpha = \alpha\sum_{k=1}^{8}(2k-1)^2 = 680\alpha = 680(16/2\cdot8^4)$$
$$= 85/64 \cong 1.3281.$$

5. Since $1/(2x^2+y^2+4) \le 1/4$, Eq.(20) yields
$$\iint_\Omega \frac{dA}{2x^2+y^2+4} \le \iint_\Omega \frac{dA}{4} = (1/4)\iint_\Omega dA = \pi/4 \text{ since } \Omega \text{ is a circle of}$$
radius 1.

7. Since $e^{-(x^2+y^2)} \leq e^0 = 1$, Eq.(20) yields
$$\iint_\Omega e^{-(x^2+y^2)}\,dA \leq \iint_\Omega dA = \pi/2 \text{ since } \Omega \text{ is a circle of radius } 1/\sqrt{2}.$$

9. The surface $z = \sqrt{9-x^2-y^2}$ is a hemisphere of radius 3. Thus
$$\iint_\Omega \sqrt{9-x^2-y^2}\,dA = (4/3)\pi(3^3)/8 = 9\pi/2, \text{ which is the volume of}$$
the hemisphere in the first octant, since $x \geq 0$, $y \geq 0$.

11. $\displaystyle\iint_\Omega (6 - 3\sqrt{x^2+y^2})\,dA = 6\iint_\Omega dA - 3\iint_\Omega \sqrt{x^2+y^2}\,dA$ by Eq.(18). Now

$\displaystyle\iint_\Omega dA = 4\pi$, since the integral represents the area of a circle

of radius 2. For the second integral consider $x^2+y^2 = 4$,
then $z = \sqrt{x^2+y^2} = 2$, and we see that the integral represents
the volume of a hemisphere (opening in the positive z
direction) of radius 2, so $\displaystyle\iint_\Omega \sqrt{x^2+y^2}\,dA = (4\pi/3)(2^3)/2 = 16\pi/3$.

Thus $\displaystyle\iint_\Omega (6 - 3\sqrt{x^2+y^2})\,dA = 6(4\pi) - 3(16\pi/3) = 8\pi$.

13. Let $g(x,y)$ be defined as in Eq.(12). Then, Eq.(13) yields
$$\iint_\Omega f(x,y)\,dA = \iint_R g(x,y)\,dA = \lim_{\|\Delta\|\to 0}\sum_{i,j} g(P^*_{ij})\Delta A_{ij}. \text{ Since}$$
$g(P^*_{ij}) = f(P^*_{ij}) \geq 0$ if P^*_{ij} is in Ω and zero otherwise and

since $\Delta A_{ij} > 0$, we conclude that $\displaystyle\sum_{i,j} g(P^*_{ij})\Delta A_{ij} \geq 0$ and hence

$\displaystyle\iint_\Omega f(x,y)\,dA \geq 0$ if f is integrable and $f(x,y) \geq 0$ on Ω.

15. As in Prob.(13),
$$\left|\iint_\Omega f(x,y)\,dA\right| = \left|\iint_R g(x,y)\,dA\right| = \left|\lim_{\|\Delta\|\to 0}\sum_{i,j} g(P^*_{ij})\Delta A_{ij}\right|$$
$$\leq \lim_{\|\Delta\|\to 0}\sum_{i,j} |g(P^*_{ij})|\Delta A_{ij},$$
since f (and hence g) is integrable and since $\Delta A_{ij} > 0$. Now,

$$\lim_{||\Delta||\to 0}\sum_{i,j}|g(P^*_{ij})|\Delta A_{ij} = \iint_R |g(x,y)|dA = \iint_\Omega |f(x,y)|dA, \text{ and hence}$$

Eq.(21) is proved.

Section 17.2, Page 903

1. From the inside limits, x varies
    from x = 0 to x = 2 and from the
    outside limits, y varies from
    y = 0 to y = 3. Thus the region
    of integration is as shown and

$$\int_0^3\int_0^2 (3x-2y)dxdy = \int_0^3 (3x^2/2 - 2xy)\Big|_0^2 dy$$

$$= \int_0^3 (6-4y)dy = 6y-2y^2\Big|_0^3 = 0.$$

3. x varies from x = -1 to x = 1 while
    y varies from y = 0 to y = 2. Thus
    the region of integration is as
    shown and

$$\int_0^2\int_{-1}^1 (1+x^2y^2)dxdy = \int_0^2 (x+x^3y^2/3)\Big|_{-1}^1 dy$$

$$= 2\int_0^2 (1+y^2/3)dy = 2(y+y^3/2)\Big|_0^2 = 52/9.$$

5. y varies from y = x^2 to y = x while
    x varies from 0 to 1. Thus the
    region is as shown and

$$\int_0^1\int_{x^2}^x (2x-5y)dydx = \int_0^1 (2xy-5y^2/2)\Big|_{x^2}^x dx$$

$$= \int_0^1 [(2x^2 - 5x^2/2) - (2x^3 - 5x^4/2)]dx$$

$$= \int_0^1 (5x^4/2 - 2x^3 - x^2/2)dx = (x^5/2 - x^4/2 - x^3/6)\Big|_0^1 = -1/6.$$

7. x varies from x = y to x = 1 while
    y varies from 0 to ln2 so the region
    is as shown and

$$\int_0^{\ln2}\int_y^1 2xe^{-y}\,dxdy = \int_0^{\ln2} x^2e^{-y}\Big|_y^1\,dy$$

$$= \int_0^{\ln2}(1-y^2)e^{-y}\,dy = (1+2y+y^2)e^{-y}\Big|_0^{\ln2}$$

$$= [(\ln2)^2 + 2\ln2 - 1]/2, \text{ where integration by}$$

    parts has been used to evaluate the last integral.

9. x varies from x = 0 to x = y while
    y varies from 0 to π/2.

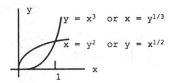

$$\int_0^{\pi/2}\int_0^y y\sin x\cos y\,dxdy = -\int_0^{\pi/2} y\cos y\cos x\Big|_0^y\,dy$$

$$= \int_0^{\pi/2}(y\cos y - y\cos^2y)\,dy = \int_0^{\pi/2}[y\cos y - y/2 - (y/2)\cos2y]\,dy$$

$$= [y\sin y + \cos y - y^2/4 - (y/4)\sin^2y - (1/8)\cos2y]_0^{\pi/2}$$

$$= \pi/2 - \pi^2/16 - 3/4.$$

11a. The region is as shown. Thus, y
     varies from $y = x^3$ to $y = \sqrt{x}$ as x
     varies from 0 to 1 and

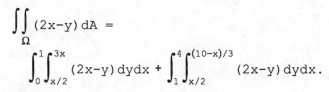

$$\iint_\Omega 2xy\,dA = \int_0^1\int_{x^3}^{\sqrt{x}} 2xy\,dydx.$$

  b. In this case x varies from $x = y^2$

     to $x = y^{1/3}$ and thus $\iint_\Omega 2xy\,dA = \int_0^1\int_{y^2}^{y^{1/3}} 2xy\,dxdy.$

  c. $\iint_\Omega 2xy\,dA = \int_0^1\int_{x^3}^{\sqrt{x}} 2xy\,dydx = \int_0^1 xy^2\Big|_{x^3}^{\sqrt{x}}\,dx = \int_0^1(x^2-x^7)\,dx = 5/24.$

13a. For 0 ≤ x ≤ 1, y varies from y = x/2
     to y = 3x and for 1 ≤ x ≤ 4, y
     varies from y = x/2 to y = (10-x)/3
     and thus

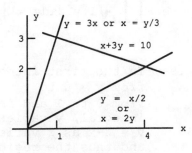

$$\iint_\Omega (2x-y)\,dA =$$

$$\int_0^1\int_{x/2}^{3x}(2x-y)\,dydx + \int_1^4\int_{x/2}^{(10-x)/3}(2x-y)\,dydx.$$

13b. For $0 \le y \le 2$, x varies from $x = y/3$ to $x = 2y$ and for
$2 \le y \le 3$, x varies from $x = y/3$ to $x = 10-3y$ and thus

$$\iint_{\Omega} (2x-y)\,dA = \int_0^2 \int_{y/3}^{2y} (2x-y)\,dx\,dy + \int_2^3 \int_{y/3}^{10-3y} (2x-y)\,dx\,dy.$$

c. From part (b),

$$\iint_{\Omega} (2x-y)\,dA = \int_0^2 (x^2-xy)\Big|_{y/3}^{2y}\,dy + \int_2^3 (x^2-xy)\Big|_{y/3}^{10-3y}\,dy$$

$$= \int_0^2 (20/9)\,y^2\,dy + \int_2^3 [100-70y+(110/9)\,y^2]\,dy$$

$$= (20/27)\,y^3\Big|_0^2 + [100y - 35y^2 + (110/27)\,y^3]_2^3 = 25/3.$$

15a. y varies from $y = 0$ to $y = 1-x^2$ as
x varies from −1 to 1 and thus the
region is as shown.

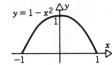

b. Reversing the order of integration,
we see that x varies from $x = -\sqrt{1-y}$ to $x = +\sqrt{1-y}$ as y varies

from 0 to 1 and hence $\displaystyle\int_0^1 \int_{-\sqrt{1-y}}^{\sqrt{1-y}} dx\,dy$ is an equivalent integral.

c. $\displaystyle\int_{-1}^1 \int_0^{1-x^2} dy\,dx = \int_{-1}^1 y\Big|_0^{1-x^2}\,dx = \int_{-1}^1 (1-x^2)\,dx = 4/3.$

17a. y varies from $y = \sqrt{x}$ to $y = 2$ as
x varies from 0 to 4 and thus the
region is as shown.

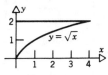

b. Reversing the order of integration,
we see that x varies from $x = 0$ to $x = y^2$ as y varies from 0

to 2 and hence $\displaystyle\int_0^2 \int_0^{y^2} \sqrt{1+y^3}\,dx\,dy$ is an equivalent integral.

c. $\displaystyle\int_0^2 \int_0^{y^2} \sqrt{1+y^3}\,dx\,dy = \int_0^2 \sqrt{1+y^3}\,x\Big|_0^{y^2}\,dy = \int_0^2 \sqrt{1+y^3}\,y^2\,dy$

$$= [(1/3)(1+y^3)^{3/2}/(3/2)]\Big|_0^2 = 52/9.$$

19a. In the first integral, y varies
from $y = 0$ to $y = \sqrt{x}$ as x varies
from 0 to 1 and in the second
integral, y varies from $y = 0$ to
$y = 2-x$ as x varies from 1 to 2
and thus the region is as shown.

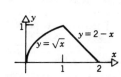

19b. Reversing the order of integration, we see that x varies from
$x = y^2$ to $x = 2-y$ as y varies from 0 to 1 and hence
$\int_0^1 \int_{y^2}^{2-y} (x-2y)\,dxdy$ is an equivalent integral.

c. $\int_0^1 \int_{y^2}^{2-y} (x-2y)\,dxdy = \int_0^1 (x^2/2 - 2xy)\,\Big|_{y^2}^{2-y}\,dy$

$$= (1/2)\int_0^1 (4 - 12y + 5y^2 + 4y^3 - y^4)\,dy = 7/30.$$

21. As in Ex.4, $M = \iint_\Omega \rho(x,y)\,dA = k\int_0^1 \int_{x^2}^{\sqrt{x}} (x^2+y^2)\,dydx$

$= k\int_0^1 (x^2 y + y^3/3)\,\Big|_{x^2}^{\sqrt{x}}\,dx = k\int_0^1 (x^{5/2} + x^{3/2}/3 - x^4 - x^6/3)\,dx = 6k/35.$

23. $M = k\int_{-\sqrt{2}}^{\sqrt{2}} \int_{x^2}^{8-3x^2} y\,dydx$

$= (k/2)\int_{-\sqrt{2}}^{\sqrt{2}} y^2\,\Big|_{x^2}^{8-3x^2}\,dx = (k/2)\int_{-\sqrt{2}}^{\sqrt{2}} (64-48x^2+8x^4)\,dx = 192k\sqrt{2}/5.$

25. The region of integration in the xy-plane is given by
$0 \le x \le 3$, $0 \le y \le 2$. Thus
$V = \int_0^3 \int_0^2 (16-x^2-y^2)\,dydx = \int_0^3 (16y-x^2y-y^3/3)\,\Big|_0^2\,dx$

$= \int_0^3 (88/3 - 2x^2)\,dx = 70.$

27. The region of integration in the xy-plane is given by
$0 \le y \le x^2$ for $0 \le x \le 1$. Thus
$V = \int_0^1 \int_0^{x^2} (4-x^2-y^2)\,dydx = \int_0^1 [(4-x^2)y-y^3/3]\,\Big|_0^{x^2}\,dx$

$= \int_0^1 (11/3 - 5x^2 + x^4 + x^6/3)\,dx = 236/105.$

29. Since the cylinders $y = z^2$ and $y = \sqrt{z}$ are parallel to the
x-axis, we project the desired volume onto the yz-plane to
obtain the region of integration $z^2 \le y \le \sqrt{z}$ for $0 \le z \le 1$.
The "top" of the volume is then given by $x = 9-2y-z$ and thus

$$V = \int_0^1 \int_{z^2}^{\sqrt{z}} (9-2y-z)\,dydz = \int_0^1 [(9-z)y-y^2]\Big|_{z^2}^{\sqrt{z}}\,dz$$

$$= \int_0^1 (9z^{1/2} - z - z^{3/2} - 9z^2 + z^3 + z^5)\,dz = 51/20.$$

31. The region of integration is given by $0 \le x \le a-y$ for $0 \le y \le a$ and thus

$$V = \int_0^a \int_0^{a-y} (a^2-x^2-y^2)\,dxdy = \int_0^a [(a^2-y^2)x-x^3/3]_0^{a-y}\,dy$$

$$= \int_0^a (2a^3/3 - 2ay^2 + 4y^3/3)\,dy = a^4/3.$$

Section 17.3, Page 915

1. From symmetry $\hat{x} = 0$ and from Eq.(21) $\hat{y} = \hat{L}_x/A$, where

$$\hat{L}_x = \int_{-a}^a \int_0^{b\sqrt{a^2-x^2}/a} y\,dydx = (b^2/2a^2)\int_{-a}^a (a^2-x^2)\,dx = 2b^2a/3.$$ Since
the area of an ellipse is πab, we then find
$\hat{y} = (2b^2a/3)/(\pi ab/2) = 4b/3\pi$.

3. From symmetry $\hat{x} = \pi/2$. In this case,

$$\hat{L}_x = \int_0^\pi \int_0^{\sin x} y\,dydx = (1/2)\int_0^\pi \sin^2 x\,dx = \pi/4 \text{ and } A = \int_0^\pi \sin x\,dx = 2$$

and thus $\hat{y} = \pi/8$.

5. From symmetry $\hat{x} = \hat{y}$, where $\hat{x} = \hat{L}_y/A = a/3$ since

$$\hat{L}_y = \int_0^a \int_0^{a-x} x\,dydx = \int_0^a (ax-x^2)\,dx = a^3/6 \text{ and } A = a^2/2.$$

7. We have $L_y = k\int_0^a \int_0^b x(2b-y)\,dydx = k(3b^2/2)\int_0^a x\,dx = 3ka^2b^2/4,$

$$L_x = k\int_0^a \int_0^b y(2b-y)\,dydx = k(2b^3/3)\int_0^a dx = 2kab^3/3, \text{ and}$$

$$M = k\int_0^a \int_0^b (2b-y)\,dydx = k(3b^2/2)\int_0^a dx = 3kab^2/2. \text{ Thus, by}$$

Eqs.(12) and (14), $\bar{x} = L_y/A = a/2$ and $\bar{y} = L_x/A = 4b/9$. Note that the centroid is at $(a/2, b/2)$.

9a. Positioning the rectangle so that $0 \le x \le b$ and $0 \le y \le h$, we
     then have by Eq.(22), $I_x = k \int_0^b \int_0^h y^2 dy dx = kbh^3/3$.

b.   By Eq.(23), $I_y = k \int_0^b \int_0^h x^2 dy dx = kb^3h/3$.

c.   Positioning the rectangle so that $-b/2 \le x \le b/2$ and
     $-h/2 \le y \le h/2$, we then have
     $$I_x = k \int_{-b/2}^{b/2} \int_{-h/2}^{h/2} y^2 dy dx = k(h^3/12) \int_{-b/2}^{b/2} dx = kbh^3/12.$$

d.   Using the position of part(a), we have by Eq.(24),
     $$I_0 = k \int_0^b \int_0^h (x^2+y^2) dy dx = kbh(b^2 + h^2)/3.$$

e.   Using the position of part(c), we have
     $$I_0 = k \int_{-b/2}^{b/2} \int_{-h/2}^{h/2} (x^2+y^2) dy dx = 2k \int_{-b/2}^{b/2} (x^2h/2 + h^3/24) dx$$
     $$= kbh(b^2 + h^2)/12.$$

11a. Orient the plate with the base on the x-axis between $-b/2$ and
     $b/2$. Then, by symmetry, $\hat{x} = 0$ and
     $$\hat{L}_x = 2 \int_0^{b/2} \int_0^{h(b-2x)/b} y \, dy dx$$
     $$= (h^2/b^2) \int_0^{b/2} (b-2x)^2 dx = -(h^2/2b^2)(b-2x)^3/3 \Big|_0^{b/2} = bh^2/6.$$
     Thus $\hat{y} = \hat{L}_x/A = h/3$ since $A = bh/2$.

b. $I_x = 2k \int_0^{b/2} \int_0^{h(b-2x)/b} y^2 dy dx = (2kh^3/3b^3) \int_0^{b/2} (b-2x)^3 dx = kbh^3/12$.

c. $I_y = 2k \int_0^{b/2} \int_0^{h(b-2x)/b} x^2 dy dx = (2kh/b) \int_0^{b/2} (bx^2-2x^3) dx = khb^3/48$.

d.   If a mass element is above the centroid, then $y-h/3$ is the
     distance to the line through the centroid. If a mass element
     is below the line, then $h/3 - y$ is the distance to the line.
     In either case the square of the distance is $(y-h/3)^2$ and
     thus, as in Ex.5 and using symmetry, we have

$$I = 2k \int_0^{b/2} \int_0^{h(b-2x)/b} (y-h/3)^2 dy dx = (2k/3) \int_0^{b/2} (y-h/3)^3 \Big|_0^{h(b-2x)/b} dx$$

$$= (2k/3) \int_0^{b/2} [8h^3(1/3 - x/b)^3 + h^3/27] dx$$

$$= (2k/3) [-2bh^3(1/3 - x/b)^4 + h^3 x/27] \Big|_0^{b/2} = kbh^3/36.$$

13a. By symmetry, $I_x = 4k \int_0^b \int_0^{a\sqrt{b^2-y^2}/b} y^2 dx dy = (4ka/b) \int_0^b y^2 \sqrt{b^2-y^2} \, dy.$

Setting $y = b\sin\theta$, we then get

$$I_x = 4kab^3 \int_0^{\pi/2} \sin^2\theta \cos^2\theta \, d\theta = 4kab^3 \int_0^{\pi/2} (\sin^2\theta - \sin^4\theta) \, d\theta$$

$= 4kab^3(\pi/4 - 3\pi/16) = \pi kab^3/4.$ Since $M = \pi abk$, we then
have, by Eq. (25), $R_x^2 = b^2/4$ and $R_x = b/2.$

  b. I_y can be evaluated as I_x was in part (a). However, from
symmetry and interchanging the role of a and b, we may
conclude that $I_y = \pi ka^3 b/4$ and that $R_y = a/2.$

  c. From Eq. (24), $I_0 = I_x + I_y = \pi kab(a^2 + b^2)/4.$

15. $d = |x-\pi/2|$ is the distance to 1 for each element of mass and
thus $I = \int_0^\pi \int_0^{\sin x} (x-\pi/2)^2 (1-y) dy dx.$

17. The two circles intersect at $(0,0)$ and $(1,1)$ and hence the
desired region is bounded above by $(x-1)^2 + y^2 = 1$, or
$y = \sqrt{1-(x-1)^2}$, and bounded below by $x^2 + (y-1)^2 = 1$, or
$y = 1-\sqrt{1-x^2}$, for $0 \le x \le 1.$ Thus, $I = k \int_0^1 \int_{1-\sqrt{1-x^2}}^{\sqrt{1-(x-1)^2}} (x^2 + y^2) x^2 dy dx.$

19. Let Ω_1 be the vertical rectangle of dimensions b and h and
and Ω_2 be the horizontal rectangle of dimensions h and b-h.
For uniform density, the center of mass and the centroid are
the same. Thus $(h/2, b/2)$ and $(h+(b-h)/2, h/2)$ are the center
of masses of Ω_1 and Ω_2 respectively and hence

$$\bar{x} = \frac{(h/2)(kbh)+(h/2+b/2)kh(b-h)}{kbh+kh(b-h)} = \frac{b^2+bh-h^2}{2(2b-h)}, \text{ and } \bar{y} = \bar{x}, \text{ from}$$

symmetry.

21. Let Ω_1 be the triangle on top of Ω_2, the square of side a. Then $\bar{x}_1 = \bar{x}_2 = 0$, $\bar{y}_1 = a+h/3$, from Prob.11, and $\bar{y}_2 = a/2$ and

thus $\bar{x} = 0$ and $\bar{y} = \dfrac{(a+h/3)(kah/2)+(a/2)(ka^2)}{kah/2+ka^2} = \dfrac{a^2+ah+h^2/3}{h+2a}$.

23a. From Fig.17.3.16,
$$I_\gamma = \iint_\Omega (y+1)^2 \rho(x,y)\,dA$$

$$= \iint_\Omega y^2\rho(x,y)\,dA + 2l\iint_\Omega y\rho(x,y)\,dA + l^2\iint_\Omega \rho(x,y)\,dA = I_\Gamma + l^2M,$$

since $\iint_\Omega y\rho(x,y)\,dA = M\bar{y} = 0$ by the choice of the axis shown.

 b. From (9a) $I_\gamma = kbh^3/3$, from (9c) $I_\Gamma = kbh^3/12$, and $l = h/2$,
$M = kbh$, so $l^2M = kbh^3/4$. Thus
$I_\Gamma + l^2M = kbh^3/12 + kbh^3/4 = kbh^3/3 = I_\gamma$.

 c. From (11b) $I_\gamma = kbh^3/12$, from (11d) $I_\Gamma = kbh^3/36$, and
$l = h/3$, $M = kbh/2$, so $l^2M = kbh^3/18$. Thus,
$I_\Gamma + l^2M = kbh^3/36 + kbh^3/18 = kbh^3/12 = I_\gamma$.

25. Position the plate so that the sides of length a are on the axis, with the vertex at the origin. Then the center of mass is at $(a/3,a/3)$, from Prob.5, and the center of mass is at a distance of $l_1 = \sqrt{a^2/36+a^2/36}$ from the hypotenuse. If γ is the hypotenuse, then $I_\gamma = ka^4/24$, from Prob.10, and thus from Prob.23, $I_\Gamma = ka^4/24 - (a^2/18)(ka^2/2) = ka^4/72$ is the moment of inertia of the triangular plate about a line parallel to the hypotenuse through the center of mass. Now, if α is a line through the vertex (at the origin) parallel to the hypotenuse, then, with $l_2 = \sqrt{a^2/9+a^2/9}$, we have $I_\alpha = I_\Gamma + l_2^2M = ka^4/72 + (2a^2/9)(ka^2/2) = ka^4/8$.

27a. Assume that the boundary of Ω is partitioned into an upper curve $y = y_2(x)$, $a \le x \le b$ and a lower curve $y = y_1(x)$, $a \le x \le b$. Then

$$\iint_{\Omega} x\,dA = \int_a^b \int_{y_1(x)}^{y_2(x)} x\,dydx = \int_a^b x[y_2(x)-y_1(x)]dx = V_y/2\pi, \text{ by Eq.(8) of}$$

Section 7.2.   Now by Eq.(20), $\hat{x} = [\iint_{\Omega} x\,dA]/A = V_y/2\pi A$.

27b. In this case partition the boundary of Ω into a right half
      $[x = x_2(y)]$ and a left half $[x = x_1(y)]$, each for $c \le y \le d$,
      and proceed as in part(a).

29. $V_x = 4\pi a^3/3$, $A = \pi a^2/2$ and thus $\hat{y} = (4\pi a^3/3)/2\pi(\pi a^2/2) = 4a/3\pi$

      and by symmetry $\hat{x} = 0$, if the origin is at the center of the
      semicircle.

Section 17.4, Page 927

1.   From the inside limits we see that
      r varies from r = 0 to r = 2 and
      from the outside limits we see that
      θ varies from $\theta = 0$ to $\theta = \pi$ and
      thus the region of integration is
      the semicircle of radius 2 shown.
      The evaluation is then given by

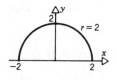

$$\int_0^\pi \int_0^2 r\sin\theta\, rdrd\theta = \int_0^\pi (r^3/3)\Big|_0^2 \sin\theta\, d\theta = (8/3)\int_0^\pi \sin\theta\, d\theta = 16/3.$$

3.   r varies from r = 0 to r = 1+sinθ as
      θ varies from 0 to $\pi/2$ and thus the
      region is in the first quadrant as
      shown and

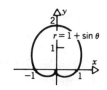

$$\int_0^{\pi/2} \int_0^{1+\sin\theta} \cos\theta\, rdrd\theta = (1/2)\int_0^{\pi/2} \cos\theta(1+\sin\theta)^2 d\theta$$

$$= (1/6)(1+\sin\theta)^3\Big|_0^{\pi/2} = 7/6.$$

5.   r varies from r = 1 to r = 2sin3θ
      while θ varies from $\pi/18$ to $5\pi/18$.
      Note that the two curves intersect
      at the θ end points and thus the
      region is as shown and

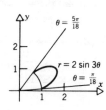

$$\int_{\pi/18}^{5\pi/18} \int_{1}^{2\sin3\theta} r\,dr\,d\theta \;=\; (1/2)\int_{\pi/18}^{5\pi/18} (4\sin^2 3\theta - 1)\,d\theta$$

$$=\; (1/2)\int_{\pi/18}^{5\pi/18} (1-2\cos6\theta)\,d\theta \;=\; [\theta/2 - (1/6)\sin6\theta]\,\Big|_{\pi/18}^{5\pi/18} \;=\; \pi/9 + \sqrt{3}/6.$$

7.   r varies from r = 2sinθ to r = 2
      while θ varies from 0 to π/2, which
      is the region shown.

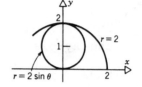

$$\int_{0}^{\pi/2} \int_{2\sin\theta}^{2} \sin\theta\, r\,dr\,d\theta$$

$$=\; (1/2)\int_{0}^{\pi/2} (4-4\sin^2\theta)\sin\theta\,d\theta \;=\; 2\int_{0}^{\pi/2} \cos^2\theta\sin\theta\,d\theta \;=\; 2/3.$$

9.   x varies from x = y to x = $\sqrt{4-y^2}$ (or
      $x^2+y^2 = 4$) as y varies from 0 to $\sqrt{2}$.
      Note that x = y is θ = π/4, that
      $x^2+y^2 = 4$ is r = 2 and that
      dxdy = dA = rdrdθ and thus

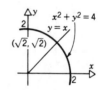

$$\int_{0}^{\sqrt{2}} \int_{y}^{\sqrt{4-y^2}} dx\,dy \;=\; \int_{0}^{\pi/4} \int_{0}^{2} r\,dr\,d\theta \;=\; 2\int_{0}^{\pi/4} d\theta \;=\; \pi/2.$$

11.   y varies from y = $\sqrt{3}$x to y = $\sqrt{9-x^2}$
      (or $x^2+y^2 = 9$) as x varies from 0
      to 3/2. Note that y = $\sqrt{3}$x is
      θ = π/3, that $x^2+y^2 = 9$ is r = 3,
      and thus

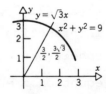

$$\int_{0}^{3/2} \int_{\sqrt{3}x}^{\sqrt{9-x^2}} e^{-x^2-y^2}dy\,dx \;=\; \int_{\pi/3}^{\pi/2} \int_{0}^{3} e^{-r^2}r\,dr\,d\theta \;=\; -(1/2)(e^{-9}-1)\int_{\pi/3}^{\pi/2} d\theta$$

$$=\; (1-e^{-9})\pi/12.$$

Note that $-x^2-y^2 = -r^2$ and
dxdy = dA = rdrdθ in the integrand.

13. The circles intersect at $a = 2a\sin\theta$ or $\theta = \pi/6, 5\pi/6$. Thus

$$A = \int_{\pi/6}^{5\pi/6}\int_{a}^{2a\sin\theta} r\,dr\,d\theta = (a^2/2)\int_{\pi/6}^{5\pi/6}(4\sin^2\theta - 1)\,d\theta$$

$$= (a^2/2)(\theta-\sin 2\theta)\Big|_{\pi/6}^{5\pi/6} = a^2[\pi/3 + \sqrt{3}/2].$$

15. The surface in polar coordinates is $z = 4-r^2$ and intersects the xy-plane in the circle $x^2+y^2 = 4$, which is $r = 2$, and

$$\text{thus } V = \int_{0}^{2\pi}\int_{0}^{2}(4-r^2)\,r\,dr\,d\theta = \int_{0}^{2\pi}(2r^2 - r^4/4)\Big|_{0}^{2}d\theta = 4\int_{0}^{2\pi}d\theta = 8\pi.$$

17. Note that $z = 4-y = 4-r\sin\theta \geq 0$ for $0 \leq r \leq 1+\cos\theta$ and thus

$$V = \int_{0}^{2\pi}\int_{0}^{1+\cos\theta}(4-r\sin\theta)\,r\,dr\,d\theta = \int_{0}^{2\pi}[2r^2 - (r^3/3)\sin\theta]\Big|_{0}^{1+\cos\theta}d\theta$$

$$= \int_{0}^{2\pi}[2(1+2\cos\theta+\cos^2\theta) - (1/3)(1+\cos\theta)^3\sin\theta]\,d\theta$$

$$= \int_{0}^{2\pi}[3+4\cos\theta+\cos 2\theta - (1/3)(1+\cos\theta)^3\sin\theta]\,d\theta$$

$$= [3\theta-4\sin\theta+(1/2)\sin 2\theta+(1/12)(1+\cos\theta)^4]\Big|_{0}^{2\pi} = 6\pi.$$

19. Subtracting the equation of one sphere from the other we find that $z = a/2$, which means the spheres intersect in the plane $z = a/2$. Setting $z = a/2$ in either one of the equations gives $x^2+y^2 = 3a^2/4$ (or $r = \sqrt{3}a/2$) as the curve of intersection. Replacing x^2+y^2 by r^2, we find that the upper sphere has equation $z = \sqrt{a^2-r^2}$, while the lower sphere has equation $z = a - \sqrt{a^2-r^2}$ and thus

$$V = \int_{0}^{2\pi}\int_{0}^{\sqrt{3}a/2}[\sqrt{a^2-r^2} - (a - \sqrt{a^2-r^2})]\,r\,dr\,d\theta$$

$$= \int_{0}^{2\pi}[(-2/3)(a^2-r^2)^{3/2} - ar^2/2]\Big|_{0}^{\sqrt{3}a/2}d\theta$$

$$= [(-2/3)(a^3/8) - 3a^3/8 + (2/3)a^3 + 0](2\pi) = 5\pi a^3/12.$$

21. From symmetry, $\hat{y} = \hat{x} = \hat{L}_y/A$, where

$$\hat{L}_y = \iint_\Omega x\,dA = \int_0^{\pi/2}\int_0^{a\sin2\theta}(r\cos\theta)\,r\,dr\,d\theta = (a^3/3)\int_0^{\pi/2}\sin^32\theta\cos\theta\,d\theta$$

$$= (8a^3/3)\int_0^{\pi/2}\sin^3\theta\cos^4\theta\,d\theta$$

$$= (8a^3/3)\int_0^{\pi/2}(\cos^4\theta\sin\theta - \cos^6\theta\sin\theta)\,d\theta = 16a^3/105$$

and $A = \int_0^{\pi/2}\int_0^{a\sin2\theta}r\,dr\,d\theta = (a^2/4)\int_0^{\pi/2}(1-\cos4\theta)\,d\theta = a^2\pi/8$. Thus

$\hat{x} = \hat{y} = 128a/105\pi$.

23a. $I_x = \iint_\Omega y^2\rho\,dA = k\int_0^\pi\int_0^a r^2\cos^2\theta\,r\,dr\,d\theta = (ka^4/4)\int_0^\pi\cos^2\theta\,d\theta = \pi ka^4/8$.

 b. $I_y = \iint_\Omega x^2\rho\,dA = k\int_0^\pi\int_0^a r^2\sin^2\theta\,r\,dr\,d\theta = (ka^4/4)\int_0^\pi\sin^2\theta\,d\theta = \pi ka^4/8$.

 c. $I_0 = I_x + I_y = \pi ka^4/4$.

25a. $\rho = k/r$ so $M = \iint_\Omega\rho\,dA = \int_0^\pi\int_a^b (k/r)\,r\,dr\,d\theta = k(b-a)\pi$.

 b. $L_y = \iint_\Omega x\rho\,dA = k\int_0^\pi\int_a^b r\cos\theta\,dr\,d\theta = (k/2)(b^2-a^2)\int_0^\pi\cos\theta\,d\theta = 0$ and

thus $\bar{x} = 0$. Note that due to symmetry we could conclude that
$\bar{x} = 0$ directly, but since ρ does vary, it's not just a simple
geometric argument. Now,

$L_x = \iint_\Omega y\rho\,dA = k\int_0^\pi\int_a^b r\sin\theta\,dr\,d\theta = (k/2)(b^2-a^2)\int_0^\pi\sin\theta\,d\theta = k(b^2-$

$a^2)$ and thus $\bar{y} = k(b^2-a^2)/k(b-a)\pi = (b+a)/\pi$.

 c. $I_x = k\int_0^\pi\int_a^b r^2\cos^2\theta\,r\,dr\,d\theta = (k/6)(b^3-a^3)\int_0^\pi(1+\cos2\theta)\,d\theta$

$= k\pi(b^3 - a^3)/6$.

 d. $I_y = k\int_0^\pi\int_a^b r^2\sin^2\theta\,r\,dr\,d\theta = (k/6)(b^3-a^3)\int_0^\pi(1-\cos2\theta)\,d\theta$

$= k\pi(b^3 - a^3)/6$.

27. Since the density of each region is constant, the center of
     mass is at the centroid in each case. For the semicircular
     region, the centroid is located $4a/6\pi$ units above the base by
     using the results of Prob.29 of Section 17.3 (substitute $a/2$
     for a in those results) or by direct calculation. Thus, if
     region 1 is the square and region 2 is the semicircle, we

     have $(\bar{x}_1,\bar{y}_1) = (0,a/2)$ and $(\bar{x}_2,\bar{y}_2) = (0,a+4a/6\pi)$ and hence by

     Prob.18 of Section 17.3, $\bar{x} = 0$, $\bar{y} = \dfrac{(a/2)(k_1a^2)+(a+4a/6\pi)(k_2\pi a^2/8)}{k_1a^2+k_2\pi a^2/8}$.

     Setting $\bar{y} = a$ and simplifying yields $k_2/k_1 = 6$.

Section 17.5, Page 941

1. $\displaystyle\int_1^2\int_0^3\int_{-1}^2 xyz\,dydzdx = \int_1^2\int_0^3 xz\,(y^2/2)\,\Big|_{-1}^2\,dzdx = (3/2)\int_1^2\int_0^3 xz\,dzdx$

    $= (3/2)\displaystyle\int_1^2 x\,(z^2/2)\,\Big|_0^3\,dx = (27/4)\int_1^2 x\,dx = 81/8.$

3. $\displaystyle\int_0^3\int_0^2\int_0^{x+y} 2xy\,dzdydx = 2\int_0^3\int_0^2 xy\,(x+y)\,dydx = 2\int_0^3 (x^2y^2/2 + xy^3/3)\,\Big|_0^2\,dx$

    $= 4\displaystyle\int_0^3 (x^2 + 4x/3)\,dx = 60.$

5. $\displaystyle\int_0^\pi\int_0^{\sin y}\int_0^{\cos y} (\pi-y)\,dxdzdy = \int_0^\pi\int_0^{\sin y} (\pi-y)\cos y\,dzdy$

    $= \displaystyle\int_0^\pi (\pi-y)\cos y\,\sin y\,dy$

    $= \pi\displaystyle\int_0^\pi \cos y\,\sin y\,dy - (1/2)\int_0^\pi y\sin 2y\,dy$

    $= [\,(\pi/2)\sin^2 y + (y/4)\cos 2y - (1/8)\sin 2y\,]\,\Big|_0^\pi = \pi/4,$

    where integration by parts has been used on the last integral.

7.   Eliminating z between the plane
      z = 3 and the sphere, we obtain
      $x^2+y^2+3^2 = 16$ or $x^2+y^2 = 7$ which is
      the projection of the curve of
      intersection of the plane and
      sphere onto the xy-plane and hence

      $x^2+y^2 \leq 7$ or $-\sqrt{7-x^2} \leq y \leq \sqrt{7-x^2}$,
      $-\sqrt{7} \leq x \leq \sqrt{7}$ is Ω_{xy}.

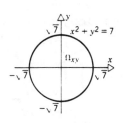

9a. $y = 4-z^2$ intersects the xz-plane
      along z = 2 and z = -2 and
      4x+2y+3z = 12 intersects the xz-plane
      along 4x+3z = 12 or x = 3-3z/4 and
      thus the region is as shown and we
      have $0 \leq x \leq 3-3z/4$ for $-2 \leq z \leq 2$.

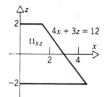

 b.   Since $y = 4-z^2$ is parallel to the
      x-axis, the region projects onto
      $0 \leq y \leq 4-z^2$, $-2 \leq z \leq 2$ in this
      case.

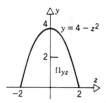

11.   Since y+2z = 2 is parallel to the x-axis, the region projects
      onto $0 \leq y \leq 2-2z$ for $0 \leq z \leq 1$ in the yz-plane. Thus
      $$M = \int_0^1 \int_0^{2-2z} \int_0^{(12-6z-4y)/3} (6-x-y)\,dxdydz.$$

13.   Let the diameter of the base be the y-axis with the vertex at
      (0,0,h). Then the cone is the surface of revolution
      $z = h - h\sqrt{x^2+y^2}/r$, which projects onto $x^2+y^2 \leq r^2$ in the
      xy-plane.   Thus

      $$I_y = k\int_{-r}^{r} \int_{-\sqrt{r^2-x^2}}^{\sqrt{r^2-x^2}} \int_0^{h-h\sqrt{x^2+y^2}/r} (x^2+z^2)\,dzdydx$$

      $$= 4k\int_0^r \int_0^{\sqrt{r^2-x^2}} \int_0^{h-h\sqrt{x^2+y^2}/r} (x^2+z^2)\,dzdydx, \text{ by symmetry.}$$

15. The projection of the region onto the xy-plane is
     $x^2/a^2 + y^2/b^2 \le 1$ and thus

$$M = k\int_{-a}^{a}\int_{-(b/a)\sqrt{a^2-x^2}}^{(b/a)\sqrt{a^2-x^2}}\int_{0}^{2b-y} (x^2+y^2+z^2)\,dzdydx$$

$$= 2k\int_{0}^{a}\int_{-(b/a)\sqrt{a^2-x^2}}^{(b/a)\sqrt{a^2-x^2}}\int_{0}^{2b-y} (x^2+y^2+z^2)\,dzdydx,$$

where we have used symmetry for the x integration but not for
y since the region is not symmetric for y.

17. $M = \int_{0}^{1}\int_{0}^{2}\int_{0}^{3} (6-x-y-z)\,dxdzdy = \int_{0}^{1}\int_{0}^{2} [(6-x-y)x - x^2/2]\Big|_{0}^{3}\,dzdy$

$$= 3\int_{0}^{1}\int_{0}^{2} (9/2 - y - z)\,dzdy = 3\int_{0}^{1} [(9/2 - y)z - z^2/2]\Big|_{0}^{2}\,dy$$

$$= 3\int_{0}^{1} (7-2y)\,dy = 18.$$

19. The projection of Ω onto the yz-plane is shown in Fig.17.5.6d
     and thus $V = \iint_{\Omega_1}\int_{0}^{2-2y} dxdA + \iint_{\Omega_2}\int_{0}^{\sqrt{4-y^2}-z} dxdA$. The curve separating

     Ω_1 and Ω_2 is the projection of the curve of intersection of
     the two surfaces and is obtained by eliminating x between the
     two equations. Thus $z = 4 - (2-2y)^2 - y^2 = 8y-5y^2$ and therefore

$$V = \int_{0}^{1}\int_{0}^{8y-5y^2}\int_{0}^{2-2y} dxdzdy + \int_{0}^{1}\int_{8y-5y^2}^{4-y^2}\int_{0}^{\sqrt{4-y^2}-z} dxdzdy.$$

21. Position the pyramid with the vertex at $(0,0,b)$ and the
     corners of its base on the xy coordinate axis at $x = \pm a/\sqrt{2}$
     and $y = \pm a/\sqrt{2}$. Then we have from symmetry that $\bar{x} = \bar{y} = 0$ and
     $\bar{z} = (4k/M)\iiint_{\Omega} z\,dzdydx$, where Ω is the portion of the pyramid
     in the first octant. Now $M = a^2bk/3$ and the pyramid's surface
     in the first octant is given by $b\sqrt{2}x + b\sqrt{2}y + az = ab$ and thus

$$\bar{z} = (4k/M) \int_0^{a/\sqrt{2}} \int_0^{a/\sqrt{2}-x} \int_0^{b-(b\sqrt{2}/a)x-(b\sqrt{2}/a)y} z \, dz dy dx$$

$$= (2k/M) \int_0^{a/\sqrt{2}} \int_0^{a/\sqrt{2}-x} [b-(b\sqrt{2}/a)x-(b\sqrt{2}/a)y]^2 dy dx$$

$$= -(2ak/3\sqrt{2}Mb) \int_0^{a/\sqrt{2}} [b-(b\sqrt{2}/a)x-(b\sqrt{2}/a)y]^3 \Big|_0^{a/\sqrt{2}-x} dx$$

$$= (2ak/3\sqrt{2}Mb) \int_0^{a/\sqrt{2}} [b-(b\sqrt{2}/a)x]^3 dx = b/4 .$$

23. From symmetry, $\bar{x} = \bar{y} = 0$ and

$$M = 4k \int_0^2 \int_0^{\sqrt{4-x^2}} \int_0^{4-x^2-y^2} dz dy dx = 4k \int_0^2 \int_0^{\sqrt{4-x^2}} (4-x^2-y^2) \, dy dx$$

$$= 4k \int_0^{\pi/2} \int_0^2 (4-r^2) \, r dr d\theta = 8k\pi ,$$

where we have switched to polar coordinates as in Ex.2. Now,

$$\bar{z} = (4k/M) \int_0^2 \int_0^{\sqrt{4-x^2}} \int_0^{4-x^2-y^2} z \, dz dy dx = (2k/M) \int_0^2 \int_0^{\sqrt{4-x^2}} (4-x^2-y^2)^2 dy dx$$

$$= (2k/M) \int_0^{\pi/2} \int_0^2 (4-r^2)^2 r dr d\theta = 4/3 .$$

25. Eliminating z between the two surfaces we see that the region projects onto $x^2 + y^2 \le 4$ in the xy-plane. Thus, using symmetry, we have

$$M = 4k \int_0^2 \int_0^{\sqrt{4-x^2}} \int_{x^2+y^2}^{8-x^2-y^2} (8-z) \, dz dy dx = 4k \int_0^2 \int_0^{\sqrt{4-x^2}} (8z-z^2/2) \Big|_{x^2+y^2}^{8-x^2-y^2} dy dx$$

$$= 4k \int_0^{\pi/2} \int_0^2 (8z-z^2/2) \Big|_{r^2}^{8-r^2} r dr d\theta = 32k \int_0^{\pi/2} \int_0^2 (4-r^2) r dr d\theta = 64k\pi \text{ and}$$

$$\bar{z} = (4k/M) \int_0^2 \int_0^{\sqrt{4-x^2}} \int_{x^2+y^2}^{8-x^2-y^2} (8z-z^2) \, dz dy dx$$

$$= (4k/M) \int_0^{\pi/2} \int_0^2 (4z^2 - z^3/3) \Big|_{r^2}^{8-r^2} r dr d\theta$$

$$= (4k/3M) \int_0^{\pi/2} \int_0^2 (256 - 24r^4 + 2r^6) r dr d\theta = 10/3 .$$

27. By symmetry, the center of mass is on the center line. For
     the vertical bar we have $\bar{z}_1$ = 1/2 and for the horizontal bar
     $\bar{z}_2$ = 1+a/2 and hence, from Prob.18 of Section 17.3, we have

$$\bar{z} = \frac{(1/2)(a^2 lm_1)+(1+a/2)(a^2 lm_2)}{a^2 lm_1+a^2 lm_2} = (lm_1+2lm_2+am_2)/2(m_1+m_2).$$

Section 17.6, Page 952

1.   The ellipsoid projects onto the circular region $x^2+y^2 \leq a^2$ or
     $0 \leq r \leq a$, $0 \leq \theta \leq 2\pi$ in the xy-plane. Thus, using symmetry,

$$V = 8\int_0^{\pi/2} \int_0^a \int_0^{(c/a)\sqrt{a^2-r^2}} r\,dz\,dr\,d\theta = (4c/a)\int_0^{\pi/2}\int_0^a \sqrt{a^2-r^2}\,r\,dr\,d\theta$$

$$= -(4c/3a)\int_0^{\pi/2} (a^2-r^2)^{3/2}\Big|_0^a d\theta = 4\pi a^2 c/3.$$

3.   The hyperboloid intersects the xy-plane in the circle
     $x^2+y^2 = a^2$ or $r = a$ and intersects the $z = h$ plane in the
     circle $x^2+y^2 = (a^2/c^2)(c^2+h^2)$ or $r = (a/c)\sqrt{c^2+h^2} > a$. Thus
     the desired volume is the sum of the volume in the cylinder
     $r = a$ between $z = 0$ and $z = h$ plus the volume between the
     cylinder and the hyperboloid. Thus, using symmetry,

$$V = \pi a^2 h + 4\int_0^{\pi/2} \int_0^{(a/c)\sqrt{c^2+h^2}} \int_{(c/a)\sqrt{r^2-a^2}}^{h} r\,dz\,dr\,d\theta$$

$$= \pi a^2 h + 4\int_0^{\pi/2} \int_0^{(a/c)\sqrt{c^2+h^2}} [h - (c/a)\sqrt{r^2-a^2}]\,r\,dr\,d\theta$$

$$= \pi a^2 h + 4[hr^2/2 - (c/3a)(r^2-a^2)^{3/2}]\Big|_a^{(a/c)\sqrt{c^2+h^2}} \int_0^{\pi/2} d\theta$$

$$= \pi a^2 h(1 + h^2/3c^2).$$

5a. Eliminating z between the two equations yields $r = 2$,
     $0 \leq \theta \leq 2\pi$, which is the projection of the volume onto the
     xy-plane. Thus, using symmetry, $\bar{x} = \bar{y} = 0$,

$$M = 4k \int_0^{\pi/2} \int_0^2 \int_r^{6-r^2} r \, dz dr d\theta = 4k \int_0^{\pi/2} \int_0^2 (6r - r^2 - r^3) \, dr d\theta$$

$$= 4(16/3)k \int_0^{\pi/2} d\theta = 32\pi k/3, \quad \text{and}$$

$$\bar{z} = (4k/M) \int_0^{\pi/2} \int_0^2 \int_r^{6-r^2} zr \, dz dr d\theta = (2k/M) \int_0^{\pi/2} \int_0^2 (36r - 13r^3 + r^5) \, dr d\theta$$

$$= (2k/M)(92/3) \int_0^{\pi/2} d\theta = 92\pi k/3M = 23/8.$$

5b. $$I_z = k \int_0^{2\pi} \int_0^2 \int_r^{6-r^2} (x^2+y^2) r \, dz dr d\theta = k \int_0^{2\pi} \int_0^2 \int_r^{6-r^2} r^3 dz dr d\theta$$

$$= k \int_0^{2\pi} \int_0^2 (6r^3 - r^4 - r^5) \, dr d\theta = k(104/15) \int_0^{2\pi} d\theta = 208\pi k/15.$$

7. $$M = k \int_0^{\pi/2} \int_0^a \int_{-\sqrt{a^2-r^2}}^{\sqrt{a^2-r^2}} (\sin 2\theta) r \, dz dr d\theta = 2k \int_0^{\pi/2} \int_0^a \int_0^{\sqrt{a^2-r^2}} r \sin 2\theta \, dz dr d\theta$$

$$= 2k \int_0^{\pi/2} \int_0^a r\sqrt{a^2-r^2} \sin 2\theta \, dr d\theta = (2ka^3/3) \int_0^{\pi/2} \sin 2\theta \, d\theta = 2ka^3/3.$$

9a. As in Ex.1, the cone is given by $z = hr/a$ and thus, by

symmetry, we have $\bar{x} = \bar{y} = 0$,

$$M = 4k \int_0^{\pi/2} \int_0^a \int_{hr/a}^h r \, dz dr d\theta = 4kh \int_0^{\pi/2} \int_0^a (r - r^2/a) \, dr d\theta$$

$$= 4kh(a^2/6) \int_0^{\pi/2} d\theta = \pi a^2 hk/3, \quad \text{and}$$

$$\bar{z} = (4k/M) \int_0^{\pi/2} \int_0^a \int_{hr/a}^h zr \, dz dr d\theta = (2h^2k/M) \int_0^{\pi/2} \int_0^a (r - r^3/a^2) \, dr d\theta$$

$$= (2h^2k/M)(a^2/4) \int_0^{\pi/2} d\theta = \pi a^2 h^2 k/4M = 3h/4.$$

b. The axis of symmetry is the z-axis and thus

$$I_z = k \int_0^{2\pi} \int_0^a \int_{hr/a}^h r^3 dz dr d\theta = hk \int_0^{2\pi} \int_0^a (r^3 - r^4/a) \, dr d\theta = \pi a^4 hk/10.$$

9c. Choose the y-axis, note that $x^2 + z^2 = r^2\cos^2\theta + z^2$, and
     integrate with respect to θ first to obtain

$$I_y = k\int_0^a\int_{hr/a}^h\int_0^{2\pi} (r^2\cos^2\theta + z^2)\,rd\theta dzdr$$

$$= k\int_0^a\int_{hr/a}^h \{r^2[\theta/2 + (1/4)\sin2\theta] + \theta z^2\}\Big|_0^{2\pi}\,rdzdr$$

$$= \pi k\int_0^a\int_{hr/a}^h (r^2 + 2z^2)\,rdzdr$$

$$= \pi kh\int_0^a [2h^2r/3 + r^3 - r^4(1/a + 2h^2/3a^3)]dr = \pi a^2hk(a^2 + 4h^2)/20.$$

11a. Place the hemisphere in the positive z range with the center
     at the origin. Then $z = \sqrt{a^2-r^2}$ is the equation of the
     hemisphere so, by symmetry, $\bar{x} = \bar{y} = 0$,

$$M = 4k\int_0^{\pi/2}\int_0^a\int_0^{\sqrt{a^2-r^2}} r\,dzdrd\theta = 2\pi ka^3/3,\text{ and}$$

$$\bar{z} = (4k/M)\int_0^{\pi/2}\int_0^a\int_0^{\sqrt{a^2-r^2}} zr\,dzdrd\theta = (2k/M)\int_0^{\pi/2}\int_0^a (a^2-r^2)\,rdrd\theta$$

$$= \pi ka^4/4M = 3a/8.$$

 b. The desired line is the z-axis and since $x^2+y^2 = r^2$, we have

$$I_z = 4k\int_0^a\int_0^{\pi/2}\int_0^{\sqrt{a^2-r^2}} r^3dzd\theta dr$$

$$= 4k\int_0^a\int_0^{\pi/2} r^3\sqrt{a^2-r^2}\,d\theta dr = 2\pi k\int_0^a r^3\sqrt{a^2-r^2}\,dr$$

$$= 2\pi k[-(r^2/3)(a^2 - r^2)^{3/2}\Big|_0^a + (2/3)\int_0^a r(a^2 - r^2)^{3/2}dr]$$

$$= 2\pi k[-(2/15)(a^2 - r^2)^{5/2}\Big|_0^a] = 4\pi ka^5/15.$$

13. The region is similar to that of Ex.3, which is shown in
     Fig.17.6.7. Although the cylinder $r = 2\cos\theta$ is traced once
     for $0 \le \theta \le \pi$, we note that r is negative for $\pi/2 \le \theta \le \pi$ and
     hence is not permissable. The cylinder is also traced once
     for $0 \le \theta \le \pi/2$ and $3\pi/2 \le \theta \le 2\pi$. However, using symmetry,

we need only consider $0 \leq \theta \leq \pi/2$ and note that $\bar{y} = 0$,

$$M = 2k \int_0^{\pi/2} \int_0^{2\cos\theta} \int_0^r r\,dz\,dr\,d\theta = (16k/3) \int_0^{\pi/2} \cos^3\theta\,d\theta$$

$$= (16k/3) \int_0^{\pi/2} (1-\sin^2\theta)\cos\theta\,d\theta$$

$$= (16k/3)[\sin\theta - (1/3)\sin^3\theta]_0^{\pi/2} = 32k/9,$$

$$\bar{x} = (2k/M) \int_0^{\pi/2} \int_0^{2\cos\theta} \int_0^r r^2\cos\theta\,dz\,dr\,d\theta = (8k/M) \int_0^{\pi/2} \cos^5\theta\,d\theta$$

$$= (8k/M) \int_0^{\pi/2} (1-\sin^2\theta)^2\cos\theta\,d\theta = 64k/15M = 6/5, \text{ and}$$

$$\bar{z} = (2k/M) \int_0^{\pi/2} \int_0^{2\cos\theta} \int_0^r zr\,dz\,dr\,d\theta = (4k/M) \int_0^{\pi/2} \cos^4\theta\,d\theta$$

$$= (k/M) \int_0^{\pi/2} [1+2\cos2\theta+(1/2)(1+\cos4\theta)]\,d\theta = 3\pi k/4M = 27\pi/128.$$

15a. The region projects onto $x^2+y^2 \leq 9$ or $0 \leq r \leq 3$, $0 \leq \theta \leq 2\pi$

in the xy-plane and thus $M = k \int_0^{2\pi} \int_0^3 \int_0^{9-r^2} (12-z)\,r\,dz\,dr\,d\theta$ or

$$M = 4k \int_0^{\pi/2} \int_0^3 \int_0^{9-r^2} (12-z)\,r\,dz\,dr\,d\theta, \text{ using symmetry.}$$

b. Since $x^2+y^2 = r^2$, we have $I_z = k \int_0^{2\pi} \int_0^3 \int_0^{9-r^2} (12-z)\,r^3\,dz\,dr\,d\theta$.

17a. Place the hemisphere as in Prob.11, then the plane $z = h$
intersects the hemisphere in the circle $r = \sqrt{a^2-h^2}$, $z = h$.
Thus the volume, when the depth is h, is given by

$$V = 2\pi a^3/3 - 4 \int_0^{\pi/2} \int_0^{\sqrt{a^2-h^2}} \int_h^{\sqrt{a^2-r^2}} r\,dz\,dr\,d\theta, \text{ which is the total}$$

volume of the hemisphere less the volume in the hemisphere
above the plane $z = h$. Hence

$$V = 2\pi a^3/3 - 4 \int_0^{\pi/2} \int_0^{\sqrt{a^2-h^2}} (r\sqrt{a^2-h^2} - rh)\,dr\,d\theta$$

$$= 2\pi a^3/3 - 4[-h^3/3 - a^2h/2 + h^3/2 + a^3/3] \int_0^{\pi/2} d\theta = \pi a^2h - \pi h^3/3.$$

17b. $\pi a^3/3$ is one half the volume of the hemisphere and thus we
want $\pi a^2 h - \pi h^3/3 = \pi a^3/3$ or $h^3 - 3a^2 h + a^3 = 0$ or
$(h/a)^3 - 3(h/a) + 1 = 0$. Using Newton's method of Section 4.5,
we find that $h/a \cong .3473$.

19.   From Eq.(24), we have $F_z = 2\pi Gm[\mu + k(\sqrt{a^2+h^2} - \sqrt{b^2+h^2})]$. Now
$\sqrt{a^2+h^2} = \sqrt{b^2+h^2} + b(b^2+h^2)^{-1/2}(a-b) + f''(\zeta)(a-b)^2/2$, where ζ is
some point between a and b and thus
$F_z = 2\pi Gm[\mu - \mu b(b^2+h^2)^{-1/2} + f''(\zeta)k\mu|a-b|/2]$ so that we have
$\lim_{a\to b} F_z = 2\pi Gm\mu[1 - b(b^2+h^2)^{-1/2}]$. This is the total

gravitational attraction since $F_x = F_y = 0$, as in Ex.4.

21a. $A = \int_0^a \int_{x^p}^{a^p} dzdx = \int_0^a (a^p - x^p)dx = pa^{p+1}/(p+1)$ and

$\hat{z}_A = (1/A)\int_0^a \int_{x^p}^{a^p} z\,dzdx = (1/2A)\int_0^a (a^{2p} - x^{2p})dx = pa^{2p+1}/A(2p+1)$
$= (p+1)a^p/(2p+1)$.
Now the surface of revolution is given by $z = (x^2+y^2)^{p/2} = r^p$

so $V = 4\int_0^{\pi/2}\int_0^a \int_{r^p}^{a^p} r\,dzdrd\theta = 4\int_0^{\pi/2}\int_0^a (a^p r - r^{p+1})drd\theta = \pi pa^{p+2}/(p+2)$

and $\hat{z}_V = (4/V)\int_0^{\pi/2}\int_0^a \int_{r^p}^{a^p} zr\,dzdrd\theta = (2/V)\int_0^{\pi/2}\int_0^a (ra^{2p} - r^{2p+1})drd\theta$
$= (\pi pa^{2p+2})/2(p+1)V = (p+2)a^p/2(p+1)$.

b. $\hat{z}_V/\hat{z}_A = \dfrac{(p+2)a^p/2(p+1)}{(p+1)a^p/(2p+1)} = \dfrac{(p+2)(2p+1)}{2(p+1)^2} = \dfrac{2p^2+5p+2}{2p^2+4p+2} = 1 + p/2(p+1)^2$.
We also have $\dfrac{d}{dp}(\hat{z}_V/\hat{z}_A) = (1-p)/2(p+1)^3$ and

$\dfrac{d^2}{dp^2}(\hat{z}_V/\hat{z}_A) = (2p-4)/2(p+1)^4$, so $\hat{z}_V/\hat{z}_A$ is a maximum at $p = 1$.

Section 17.7, Page 962

1.   From Eqs.(3a), we have $x = 2\sin(\pi/3)\cos(\pi/4) = \sqrt{3/2}$,
$y = 2\sin(\pi/3)\sin(\pi/4) = \sqrt{3/2}$, and $z = 2\cos(\pi/3) = 1$.

3. $x = 2\sin(2\pi/3)\cos(\pi/6) = 3/2$, $y = 2\sin(2\pi/3)\sin(\pi/6) = \sqrt{3}/2$,
      and $z = 2\cos(2\pi/3) = -1$.

5. $x = 2\sin(3\pi/4)\cos(3\pi/4) = -1$, $y = 2\sin(3\pi/4)\sin(3\pi/4) = 1$,
      and $z = 2\cos(3\pi/4) = -\sqrt{2}$.

7.   From Eqs. (3b), we have $\rho = [1^2 + (-1)^2 + (\sqrt{2})^2]^{1/2} = 2$,
      $\tan\theta = -1/1 = -1$ or $\theta = 7\pi/4$, and $\cos\phi = \sqrt{2}/2$ or $\phi = \pi/4$.

9. $\rho = (3/4 + 9/4 + 1)^{1/2} = 2$, $\tan\theta = \dfrac{3/2}{\sqrt{3}/2} = \sqrt{3}$ or $\theta = \pi/3$, and

      $\cos\phi = -1/2$ or $\phi = 2\pi/3$.

11. $\rho = (27/8 + 9/8 + 9/2)^{1/2} = 3$, $\tan\theta = \dfrac{3/2\sqrt{2}}{3\sqrt{3}/2\sqrt{2}} = 1/\sqrt{3}$ or

      $\theta = \pi/6$, and $\cos\phi = \dfrac{3/\sqrt{2}}{3} = 1/\sqrt{2}$ or $\phi = \pi/4$.

13.   Expanding $(z-c)^2$, we have $x^2 + y^2 + z^2 = 2cz$ which by Eqs. (3a) is
      $\rho^2 = 2c\rho\cos\phi$ or $\rho = 2c\cos\phi$.

15.   Multiplying by 36 and using Eqs. (3a), we have
      $9(\rho^2\sin^2\phi\cos^2\theta + \rho^2\sin^2\phi\sin^2\theta) - 4\rho^2\cos^2\phi = 0$ or $\tan^2\phi = 4/9$ so
      that $\phi = \arctan(2/3)$.

17. $\rho\sin\phi\cos\theta = 2\rho\sin\phi\sin\theta$ or $\tan\theta = 1/2$ so that $\theta = \arctan(1/2)$.

19.   Since $r = \rho\sin\phi$ from Eq. (1a), we have $\rho\sin\phi = 3\rho\cos\phi$ so
      $\phi = \arctan 3$.

21.   Since $z = \rho\cos\phi$, we have $z = b$, which is a plane parallel to
      and b units from the xy-plane.

23.   Since $r = \rho\sin\phi$, we have $r = b$, which is a circular cylinder
      of radius b about the z-axis.

25.   We have $\rho\sin\phi\cos\theta + \rho\sin\phi\sin\theta = a$ or $x+y = a$, which is a plane
      parallel to the z-axis.

27a. Position the hemisphere as in Fig. 17.7.11. Then its volume
      satisfies $0 \le \rho \le a$, $0 \le \theta \le 2\pi$, and $0 \le \phi \le \pi/2$. Now
      $x^2 + y^2 = \rho^2\sin^2\phi$, and thus

$$I_z = k \int_0^{\pi/2} \int_0^{2\pi} \int_0^a (\rho^2 \sin^2\phi)\rho^2 \sin\phi \, d\rho d\theta d\phi = (ka^5/5) \int_0^{\pi/2} \int_0^{2\pi} \sin^3\phi \, d\theta d\phi$$

$$= (2\pi ka^5/5) \int_0^{\pi/2} \sin^3\phi \, d\phi = (2\pi ka^5/5) [-\cos\phi + (1/3)\cos^3\phi] \Big|_0^{\pi/2}$$

$$= 4\pi ka^5/15.$$

27b. $I_y = k \int_0^{\pi/2} \int_0^{2\pi} \int_0^a (\rho^2 \sin^2\phi\cos^2\theta + \rho^2\cos^2\phi)\rho^2\sin\phi \, d\rho d\theta d\phi$

$$= (ka^5/5) \int_0^{\pi/2} \int_0^{2\pi} \{\sin^3\phi[1/2 + (1/2)\cos 2\theta] + \cos^2\phi\sin\phi\} d\theta d\phi$$

$$= (ka^5/5) \int_0^{\pi/2} \{\sin^3\phi[\theta/2 + (1/4)\sin 2\theta] + \theta\cos^2\phi\sin\phi\} \Big|_0^{2\pi} d\phi$$

$$= (ka^5/5) \{\pi[-\cos\phi + (1/3)\cos^3\phi] - (2\pi/3)\cos^3\phi\} \Big|_0^{\pi/2} = 4\pi ka^5/15.$$

29a. Locate the origin at the center of the shell. Then the mass is located such that $a \le \rho \le b$, $0 \le \theta \le 2\pi$, $0 \le \phi \le \pi$. Thus

$$M = \int_0^\pi \int_0^{2\pi} \int_a^b (k/\rho)\rho^2\sin\phi \, d\rho d\theta d\phi = [k(b^2 - a^2)/2] \int_0^\pi \int_0^{2\pi} \sin\phi d\theta d\phi$$

$$= 2\pi k(b^2 - a^2).$$

b. $I_z = \int_0^\pi \int_0^{2\pi} (k/\rho)(\rho^2\sin^2\phi)\rho^2\sin\phi \, d\rho d\theta d\phi$

$$= [k(b^4 - a^4)/4] \int_0^\pi \int_0^{2\pi} \sin^3\phi \, d\theta d\phi$$

$$= [\pi k(b^4 - a^4)/2] [-\cos\phi + (1/3)\cos^3\phi] \Big|_0^\pi = 2\pi k(b^4 - a^4)/3.$$

31. $\rho = 2a\cos\phi \Rightarrow \rho^2 = 2a\rho\cos\phi \Rightarrow x^2+y^2+z^2 = 2az$, which is a sphere with center at $(0,0,a)$ and radius a. Now the density is $kz = k\rho\cos\phi$ and thus

$$M = \int_0^\alpha \int_0^{2\pi} \int_0^{2a\cos\phi} (k\rho\cos\phi)\rho^2\sin\phi \, d\rho d\theta d\phi = (4ka^2) \int_0^\alpha \int_0^{2\pi} \cos^5\phi\sin\phi \, d\theta d\phi$$

$$= (-8\pi ka^4/6)\cos^6\phi \Big|_0^\alpha = 4k\pi a^4(1 - \cos^6\alpha)/3.$$

33a. $M = \int_0^{\pi/2} \int_0^{2\pi} \int_{a/2}^a (k/\rho)\rho^2\sin\phi \, d\rho d\theta d\phi = (k/2)(a^2 - a^2/4) \int_0^{\pi/2} \int_0^{2\pi} \sin\phi \, d\theta d\phi$

$$= 3\pi ka^2/4.$$

33b. $\bar{x} = \bar{y} = 0$ by symmetry, and

$$\bar{z} = (1/M) \int_0^{\pi/2} \int_0^{2\pi} \int_{a/2}^{a} (\rho\cos\phi)(k/\rho)\rho^2\sin\phi \, d\rho d\theta d\phi$$

$$= (k/3M)(a^3 - a^3/8) \int_0^{\pi/2} \int_0^{2\pi} \cos\phi\sin\phi \, d\theta d\phi = (7\pi ka^3/24M)\sin^2\phi \Big|_0^{\pi/2}$$

$$= 7\pi ka^3/24M = 7a/18.$$

c. $$I_z = \int_0^{\pi/2} \int_0^{2\pi} \int_{a/2}^{a} (\rho^2\sin^2\phi)(k/\rho)\rho^2\sin\phi \, d\rho d\theta d\phi$$

$$= (15ka^4/64) \int_0^{\pi/2} \int_0^{2\pi} \sin^3\phi \, d\theta d\phi$$

$$= (15\pi ka^4/32)[-\cos\phi + (1/3)\cos^3\phi] \Big|_0^{\pi/2} = 5\pi ka^4/16.$$

35. Using the hint, the center of the sphere is at $(0,0,a)$ and, from Prob.31, the equation of the sphere is $\rho = 2a\cos\phi$. Thus the mass lies in the region $0 \le \rho \le 2a\cos\phi$, $0 \le \theta \le 2\pi$, $0 \le \phi \le \pi/2$ and

$$M = \int_0^{\pi/2} \int_0^{2\pi} \int_0^{2a\cos\phi} (k\rho)\rho^2\sin\phi \, d\rho d\theta d\phi = 4ka^4 \int_0^{\pi/2} \int_0^{2\pi} \cos^4\phi\sin\phi \, d\theta d\phi$$

$$= -(8\pi ka^4/5)\cos^5\phi \Big|_0^{\pi/2} = 8\pi ka^4/5.$$

37a. From Fig.17.7.14, F_z will be in the negative z direction since m is above the sphere. Now dV is a distance R from m and with Ψ as shown in Fig.17.7.14, we have, as in Ex.5, $F_z = -Gmk\iiint_\Omega (\cos\Psi/R^2) dV$. In this case the volume of the sphere satisfies $0 \le \rho \le a$, $0 \le \theta \le 2\pi$, and $0 \le \phi \le \pi$ and thus we may integrate in any order. Hence

$$\|\mathbf{F}\| = |F_z| = Gmk \int_0^a \int_0^\pi \int_0^{2\pi} (\cos\Psi/R^2)\rho^2\sin\phi \, d\theta d\phi d\rho$$

$$= 2\pi Gmk \int_0^a \int_0^\pi (\cos\Psi/R^2)\rho^2\sin\phi \, d\phi d\rho,$$

since Ψ and R are independent of θ for fixed ϕ and ρ.

b. Eq.(iii) $\Rightarrow \cos\Psi = (1-\rho\cos\phi)/R = (2l^2 - 2\rho l\cos\phi)/2Rl$.

Eq.(ii) $\Rightarrow 2l^2 - 2\rho l\cos\phi = l^2 + R^2 - \rho^2$ and thus $\cos\Psi = (l^2 + R^2 - \rho^2)/2Rl$.

37c. Taking the differential of both sides of Eq.(ii), holding ρ
      constant, yields $2RdR = 2l\rho\sin\phi\,d\phi$ or $\sin\phi\,d\phi = (R/l\rho)\,dR$.
      Therefore we have

$$\|\mathbf{F}\| = 2\pi Gmk\int_0^a\int_{l-\rho}^{l+\rho}\frac{(l^2+R^2-\rho^2)}{2lR^3}\rho^2\left(\frac{R}{l\rho}\right)dRd\rho$$

$$= (\pi Gmk/l^2)\int_0^a\int_{l-\rho}^{l+\rho}\rho[1+(l^2-\rho^2)/R^2]dRd\rho.$$

 d. $$\|\mathbf{F}\| = (\pi Gmk/l^2)\int_0^a\rho[R-(l^2-\rho^2)/R]\,\Big|_{l-\rho}^{l+\rho}\,d\rho$$

$$= (4\pi Gmk/l^2)\int_0^a\rho^2 d\rho = 4\pi a^3 kGm/3l^2 = GMm/l^2,$$

      where $M = 4\pi a^3 k/3$ is the mass of the sphere.

 e. In this case Eq.(vi) is replaced by

$$\|\mathbf{F}\| = (\pi Gm/l^2)\int_0^a\int_{l-\rho}^{l+\rho}\rho(1+\frac{l^2-\rho^2}{R^2})f(\rho)\,dRd\rho,\quad\text{since the earlier}$$

      constant density is replaced by $f(\rho)$ and all other steps
      leading up to Eq.(vi) remain the same. Integrating with
      respect to R, as in part(d), we get

$$\|\mathbf{F}\| = (4\pi Gm/l^2)\int_0^a\rho^2 f(\rho)\,d\rho = GMm/l^2,\quad\text{since}$$

$$M = \int_0^a\int_0^\pi\int_0^{2\pi}\rho^2 f(\rho)\sin\phi\,d\theta d\phi d\rho = 4\pi\int_0^a\rho^2 f(\rho)\,d\rho.$$

39.   As in Ex.5, the x and y components of the attractive force
      are zero and $F_z = Gmk\iiint_\Omega\frac{\cos\psi}{R^2}dV$. Since $\psi = \phi$ and $R = \rho$ (as in
      Ex.5), we have

$$F_z = Gmk\int_0^{\pi/2}\int_0^{2\pi}\int_0^a(\cos\phi/\rho^2)\rho^2\sin\phi\,d\rho d\theta d\phi = \pi Gmka\sin^2\phi\,\Big|_0^{\pi/2} = \pi Gmka.$$

Chapter 17 Review, Page 964

1.   The centers are at $(1,3)$, $(3,3)$, $(1,5)$, $(3,5)$, $(1,7)$, $(3,7)$,
      and $\Delta A_{ij} = 4$ for all i,j. Thus

$$\sum_{i=1,2,\,j=1,2,3} f(P_{ij})\Delta A_{ij} = (2+2+2+2+4+4)(4) = 64.$$

3.   With the centers and ΔA_{ij} as in Prob.1,

$$\sum_{i,j} f(P_{ij})\Delta A_{ij} = [e^3+3e^3+e^5+3e^5+e^7+3e^7](4) = 16(e^3+e^5+e^7).$$

5. $I = \displaystyle\int_0^2 (e^{2-x} - 1)\,dx = [-e^{(2-x)} - x]\Big|_0^2 = e^2 - 3.$

7. $I = \displaystyle\int_0^3\int_0^3 (x^2+4y^2)\,dy\,dx = \int_0^3 (3x^2+36)\,dx = 135.$

9. $I = \displaystyle\int_{\pi/6}^{\pi/2} \cot x \ln(\sin x)\,dx$

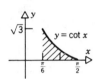

$= (1/2)[(\ln 1)^2 - (\ln(1/2))^2]$

$= -(\ln 2)^2/2.$

11. $I = \displaystyle\int_0^1 [x^3/2 + x^2/2 - x^4]\,dx$

$= 11/120.$

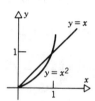

13a. $x = \ln y \Rightarrow y = e^x$ so $I = \displaystyle\int_0^1\int_1^{e^x} (x-2y)\,dy\,dx.$

  b. $I = \displaystyle\int_1^e\int_{\ln y}^1 (x-2y)\,dx\,dy.$

  c. $I = \displaystyle\int_0^1 [(xe^x-e^{2x}) - (x-1)]\,dx$ [from part(a)]

      $= [(x-1)e^x - (1/2)e^{2x} - x^2/2 + x]\Big|_0^1 = 2 - e^2/2.$

15a. $\displaystyle\int_0^1\int_{x^3}^{x^2} \frac{x+y}{xy}\,dy\,dx.$ b. $\displaystyle\int_0^1\int_{y^{1/2}}^{y^{1/3}} \frac{x+y}{xy}\,dx\,dy.$

  c. $I = \displaystyle\int_0^1 (x-x^2-\ln x)\,dx = 1/6 - \lim_{h\to 0+}(x\ln x - x)\Big|_h^1 = 1/6 + 1 = 7/6.$

17. $I = (1/2) \displaystyle\int_0^{2\pi} (\sin 4 - \sin 1) \, d\theta$

$= \pi (\sin 4 - \sin 1).$

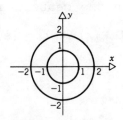

19. $I = (1/3) \displaystyle\int_0^{3\pi/2} \cos^3\theta \, d\theta$

$= (1/3) \displaystyle\int_0^{3\pi/2} (\cos\theta - \cos\theta \sin^2\theta) \, d\theta$

$= -2/9.$ The top half of the region
is covered twice.

21. $x = 2\cos(\pi/6) = \sqrt{3},\ y = 2\sin(\pi/6) = 1,$ and $z = 1.$

23. $r = \rho\sin\phi = 2\sin(\pi/2) = 2,\ \theta = \pi/3,\ z = \rho\cos\phi = 2\cos(\pi/2) = 0.$

25.   In spherical coordinates, $r = \rho\sin\phi$
and $z = \rho\cos\phi$ so $r^2 + z^2 = \rho^2 = 4,$
which is a sphere, radius 2,
center at origin.

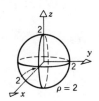

27.   Writing the equation as $\rho\cos\phi = 3,$
it becomes $z = 3$ in rectangular
coordinates, which is a plane.

29. $I = -(8/3) \displaystyle\int_{4\pi/3}^{5\pi/4} \sec^2\theta \, d\theta = (-8/3)\tan\theta \Big|_{4\pi/3}^{5\pi/4} = (8/3)(\sqrt{3} - 1).$

31.   In cylindrical coordinates,

$$V = \int_0^{2\pi}\int_0^{\sqrt{2}}\int_r^{\sqrt{4-r^2}} r \, dz\,dr\,d\theta = \int_0^{2\pi}\int_0^{\sqrt{2}} (r\sqrt{4-r^2} - r^2) \, dr\,d\theta$$

$$= (-1/3)\int_0^{2\pi} [(\sqrt{4-r^2})^{3/2} + r^3] \Big|_0^{\sqrt{2}} \, d\theta = (8\pi/3)(2 - \sqrt{2}).$$

33. $M = \int_0^1 \int_{2x}^2 kxy \, dydx = k\int_0^1 (2x-2x^3) \, dx = k/2.$

35. With $\rho(r,\theta,z) = kr^2$,

$$M = k\int_0^{2\pi}\int_0^1\int_0^{\sqrt{4-r^2}} r^3 \, dzdrd\theta = k\int_0^{2\pi}\int_0^1 r^3\sqrt{4-r^2} \, drd\theta$$

$$= k\int_0^{2\pi} (-r^2/5 - 8/15)(4-r^2)3/2\Big|_0^1 \, d\theta = k\pi(128 - 66\sqrt{3})/15.$$

37. $\bar{x} = L_y/M = (\int_0^1\int_1^{e^x} kxdydx)/(\int_0^1\int_1^{e^x} kdydx) = 1/2(e-2).$

39. $\bar{x} = (k\int_0^4\int_0^x\int_0^{\sqrt{16-x^2}/2} x \, dzdydx)/(k\int_0^4\int_0^x\int_0^{\sqrt{16-x^2}/2} dzdydx)$

$$= (8k\pi)/(32k/3) = 3\pi/4.$$

41. $R_y^2 = I_y/M = (k\int_0^1\int_1^{e^x} x^2dydx)/k(e-2) = [\int_0^1 (x^2e^x - x^2) \, dx]/(e-2)$

$$= (e - 7/3)/(e-2).$$

43. $R_y^2 = \iiint_\Omega k(x^2+z^2) \, dV/M = (k\int_0^4\int_0^x\int_0^{\sqrt{16-x^2}/2} (x^2+z^2) \, dzdydx)/(32k/3)$

$$= (3/32)\int_0^4 [(1/2)x^3\sqrt{16-x^2} + (x/24)\sqrt{(16-x^2)^3}] \, dx = 36/5.$$

CHAPTER 18

1. From $\mathbf{r}(t)$, we see that the parametric equations of C are $x = t$, $y = \sin\pi t$, and thus $dx = dt$ and $dy = \pi\cos\pi t\ dt$. Hence

$$\int_C \mathbf{F}\cdot d\mathbf{r} = \int_C x dx + y dy = \int_0^2 [t dt + (\sin\pi t)(\pi\cos\pi t dt)]$$

$$= \int_0^2 (t + \pi\sin\pi t\cos\pi t) dt = [t^2/2 + (1/2)\sin^2\pi t]\Big|_0^2 = 2.$$

3. We have $x = t$, $y = -2t$, and $z = -\ln t$ so that $dx = dt$, $dy = -2dt$, $dz = -dt/t$ and thus

$$\int_C \mathbf{F}\cdot d\mathbf{r} = \int_C xy dx + y^2 dy - zx dz$$

$$= \int_1^3 [t(-2t) dt + (4t^2)(-2dt) - (-\ln t)(t)(-dt/t)]$$

$$= -\int_1^3 (10t^2 + \ln t) dt = -[10t^3/3 + (t\ln t - t)]\Big|_1^3 = -(254/3 + 3\ln 3).$$

5. The arc is given by $x = a\cos t$, $y = a\sin t$ for $0 \le t \le \pi/2$. Thus $dx = -a\sin t\ dt$, $dy = a\cos t\ dt$, and

$$\int_C \mathbf{F}\cdot d\mathbf{r} = \int_C e^{x+y}(x dx + y dy)$$

$$= \int_0^{\pi/2} e^{a(\cos t + \sin t)}[(a\cos t)(-a\sin t) dt + (a\sin t)(a\cos t) dt] = 0.$$

7. Let x be the parameter, so $x = x$, $y = x^2$ and $dx = dx$, $dy = 2x dx$. Thus $\int_C \mathbf{F}\cdot d\mathbf{r} = \int_C 2xy^2 dx + 3\sin y dy$

$$= \int_0^2 (2x)(x^4) dx + (3\sin x^2)(2x dx) = x^6/3 - 3\cos x^2\Big|_0^2 = 73/3 - 3\cos 4.$$

9. Let C_1 be the line from $(1,0)$ to $(0,1)$, C_2 be the line from $(0,1)$ to $(-1,0)$, and C_3 be the line from $(-1,0)$ to $(1,0)$. On C_1, $x = t$, $y = -t+1$, $dx = dt$, $dy = -dt$ for $1 \ge t \ge 0$; on C_2, $x = t$, $y = t+1$, $dx = dt$, $dy = dt$ for $0 \ge t \ge -1$; and on C_3, $y = 0$ and $dy = 0$. Thus, by Eq.(15),

$$\int_C \mathbf{F} \cdot d\mathbf{r} = \int_{C_1} y dx - x dy + \int_{C_2} y dx - x dy + \int_{C_3} y dx - x dy$$

$$= \int_1^0 [(1-t) dt + t dt] + \int_0^{-1} [(t+1) dt - t dt] + 0 = \int_1^0 dt + \int_0^{-1} dt = -2.$$

11.   The desired arc may be written as $x = a \cos t$, $y = b \sin t$ for $0 \le t \le \pi/2$. Hence $dx = -a \sin t\, dt$ and $dy = a \cos t\, dt$ and thus

$$\int_C \mathbf{F} \cdot d\mathbf{r} = \int_C y^2 dx + x^2 dy = \int_0^{\pi/2} (-ab^2 \sin^3 t + a^2 b \cos^3 t) dt$$

$$= \{-ab^2 [-\cos t + (1/3) \cos^3 t] + a^2 b [\sin t - (1/3) \sin^3 t]\} \Big|_0^{\pi/2}$$

$$= 2ab(a-b)/3.$$

13.   The straight line is given by $y = 5x/3$ so, with x as the parameter, $\int_C (x+y)^2 dx = \int_0^3 (x+5x/3)^2 dx = (64/9) \int_0^3 x^2 dx = 64.$

15.   Let $x = a \cos t$ and $y = a \sin t$, then $dx = -a \sin t\, dt$,

$$dy = a \cos t\, dt, \text{ and } \int_C \frac{xy dy}{\sqrt{x^2+y^2}} = \int_0^{\pi/2} \frac{(a \cos t)(a \sin t)(a \cos t dt)}{a}$$

$$= -(a^2/3) \cos^3 t \Big|_0^{\pi/2} = a^2/3.$$

17a. $y = bx/a$, $0 \le x \le a$, and thus

$$\int_C x^2 y^3 dx = (b^3/a^3) \int_0^a x^5 dx = a^3 b^3/6.$$

   b. $dy = b dx/a$ and thus $\int_C x^2 y^3 dy = (b^4/a^4) \int_0^a x^5 dx = a^2 b^4/6.$

19a. $y = x$, $0 \le x \le 1 \Rightarrow \int_C (2y-x) dx + (3x+y) dy = \int_0^1 (x dx + 4x dx) = 5/2.$

   b. $y = x^2$, $dy = 2x dx$, $0 \le x \le 1 \Rightarrow$

$$\int_C (2y-x) dx + (3x+y) dy = \int_0^1 (2x^2-x) dx + (3x+x^2)(2x dx)$$

$$= \int_0^1 (2x^3+8x^2-x) dx = 8/3.$$

19c. From $(0,0)$ to $(1,0)$ we have $y = 0$, $dy = 0$, $0 \le x \le 1$, and from $(1,0)$ to $(1,1)$ we have $x = 1$, $dx = 0$, $0 \le y \le 1$. Hence

$$\int_C (2y-x)\,dx + (3x+y)\,dy = \int_0^1 -x\,dx + \int_0^1 (3+y)\,dy = 3.$$

d. From $(0,0)$ to $(0,1)$ we have $x = 0$, $dx = 0$, $0 \le y \le 1$, and from $(0,1)$ to $(1,1)$ we have $y = 1$, $dy = 0$, $0 \le x \le 1$. Hence

$$\int_C (2y-x)\,dx + (3x+y)\,dy = \int_0^1 y\,dy + \int_0^1 (2-x)\,dx = 2.$$

e. $y = \sqrt{x}$ and $dy = dx/2\sqrt{x}$ so

$$\int_C (2y-x)\,dx + (3x+y)\,dy = \int_0^1 (2\sqrt{x}-x)\,dx + (3\sqrt{x}/2 + 1/2)\,dx = 7/3.$$

21a. If $x = 2\cos t$ and $y = 2\sin t$, then t goes from π to $\pi/2$. Thus

$$\int_C \mathbf{F} \cdot d\mathbf{r} = \int_C y\,dx + x\,dy = \int_\pi^{\pi/2} [(2\sin t)(-2\sin t) + (2\cos t)(2\cos t)]\,dt$$

$$= 4\int_\pi^{\pi/2} (\cos^2 t - \sin^2 t)\,dt = 4\int_\pi^{\pi/2} \cos 2t\,dt = 0.$$

b. Use the same parametric equations as in part (a) except t goes from π to $5\pi/2$. Then $\displaystyle\int_C \mathbf{F} \cdot d\mathbf{r} = 4\int_\pi^{5\pi/2} \cos 2t\,dt = 0.$

23a. Let $\mathbf{r} = t\mathbf{i}+t\mathbf{j}+t\mathbf{k}$, $0 \le t \le 1$. Then

$$\int_C \mathbf{F} \cdot d\mathbf{r} = \int_C (e^x\,dx + e^y\,dy + e^z\,dz) = \int_0^1 (e^t\,dt + e^t\,dt + e^t\,dt) = 3(e-1).$$

b. On the first segment $y = z = 0$, $dy = dz = 0$, $0 \le x \le 1$; on the second, $x = 1$, $z = 0$, $dx = dz = 0$, $0 \le y \le 1$; and on the third, $x = y = 1$, $dx = dy = 0$, $0 \le z \le 1$. Thus

$$\int_C \mathbf{F} \cdot d\mathbf{r} = \int_0^1 e^x\,dx + \int_0^1 e^y\,dy + \int_0^1 e^z\,dz = 3(e-1).$$

c. We have $dx = dt$, $dy = 2t\,dt$, $dz = 3t^2\,dt$ and thus

$$\int_C \mathbf{F} \cdot d\mathbf{r} = \int_0^1 (e^t\,dt + e^{t^2}2t\,dt + e^{t^3}3t^2\,dt) = (e^t+e^{t^2}+e^{t^3})\Big|_0^1 = 3(e-1).$$

25. The length of the wire is $a\pi$ and we have $\rho = ks$, where k is the constant of proportionality. Thus $M = \displaystyle\int_0^{a\pi} (ks)\,ds = ka^2\pi^2/2.$

27. We have $\rho = k|\mathbf{r}|^2 = k(4\cos^2 t + 4\sin^2 t + 9t^2) = k(4+9t^2)$ and
     $ds = (ds/dt)dt = |\mathbf{r}'|dt = (4\sin^2 t + 4\cos^2 t + 9)^{1/2}dt = \sqrt{13}dt$.
     Thus $M = k\displaystyle\int_0^{4\pi}(4+9t^2)\sqrt{13}\,dt = 16\sqrt{13}\pi(1 + 12\pi^2)k$.

29. We have $\rho = k\theta$ as the density and $\mathbf{r} = \theta\cos\theta\mathbf{i} + \theta\sin\theta\mathbf{j}$ as the
     position vector. Thus $ds = |d\mathbf{r}/d\theta|d\theta = (1+\theta^2)^{1/2}d\theta$ [which is
     equivalent to Eq.(21) of Section 14.3] and hence
     $$M = k\int_0^{2\pi}\theta(1+\theta^2)^{1/2}d\theta = [(1+4\pi^2)^{3/2} - 1]k/3.$$

31. We have $ds = (ds/dt)dt = |\mathbf{r}'|dt = adt$ and $\rho = k$. Thus
     $$I_x = k\int_0^{\pi}(a^2\sin^2 t)adt = \pi ka^3/2, \quad I_y = k\int_0^{\pi}(a^2\cos^2 t)adt = \pi ka^3/2.$$

33. By symmetry $\hat{x} = 0$ and $\hat{y} = (1/L)\displaystyle\int_C yds = (1/3a)[2\int_{C_1} yds + \int_{C_2} yds]$,
     where C_1 is the segment from $(0,0)$ to $(a/2,\sqrt{3}a/2)$ and C_2 is
     the segment from $(a/2,\sqrt{3}a/2)$ to $(-a/2,\sqrt{3}a/2)$. On C_1 $x = t$,
     $y = \sqrt{3}t$, $0 \le t \le a/2$, and $ds = 2dt$, while on C_2 $x = t$,
     $y = \sqrt{3}a/2$, where t goes from $a/2$ to $-a/2$, and $ds = -dt$ (since
     ds must be positive). Thus
     $$\hat{y} = (1/3a)[2\int_0^{a/2}(\sqrt{3}t)2dt + \int_{a/2}^{-a/2}(\sqrt{3}a/2)(-dt)] = (1/3a)(a^2\sqrt{3})$$
     $= a/\sqrt{3}$.

     This result can also be obtained by using the composite body
     concept of Prob.18, Section 17.3.

35. $I_x = k\displaystyle\int_0^a y^2 ds + k\int_0^b y^2 ds = k\int_0^b s^2 ds = kb^3/3$ since $y = 0$ on the
     horizontal portion and $y = s$ on the vertical portion.
     Likewise, $I_y = k\displaystyle\int_0^a x^2 ds + k\int_0^b x^2 ds = k\int_0^a s^2 ds = ka^3/3$ since $x = 0$
     on the vertical portion.

Section 18.2, Page 991

1.   Let $P = 2xy+6$ and $Q = x^2-3$, then $P_y = Q_x = 2x$ so Eqs.(21) are
      satisfied and **F** is therefore conservative. From Eq.(9), we
      have $\phi_x = 2xy+6$ so $\phi = x^2y + 6x + h(y)$ and thus
      $\phi_y = x^2 + h'(y) = x^2 - 3$ so $h(y) = -3y+k$ and hence
      $\phi = x^2y + 6x - 3y + k$.

3.   Let $P = y\cos x + 2xe^y$ and $Q = \sin x + x^2e^y + 2$, then
      $P_y = Q_x = \cos x + 2xe^y$ and hence **F** is conservative. Now,
      $\phi_x = y\cos x + 2xe^y \Rightarrow \phi = y\sin x + x^2e^y + h(y)$ and thus
      $\phi_y = \sin x + x^2e^y + h'(y) = \sin x + x^2y + 2$ so $h(y) = 2y+k$ and
      $\phi = y\sin x + x^2e^y + 2y + k$.

5.   Let $P = 2x+4y$ and $Q = 2x-2y$, then $P_y = 4$ and $Q_x = 2$ and hence
      F is not conservative.

7.   Let $P = y/z - e^z$, $Q = x/z + 3$, and $R = -(xe^z + xy/z^2)$, then
      $R_y = Q_z = -x/z^2$, $P_z = R_x = -y/z^2 - e^z$ and $Q_x = P_y = 1/z$ so
      Eqs.(22) are satisfied and **F** is therefore conservative. Now
      $\phi_x = y/z - e^z \Rightarrow \phi = xy/z - xe^z + h(y,z)$ and thus
      $\phi_y = x/z + \partial h/\partial y = x/z + 3$ so $h(y,z) = 3y + g(z)$. Now
      $\phi = xy/z - xe^z + 3y + g(z)$ and thus
      $\phi_z = -xy/z^2 - xe^z + g'(z) = -xe^z - xy/z^2$ so $g(z) = k$ and
      finally $\phi = xy/z - xe^z + 3y + k$.

9.   Let $P = 2\sin x + y^2$ and $Q = 2xy + e^y$, then $P_y = Q_x = 2y$ and thus
      the line integral is independent of the path. To evaluate the
      integral we need to find ϕ so that $\phi_x = P$ and $\phi_y = Q$. Now
      $\phi_x = 2\sin x + y^2 \Rightarrow \phi = -2\cos x + xy^2 + h(y)$ so
      $\phi_y = 2xy + h'(y) = 2xy + e^y$ and thus $h = e^y + k$ and
      $\phi = -2\cos x + xy^2 + e^y + k$. Hence,
      $$\int_C Pdx + Qdy = (-2\cos x + xy^2 + e^y + k) \Big|_{(0,1)}^{(\pi,2)} = 4(1+\pi) + e^2 - e.$$
      Note that k does not affect the value of the integral and
      will not be used in the rest of the integration problems.

11. Let $P = z^2 - y\sin x$, $Q = \cos x - 2z$, $R = 2zx-2y+z$, then $R_y = Q_z$,
     $P_z = R_x$, and $Q_x = P_y$ and thus the integral is independent of
     the path. Now, $\phi_x = P \Rightarrow \phi = xz^2 + y\cos x + h(y,z)$,
     $\phi_y = Q \Rightarrow \cos x + h_y = \cos x - 2z \Rightarrow h = -2yz + g(z) \Rightarrow$
     $\phi = xz^2 + y\cos x - 2yz + g(z)$, and
     $\phi_z = R \Rightarrow 2xz - 2y + g'(z) = 2zx-2y+z \Rightarrow g(z) = z^2/2$ and hence
     $\phi = xz^2 + y\cos x - 2yz + z^2/2$. Thus
     $$\int_C Pdx+Qdy = (xz^2 + y\cos x - 2yz + z^2/2)\Big|_{(\pi/4,1,1)}^{(\pi,0,2)} = 15\pi/4 + (7-\sqrt{2})/2.$$

13. Let $P = xe^{r^2}$, $Q = ye^{r^2}$, and $R = ze^{r^2}$, then Eqs.(22) are
     satisfied and the line integral is path independent. Now
     $\phi_x = xe^{r^2} \Rightarrow \phi = e^{r^2}/2 + h(y,z)$,
     $\phi_y = ye^{r^2} \Rightarrow ye^{r^2} + h_y = ye^{r^2} \Rightarrow h(y,z) = g(z)$ and $\phi = e^{r^2} + g(z)$,
     and $\phi_z = ze^{r^2} \Rightarrow ze^{r^2} + g' = ze^{r^2} \Rightarrow g(z) = k$ and $\phi = e^{r^2}/2$.
     Thus $\int_C Pdx+Qdy+Rdz = (1/2)e^{r^2}\Big|_{(0,0,0)}^{(a,b,c)} = (1/2)[\exp(a^2+b^2+c^2) - 1]$.

15. If $P = xy^2 + \alpha x^2 y$ and $Q = x^3 + x^2 y$, then
     $P_y = Q_x \Rightarrow 2xy + \alpha x^2 = 3x^2 + 2xy$, so $\alpha = 3$. Now, with $\alpha = 3$,
     $\phi_x = P \Rightarrow \phi = x^2 y^2/2 + x^3 y + h(y)$,
     $\phi_y = Q \Rightarrow x^2 y + x^3 + h' = x^3 + x^2 y \Rightarrow h = k$ and $\phi = x^2 y^2/2 + x^3 y$.
     Thus $\int_C Pdx+Qdy = (x^2 y^2/2 + x^3 y)\Big|_{(0,0)}^{(1,-1)} = -1/2$.

17. Since $(\cos x\cos y)_y = (-\sin x\sin y)_x$, the integral is path
     independent and thus its value is 0 by Theorem 18.2.3.

19. If $P = (y/x + z - 4x)$, $Q = 3+\ln x$, and $R = x$, then $R_y = Q_z$,
     $P_z = R_x$, and $Q_x = P_y$ so the integral is path independent.
     $\phi_x = P \Rightarrow \phi = y\ln x + xz - 2x^2 + h(y,z)$, $\phi_y = Q \Rightarrow \ln x + h_y = 3 + \ln x$
     or $h = 3y + g(z)$ and $\phi = y\ln x + xz - 2x^2 + 3y + g(z)$, and
     $\phi_z = R \Rightarrow x + g'(z) = x$ or $g(z) = k$ and $\phi = y\ln x + xz - 2x^2 + 3y$.
     Thus $\int_C Pdx+Qdy+Rdz = (y\ln x + xz - 2x^2 + 3y)\Big|_{(1,0,0)}^{(e,2,1)} = 10 + e - 2e^2$.

21. Since $d(r^2/2) = d(\mathbf{r}\cdot\mathbf{r}/2) = \mathbf{r}\cdot d\mathbf{r}$, the integral is path

independent   and $\int_C \mathbf{r}\cdot d\mathbf{r} = \int_C d(r^2/2) = r^2/2\Big|_{r_a}^{r_b} = (r_b^2 - r_a^2)/2$.

23. Let $P = y\sin(x^2y^2)$ and $Q = x\sin(x^2y^2)$, then $P_y = Q_x$ and the
integral is independent of the path. In this case, however,
neither $\phi_x = P = y\sin(x^2y^2)$ nor $\phi_y = Q = x\sin(x^2y^2)$ is
integrable in terms of elementary functions so we can't find
$\phi(x,y)$ easily. In order to evaluate the given integral we
must seek another path which simplifies the calculations.
Thus we choose straight lines from $(0,2)$ to $(0,0)$ (C_1) and
then from $(0,0)$ to $(2,0)$ (C_2). On C_1 $x = 0$ so $P = Q = 0$ and
on C_2 $y = 0$ so $P = Q = 0$, and thus the given integral is 0.

25. Since $\nabla(\phi\psi) = \psi\nabla\phi + \phi\nabla\psi$, we have

$\int_C \nabla(\phi\psi)\cdot d\mathbf{r} = \int_C (\psi\nabla\phi + \phi\nabla\psi)\cdot d\mathbf{r} = \int_C \psi\nabla\phi\cdot d\mathbf{r} + \int_C \phi\nabla\psi\cdot d\mathbf{r}$ and

$\int_C \nabla(\phi\psi)\cdot d\mathbf{r} = \psi\phi\Big|_{\mathbf{r}_a}^{\mathbf{r}_b} = \psi(\mathbf{r}_b)\phi(\mathbf{r}_b) - \psi(\mathbf{r}_a)\phi(\mathbf{r}_a)$. Equating the

two right sides yields the desired identity.

27a. Let $P = -kx$ and $Q = 0$ then $P_y = Q_x$ and thus $\mathbf{F} = -kx\mathbf{i}$ is
conservative. Now $\phi_x = kx \Rightarrow \phi = kx^2/2 + h(y)$, $\phi_y = 0 \Rightarrow$
$h' = 0$, and hence $\phi = kx^2/2$ is a potential function and
$\mathbf{F} = -\nabla\phi$. In scalar form we have $F = -kx = d(-kx^2/2)/dx$.

b. $W = \int_a^b \mathbf{F}\cdot d\mathbf{r} = \int_a^b Fdx = \int_a^b d(-kx^2/2) = k(x_a^2 - x_b^2)/2$.

Section 18.3, Page 1000

1. Let $P = y$, $Q = x^2$, $x = a\cos t$, $y = a\sin t$ for $0 \le t \le 2\pi$.

Then $\oint_C Pdx+Qdy = \int_0^{2\pi} [(a\sin t)(-a\sin t\, dt) + (a^2\cos^2 t)(a\cos t\, dt)]$

$= -(a^2/2)\int_0^{2\pi} (1-\cos 2t)dt + a^3\int_0^{2\pi} (1-\sin^2 t)\cos t\, dt = -\pi a^2$,

and $\iint_\Omega (Q_x-P_y)dA = \iint_\Omega (2x-1)dA = \int_0^{2\pi}\int_0^a (2r\cos\theta - 1)rdrd\theta$

$= \int_0^{2\pi} [(2a^3/3)\cos\theta - a^2/2]d\theta = -\pi a^2$.

3.   Let $P = 2y-x$, $Q = 3x+y$, C_1 be $y = x^2$, $0 \le x \le 1$, and C_2 be $x = y^2$ for y going from 1 to 0. Then

$$\oint_C Pdx+Qdy = \int_0^1 [(2x^2-x)dx + (3x+x^2)2xdx]$$

$$+ \int_1^0 [(2y-y^2)2ydy+(3y^2+y)dy]$$

$$= \int_0^1 (2x^3+8x^2-x)dx + \int_1^0 (-2y^3+7y^2+y)dy = 1/3,$$

and $\iint_\Omega (Q_x-P_y)dA = \iint_\Omega (3-2)dA = \int_0^1 \int_{x^2}^{\sqrt{x}} dydx = 1/3.$

5. $\oint_C 2dx-3dy = \iint_\Omega (0-0)dA = 0.$

7. $\oint_C xy\, dy = \int_{-a}^a \int_0^{\sqrt{a^2-x^2}} (y-0)dydx = (1/2)\int_{-a}^a (a^2-x^2)dx = 2a^3/3.$

9. $\oint_C 3xy^2dx - 5x^2ydy = \int_{-1}^1\int_{-1}^1 (-10xy-6xy)dydx = -8\int_{-1}^1 xy^2 \big|_{-1}^1 dx = 0.$

11. $A = \oint_C xdy = \int_0^{2\pi} (a\cos t)(a\cos t\, dt) = \pi a^2.$

13.   Let C_1 be $y = 0$, $0 \le x \le 2\pi$ and C_2 be the cycloid starting at $(2\pi,0)$ and going counterclockwise to $(0,0)$. Then

$$A = \oint_C xdy = \int_{C_1} x\cdot 0 + \int_{2\pi}^0 a(t-\sin t)(a\sin t\, dt)$$

$$= a^2(-t\cos t + \sin t)\big|_{2\pi}^0 - a^2[t/2 - (1/4)\sin 2t]\big|_{2\pi}^0 = 3\pi a^2.$$

15.   Let $P = \phi_y$, $Q = -\phi_x$, so $P_y = \phi_{yy}$, $Q_x = -\phi_{xx}$, and $Q_x - P_y = -\phi_{xx} - \phi_{yy} = 0$. Thus Eq.(1) yields the desired result.

17.   We have $\hat{x} = (1/A)\iint_\Omega xdA$ and hence we wish to choose $Q_x-P_y = x$.

Take $P = 0$ (so $P_y = 0$) and $Q = x^2/2$ (so $Q_x = x$) and thus

$\hat{x} = (1/A)\iint_\Omega xdA = (1/A)\oint_C Qdy = (1/2A)\oint_C x^2dy.$ Likewise,

$\hat{y} = (1/A)\iint_\Omega y\,dA$ so we want $Q_x - P_y = y$.   Take $Q = 0$, $P = -y^2/2$

and thus $\hat{y} = (1/A)\iint_\Omega y\,dA = (1/A)\oint_C P\,dx = (-1/2A)\oint_C y^2 dx$.

19a. $P_y = (-2xy)/(x^2+y^2)^2$, $Q_x = (-2xy)/(x^2+y^2)^2$ and thus $P_y = Q_x$
   for $(x,y) \neq (0,0)$, which is excluded in D.

  b. On the circle $P\,dx = (a\cos t)(-a\sin t\, dt)/a^2$ and
   $Q\,dy = (a\sin t)(a\cos t\, dt)/a^2$ and thus $P\,dx + Q\,dy = 0$.

  c. If C does not enclose the origin then Eq.(1) applies and
   $$\oint_C P\,dx + Q\,dy = \iint_\Omega (Q_x - P_y)\,dA = 0 \text{ by part (a). If C encloses the}$$
   origin, choose an $a \geq \varepsilon$ such that the circle of part(b) lies
   entirely within C.   Then, by Eq.(12),
   $$\oint_C P\,dx + Q\,dy + \oint_{x^2+y^2=a^2} P\,dx + Q\,dy = \iint_\Omega (Q_x - P_y)\,dA = 0 \text{ by part (a), so}$$
   $$\oint_C P\,dx + Q\,dy = \oint_{x^2+y^2=a^2} P\,dx + Q\,dy = 0 \text{ by part (b).}$$

  d. $\phi_x = P \Rightarrow \phi = (1/2)\ln(x^2+y^2) + h(y)$ and
   $\phi_y = Q \Rightarrow y/(x^2+y^2) + h'(y) = y/(x^2+y^2)$, so $h = 0$ and
   $\phi(x,y) = (1/2)\ln(x^2+y^2)$.

21a. $\dfrac{\partial u}{\partial n} = \nabla u \cdot \mathbf{N} = u_x\dfrac{dy}{ds} - u_y\dfrac{dx}{ds}$ and thus

   $$\oint_C \frac{\partial u}{\partial n}\,ds = \oint_C -u_y\,dx + u_x\,dy = \iint_\Omega (u_{xx} + u_{yy})\,dA = \iint_\Omega \nabla^2 u\,dA.$$

  b. $\dfrac{\partial v}{\partial n}\,ds = -v_y\,dx + v_x\,dy$ and hence

   $$\oint_C (u\frac{\partial v}{\partial n} - v\frac{\partial u}{\partial n})\,ds = \oint_C [(vu_y - uv_y)\,dx + (uv_x - vu_x)\,dy]$$
   $$= \iint_\Omega [u(v_{xx}+v_{yy}) - v(u_{xx}+u_{yy})]\,dA = \iint_\Omega (u\nabla^2 v - v\nabla^2 u)\,dA.$$

  c. $\displaystyle\oint_C u\frac{\partial v}{\partial n}\,ds = \oint_C -uv_y\,dx + uv_x\,dy = \iint_\Omega (u_xv_x+uv_{xx}+u_yv_y+uv_{yy})\,dA$
   $$= \iint_\Omega (u\nabla^2 v + \nabla u \cdot \nabla v)\,dA.$$

21d. $\oint_C u\dfrac{\partial u}{\partial n}ds = \oint_C -uu_ydx + uu_xdy = \iint_\Omega (u_x{}^2 + uu_{xx} + u_y{}^2 + uu_{yy})\,dA$

$\qquad\qquad = \iint_\Omega (\|\nabla u\|^2 + u\nabla^2 u)\,dA$. The same result may also be

obtained by setting $v = u$ in part(c).

Section 18.4, Page 1013

1. One eighth of the surface area projects onto Ω_{xy} given by

    $0 \le y \le \sqrt{a^2-x^2}$, $0 \le x \le a$. Let $F(x,y,z) = x^2+y^2+z^2-a^2$, then

    by Eq.(21), we have $A(S) = 8\iint_{\Omega_{xy}} \{[(4x^2+4y^2+4z^2)^{1/2}]/2z\}dA_{xy}$,

    where $z = \sqrt{a^2-x^2-y^2}$. Thus

    $A = 8a\iint_{\Omega_{xy}} (a^2-x^2-y^2)^{-1/2}dA_{xy} = 8a\int_0^{\pi/2}\int_0^a (a^2-r^2)^{-1/2}rdrd\theta = 4\pi a^2.$

3. The plane projects onto $0 \le y \le b-bx/a$, $0 \le x \le a$. Let
    $F(x,y,z) = x/a + y/b + z/c - 1$ then

    $A(S) = \int_0^a\int_0^{b-bx/a} [(1/a^2 + 1/b^2 + 1/c^2)^{1/2}/(1/c)]dydx$

    $\qquad = (c^2/a^2 + c^2/b^2 + 1)^{1/2}\int_0^a (b-bx/a)\,dx = (b^2c^2 + a^2c^2 + a^2b^2)^{1/2}/2.$

5. The total surface area is four times that lying in the first
    octant, and $x = yz$ intersects the yz-plane along $y = 0$ and

    $z = 0$. Thus Ω_{yz} is given by $0 \le z \le \sqrt{a^2-y^2}$, $0 \le y \le a$, and

    hence Eq.(23) yields $A(S) = 4\iint_{\Omega_{yz}} [(1^2+z^2+y^2)^{1/2}/(1)]dA_{yz}$,

    where we have chosen $F(x,y,z) = x-yz$. Using polar
    coordinates we then obtain

    $A(S) = 4\int_0^{\pi/2}\int_0^a (1+r^2)^{1/2}rdrd\theta = (2\pi/3)[(1+a^2)^{3/2} - 1].$

7. Projecting the surface onto the yz-plane we get
    $0 \le z \le 2y+3$, $0 \le y \le a$. If $F(x,y,z) = x^2+y^2-a^2$, then

    $A(S) = \iint_{\Omega_{yz}} [(4x^2+4y^2)^{1/2}/2x]dA_{xy}$, where $x = \sqrt{a^2-y^2}$. Thus

$$A(S) = a\int_0^a\int_0^{2y+3}(a^2-y^2)^{-1/2}dzdy = a\int_0^a(2y+3)(a^2-y^2)^{-1/2}dy$$

$$= [-2a(a^2-y^2)^{1/2} + 3a\,\arcsin(y/a)]\Big|_0^a = 2a^2 + 3\pi a/2.$$

9.   The surface projects onto $0 \leq z \leq x^3$, $0 \leq x \leq 2$. Letting $F(x,y,z) = x^3 + y - 8$, Eq.(22) then yields

$$A(S) = \int_0^2\int_0^{x^3}(9x^4+1)^{1/2}dzdx = \int_0^2(9x^4+1)^{1/2}x^3dx = (145^{3/2} - 1)/54.$$

11. $z = Hr/R = H(x^2+y^2)^{1/2}/R$ is the equation of the cone if the vertex is at the origin. Thus the cone projects onto a circle of radius R in the xy-plane. With

$F(x,y,z) = H\sqrt{x^2+y^2} - Rz$, we then obtain

$$A(S) = \int\!\!\int_{\Omega_{xy}}\{[(H^2x^2 + H^2y^2)/(x^2+y^2) + R^2]^{1/2}/R\}dAxy$$

$$= [(H^2 + R^2)^{1/2}/R]\int\!\!\int_{\Omega_{xy}}dAxy = 2\pi R(H^2 + R^2)^{1/2},$$

where the last integral is simply the area of the circle.

13. $F(x,y,z) = x^2+y^2+z^2-a^2$ and $z = \sqrt{a^2-x^2-y^2}$ on the surface, so

$$M = k\int\!\!\int_{\Omega_{xy}}(x^2+y^2)^{1/2}[(4x^2+4y^2+4z^2)^{1/2}/2z]dAxy$$

$$= ak\int\!\!\int_{\Omega_{xy}}(x^2+y^2)^{1/2}(a^2-x^2-y^2)^{-1/2}dAxy = ak\int_0^{2\pi}\int_0^a r^2(a^2-r^2)^{-1/2}drd\theta$$

$$= ak[-r(a^2-r^2)^{1/2}/2 + (a^2/2)\arcsin(r/a)]\Big|_0^a\int_0^{2\pi}d\theta = \pi^2ka^3/2.$$

15.   The projection of the surface on the xz-plane is

$0 \leq z \leq \sqrt{a^2-x^2}$, $0 \leq x \leq a$. Thus

$M = \int\!\!\int_{\Omega_{xz}}(a-x)[(4x^2+4y^2)^{1/2}/2y]dAxz$, where $y = \sqrt{a^2-x^2}$. Hence

$$M = a\int_0^a\int_0^{\sqrt{a^2-x^2}}(a-x)(a^2-x^2)^{-1/2}dzdx = a\int_0^a(a-x)dx = a^3/2.$$

17. One eighth of the surface projects onto the quarter circle
$0 \leq y \leq \sqrt{a^2-x^2}$, $0 \leq x \leq a$. With $z = (1 - x^2/4 - y^2/4)^{1/2}$, then

$$A(S) = 8\iint_{\Omega_{xy}} [(x^2/4 + y^2/4 + 4z^2)^{1/2}/(2z)]dA_{xy}$$

$$= 4\iint_{\Omega_{xy}} (16-3x^2-3y^2)^{1/2}(4-x^2-y^2)^{-1/2}dA_{xy}$$

$$= 4\int_0^a \int_0^{\pi/2} (16-3r^2)^{1/2}(4-r^2)^{-1/2}r d\theta dr$$

$$= 2\pi \int_0^a [(16-3r^2)^{1/2}/(4-r^2)]^{1/2}r dr.$$

19. The surface projects onto $0 \leq x \leq 1-y^2$, $0 \leq y \leq 1$, and
$\rho = 1-z = x^2$ on the surface $x^2 + z - 1 = 0$. Thus

$$M = \int_0^1 \int_0^{1-y^2} x^2(4x^2+1)^{1/2}dxdy.$$

Section 18.5, Page 1023

1. Solving the first two equations for u and v, we get
$u = (2x+y)/5$ and $v = (-x+2y)/5$, which when substituted into
the equation for z yields $3x-y-5z = 0$. This is a plane
through the origin, and the whole plane is generated by the
given u and v ranges, since all x, y, z values are obtained.

3. We have $x^2 - y^2 = -u^2 - v^2u^2$ and hence $x^2-y^2+z^2 = 0$, which is a
cone. Since $v \geq 0$, we see that $x \geq 0$ and $y \geq 0$ only.

5. We have $x^2 + (y/2)^2 = \sinh^2 v$, so $z = x^2 + y^2/4$, which is an
elliptic paraboloid. Since $0 \leq u \leq \pi$ and $v \geq 0$, we have $x \geq 0$.

7. Let $\mathbf{r} = (2u-v)\mathbf{i} + (u+2v)\mathbf{j} + (u-v)\mathbf{k}$, then $(u,v) = (2,3)$ gives
the desired point and $\partial\mathbf{r}/\partial u = 2\mathbf{i}+\mathbf{j}+\mathbf{k}$ and $\partial\mathbf{r}/\partial v = -\mathbf{i}+2\mathbf{j}-\mathbf{k}$. By
Eq.(14), $\mathbf{n} = (\partial\mathbf{r}/\partial u)\times(\partial\mathbf{r}/\partial v) = -3\mathbf{i}+\mathbf{j}+5\mathbf{k}$ and thus
$\dfrac{x-1}{-3} = \dfrac{y-8}{1} = \dfrac{z+1}{5}$ is the normal line and
$-3(x-1)+(y-8)+5(z+1) = 0$ or $-3x+y+5z = 0$ is the tangent plane.

9. Let $\mathbf{r} = v(1-u^2)^{1/2}\mathbf{i} + (u^2+v^2)^{1/2}\mathbf{j} + u(1+v^2)^{1/2}\mathbf{k}$, then
$(u,v) = (0,1)$ gives the desired point, and
$\partial\mathbf{r}/\partial u = [-uv/(1-u^2)^{1/2}]\mathbf{i} + [u/(u^2+v^2)^{1/2}]\mathbf{j} + (1+v^2)^{1/2}\mathbf{k} = \sqrt{2}\mathbf{k}$
at $(0,1)$ and
$\partial\mathbf{r}/\partial v = (1-u^2)^{1/2}\mathbf{i} + [v/(u^2+v^2)^{1/2}]\mathbf{j} + [uv/(1+v^2)^{1/2}]\mathbf{k} = \mathbf{i}+\mathbf{j}$

at $(0,1)$ so $\mathbf{n} = (\sqrt{2}\mathbf{k}) \times (\mathbf{i}+\mathbf{j}) = -\sqrt{2}(\mathbf{i}-\mathbf{j})$. Since any vector parallel to $\mathbf{n}$ may be used, we choose $\mathbf{i}-\mathbf{j}$ and thus $x-1 = -(y-1)$, $z = 0$ is the normal line and $x-y = 0$ is the tangent plane.

11. Let $\mathbf{r} = \sin u \sinh v \mathbf{i} + 2\cos u \sinh v \mathbf{j} + \sinh^2 v \mathbf{k}$. Now $\sinh^2 v = 9/16 \Rightarrow \sinh v = 3/4 \Rightarrow e^v - 1/e^v = 3/2 \Rightarrow e^v = 2$ or $v = \ln 2$, and $\sinh v = 3/4 \Rightarrow \sin u = 1$, $\cos u = 0$ or $u = \pi/2$, hence $(u,v) = (\pi/2, \ln 2)$ yields the given point. Now $\partial\mathbf{r}/\partial u = (-3/2)\mathbf{j}$ and $\partial\mathbf{r}/\partial v = (5/4)\mathbf{i}+(15/8)\mathbf{k}$ at $(\pi/2, \ln 2)$ and thus a normal vector is given by $\mathbf{n} = \mathbf{j} \times [\mathbf{i}+(3/2)\mathbf{k}] = (3/2)\mathbf{i}-\mathbf{k}$. The normal line can then be written as
$$\frac{x-3/4}{3} = \frac{z-9/16}{-2}, \quad y = 0$$ and the tangent plane as $3x-2z = 9/8$.

13a. From Ex.4, $A = R^2(\theta_2 - \theta_1)(\cos\phi_1 - \cos\phi_2)$ and $R = 3960$ miles, and, from Fig.18.5.12, $\theta_2 - \theta_1 = 7° \cong 0.12217305$ radians, $\phi_1 = 45° = \pi/4$ radians, and $\phi_2 = 49° \cong .85521133$ radians. Thus $A = 97,801$ mi^2.

 b. $5° = .08726646$ radians and $2° = .03490659$ radians, so we have
$$A = 3960^2[.08726646(\cos 48° - \cos 53°)$$
$$- .03490659(\cos 48° - \cos 49°)] = 84,965 \text{mi}^2.$$

15a. We have $x^2+y^2 = \rho^2\sin^2\phi_0$ and thus $(x^2+y^2)/z^2 = \tan^2\phi_0$ or $x^2+y^2-c^2z^2 = 0$ $(c^2 = \tan^2\phi_0)$, which is a right circular cone.

 b. $\partial\mathbf{r}/\partial\rho = \sin\phi_0\cos\theta\mathbf{i} + \sin\phi_0\sin\theta\mathbf{j} + \cos\phi_0\mathbf{k}$ and
$\partial\mathbf{r}/\partial\theta = -\rho\sin\phi_0\sin\theta\mathbf{i} + \rho\sin\phi_0\cos\theta\mathbf{j}$ so
$\mathbf{r}_\rho \times \mathbf{r}_\theta = -\rho\sin\phi_0\cos\phi_0\cos\theta\mathbf{i} - \rho\sin\phi_0\cos\phi_0\sin\theta\mathbf{j} + \rho\sin^2\phi_0\mathbf{k}$.
Thus $d\sigma = \|(\partial\mathbf{r}/\partial\rho) \times (\partial\mathbf{r}/\partial\theta)\|d\rho d\theta = \rho\sin\phi_0 d\rho d\theta$.

 c. For the given cone we have $\sin\phi_0 = r/\sqrt{r^2+h^2}$, $0 \leq \rho \leq \sqrt{r^2+h^2}$, and $0 \leq \theta \leq 2\pi$. Thus
$$A(S) = \int_0^{2\pi}\int_0^{\sqrt{r^2+h^2}} \rho\sin\phi_0 d\rho d\theta = (r/\sqrt{r^2+h^2})(\rho^2/2)\Big|_0^{\sqrt{r^2+h^2}}\int_0^{2\pi} d\theta$$
$$= \pi r\sqrt{r^2+h^2}.$$

17. $\partial r/\partial u = \cos v\mathbf{i} + \sin v\mathbf{j}$ and $\partial r/\partial v = -u\sin v\mathbf{i} + u\cos v\mathbf{j} + c\mathbf{k}$ so

$(\partial r/\partial u)\times(\partial r/\partial v) = c\sin v\mathbf{i} - c\cos v\mathbf{j} + u\mathbf{k}$. Thus $d\sigma = \sqrt{u^2+c^2}\,dudv$

and $A(S) = \int_a^b\int_0^{2\pi}\sqrt{u^2+c^2}\,dvdu = 2\pi\int_a^b\sqrt{u^2+c^2}\,du$

$= \pi[u\sqrt{u^2+c^2} + c^2\ln(u + \sqrt{u^2+c^2})]\,|_a^b$

$= \pi[b\sqrt{b^2+c^2} - a\sqrt{a^2+c^2} + c^2\ln(b+\sqrt{b^2+c^2}) - c^2\ln(a+\sqrt{a^2+c^2})].$

19. $\partial r/\partial u = a\cos u\cos v\mathbf{i} + b\cos u\sin v\mathbf{j} - c\sin u\mathbf{k}$ and

$\partial r/\partial v = -a\sin u\sin v\mathbf{i} + b\sin u\cos v\mathbf{j}$ so

$(\partial r/\partial u)\times(\partial r/\partial v) = bc\sin^2 u\cos v\mathbf{i} + ac\sin^2 u\sin v\mathbf{j} + ab\cos u\sin u\mathbf{k}$.

Thus $d\sigma = \sin u(b^2c^2\sin^2 u\cos^2 v + a^2c^2\sin^2 u\sin^2 v + a^2b^2\cos^2 u)dvdu$.

From $z = c\cos u$ we see that $0 \le u \le \pi$, and from

$x = a\sin u\cos v$, $y = b\sin u\sin v$ we see that we need $0 \le v \le 2\pi$

to give all x and y values. Hence $A(S) = \int_0^\pi\int_0^{2\pi} d\sigma$.

Section 18.6, Page 1030

1. By Eq.(1), $\operatorname{div}\mathbf{v} = \nabla\cdot\mathbf{v} = \partial(xy\sin z)/\partial x + \partial(x^2 z)/\partial y + \partial(-y\cos z)/\partial z$
$= 2y\sin z$.

From Eq.(5), $\operatorname{curl}\mathbf{v} = \nabla\times\mathbf{v} = \begin{vmatrix} \mathbf{i} & \mathbf{j} & \mathbf{k} \\ \partial/\partial x & \partial/\partial y & \partial/\partial z \\ xy\sin z & x^2 z & -y\cos z \end{vmatrix}$

$= -(\cos z+x^2)\mathbf{i} + xy\cos z\mathbf{j} + (2xz-x\sin z)\mathbf{k}$.

3. We have $\partial(x/r)/\partial x = 1/r - x^2/r^3$, $\partial(y/r)/\partial y = 1/r - y^2/r^3$,
$\partial(z/r)/\partial z = 1/r - z^2/r^3$ and thus
$\nabla\cdot\mathbf{v} = 3/r - (x^2+y^2+z^2)/r^3 = 3/r - 1/r = 2/r$. For the curl we

have $\nabla\times\mathbf{v} = \begin{vmatrix} \mathbf{i} & \mathbf{j} & \mathbf{k} \\ \partial/\partial x & \partial/\partial y & \partial/\partial z \\ x/r & y/r & z/r \end{vmatrix}$

$= (-zy/r^3 + yz/r^3)\mathbf{i} + (-xz/r^3 + zx/r^3)\mathbf{j} + (-yx/r^3 + xy/r^3)\mathbf{k} = \mathbf{0}$.

5. $\nabla\cdot\mathbf{v} = \partial(-ry)/\partial x + \partial(rx)/\partial y = -yx/r + xy/r = 0$ and, by Eq.(4),
$\nabla\times\mathbf{v} = 0\mathbf{i} + 0\mathbf{j} + [\partial(rx)/\partial x + \partial(ry)/\partial y]\mathbf{k} = [2r+(x^2+y^2)/r]\mathbf{k} = 3r\mathbf{k}$.

7. $\nabla\cdot\mathbf{v} = (2xyz)_x - (x^2+y^2)_y + (xyz^2)_z = 2yz-2y+2xyz$ and

   $\nabla\times\mathbf{v} = (xyz^2)_y\mathbf{i} + [(2xyz)_z - (xyz^2)_x]\mathbf{j} - [(x^2+y^2)_x + (2xyz)_y]\mathbf{k}$
      $= xz^2\mathbf{i} + (2xy-yz^2)\mathbf{j} - (2x+2xz)\mathbf{k}.$

9. $\nabla\cdot\mathbf{v} = (\sin x \sin y)_x + (\cos y \sin z)_y + (\cos z \cos x)_z$
      $= \cos x \sin y - \sin y \sin z - \sin z \cos x.$

   $\nabla\times\mathbf{v} = -(\cos y \sin z)_z\mathbf{i} - (\cos z \cos x)_x\mathbf{j} - (\sin x \sin y)_y\mathbf{k}$
      $= -\cos y \cos z\mathbf{i} + \cos z \sin x\mathbf{j} - \sin x \cos y\mathbf{k}.$

11a. $\nabla\cdot(c_1\mathbf{u}+c_2\mathbf{v}) = (c_1u_1 + c_2v_1)_x + (c_1u_2 + c_2v_2)_y + (c_1u_3 + c_2v_3)_z$
      $= c_1[(u_1)_x + (u_2)_y + (u_3)_z] + c_2[(v_1)_x + (v_2)_y + (v_3)_z]$
      $= c_1(\nabla\cdot\mathbf{u}) + c_2(\nabla\cdot\mathbf{v}).$

   b. $\nabla\times(c_1\mathbf{u}+c_2\mathbf{v}) = \begin{vmatrix} \mathbf{i} & \mathbf{j} & \mathbf{k} \\ \partial/\partial x & \partial/\partial y & \partial/\partial z \\ c_1u_1+c_2v_1 & c_1u_2+c_2v_2 & c_1u_3+c_2v_3 \end{vmatrix}$

      $= c_1\begin{vmatrix} \mathbf{i} & \mathbf{j} & \mathbf{k} \\ \partial/\partial x & \partial/\partial y & \partial/\partial z \\ u_1 & u_2 & u_3 \end{vmatrix} + c_2\begin{vmatrix} \mathbf{i} & \mathbf{j} & \mathbf{k} \\ \partial/\partial x & \partial/\partial y & \partial/\partial z \\ v_1 & v_2 & v_3 \end{vmatrix}$

      $= c_1(\nabla\times\mathbf{u}) + c_2(\nabla\times\mathbf{v}).$ Note that we have used
   properties of determinants to perform the middle step here.

13. $\mathbf{A}\times\mathbf{r} = (zA_2-yA_3)\mathbf{i} + (xA_3-zA_1)\mathbf{j} + (yA_1-xA_2)\mathbf{k}.$ Thus
   $\nabla\times(\mathbf{A}\times\mathbf{r}) = (A_1+A_1)\mathbf{i} + (A_2+A_2)\mathbf{j} + (A_3+A_3)\mathbf{k} = 2\mathbf{A}$ and
   $\nabla\cdot(\mathbf{A}\times\mathbf{r}) = 0,$ since A_1, A_2, and A_3 are constants.

15. $\nabla\times(\phi\mathbf{u}) = \begin{vmatrix} \mathbf{i} & \mathbf{j} & \mathbf{k} \\ \partial/\partial x & \partial/\partial y & \partial/\partial z \\ \phi u_1 & \phi u_2 & \phi u_3 \end{vmatrix} = [\phi_y u_3+\phi(u_3)_y-\phi_z u_2-\phi(u_3)_x]\mathbf{i}$

      $+ [\phi_z u_1+\phi(u_1)_z-\phi_x u_3-\phi(u_3)_x]\mathbf{j} + [\phi_x u_2+\phi(u_2)_x-\phi_y u_1-\phi(u_1)_y]\mathbf{k}$
   $= (\phi_y u_3-\phi_z u_2)\mathbf{i} + (\phi_z u_1-\phi_x u_3)\mathbf{j} + (\phi_x u_2-\phi_y u_1)\mathbf{k}$
      $+ \phi\{[(u_3)_y-(u_2)_z]\mathbf{i} + [(u_1)_z-(u_3)_x]\mathbf{j} + [(u_2)_x-(u_1)_y]\mathbf{k}\}$
   $= (\nabla\phi)\times\mathbf{u} + \phi(\nabla\times\mathbf{u}).$

17. $\nabla \cdot (\phi \nabla \chi) = (\phi \chi_x)_x + (\phi \chi_y)_y + (\phi \chi_z)_z$

$\qquad = \phi_x \chi_x + \phi_y \chi_y + \phi_z \chi_z + \phi (\chi_{xx} + \chi_{yy} + \chi_{zz}) = \nabla \phi \cdot \nabla \chi + \phi \nabla^2 \chi.$

19a. We have $\partial R / \partial y = 0$, $\partial Q / \partial z = -L$ (by the Fundamental Theorem of
Calculus), and hence $\partial R / \partial y - \partial Q / \partial z = L$. Likewise $\partial R / \partial x = 0$,
$\partial P / \partial z = M$, and hence $\partial P / \partial z - \partial R / \partial x = M$. Thus the first two of
Eqs.(ii) are satisfied.

  b. Now $\partial Q / \partial x = -\int_{z_0}^{z} L_x (x.y, \zeta) \, d\zeta$ and $\partial P / \partial y = \int_{z_0}^{z} M_y (x, y, \zeta) \, d\zeta + \partial g / \partial y$,

each of which makes use of the equation appearing just before
part(a). Hence

$$\partial Q / \partial x - \partial P / \partial y = -\int_{z_0}^{z} (\partial L / \partial x - \partial M / \partial y) \, d\zeta - \partial g / \partial y = \int_{z_0}^{z} N_\zeta (x, y, \zeta) \, d\zeta - \partial g / \partial y$$

$$= N(x, y, z) - N(x, y, z_0) - \partial g / \partial y. \text{ Thus the third of}$$

Eqs.(ii) is satisfied if $\partial g / \partial y = -N(x, y, z_0)$ or

$$g(x, y) = -\int_{y_0}^{y} N(x, \eta, z_0) \, d\eta.$$

21. $\nabla \cdot \mathbf{u} = 2+1-3 = 0$ so $\mathbf{u} = \nabla \times \mathbf{v}$.   From Prob.19 we have
$L = 2x$, $M = y$, and $N = -3z$ so that

$$P(x, y, z) = \int_{0}^{z} y \, d\zeta - \int_{0}^{y} (-3)(0) \, d\eta = yz \quad \text{and}$$

$$Q(x, y, z) = -\int_{0}^{z} 2x \, d\zeta = -2xz, \text{ where we have chosen}$$

$(x_0, y_0, z_0) = (0, 0, 0)$.   Thus $\mathbf{v} = P\mathbf{i} + Q\mathbf{j} = yz\mathbf{i} - 2xz\mathbf{j}.$

23. $\nabla \cdot \mathbf{u} = y \sin z + 0 + y \sin z \neq 0$, so no $\mathbf{v}$ exists for which $\mathbf{u} = \nabla \times \mathbf{v}$.

25. $\nabla \cdot \mathbf{u} = y \sin z + 0 - y \sin z = 0$ so $\mathbf{u} = \nabla \times \mathbf{v}$.   Now $L = xy \sin z$,
$M = x^2 z$, $N = y \cos z$ so, if $(x_0, y_0, z_0) = (0, 0, 0)$, we have

$N(x, \eta, 0) = \eta$.   Thus $P = \int_{0}^{z} x^2 \zeta \, d\zeta - \int_{0}^{y} \eta \, d\eta = x^2 z^2 / 2 - y^2 / 2,$

$Q = -\int_{0}^{z} xy \sin \zeta \, d\zeta = xy (\cos z - 1)$, and

$\mathbf{v} = [(x^2 z^2 - y^2) / 2] \mathbf{i} + xy (\cos z - 1) \mathbf{j}.$

1a.   Let $\mathbf{v} = 2x\mathbf{i}-3y\mathbf{j}+4z\mathbf{k}$ then $\iint_S \mathbf{v}\cdot\mathbf{N}d\sigma = \iiint_D \nabla\cdot\mathbf{v}dV = \iiint_D 3dV = 3.$

  b.   S consists of a top ($z = 1$, $\mathbf{N} = \mathbf{k}$ so $\mathbf{v}\cdot\mathbf{N} = 4$, $d\sigma = dxdy$), a
       bottom ($z = 0$, $\mathbf{N} = -\mathbf{k}$ so $\mathbf{v}\cdot\mathbf{N} = 0$), a right side ($y = 1$,
       $\mathbf{N} = \mathbf{j}$ so $\mathbf{v}\cdot\mathbf{N} = -3$, $d\sigma = dxdz$), a left side ($y = 0$, $\mathbf{N} = -\mathbf{j}$,
       so $\mathbf{v}\cdot\mathbf{N} = 0$), a front ($x = 1$, $\mathbf{N} = \mathbf{i}$ so $\mathbf{v}\cdot\mathbf{N} = 2$, $d\sigma = dydz$),
       and a back ($x = 0$, $\mathbf{N} = -\mathbf{i}$ so $\mathbf{v}\cdot\mathbf{N} = 0$).   Thus

$$\iint_S \mathbf{v}\cdot\mathbf{N}d\sigma = \int_0^1\int_0^1 xdxdy + \int_0^1\int_0^1 (-3)dxdz + \int_0^1\int_0^1 2dydz = 3.$$

3a. $\iint_S (yz\mathbf{i}+xz\mathbf{j}+xy\mathbf{k})\cdot\mathbf{N}d\sigma = \iiint_D (0+0+0)dV = 0.$

  b. $\iint_S (yz\mathbf{i}+xz\mathbf{j}+xy\mathbf{k})\cdot\mathbf{N}d\sigma = \underset{\text{Top}}{\int\int} xy\,dxdy - \underset{\text{Bottom}}{\int\int} xy\,dxdy + \underset{\text{Right}}{\int\int} xz\,dxdz$

$$- \underset{\text{Left}}{\int\int} xz\,dxdz + \underset{\text{Front}}{\int\int} yz\,dydz - \underset{\text{Back}}{\int\int} yz\,dydz = 0,$$

       where the various sides and their normal vectors are the same
       as those given in Prob.1.

5. $\iint_S \mathbf{v}\cdot\mathbf{N}d\sigma = \iiint_D (2+1-1)dV = 2\iiint_D dV = 8\pi a^3/3$, since the volume of a

     sphere is $4\pi a^3/3$.

7. $\iint_S \mathbf{v}\cdot\mathbf{N}d\sigma = \iiint_D (2x+2y)dV = 2\int_{-1}^1\int_{-\sqrt{1-x^2}}^{\sqrt{1-x^2}}\int_0^2 (x+y)dzdydx$

$$= 4\int_{-1}^1\int_{-\sqrt{1-x^2}}^{\sqrt{1-x^2}} (x+y)dydx = 8\int_{-1}^1 x\sqrt{1-x^2}dx = 0.$$

9. $\iint_S \mathbf{v}\cdot\mathbf{N}d\sigma = \iiint_D (2x+2y+2z)dV = 2\int_{-2}^2\int_{-\sqrt{4-x^2}}^{\sqrt{4-x^2}}\int_0^3 (x+y+z)dzdydx$

$$= 2\int_{-2}^2\int_{-\sqrt{4-x^2}}^{\sqrt{4-x^2}} (3x+3y+9/2)dydx = 4\int_{-2}^2 (3x+9/2)\sqrt{4-x^2}dx$$

$$= \{-8(4-x^2)^{3/2}/3 + 9[x\sqrt{4-x^2} + 4\arcsin(x/2)]\}\Big|_{-2}^2 = 36\pi,$$

     where the last term is obtained using $x = \sin\theta$.

11.   From Ex.4, we have $\iint_S \mathbf{f} \cdot \mathbf{N} d\sigma = (-K/a^2)\iint_S d\sigma = -2\pi K$, where we have

replaced Σ by S and used the fact that the surface area of a
hemisphere is $2\pi a^2$.   Note that we can not "close" the
hemisphere by including the disk $0 \leq x^2+y^2 \leq a^2$ for $z = 0$
since this portion of the surface includes the origin and $\mathbf{f}$
is not continuous there.

13.   Let Σ be the disk $0 \leq x^2+y^2 \leq a^2$ for $z = 0$.   Then S and Σ
enclose a region D for which the divergence theorem holds,
$\iint_S \mathbf{v} \cdot \mathbf{N} d\sigma + \iint_\Sigma \mathbf{v} \cdot \mathbf{N} d\sigma = \iiint_D U dv$.   Now $\iint_\Sigma \mathbf{v} \cdot \mathbf{N} d\sigma = 0$ since $\mathbf{v} = U z \mathbf{k} = \mathbf{0}$

on $z = 0$ and thus $\iint_S \mathbf{v} \cdot \mathbf{N} d\sigma = 4U \int_0^{\pi/2} \int_0^a \int_0^{a^2-r^2} r \, dz \, dr \, d\theta = \pi U a^4/2$.

15.   Let D be the region bounded by $z = \sqrt{1+x^2+y^2}$ and $z = 2$.   Then
we have $\iint_S \mathbf{f} \cdot \mathbf{N} d\sigma + \iint_\Sigma \mathbf{f} \cdot \mathbf{N} d\sigma = -\iiint_D (\nabla \cdot \mathbf{f}) \, dV = -3\iiint_D dV$, where the

negative value comes from the fact that $\mathbf{N}$ is the inward
normal (since $\mathbf{N}$ is upward on S) and Σ is the disk
$0 \leq x^2+y^2 \leq a^2$, $z = 2$. On Σ, $\mathbf{N} = -\mathbf{k}$, $z = 2$, $d\sigma = dxdy = rdrd\theta$,
$\mathbf{f} \cdot \mathbf{N} = -2$ and hence

$$\iint_S \mathbf{f} \cdot \mathbf{N} d\sigma = -3\int_0^{2\pi} \int_0^{\sqrt{3}} \int_{\sqrt{1+r^2}}^2 r \, dz \, dr \, d\theta + 2\int_0^{2\pi} \int_0^{\sqrt{3}} r \, dr \, d\theta$$

$$= -3\int_0^{2\pi} \int_0^{\sqrt{3}} (2r - r\sqrt{1+r^2}) \, dr \, d\theta + 6\pi = 2\pi.$$

17.   Since $\mathbf{f}$ has only a vertical component, Eq.(8) yields
$F_z = \alpha \iint_S z d\sigma$.   Now $\mathbf{N} = (x\mathbf{i}+y\mathbf{j}+z\mathbf{k})/a$,   so

$d\sigma = dxdy/|\mathbf{N} \cdot \mathbf{k}| = (a/z)dxdy$ and $F_z = \alpha a \iint_R dxdy = \pi \alpha a^3$.

19.   Since $\nabla \cdot \mathbf{v} = 0$ for $(x,y) \neq (0,0)$, we have $\iint_S \mathbf{v} \cdot \mathbf{N} d\sigma = \iint_\Sigma \mathbf{v} \cdot \mathbf{N} d\sigma$ as

in Ex.4, where $\mathbf{N}$ is the outward normal in each case.   Now the
top of Σ has $\mathbf{N} = \mathbf{k}$, and the bottom has $\mathbf{N} = -\mathbf{k}$ so
$\mathbf{v} \cdot \mathbf{N} = 0$ in both cases and thus the integral over Σ is simply
an integral over the cylinder $x^2+y^2 = 1$, $-2 \leq z \leq 2$.   In this
case, $\mathbf{N} = x\mathbf{i}+y\mathbf{j}$ so $\mathbf{v} \cdot \mathbf{N} = x^2+y^2 = 1$ and hence

$\iint\limits_{S} \mathbf{v} \cdot \mathbf{N} d\sigma = \int\limits_{\text{cylinder}} \int d\sigma = 8\pi$, since the surface area of the

cylinder is $2\pi h = 8\pi$.

21. $\iint\limits_{S} \mathbf{F} \cdot \mathbf{N} d\sigma = \iiint\limits_{D} \nabla \cdot \mathbf{F} dV = 3\iiint\limits_{D} dV = 3V(D)$, or $V(D) = (1/3)\iint\limits_{S} \mathbf{F} \cdot \mathbf{N} d\sigma$.

23a. From Prob.17 Section 18.6, we have $\nabla \cdot (g\nabla f) = \nabla g \cdot \nabla f + g\nabla^2 f$.

  b. From Section 16.4, $\partial f/\partial n = \nabla f \cdot \mathbf{N}$ so $g\partial f/\partial n = g\nabla f \cdot \mathbf{N}$. Thus
      $\iint\limits_{S} g(\partial f/\partial n) d\sigma = \iint\limits_{S} g\nabla f \cdot \mathbf{N} d\sigma = \iiint\limits_{D} (g\nabla^2 f + \nabla f \cdot \nabla g) dV$.

  c. Interchange f and g in part(b) to obtain
      $\iint\limits_{S} f(\partial g/\partial n) d\sigma = \iiint\limits_{D} (f\nabla^2 g + \nabla g \cdot \nabla f) dV$. Thus

      $\iint\limits_{S} [g(\partial f/\partial n) - f(\partial g/\partial n)] d\sigma = \iiint\limits_{D} (g\nabla^2 f - f\nabla^2 g) dV$.

Section 18.8, Page 1051

1.   The path lies in the plane $3x+3y+2z = 6$ so
     $\mathbf{N} = (3\mathbf{i}+3\mathbf{j}+2\mathbf{k})/\sqrt{22}$ and $d\sigma = dxdy/|\mathbf{N}\cdot\mathbf{k}| = (\sqrt{22}/2)dxdy$. Note
     that $\mathbf{N}$ has the correct orientation for the path Γ. Now
     $\nabla x\mathbf{v} = 3\mathbf{i}+2\mathbf{j}-\mathbf{k}$, so $\nabla x\mathbf{v}\cdot\mathbf{N} = 13/\sqrt{22}$ or $(\nabla x\mathbf{v}\cdot\mathbf{N}) d\sigma = (13/2)dxdy$
     and thus $\oint\limits_{\Gamma} \mathbf{v}\cdot\mathbf{T} ds = \iint\limits_{S} (\nabla x\mathbf{v}\cdot\mathbf{N}) d\sigma = (13/2)\int\limits_{\Omega_{xy}}\int dxdy = 13$ since S

     projects onto the triangle Ω_{xy} with vertices $(0,0),(2,0)$, and
     $(0,2)$, which has area 2.

3.   Let $\mathbf{a} = 3\mathbf{i}+\mathbf{j}+\mathbf{k}$ and $\mathbf{b} = \mathbf{i}+3\mathbf{j}+3\mathbf{k}$ be the position vectors of
     the respective points. Then $\mathbf{N} = (\mathbf{a}x\mathbf{b})/\|\mathbf{a}x\mathbf{b}\| = (-\mathbf{j}+\mathbf{k})/\sqrt{2}$ is
     the normal with the correct orientation to the plane that
     contains Γ. Now $\nabla x\mathbf{v} = -2\mathbf{i}+\mathbf{j}-3\mathbf{k}$ and $d\sigma = \sqrt{2}dxdy$ so
     $\oint\limits_{\Gamma} \mathbf{v}\cdot\mathbf{T} ds = \iint\limits_{S} (-4/\sqrt{2}) d\sigma = -4\int\limits_{\Omega_{xy}}\int dxdy = -16$ since S projects

     onto the triangle Ω_{xy} with vertices $(0,0),(3,1)$, and $(1,3)$,
     which has area 4.

5.   We have $\nabla x\mathbf{v} = 3\mathbf{i}-\mathbf{j}+2\mathbf{k}$ and the normal to $y = x$ with the
     correct orientation is $\mathbf{N} = (-\mathbf{i}+\mathbf{j})/\sqrt{2}$ so
     $d\sigma = dxdz/|\mathbf{N}\cdot\mathbf{j}| = \sqrt{2}dxdz$. Thus

$$\oint_\Gamma \mathbf{v}\cdot\mathbf{T}\,ds = \iint_S (-4/\sqrt{2})\,d\sigma = -4\iint_{\Omega_{xz}} dxdz = -16\pi, \text{ since } \Omega_{xz} \text{ is the}$$

circle $0 \le x^2+z^2 \le 4$, $y = 0$, which has area 4π.

7. Since $\nabla\times\mathbf{v} = \mathbf{0}$, we have $\oint_\Gamma \mathbf{v}\cdot\mathbf{T}\,ds = 0$ by Stokes' Theorem.

9. We have $\nabla\times\mathbf{v} = 3\mathbf{j}$ and the normal to the plane $z = x+2$, with
    the correct orientation, is $\mathbf{N} = (\mathbf{i}-\mathbf{k})/\sqrt{2}$. Thus $\nabla\times\mathbf{v}\cdot\mathbf{N} = 0$
    and $\oint_\Gamma \mathbf{v}\cdot\mathbf{T}\,ds = 0$ by Stokes' Theorem.

11. The boundary Γ of S is the circle $x^2+y^2 = a^2$, $z = 0$ and thus
    $\iint_S \nabla\times\mathbf{v}\cdot\mathbf{N}\,d\sigma = \oint_\Gamma \mathbf{v}\cdot\mathbf{T}\,ds$. Note that Γ lies completely in the
    plane $z = 0$ (call the disk $0 \le x^2+y^2 \le a^2$, $z = 0$ the surface
    S_1), which has normal $\mathbf{N} = \mathbf{k}$ for the correct orientation.
    Thus, using Stoke's Theorem again, we have
    $\oint_\Gamma \mathbf{v}\cdot\mathbf{T}\,ds = \iint_{S_1} \nabla\times\mathbf{v}\cdot\mathbf{k}\,dxdy$. [Note that since $dz = 0$ on Γ, this
    is really Green's Theorem as expressed in Eq.(5)]. Now
    $\nabla\times\mathbf{v} = 2x\mathbf{i}-y\mathbf{j}-z\mathbf{k}$ so $\nabla\times\mathbf{v}\cdot\mathbf{k} = -z = 0$ on S_1 so we conclude that
    $\iint_S \nabla\times\mathbf{v}\cdot\mathbf{N}\,d\sigma = \iint_{S_1} (-z)\,dxdy = 0$.

13. As in Prob.11, $\iint_S \nabla\times\mathbf{v}\cdot\mathbf{N}\,d\sigma = \iint_{S_1} \nabla\times\mathbf{v}\cdot\mathbf{k}\,dxdy = \iint_{S_1} dxdy = \pi a^2$ since
    $\nabla\times\mathbf{v}\cdot\mathbf{k} = (1-xz) = 1$ on S_1 and the area of S_1 (a circle of
    radius 1) is πa^2.

15. The boundary Γ of S is the circle $x^2+y^2 = 3a^2/4$, $z = a/2$.
    Now let S_1 be the disk $0 \le x^2+y^2 \le 3a^2/4$, $z = a/2$ with normal
    $\mathbf{N} = \mathbf{k}$ so that S_1 also has boundary Γ and thus as in Prob.11,
    we have $\iint_S \nabla\times\mathbf{v}\cdot\mathbf{N}\,d\sigma = \oint_\Gamma \mathbf{v}\cdot\mathbf{T}\,ds = \iint_{S_1} \nabla\times\mathbf{v}\cdot\mathbf{k}\,d\sigma$. Now $d\sigma = dxdy$ and
    $\nabla\times\mathbf{v}\cdot\mathbf{k} = z^2 = a^2/4$ on S_1, so $\iint_{S_1} \nabla\times\mathbf{v}\cdot\mathbf{k}\,d\sigma = \iint_{S_1} (a^2/4)\,dxdy = 3\pi a^4/16$,
    since the area of S_1 is $3\pi a^2/4$.

17. $\nabla \times \mathbf{v} = \mathbf{0}$ and thus $\oint_\Gamma \mathbf{v} \cdot \mathbf{T}\, ds = \iint_S \nabla \times \mathbf{v} \cdot \mathbf{N}\, d\sigma = 0$.

19.   Choose any simple closed curve Γ on S. Γ then divides S into
      two portions S_1 and S_2 and thus we have

$$\iint_S \nabla \times \mathbf{v} \cdot \mathbf{N}\, d\sigma = \iint_{S_1} \nabla \times \mathbf{v} \cdot \mathbf{N}\, d\sigma + \iint_{S_2} \nabla \times \mathbf{v} \cdot \mathbf{N}\, d\sigma = \oint_\Gamma \mathbf{v} \cdot \mathbf{T}_1\, ds + \oint_\Gamma \mathbf{v} \cdot \mathbf{T}_2\, ds,$$

      where we must have $\mathbf{T}_1 = -\mathbf{T}_2$ in order to have the correct

      orientation.   Hence $\iint_S \nabla \times \mathbf{v} \cdot \mathbf{N}\, d\sigma = \oint_\Gamma \mathbf{v} \cdot \mathbf{T}_1\, ds - \oint_\Gamma \mathbf{v} \cdot \mathbf{T}_1\, ds = 0$.

Chapter 18 Review, Page 1052

1.   With $d\mathbf{r} = \mathbf{i} + t\mathbf{j}$, $\int_C \mathbf{F} \cdot d\mathbf{r} = \int_1^2 (\ln t + te^{t^2/2})\, dt = t\ln t - t + e^{t^2/2}\Big|_1^2$
      $$= e^2 - e^{1/2} + \ln 4 - 1.$$

3. $d\mathbf{r} = \cos t\, \mathbf{i} + \sin t\, \mathbf{j}$, so $\int_C \mathbf{F} \cdot d\mathbf{r} = \int_0^{\pi/2} (\sin^2 t - \sin t)\cos t\, dt = -1/6$.

5. $\int_C (x^2 + y^2)\, dx = \int_0^2 (t^2 + t^6)\, dt = 8/3 + 128/7 = 440/21$.

7. $\int_C (x/y)\, dx = \int_0^3 [x/(x^2+1)]\, dx = (1/2)\ln(x^2+1)\Big|_0^3 = \ln\sqrt{10}$.

9.   Since $P_y = e^x + xye^{xy} + 1/x^2y^2 = Q_x$, $\mathbf{F}$ is conservative.   Thus
      $\phi_x = P = ye^{xy} + 1 - 1/x^2y$ and $\phi = e^{xy} + x + 1/xy + h(y)$.   As
      $\phi_y = Q$, we have $h'(y) = 0$, so $h(y) = c$ and
      $\phi(x,y) = e^{xy} + x + 1/xy + c$.

11. $P_z = 0 = R_x$, $P_y = 0 = Q_x$, and $Q_z = \sin y \cos z = R_y$ so $\mathbf{F}$ is
      conservative.   Thus $\phi_x = 2$ and $\phi = 2x + h(y,z)$, and thus
      $\phi_y = h_y = \sin y \sin z$, so $h(y,z) = -\cos y \sin z + g(z)$, giving
      $\phi_z = -\cos y \cos z + g'(z) = -\cos y \cos z$, or $g(z) = c$.   Therefore
      $\phi(x,y,z) = 2x - \cos y \sin z + c$.

13. $P_y = x+y = Q_x$, so the integral is independent of path. Since
     $\phi_x = xy + y^2/2$, $\phi = x^2y/2 + xy^2/2 + g(y)$ and we have that
     $\phi_y = x^2/2 + xy + g'(y) = x^2/2 + xy$. Choosing $g(y) = c = 0$, the
     integral is $\int_{(0,1)}^{(1,0)} d[x^2y/2 + xy^2/2] = [x^2y/2 + xy^2/2]_{(0,1)}^{(1,0)} = 0$.

15. $P_y = yz + z^2/2 = Q_x$, $P_z = y^2/2 + zy = R_x$, and $Q_z = xy + xz = R_y$,
     so the integral is independent of path. The integral is
     $(1/2)\int_{(0,0,0)}^{(1,1,1)} d(xy^2z + xz^2y) = (1/2)(xy^2z + xz^2y)\big|_{(0,0,0)}^{(1,1,1)} = 1$.

17. $Q_x - P_y = 2(x-y)$, so $I = 2\iint_\Omega (x-y)\,dA = 2\int_0^{2\pi}\int_0^1 r^2(\cos\theta - \sin\theta)\,dr\,d\theta$

     $= (2/3)\int_0^{2\pi} (\cos\theta - \sin\theta)\,d\theta = 0$.

19. $I = \iint_\Omega (2\sin x \cos x + 2\sin x \cos x)\,dx = 2\int_0^\pi \int_0^\pi \sin 2x\,dx = 0$.

21a. The area of the ellipse is $\pi ab = 6\pi$.

   b. $\hat{x} = (1/2A)\oint_C x^2\,dy$, where $x = 2\cos t$, $y = 3\sin t$, $0 \le t \le 2\pi$. Thus

      $\hat{x} = (1/12\pi)\int_0^{2\pi} 12\cos^3 t\,dt = 0$. Also, $\hat{y} = (1/12\pi)\int_0^{2\pi} 18\sin^3 t\,dt = 0$.

   c. $I_0 = (1/3)\oint_C (-y^3 dx + x^3 dy) = (1/3)\int_0^{2\pi} (54\sin^4 t + 24\cos^4 t)\,dt$

      $= (1/3)\int_0^{2\pi} [(27/2)(1-2\cos 2t + \cos^2 2t) + (12/2)(1+2\cos 2t + \cos^2 2t)]\,dt$

      $= (1/3)\int_0^{2\pi} [39/2 - 15\cos 2t + (39/2)\cos^2 2t]\,dt = 39\pi/2$.

23a. $A = \int_0^1 e^x dx = e-1$.

   b. $\hat{x} = [1/2(e-1)][\int_0^e dy + \int_1^0 x^2 e^x dx] = [1/2(e-1)][e+2-e] = (e-1)^{-1}$,

      $\hat{y} = [-1/2(e-1)][\int_1^0 e^{2x} dx] = (e+1)/4$.

23c. $I_0 = (1/3) [\int_0^e dy + \int_1^0 -e^{3x}dx + \int_1^0 x^3 e^x dx]$

$= (1/3)[(1/3)(e^3-1) + (e) + (2e-6)] = (e^3 + 9e - 19)/9.$

25. $A = \int\int_{\Omega_{xy}} [4/9 + 1/9 + 1]^{1/2} dA_{xy} = (\sqrt{14}/3)\int\int_{\Omega_{xy}} dA_{xy} = 2\sqrt{14}/3.$

27. $A = \int\int_{\Omega_{xy}} \{[64x^2 + 64y^2 + 4z^2]^{1/2}/2z\} dA_{xy}$

$= \int\int_{\Omega_{xy}} [\sqrt{80(x^2+y^2)}/2\sqrt{4(x^2+y^2)}] dA_{xy} = \sqrt{5}\int\int_{\Omega_{xy}} dA_{xy} = \sqrt{5}/2.$

29a. $\nabla \cdot \mathbf{v} = 2xy + 2yz + 2zx.$

b. $\nabla \times \mathbf{v} = \begin{vmatrix} \mathbf{i} & \mathbf{j} & \mathbf{k} \\ \dfrac{\partial}{\partial x} & \dfrac{\partial}{\partial y} & \dfrac{\partial}{\partial z} \\ x^2 y & y^2 z & z^2 x \end{vmatrix} = -(y^2\mathbf{i} + z^2\mathbf{j} + x^2\mathbf{k}).$

31a. $\nabla \cdot \mathbf{v} = -[xy^2 (x^2+y^2)^{-3/2} + yz^2 (y^2+z^2)^{-3/2} + zx^2 (x^2+z^2)^{-3/2}].$

b. $\nabla \times \mathbf{v} = -\{[(z^3+2zy^2)/(z^2+y^2)^{3/2}]\mathbf{i} + [(x^3+2xz^2)/(x^2+z^2)^{3/2}]\mathbf{j}$

$+ [(y^3+2yx^2)/(y^2+x^2)^{3/2}]\mathbf{k}\}.$

33. $\int\int_S \mathbf{v}\cdot\mathbf{N} \, d\sigma = \int\int\int_D \nabla\cdot\mathbf{v} \, dV - \int\int_{S_1} \mathbf{v}\cdot\mathbf{N} \, d\sigma = \int\int\int_D \nabla\cdot\mathbf{v} \, dV$ (where S_1 is $z = 0$)

$= \int\int\int_D 4z \, dV = \int_0^{2\pi}\int_0^{\pi/2}\int_0^1 4\rho^3 \sin\phi \cos\phi \, d\rho d\phi d\theta = \pi.$

35. $\int\int\int_D \nabla\cdot\mathbf{v} \, dV = \int_0^3\int_0^4\int_0^2 (x+y+z) \, dz dy dx = 108.$

37. $\int\int_S \nabla\times\mathbf{v}\cdot\mathbf{N} \, d\sigma = \oint_C \mathbf{v}\cdot\mathbf{T} \, ds = \oint_C (2y+z)dx + (2z+x)dy + (2x+y)dz$

$= \int_{-2}^2 (-2)dx + \int_{-2}^2 6dy + \int_2^{-2} 6dx + \int_2^{-2} 2dy = -16.$ The

surface integral may also be evaluated directly since

$\nabla\times\mathbf{v} = -(\mathbf{i}+\mathbf{j}+\mathbf{k})$ and $\mathbf{N} = \mathbf{k}$, so $\int\int_S \nabla\times\mathbf{v}\cdot\mathbf{N} \, d\sigma = -\int_{-2}^2\int_{-2}^2 dx dy = -16.$

39.   Since $\mathbf{v} \cdot \mathbf{T} = 0$ on each side of the triangle,

$$\iint_{S} \nabla \times \mathbf{v} \cdot \mathbf{N} \, d\sigma = \oint_{C} \mathbf{v} \cdot \mathbf{T} \, ds = 0.$$

41.   With $\mathbf{r} = t\mathbf{i} + t^2\mathbf{j}$ and $\mathbf{F}(t) = (t^5 - t^4)\mathbf{j}$, we have

$$W = \int_{C} \mathbf{F} \cdot \mathbf{T} \, ds = \int_{C} \mathbf{F} \cdot d\mathbf{r} = \int_{0}^{2} (2t^6 - 2t^5) \, dt = 320/21 \text{ N-m.}$$